Matthew Murray
(1765-1826)
and
The firm of
Fenton Murray & Co
(1795-1844)

by

Paul Murray Thompson

Contents

List of Illustrations

Including acknowledgements

vol.37 1814.

51. Murray's design for a hydraulic rope tester 1814 as sent to Goodrich.

Goodrich Papers A545. Reproduced with the permission of the Science & Society Picture Library.

52 Distinctive column capital. Campion's Mill Whitby. Author's photograph.

Chapter 11.   **53**. Fenton Murray & Co proposed 'A' frame steam engine for Henry Heth in Virginia U.S.A. Note firm's paper. Reproduced with the permission of the University of Virginia Library.

54. Author's redraw of Goodrich sketch of Fenton Murray & Co lathe 1819. Goodrich Journal B40, Science Museum Library.

55. Author's redraw of Goodrich sketch of Fenton Murray & Co shaft connector 1819. Goodrich Journal B40, Science Museum Library.

56. Murray's smoke burning apparatus. Newton's London Journal of Arts and Sciences 1821.

Chapter 12.   **57**. Author's redraw of Goodrich sketch of the P.S. Hero's boiler. Goodrich Journal B47, Science Museum Library.

58. Fenton Murray & Co engine at the Westminster Gas Works 1823. Mechanic's Magazine vol. 7 1827.

59. Pode Hole Pumping Station. Author's photograph.

60. 3 views of Kesteven (Fenton Murray & Co) Engine at Pode Hole Station. Reproduced from Holland and Kesteven Pumping Engines at Pode Hole Spalding 1952 by permission of Welland and Deeping Internal Drainage Board.

61. Murray's 1825 design for a locomotive engine. Newton's London Journal of Arts and Science 1826.

Chapter 13.   **62**. Typical 'A' frame engine as proposed by Fenton Murray & Co to John Watson for Garforth Colliery near Leeds. Reproduced with permission of NEIMME, Wat2/3.

63. Drawing of Vulcan, One of Liverpool and Manchester Railway locomotives by Fenton Murray & Co 1831. NRM 1998-11384/70. Reproduced with the permission of the Science & Society Picture Library.

Chapter 14.   **64**. 20 H.P marine engines for use on the Saone/Rhone. Archives du Musee des arts et metiers, CNAM S 87.

65. Fenton Murray & Co locomotive for the Andrezieux-Roanne Railway. Flachat and Petier 1849.

66. Fenton Murray & Co locomotive for the Lyons-St Etienne railway 1833. Archives du Musee des arts et metiers, CNAM S 183.

67. Fenton Murray & Co locomotive for the Paris- St Germain en Laye Railway c. 1836. L'Industrie des Chemins de fer, Armengaud.

Foreword

Having grown up with the hazy knowledge of Matthew Murray in the background, although we did not know all that much about him, a time came when I felt I would attempt to find out what I could about my great x4 grandfather. I started in the Science Museum Library in London and kept finding out more and more leading me on to other places. Having acquired quite a large amount of information I felt it should be recorded, as many of the facts as currently on record are fleeting or imprecise. The result is this book, which due to my start point, is more of a source history of the records. This has taken over 10 years and while I have made some comparison with Murray's contemporaries, I am aware that more comparisons with other engineers would have been better. But at least if the available facts are published, others may use the data to understand more where the man and the firm fits into the overall story of the Industrial Revolution. As will be seen Murray was into technology and seems to have looked at most things new, some he did not pursue, others such as the construction of fire proof mills became pieces of important business for the firm. While his business practices at times seem unethical in today's world, they seem to have been fairly typical of those practised by many major firms at the time. Patents on steam engines and textile machinery proliferated at such a great extent that it must have been almost impossible to absorb them all or to find that part which made them unique!

It is also interesting to note that Murray really only gets a bad press from Boulton and Watt and their cronies, or that is what has survived, and it should be remembered that it was in their interests to cast him as a blackguard.

There are areas, such as the impact on the Russian market, which have only been touched upon using sources in this country and may be far more important than the credit given here. There is probably more to be found in France, which was a major market in the last decade of the firm.

Some of the data is repetitive, for which I do not apologise, because I hope it shows the amount of business being undertaken, e.g. the purchase of boiler plates from Kirkstall Forge. The whole mix of the book is uneven, with far more on the early days for which there seem to be far more records, mainly due to Murray coming under the eagle eye of Boulton and Watt. But each part is dependent on the quantity of surviving records.

Repetition is also unavoidable between the sequential history of the firm and the subject chapters 16 to 18, but it is my belief that confining everything to subjects loses the impact of the amount of work that the firm was dealing with at times, especially between about 1810 and 1814.

Acknowledgements.

First I have to render an enormous thank you to Lesley not only for her patience over the years, but her active encouragement and interest.

For encouragement and crash courses in many technical matters over time my great thanks are due to Ron Fitzgerald, who I first met on a sleety November day at the Round Foundry in Leeds. Ron has also shared with me much of the data on the Round Foundry and other places he has worked for which I also give my thanks. He has also provided encouragement and a discussion board.

Sheila Bye has also given me much help and encouragement while I have been researching and writing and has been particularly helpful in tracking down some early parish records and we also eventually managed to track down Richard Jackson's movements in outline after the bankruptcy.

I would also like to express my appreciation of the help and assistance provided by many people working in archives which I have visited either physically or electronically.

All mistakes remain my own.

A note on sources.

I have looked at so many books that I cannot provide a bibliography, some I cannot even remember. However I have tried to provide proper referencing for the major points.

Any study of this subject probably begins with two works,' Matthew Murray, Pioneer Engineer' edited by Kilburn Scott which includes the reprint of G F Tyas's 1926 presentation to the Newcomen Society; and secondly Trevor Turner's thesis from the 1960's 'Fenton Murray and Wood' (unpublished). The sources listed by Turner gave a very sound basis for starting on the road to this book.

The most important archive sources are:

The Boulton and Watt Archive in The Library of Birmingham.

The Goodrich Archive in the Science Museum Library.

The Watson Papers and others at the North of England Institute of Mining and Mechanical Engineers in Newcastle upon Tyne.

West Yorkshire Archive Service Leeds, for the records of Kirkstall Forge, and Middleton Colliery.

National Archives at Kew, particularly those for The Admiralty and the Early Railway Companies.

Archives of the National Maritime Museum, the Caird Library, for Brunel's letter Book and Admiralty papers.

London Metropolitan Archives for Waterworks Companies and Insurance policies.

The Curwen Archive in the Cumbrian Archives at Whitehaven.

The Archives du Musee des arts et metiers Paris.

French National Archives, Paris.

Service historique de la marine, Vincennes Paris.

And the

British Library's Newspaper Archive on line.

Sources and copyright.

Most of the original sources used are out of copyright, all references within and without are intended to be properly acknowledged in the notes.

Whilst every effort has been made to acknowledge sources and copyright if any holder of same feels any acknowledgement or copyright has been incorrectly made or omitted, then please contact the author.

Abbreviations

[ ]	denotes that part of a text is indecipherable.
[word]	possible meaning for unclear text.
....	omission of text for irrelevance or brevity.
H.P.	Horse Power.
£ s d	Pounds Shillings Pence pre decimal currency.
c.w.t	hundredweight (112 lb).
qtr.	a quarter of a c.w.t or 28lbs.
B & W	Boulton and Watt.
NMM	National Maritime Museum Greenwich.
NEIMME	North of England Institute of Mining and Mechanical Emgineers.
WYAS	West Yorkshire Archive Service.

Chapter 1

Origins, Newcastle upon Tyne and Northumberland

If a traveller today arrives at Leeds Railway Station and leaves it at the north end to cross Wellington Street they arrive in City Square where among other statues is one of James Watt who, with a Parliamentary extension, was granted a patent on his main inventions on the steam engine for 31 years and not surprisingly in such a period found many engineers and mechanics trying to circumvent the patent and put into practice many good improvements. To enforce his rights James Watt and his partner Matthew Boulton had recourse to the Courts of Law, in more than one famous case, before the patent expired in 1800.

If on the other hand the traveller turned south down Neville Street, crossed the River Aire and then turned right down Water Lane and followed this road to its junction with Globe Road, there is the Steam Engine Manufactory, better known as the Round Foundry, the earliest parts of which were built by Matthew Murray and David Wood and were contemporary with the Boulton and Watt Soho Foundry in Birmingham. Murray has been called the Father of Leed's Engineering and it has often been mooted in the past that it should be Murray who has a statue in City Square rather than James Watt. It can arguably be stated that Boulton and Watt were at the very least equalled in eminence in the steam engine business by Matthew Murray within a very short time of the expiry of James Watt's patent. This was no doubt due in some part to the fact that Murray was building steam engines before Watt's patent expired and was considered by Boulton and Watt as one of "the pirates"; which resulted in Fenton Murray and Wood receiving a lot of attention from Soho for a number of years. The strength of their feeling can be summarised by quoting from their response to a request from Messrs Benyon Marshall and Bage of Shrewsbury for a license to apply their patent to the new Fenton Murray and Wood steam engine they had installed in their Shrewsbury Mill (Ditherington). The Boulton and Watt reply, dated the 22nd February 1799, stated:

"We see no reason why Messrs Murray & Wood should be made partners to our Parliamentary Privilege" [1].

It is known that Murray was the main mechanical mind behind the Leeds partnership's textile machinery and steam engines but who was he?

Dr. Samuel Smiles in his book 'Industrial Biography', states that:

"Matthew Murray was born in Stockton-on -Tees in the year 1763. His parents were of the working class, and Matthew, like the other members of the family, was brought up with the ordinary career of labour before him" [2].

Smiles lived in Leeds for some time and from 1838 to 1845 was editor of the Leeds Times, and in 1842 set up a surgery in Holbeck and knew Murray's surviving family. In his book he acknowledges assistance from Murray's son in law Mr J O March.

Stockton also lays claim to Murray in an entry in "The Local records of Stockton and Neighbourhood" for which the first part of an entry for December 1825 reads:

"Matthew Murray, a native of Stockton, died at Leeds this year, aged 62. His parents were of the working class, and he was apprenticed to a blacksmith, in which trade he very soon acquired considerable expertness" [3].

As shall be seen Murray's first child a daughter Margaret was born in Stockton on August 27th 1786, and this event may well have given rise to the idea that Matthew Murray himself was born in Stockton.

The "Biographia Leodiensis" provides some wider alternatives:

"A celebrated engineer (according to Smile's Industrial Biography),* was born at Stockton-on -Tees, in the year 1763."

The footnotes then explain

"The following Notes, in correction of, and in addition to, Dr Smile's interesting account, have been kindly contributed by Mr J. O. March, one of the senior aldermen of Leeds, and an eminent machine-maker, who married one of Mr Matthew Murray's daughters; his late partner, Mr Charles Gascoigne Maclea, having married another.
We believe he was born in or near Newcastle, where he also served his apprenticeship, and afterwards for a year or two worked in a mechanics' shop at Stockton. Mrs Murray's family lived at Whickham [on the south bank of the Tyne west of Gateshead], near Newcastle, where he became acquainted with her.
Mr Murray died in February, 1826, in his sixty-first year, and was, therefore, born in 1765" [4].

It has been established, from the parish records, that Matthew Murray was married in St Mary's Gateshead, and from this church looking across the Tyne one can see the parish church of All Saints (also known as All Hallows) Newcastle (albeit this is the church as splendidly rebuilt in 1796) and if one looks in the Parish Records of All Saints there is a baptismal entry for November 24th 1765;

"Matthew son of Reynard Morrow W.smith" (whitesmith).

Can Morrow be taken to be synonymous with Murray? The evidence here is satisfactory as Murray's name in Leeds is variously spelt anything from Mirror to Murrow, so a Morrow is certainly not far-fetched in an era when the oral tradition was a majority and the clerks wrote down what they heard. The additional confirmation comes from further exploration of The All Saint's registers, as it is known that Matthew Murray had two sisters and searching forward in the register uncovers additional family entries:

"Baptism 13th September 1767
Margaret daughter of Renold Murray blacksmith;

Burial on 20th December 1767
Margaret daughter of Leonard Murray blacksmith;

Baptism on September 13th 1772
Alice daughter Reginald Murray whitesmith;

Baptism on 16th May 1777
Reginald son Of Reginald Murray whitesmith;

Baptism on 18th September 1779
Margaret daughter of Reynard Murray smith;"

The recorded names of Matthew Murray's sisters were Alice and Margaret. The name Reginald and Reynard are interchangeable, and it is possible that during times of tension with France that Mr Murray preferred the more English Reginald.

A search back in time in these parish Registers reveals that:

"Reynard Murrey of the Parish of Walls End & Margaret Price of this Chapelry were married in this Chapel by Licence the 30th Day of Sept 1764 by me Geo Stephenson Curate. This marriage was solemnized between us Reynard Murrey His X Mark margrat price (signature). In the presence of us Edward Kell Henry Swan."

So Matthew Murray's mother could sign her name, while at the time of the marriage Reynard used the mark. The reference to the chapelry indicates that the marriage service was carried out at the chapel of ease St Anns in the Sandgate area. The current building was consecrated in 1768 a few years later, but has been the site of a church since mediaeval times.

Further study of these registers reveal much of the probable family of Matthew Murray's mother; Margaret Price was baptised on the 23rd November 1736 and had two elder brothers Thomas, baptised 3 April 1732, and John baptised on 23rd April 1734. She had two younger siblings but they both died in infancy. Her parents were Thomas Price and Margaret Currey who were married in the same parish on 20th July 1731. Thomas Price was a smith, who died in 1741 and was buried on 5th February 1741, and his eldest son Thomas took over the business.

From this family data it can begin to be understood where the seeds of Matthew Murray's mechanical ability were sown. His father was a smith and while he was growing up his uncle Thomas Price was practising as a blacksmith in Newcastle. Tracing the origins of Reynard Murray is not so clear cut. As seen from the marriage lines, at the time Reynard was shown as being from the parish of Walls End. Walls End even at this time was an area of boat building that would give employment to a smith, but to date no record of a Reynard Murray has come to light in this parish. A

Matthew Murray married an Alice Purvis in February 1743, but there is little else to do with Murrays.

Reynard and Margaret Murray were married by licence, and the licence survives in the records of Durham Diocese and in this document Reynard Murrey is described as a mariner and his witness Henry Swan as a farmer [2]. Reynard however was at the time an unusual name and tracking through the Northumberland and Durham parishes has produced only a very few Reynard Murrays and they are confined to the parish of Kirkwhelpington to the north west of Newcastle, roughly west of Morpeth. The only entry to fit Matthew Murray's father is a baptism, the page upon which it is written is faded almost to the point of illegibility, but it reads:

"30th June 1728 Reynold the son of Reynold Murrah of Cornhills was baptised".

Cornhills lies to the west of Kirkwhelpington village in an area that contains an abandoned village named West Whelpington, the surprising fact concerning this abandonment was that it took place as late as around 1720, some fifteen farmers being moved on by the landlord Stott. Similar occurrences are reported in the area of Ray, to the immediate north on the other side of the river. It would also appear unusually that the clearance was to promote more efficient arable farming rather than to turn the land over to sheep. Stott leased one of the nearby farms, Cornhills, and the steading appears in the registers for 1689/1690. The farm was one of the four that supplanted the area previously under the sway of West Whelpington.

In 1748 it is listed as having:
Three lofted & thatched dwellings each with its byre and barn
A Milk House
A Herd's house
A Cottage [6].
The area is rural and agricultural, but in the period in question the number of jobs was decreasing fairly sharply and this would explain a migration towards Tyneside and its relatively better prospects, and indeed the number of Murray entries in the parish register declines. A few do still take place. A Reynard Murray married in Kirkwhelpington in 1740, but would be too young to be the one born at Cornhills. There were Murrahs and Murrays in West Whelpington, Whelpington and Ray and some examples from the parish register are the death of Margaret the wife of Matthew Murray of West Whelpington on the 11th February 1717, and an Edward the son of Reynard Murray of Ray was buried in 1723, and he was probably the same person for whom the birth was recorded in 1720 as Edward son of Reynard Murrah

of Whelpington. There is not necessarily any contradiction in the changing place names. They were probably agricultural labourers and would when necessary shift farms for the annual hirings. That they were not wealthy in any way is supported by there being no Murrah gravestones in the churchyard despite the number of recorded deaths.

Other data is available from Newcastle records, which it appears probably relates to the Murray and Price families. Some of the Poll Lists for this period survive, and in the 1741 register, listed under Smiths is Thomas Price (probably the senior) and again in 1774 an entry Price Thomas Smith (the son). The Price's as smiths were wealthy enough to own the property necessary to qualify, and teach their daughter to read and write. Reynard Murray, whose family had probably failed to survive in the reduced employment available in the Cornhills area, had no doubt moved to Wallsend Parish or nearby where Reynard had served apprentice to a smith. But for whatever reason he had ended up making a living as a mariner. Reynard then married into a family of smiths, which possibly gave him the impetus to return to his trade However as a smith just starting up he will have rented his property and would therefore not have qualified for a vote.

The Rate Assessments for the Poor Law Tax in the Sandgate Ward survive in the records of the Parish of All Saints. In these there are references starting in October 1767 to a Mr Murry, Smith paying a rent of £7 leading to an assessment for tax of 7d (old pence). The assessment being one farthing per week or One Penny per pound per month. This increased to 10 1/2d assessment in August 1772 by which time the spelling of the name had changed to Mr Murray. From April to September the assessment is made on Mr Murray Ho[use] and then ceases in October 1773. It is interesting to note that in 1773 these records show a Mr Edw Kells living in Sandgate, and a gentleman of this name was a witness at Reynard Murray's marriage to Margaret Price; a further indicator that the above records relate to Reynard Murray's Poor Tax assessments [7].

In 1778, Whiteheads Newcastle Directory was published and it contains two entries of interest. The first is found under the heading of WHITE-SMITHS and reads:
"MURRAY REN Painter-heugh (foot of the street)".
The other is under BLACK-SMITHS and reads:
"PRICE THOMAS High-bridge (middle of the street)".
High Bridge (formerly Upper Dean Bridge) still survives as a street in the much changed Newcastle, although now accommodating Grey Street crossing it about half way along its length. Off Dean Street there is an alley that goes off before The Side is reached which leads to and past a multi storey car park. The alley still carries a name

plate calling it Painter Heugh. In Brand's History of Newcastle there is a short description:

"The Painter-Hugh, by a steep descent, conducted from the bottom of Pilgrim Street to the middle of the Side. There is a flight of stone steps on one side for the convenience of foot passengers" [8].

Bourne derives the name from 'painter' or rope by which boats are moored; and hugh or heugh a steep hill or bank [9]. There do not appear to be many surviving photographs of Painter Heugh, one in the Northumberland Archives obviously taken at the time the street was dismantled, still shows the steps to one side.

So it would appear that Matthew Murray grew up in the Quays area of Newcastle close to the river and that by the time he was in his teens his father was advertising himself as a white smith.

It is quite difficult today to find a satisfactory definition of a whitesmith, other than 'a worker in white metal, a tinsmith', but the following may help again with understanding Matthew Murray's skill-base although it was written a little later than our period

'*A modern whitesmithery establishment generally comprises the conveniences requisite for the production of every description of work from what is called blacksmithing to machine making or engineering. A first rate whitesmith is not only required to understand generally the qualities of common iron and steel, and the methods of working them, he must likewise have a competent knowledge of the principles of the mechanical side'*

The attractions of the Sandgate area as the initial place to live are fairly clear in that it is very close to the River Tyne and near the point at which it's tributary the Ouse Burn flows into it. The area was quite industrial having roperies, glassworks and the iron works or smithery at some time owned by Messrs Hawks and Ridley. So Sandgate would have offered employment at some iron working concern such as Hawks and Ridley, and plenty of affordable accommodation. Later on setting up as a whitesmith Reynard Murray would probably have gone in for the smaller end of the business such as weighing machines and other goods with a wider market and hence a move closer to the city centre into Painters Heugh.

As a 6 year old child growing up near the Tyne, Matthew Murray would in some form or other have experienced the floods and storms that saw the Tyne Bridge destroyed in 1771.

It is clear that Matthew Murray received somewhat more than a basic education. He could read and write, he had a good grasp of mathematics, as one critic of his mathematics pointed out he was adept at using a slide rule [10]. An instrument that is rapidly disappearing from society, and doing so possibly because

6

using one requires a firm grasp of arithmetic and an understanding of exactly what it is that needs to be done in order to carry out a successful calculation. Schools were available in Newcastle from the Royal Free Grammar School to the Parish Charity Schools. Henry Bourne records that All-Saints Charity School was set up by voluntary subscription in 1709 and has continued ever since on the same footing.

"There are 41 boys taught to read, write, and cast accounts…, and 17 girls are taught to read, knit, sew make and mend their own clothes….. These children have Coats and Caps once a year, and shoes, stockings, shirts and bands twice a year, and at their leaving the School they have Forty Shillings each to put them out apprentice or equip them for services and each of them a bible, with the Common-Prayer, a whole Duty of Man, and Lewis's catechism" [11].

So Matthew Murray may have been educated at this level or it is possible that if he had shown aptitude early on that his mother's family may have helped with schooling costs, maybe covering extra tuition. As his mother had been taught to write one might expect her to have wanted her firstborn to have as good an education as possible.

On reaching the age of fourteen Matthew Murray would have been apprenticed, and his son in law Joseph March believed that apprenticeship was in Newcastle. In this period in describing Newcastle as an area, the description was generally held to include Gateshead. This enlarges the field in terms of with whom he was apprenticed, and in truth no surviving clues have yet been found. He could have become apprenticed to his father or his uncle Thomas Price. There is one cogent argument that might support this, Matthew Murray was married in Gateshead when he was one month short of the twentieth anniversary of his baptism. With the delays common in having children baptised this could easily mean that he was twenty years old. However at the time it was certainly not normal for apprentices to marry (in fact it was usually prohibited by the indentures) and apprenticeships normally lasted until the twenty first birthday. If he was apprenticed to a family member there may have been some latitude allowed. There is additional evidence that may support Matthew Murray having been an apprentice to his father, it is certainly not conclusive, but adds to the pointers above. In his first patent number 1752 of 1790 Matthew Murray describes himself as a whitesmith, and in the second number 1971 of 1793 he uses the terms whitesmith and mechanic.

Alternatively he may have been apprenticed to any number of smiths or indeed with someone such as Hawkes of Ouseburn and Gateshead. However at this point it remains speculation.

In 1782 when Matthew Murray was seventeen his mother died and was buried in All Saints parish on 5th December.

The next certain event in Matthew Murray's life is his marriage. This took place by the normal route of Banns being published at St Mary's Gateshead on 28th August 1785 and the 4th September 1785. The marriage itself was celebrated on the 25th September 1785 and is recorded as follows:

"Matt^w Murray & Mary Thompson both of this Parish were married in this Church by Banns this 25th day of Sept 1785 by me Thos Bowman Curate. This marriage was solemnized between us Matthew Murray Mary Thompson in the presence of us Geo Haggerston his mark John Bowman."

Mary Thompson was the daughter of John and Ann Thompson and was born in Whickham on 26th February 1764. Whickham is the next village west of Gateshead, but is a separate parish.

As stated at the beginning of this chapter the parish church of St Mary's Gateshead is easily in view of the parish church of All Saints Newcastle and around a twenty minute walk depending on one's state of fitness. However this does not explain why the marriage took place in Gateshead. The first reason is probably that Mary Thompson was by then living there and qualifying for the description in the marriage lines as "of this parish". The same description is applied to Matthew Murray which is quite surprising, unless the inclusion of Gateshead in Newcastle went further than thought, but this presents difficulties as the Tyne divided the diocese of Northumberland and the diocese of Durham.

Matthew and Mary's first child a daughter, Margaret, was born in Stockton on Tees on August 27th 1786 over nine months later. Matthew Murray was still probably under twenty-one.

Three explanations appear credible but there may be something completely different which has become buried over time. Firstly Matthew Murray may have been apprenticed to a whitesmith in Gateshead, and living on the premises; Gateshead directories show four or five iron workers in the areas of Battlebank and Pipewellgate. Secondly his employer may have gone bankrupt or ceased trading. Thirdly Matthew Murray may have been born some months or even a year before he was baptised. Joseph March in his notes for Leeds Worthies works back from Matthew Murray's age to the fact that he was born in 1765. Joseph March is clear that Matthew Murray spent his journeyman years in Stockton and it is evidenced by Margaret's baptism on October 22nd 1786 (two months after her birth). There would therefore have needed to be some fudging of the apprenticeship issue.

Notes

1.   B & W (Boulton and Watt) papers Birmingham Central Library, Letter Book MS 3147/3/93. Industrial Biography by Samuel Smiles, John Murray 1863.

2. The Local Records of Stockton and Neighbourhood by Thomas Richmond published 1868, reissued 1972 by Patrick & Shotton.

3. The Biographia Leodiensis by Rev R V Taylor, John Hamer 1865.

4. Durham Diocese Marriage Bonds and Allegations 1692-1900, DDR EJ MLA 1 1764 30711. I am grateful to Sheila Bye for telling me of this record.

5. The Deserted Village of West Whelpington, Northumberland; Second Report by Michael G Jarrett, Society of Antiquaries of Newcastle upon Tyne 1970.

6. Tyne & Wear Archives, All Saints Rate Assessment Books.

7. The History and Antiquities of the Town and County of the Town of Newcastle upon Tyne by John Brand 1789.

8. The History of Newcastle upon Tyne by Henry Bourne, John White 1736 and Frank Graham 1980. Page 88.

9. Samuel Owen, who worked for both Boulton and Watt and Fenton Murray and Wood before establishing himself in Sweden in 1806.

10. The History of Newcastle upon Tyne by Henry Bourne, John White 1736 and Frank Graham 1980. Page 106.

Chapter 2

Stockton on Tees, Developments in Cotton Spinning and the first attempts at flax spinning.

At some point after his marriage, Matthew Murray and his wife Mary left the Newcastle area for Stockton on Tees. This in itself would on the surface appear to be a questionable move for a whitesmith newly out of indentures, as it would be expected that there would be more job opportunities in the Newcastle / Gateshead area than in the much smaller and quieter Stockton.

In 1784 the Gentleman's Magazine published an account of Stockton on Tees, from which the following extracts are taken:

"At the end of this road the principal street affords a prospect which can scarcely be excelled in beauty in any town view. The whole town is well paved, and kept extremely clean; and, though there is little doubt of its being a very ancient place, it is so full of elegant modern houses that scarce a building in it bears the stamp of antiquity except an old house in the market place. Here are two markets a week, on Wednesday and Saturday, well supplied with provisions of every kind, which are generally sold at more moderate prices than in any other market town in the neighbourhood.
The ale brewed here is highly esteemed by the lovers of that liquor. Much sail-cloth is manufactured; and many ships, greatly admired for their beauty and strength, are built here; a company of Gentlemen are likewise engaged in the business of sugar refining. Several ships are constantly employed by the merchants of this place in the London Trade; they also carry on a traffic with Holland, Norway &c; their exports consisting chiefly of lead, corn, butter, pork &c are very considerable. A charity school for the educating and cloathing 20 boys and 15 girls, is supported by the donations and subscriptions of the inhabitants" [1].

Joseph March in his information supplied to the Reverend R V Taylor states that Matthew Murray

"for a year or two worked in a mechanic's shop at Stockton" [2].

It was during the Murray's time at Stockton that their first child Margaret was born on 27th August 1786, being baptised in the parish church of St Thomas on October 22nd of the same year. This short section states all the facts known of Matthew Murray's time at Stockton. However as will be seen there are possibilities, if not probabilities as to the work he did.

About 10 miles from Stockton on Tees lies the town of Darlington, and in this period some important developments were taking place in the mechanisation of the process of spinning flax. However before these are examined it would be helpful to examine the background to the explosion of inventions in the textile trade by examining the most pertinent inventions which were made in the cotton industry and upon which the other inventions in the main were based.

Part of the impetus to invent machines for spinning cotton came through the use of John Kay's fly shuttle which although invented in 1733 did not become widely adopted for weaving until the late 1750s to 1760 once its advantages were demonstrated. Such was the improvement in efficiency that resulted from its use in weaving that there began to be a shortage of yarn to be woven. By the 1760s some estimates put the number of spinners or yarn producers needed to satisfy one cotton weaver at around eight.

A Mr Paul and a Mr Wyatt invented a spinning machine c 1740 which operated groups consisting of a pair of rollers, a flyer spindle and bobbin. The machine worked from a central driving shaft with the groups of rollers, spindles etc arranged around it in a circle. The utilisation of only one pair of rollers to each spindle and flyer will have meant that the sliver sent for spinning was unlikely to be fine enough to produce a quality yarn. While a number of concerns persevered with the system until the 1750s it was dropped around that time.

James Hargreaves came up with his well-known spinning jenny around 1767 (Rees Cyclopaedia) which was not patented until 1770. The jenny relies on drawing and twisting only and did not in its nascent form use rollers, and primarily was aimed at mimicking the action of a hand weaver but with a machine capable of producing more than one yarn at a time. The jenny did work and was fairly widely used and read across into the woollen trade. It was extended to produce over 50 yarns, but remained difficult to use, required a skilled operative and was never adapted to a factory system.

Then in 1768 Richard Arkwright with a John Kay (no relation to the fly shuttle inventor) began working to produce their machine for 'roller spinning'. The basic construction of the drawing-rollers, spindles, flyers and bobbins were fundamentally the same as those of Paul and Wyatt. The difference was that Arkwright's machine was rectangular with a completely different shafting arrangement and used a number of sequential weighted rollers. Also Arkwright's bobbins were turned by the yarn and not mechanically as were Paul and Wyatt's. The initial trials were at a small horse powered mill in Nottingham. Arkwright had originally secured some funding through a partnership with a John Smalley and a David Thornley. Arkwright's first patent was granted in 1769. The horse mill must have produced good enough quality yarns to be considered of use in the manufacture of stockings in the successful firms of Jedediah Strutt and Samuel Need. In 1770 the above partnership was extended to include both Jedediah Strutt and Samuel Need. In 1771 the large water powered mill at Cromford was erected and larger versions of the roller spinning machines installed. Due to the source of power they now became known as 'water frames'. Within the early mills the processes to take in bales of raw cotton and output yarn were all established. The cotton had to be cleaned and seeds removed. The cotton then had to be carded into a roving or sliver. The carded cotton was too broad to immediately enter the spinning process, and had not enough strength to stop it braking unless it was twisted. Any twist had to very light as if there were too much it would prevent the fibres being drawn out finer in the spinning machine. The carded sliver also still consisted of fibres lying in all directions and to achieve a good spun yarn they needed to be as parallel as possible. Arkwright introduced 'drawing frames' which drew out or drafted the slivers through 3 or 4 pairs of drawing rollers. The drawing out on each frame was dependent on each set of rollers moving successively at a higher speed. The back rollers that admitted the roving being the slowest and the front ones the quickest. The amount of drawing out was usually matched by a requisite input of slivers so that the end sliver did not need a twist, so if the slivers were drawn out 4 times finer, the input was 4 slivers. In terms of putting in the slight twist for strength various methods were tried including lantern frames and jack in the boxes. It became the practice in Arkwright's mills to use three drawing frames in succession and to merge 6 slivers in each drawing meaning an input of 216 slivers, before one was put to the final spinning. It had been found that combining irregularities in the slivers by drawing tended to eliminate them and the above formula gave good results. Arkwright's patent included the words:

"A *new piece of machinery never before found out, practised, or used, for the making of weft or yarn from cotton, flax, and wool, which would be of great utility to a great many manufacturers in this his Kingdom of England, as well as to his subjects in general, by employing a great number of poor people in working the said machinery, and by making the said weft or yarn much superior in quality to any heretofore manufactured or made"* [3].

Arkwright's patents were challenged, one of the challengers being a Mr High who claimed to have been making a roller spinner with John Kay when Arkwright had persuaded Kay to go and work for him. Kay supported High's claim. This culminated in concerted action in 1785, and Arkwright lost the patent. However there were two important technological advances in Arkwright's patent that probably distinguished his machine from Paul and Wyatt's and Thomas High's. Arkwright realised that the rollers had to be set apart at the right distances for the cotton fibres. The fibres also need to be held tightly in the rollers and to this end Arkwright used weights to hold his rollers together with the necessary tightness. So far as can be ascertained with hindsight neither High nor Paul and Wyatt had understood these principles.

As stated in the quote from the patent above, Arkwright's patent, as would most of the textile spinning patents of the era, claimed that the machine would spin cotton, flax or wool. As initially constructed it would not have worked for either wool or flax, because their fibre lengths are different to cotton and to each other. The average length of cotton fibres is about an inch, with the best cotton fibres being 1 ½ inches; most wool fibres are 2 to 6 inches long and flax tends to end up with two lengths, which are incompatible. If the distance between the rollers is shorter than the fibre length then the fibres will be broken by being held by two sets of rollers moving at different speeds, and if the distance between the rollers exceeds the fibre length by too much the fibres end up not going uniformly forward, drift across the yarn and make it lumpy and uneven. Additionally in the latter case the yarn may well break. The weights ensure that the rollers have a good enough grip on the fibres and this requirement can also vary from textile to textile. So a machine set up for one textile will not work on those settings for other textiles. These factors explain much of the work that had to be undertaken by John Marshall and Matthew Murray in obtaining satisfactory flax spinning machinery, made more complicated by the flax being divided into tow (short fibres) and line(long fibres), this will be described in the next chapter.

About 1773 it was suggested the yarns would serve as warp yarns for weaving calicos, and this was successful. Belper mill (a Strutt mill) began to concentrate on yarns for hosiery and Cromford on yarns for cotton goods. Other mills were established at Milford Derbyshire, Wirksworth, Matlock Bath, and Bakewell. During this period Need died and in 1781, Arkwright and Strutt dissolved their partnership. Strutt had Belper Mill, and Arkwright had Cromford Mill. Many other factories were started under license.

Arkwright does not seem to have lost financially as a result of losing his patent. Among many other successful ventures he formed a partnership with David Dale and in 1784 they established New Lanark.

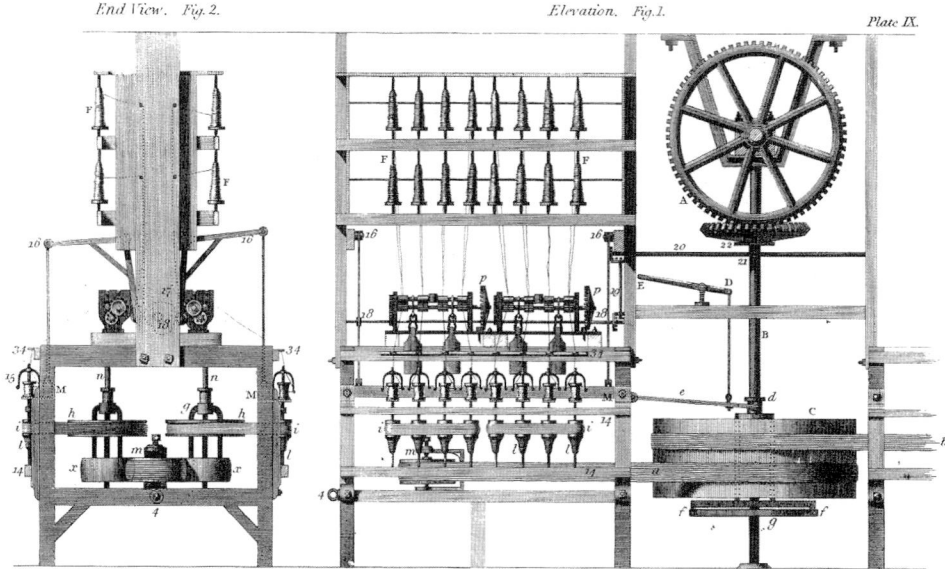

1 Arkwright Water Frame as developed.

The following side view clearly shows the roving entering the rollers from the upper side, the three sets of rollers with their weights hanging down and the resultant yarn passing to the flyer where it was twisted and wound onto a bobbin.

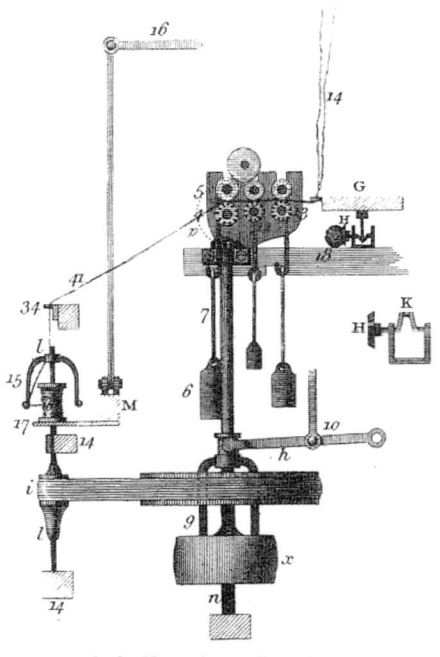

2 Detail of spindle

The next fundamental invention was Samuel Crompton's spinning mule. Crompton was born near Bolton 1753. He worked in a spinning family using an early jenny. He was not satisfied with the wefts produced and set out to improve the machine. He replaced the clasp/drawbar used by Hargreaves to release and hold the slubbings to control the input to the jenny, with a pair of rollers pressed together by springs and also a carriage to carry the spindles so that the combined spindles and carriage became the moveable part that drew the textile out. Crompton was using his machine to spin by 1778/9, some 8 -10 years after Arkwright's water frame had commenced, but Crompton always denied knowing of Arkwright's machine or even of Paul and Wyatt's earlier use of rollers. It has been opined that Crompton could not have been familiar with Arkwright's machine as if he had been he would have developed gearing to drive his rollers and not used bands and pulleys. Crompton did not patent his ideas and was persuaded to give away the details in exchange for an unstated payment, which in the end was around £60. Crompton's machine spun more yarn than the other systems and because it combined rollers and the stretching or pulling out process the yarns were much finer and softer than those otherwise produced.

Crompton's machines were all man powered by turning a wheel and it was not until 1825 that Richard Roberts produced the fully automatic mule. Many other improvements took place as this type of machine and its derivatives remained one of the main cotton yarn producers into the 20th century.

Ree's Cyclopaedia provides illustrations of the mule, although of one somewhat later than Crompton's initial machines:

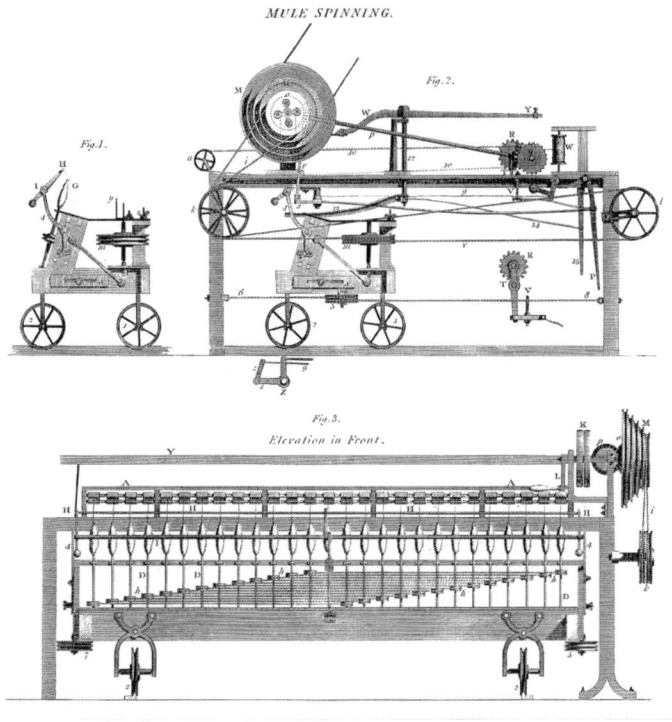

3 Mule Spinner

Alongside, and following the major impact the above machines were having on the mechanisation of the cotton industry, other mechanics and inventors were patenting machines for the preparation and spinning of other textiles [4].

In 1783 a Samuel Barber of Derby took out patent number 1365 for:

"an *engine or machine for the reducing and spinning of wove flax, cotton, wool etc*" [5].

In the same year and entered just days after Barber's patent there came a combined carding and spinning frame patented by Thomas Oldham and George Prestwidge. They lived at Shirland just outside Alfreton north of Derby [5].

Also in 1783, a Benjamin Partridge of Halesowen, a wick yarn manufacturer, patented another carding and spinning machine, specifically for "*the spinning of short tow, commonly called hurds*" [5].

In 1796 a John Royds of Falinge, close to Rochdale, took a patent for a frame to draw double or spin various fibres [5].

Then on June 19th 1787 a patent was granted to:

"*Mr. John Kendrew, of Darlington, in the County of Durham, Optic Glass-grinder, and Mr. Thomas Porthouse of the same place, clockmaker; for a new Mill or Machine, upon new principles, for spinning Yarn from Hemp, Tow, Flax, or Wool*" [6].

Darlington had developed a reputation for linen bleaching in the early part of the 18th century even, according to Defoe, having cloth brought from Scotland there for bleaching. By the middle of the century it was a centre for linen manufacture of the sort used in making napkins and tablecloths.

John Kendrew, a Quaker, was brought up as a handloom weaver to make 'checked tammys', a common local worsted product, but was obviously keen to move on as he developed a small machine for grinding and polishing concave and convex glasses for spectacles and optical instruments and he progressed to running a sizeable business from Low Mill and traded his optical wares as far as Birmingham. However his machinery seems to have been copied in Birmingham and Sheffield, and trade dropped off. While visiting Lancashire he saw cotton machnery in operation and considered applying the principles to spinning flax. He enlisted Thomas Porthouse, a watchmaker with some money and mechanical skill, and they worked together to produce their machinery. They probably had to spend some considerable time in development due to the peculiarities of the fibre, but eventually with the financial assistance of Mr Backhouse the banker they obtained their royal patent [7].

The machinery put forward and described in the patent fundamentally consists of a drawing machine / spreadboard and a spinning machine as follows:

Description of the spreadboard and drawing frame figure 1

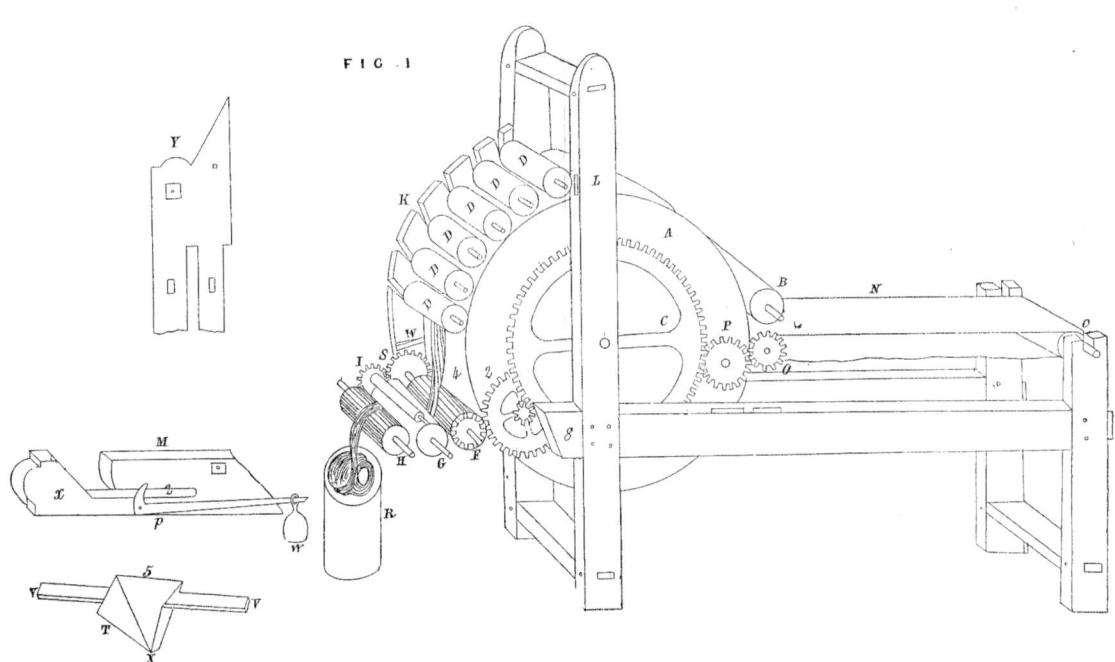

4. Kendrew & Porthouse's Spreadboard & Drawing machine

A cylinder ten inches (25.4 cm) wide and three foot (91.44 cm) in diameter (shown as A) made of wood or metal is covered in smooth leather. Set in a slotted timber frame on the upper quarter of this cylinder on the other side to the spreadboard are a number of rollers (D), also covered in leather, which press down onto the main cylinder (A). These rollers (D) are of different weights, ranging from the uppermost at two stone (12.7 kilos) to two pounds (0.907 kilos). Then underneath the rollers (D) and moving out from the main cylinder (A) there are in succession three further different rollers. First is an iron fluted roller (F) again rotating in a slit in the frame and having a toothed wheel at each end to drive it, followed by a wooden roller (G) covered in cloth and then smooth leather , and thirdly a further iron fluted roller (H). These rollers fit into the slit marked (Z) in the frame (M) shown in figure 1. The rollers G and F are squeezed together by the lever p and its weight w (fig 1). The roller H is pressed onto the roller G by virtue of being on the inclined slope (plane) shown at x in figure 1. A subsidiary rubbing roller, covered in woollen cloth sits upon G and prevents dirt or fibres sticking to G. It is operated by the cogged wheels I and S.

There is a cloth N, over the spread board proper, turning on two rollers O, O which are driven from the wheels C and P. This cloth and the cylinder A move at the same speed. The wheels O,P and the axle of the roller shown at B are supported by a frame Y ,on either side of the machine, shown in figure 1 which fits in the slots on the main frame ZZ.
At the side of the moving cloth N is a table (not shown) of the same dimensions, having two smooth cloths or leathers the size of the table.
The power source, water horse or other kind of mill, drives mill work which

is attached to and drives the roller F, which in turn drives Q and C and so the cylinder A and all the rollers, including O which are those that drive the cloth N.

The workman will spread a quantity of hemp, flax, tow, or wool (the material) evenly on the table, the quantity to depend on the fineness of the thread to be made. The material will then be placed on the cloth N which will move it to the cylinder A, and as N revolves to continue to accept the material, the material itself will pass under the roller B onto A and pass under the rollers D and then through a gathering device ab shown at figure 1c (which fits in the slots in the frame marked W) and then falling between the rollers F and G under G and over H. The system of rollers at F G H , due to its holding the material at two points manages to drag the long fibres forward , and the rollers here by moving at three times the rate of the remainder of the machine extend the length of the material by three times . The material at this point is referred to as the sliver, and the process of the rollers pulling the fibres forward as drawing. As the workman continuously lays further quantities of the material on to N, which overlap the previous quantities of material, it is drawn through the machine and a continuous sliver is produced which is deposited in the collecting can R. The device ab works by the narrow exit marked X being so small as to press the individual fibres close to each other such that they remain so in the sliver passing through the rollers F G H and do not entangle.

This first machine produces a thick sliver, which needs further reduction which is achieved by passing it through succeeding machines that operate the same process but do not have the moving cloth N as the sliver is presented directly under the roller B and drawn into a finer sliver until it is small enough for the spinning machine. The subsequent drawing machines have smaller cylinders, the cylinder diameter on the last in this process being two feet (60.96 cm). The aperture X in the device ab is reduced in succession as are the distances between the rollers D and the rollers F G H. These distances are not absolute and will depend on the material being processed and the length of the fibres [6].

Description of the Spinning Machine figure 2.

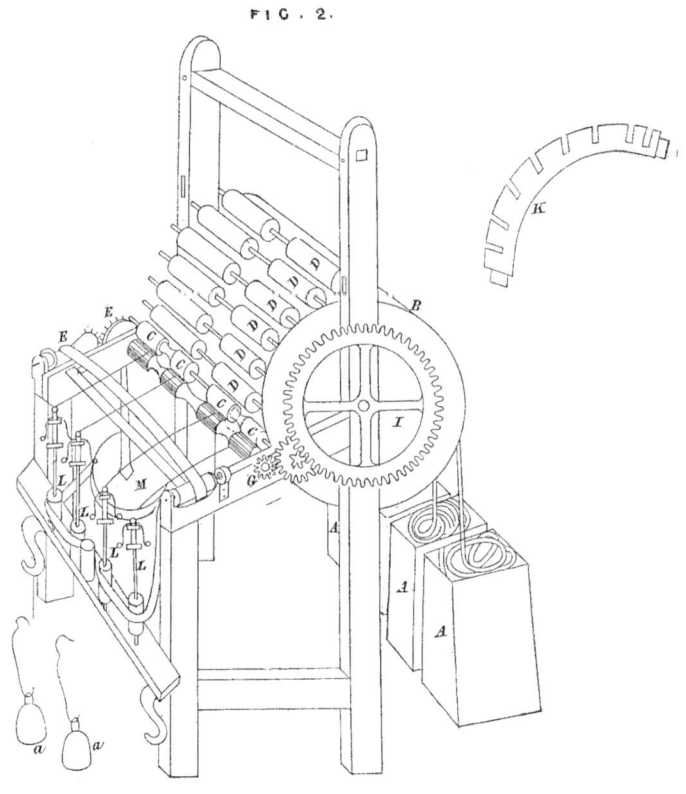

FIG. 2.

5 Kendrew & Porthouse's Spinning Machine

18

The spinning machine in its operation and its process of drawing the material through, works on the same principles as the drawing or spread board machine. The slivers are presented to the spinning machine from canisters shown as A and drawn over a cylinder B with rollers D. The fibres are drawn by the rollers marked C, where the lower roller is made of fluted iron. The upper one of wood and covered in leather. These rollers C rotate six to eight times faster than the cylinder B and are able to draw the material forward under the pressing cylinders D being squeezed together by the weights a,a, which hang off the rollers C by the wire hangers shown. The power source is applied to drive at point F a wheel on the line of the fluted roller (lower C) which drives:

> *two rollers moving a belt of smooth cloth from side to side*
> *by a small cogged wheel on the other end at G the wheel H which in turn*
> *drives the wheel I and hence the cylinder B, with its rollers.*

The rollers on the cylinder are held by metal plates K having slits to hold the axles and located within the main frame of the machine. These rollers D press upon the cylinder B at all times and are covered in cloth and leather. The top roller D weighs 10 lbs (4.53 kilos) and the bottom one about one lb (0.45 kilos).

After passing the rollers C the yarn commences to be turned by the revolving spindles L, and if linen, is rubbed over the wet belt cloth (for wool this must be removed) and then descends to collect on the spindles as fast as the rollers let it. The spindles L are turned by a belt operated by a wheel M driven by a shaft descending from the wheel F [6].

T Turner in his 1964 thesis on Fenton Murray and Wood describes this machinery as:

"of the utmost crudity but it was the first to succeed, albeit only partially, in the processing of long flax fibre. The conditions required were that sufficient drag was provided for the drawing operation, and also that the long fibres were supported over their whole length, and at the same time allowed to slide over one another" [5].

Kendrew and Porthouse's machinery attracted attention. Between 1788 and 1792 it was licensed to four Scottish firms, including Sims and Thom on the Haughs of Bervie and James Ivory and Co at Douglas town near Forfar. The machinery was also copied illegally with the party being successfully sued for patent infringement, and it may well be that the advert reproduced below, which appeared widely over a number of months, gives the names of the defendants.

To the MANUFACTURERS of LINEN and WORSTED Yarn
George Parkinson and Robert Eastgate, manufacturers, of Darlington, in the county of Durham, respectively beg leave to inform their friends and the public that machines for the spinning of hemp, lint, tow, and wool into yarn, are now fitting up, and will soon be ready for inspection, when public notice will be given.
The machines are simple in construction, making exceeding good yarn with dispatch, and are erected at a moderate expence, which it is apprehended will recommend them to the notice of every manufacturer.
Samples of yarn, on the shortest notice, will be sent by the readiest conveyance to any part of the kingdom and application in person, or by letter will meet with due attention.
Darlington 12th Nov. 1788. [8].

This advert received a prompt response from Kendrew and Porthouse which alleged that Parkinson and Eastgate had used a whitesmith employed to work on Kendrew and Porthouse's patent machines, because they were unable to replicate some parts themselves, and that anyone using Parkinson and Eastgate's machinery would: "*be prosecuted for damages under the authority of the patent*" [9].

By February 1789, despite Parkinson and Eastgate still pushing their advert, a further piece in the papers from Kendrew, Porthouse and Backhouse (their financial partner from the Backhouse banking family) stated that Parkinson and Eastgate had applied for a patent, the Attorney General had called for a model and upon examining it stated it to be the same in principle and declining to go forward with the patent. This series of adverts also provided the outline terms of license for using Kendrew & Co's patent machinery. By March Kendrew & Co's advert was repeated with an additional piece on the end that the offending machine could be examined at their mill. The name of the whitesmith is not given. Many authorities, originating with a reference in Smiles, quote as a fact that Murray worked for Kendrew, but there is no surviving evidence. This case does raise some additional speculation, however at this distance in time it is unlikely that the position can be proven one way or the other.

The most interesting person to approach Kendrew and Porthouse for a license was a Mr John Marshall of Leeds. John Marshall was born in 1765, the same year as Matthew Murray, but Marshall's father was a successful Linen Merchant with a shop at 1 Briggate Leeds. Early in December 1787 Jeremiah Marshall, John's father died suddenly. John Marshall at the age of twenty two found himself controlling partner in a business and an estate worth £9000. Marshall had seen the immense profits made from cotton spinning and had been contemplating how to spin flax by machinery, in the hope that it would be equally advantageous. He was now in a position to try it out.

Marshall went to Darlington with great haste, and presumably reached an agreement for a license at least in principle, for he returned to Leeds [9]. Marshall formed a partnership and took a lease from James Whitely of 'a New Erected Water Mill called Scotland Mill' in the parish of Adel to the north of Leeds on 5th January 1788. This, Marshall's first partnership which traded as Marshall Fenton and Dearlove actually consisted of John Marshall, Samuel Fenton junior, James Fenton junior, Mary Marshall (John's mother), and Ralph Dearlove. Samuel Fenton had been in partnership with Jeremiah Marshall. James Fenton, a member of the same family, was later to join Matthew Murray and David Wood in the firm of Fenton, Murray and Wood. Ralph Dearlove was a Knaresborough linen merchant. Mary Marshall and James Fenton were sleeping partners.[11]

Notes

1. Gentlemens Magazine Volume 2 1784.
2. The Biographia Leodiensis by Rev R V Taylor, John Hamer 1865.
3. Patent number 931 of 1769.
4. For this period see: The Textile Industry by W. English Longmans 1969; Richard Arkwright by Richard L Hills Priory Press 1973; The Strutts and the Arkwrights by Fitton and Wadsworth M.U.P. 1958.
5. Fenton Murray & Wood, unpublished thesis by Trevor Turner
6. Patent number 1613 granted to John Kendrew and Thomas Porthouse for a new mill or machine, upon new principles, for spinning yarn from hemp, tow, flax, or wool. June 19th 1787.
7. Victoria County history of Durham – Linen.   The History and Antiquities of the Parish of Darlington by W.   Longstaffe, 1854. Page 311.
8. One of the later adverts which appeared in the Newcastle Courant Saturday November 15th 1788.
9. The Caledonian Mercury Monday October 13 1788.
10. Marshall's of Leeds 1788-1886 by W G Rimmer, Cambridge University Press 1960. Also John Marshall's sketch of his own Life, Marshall Papers MS200, Brotherton Library Leeds University.
11. London Gazette 6th April 1793 notice of dissolution of partnership.

6. Portrait of John Marshall from Hallsteads as High Sherrif of Cumberland in 1821, courtesy of the Outward Bound Trust.

Chapter 3

The Marshall Years

"I longed for an employment where there was a field for exertion and improvement, where difficulties were to be encountered, and distinction and riches to be obtained by overcoming them. My attention was accidentally turned to spinning of flax by machinery, it being a thing much wished for by the linen manufacturers" [1]

So wrote John Marshall in his recollections dated the 18th September 1796 describing how he felt after his father's death. He goes on to record

"My partner S. Fenton was as eager to undertake it as myself, and in partnership with Ralph Dearlove of Knaresborough a linen manufacturer of great probity and good sense, we took a small water mill at Scotland four miles from Leeds, and worked some machinery on the patent plan of Kendrew and Co" [1]

It is interesting that John Marshall does not include either his mother or James Fenton among his partners, hence an assumption that they were sleeping partners. John Marshall had presumably been contemplating what he wished to make of his life for some time prior to his father's death in December 1787 as it was immediately afterwards in January 1788 that the new partnership of Marshall, Fenton and Dearlove agreed the lease of Scotland Mill, described as a "New Erected Water Mill" in the parish of Adel to the north of Leeds [2]. The lease document states that the

"Lessees...have erected a very large Water Wheel at the side of the Mill & have also erected within the Mill several machines, Benches & other things for carrying on their manufactory of Linen Cloth and the spinning of Linen yarn some of which are let into the Outer Wall of the said Mill" [3].

The lease on the Mill had a break clause allowing the lease to be terminated after three years.

With Mill and machinery set up, the partners under John Marshall's drive attempted to get it to work. Marshall recorded simply that Kendrew and Porthouse's machinery "did not answer" and Rimmer states that the drawing frames could not reduce

" 'tow', and only heavy (5-7 lea)' line' yarns could be spun. There were frequent breakages and the yarn was lumpy and hairy" [2].

Line was the term applied to the long fibres produced through the preparatory stages. Tow was the term given to a large proportion of the flax emerging from the preparatory processes of scutching and hackling and consisted of the shorter fibres. Within tow there were divisions into long tow and shorts

By the middle of 1788 John Marshall had realised that his machinery had to be improved in order to make yarns that could compete with hand spun work, which produced much finer linens, and so make the partnership a profit. John Marshall knew that to realise the maximum commercial potential it would be necessary to machine spin tow as well as line.

So John Marshall turned his attention to spinning tow, and being of a fairly systematic frame of mind, as he started conducting a series of experiments, he kept a record. His experiment books survive among the Marshall papers at Leeds University and from these it can be seen that between June and October 1788 John Marshall conducted a series of seventeen experiments. The descriptions of the first thirteen of these experiments are quite brief and none of them gives any indication of the team performing the experiments. We do not know whether John Marshall's partners were involved or the names of any other person who may have participated.

As with all the attempts to mechanise linen production the impetus was coming from the successful development of machinery for cotton, wool and other textiles. The example of men such as Arkwright, with major mills in Derbyshire not far south of Leeds fired a few persons to find a way of applying the principles of the cotton machinery to flax and other textiles, amongst whom was John Kendrew and John Marshall with the possibilities of achieving commercial success. John Marshall seems therefore to have kept an ear to the ground for developments in other textiles, and by his fourteenth experiment used some knowledge of developments happening in the neighbourhood where Arkwright's methods were being tried for worsted spinning. Marshall introduced the use of a jack frame in his experiments. Arkwright had used jacks to try and sort out a method of winding soft sliver onto a bobbin without breakage and with the correct twist. Marshall obtained the 'jack' from Addingham, between Ilkley and Skipton, where the first mechanised worsted mill had recently started up. Marshall tried long tow in experiment 14 and short tow in experiment 15 both with fairly unpromising results in terms of obtaining yarn to the necessary strength, and this appeared to be due to issues with getting enough twine [twist] in the yarn and a compact enough sliver.

By the middle of September 1788 he was trying out the machinery at the mill and after a number of changes and adjustments he observed

" *the sliver and carding were got very even and the yarn tolerable so, but it was continually breaking in spinning so that it was one persons work to attend to six spindles, which we supposed must be owing to its being spun off the side, the fibres being all jumbled together and not drawn lengthways*" [4].

By 27th September they had established that one of the major issues with this set up was that the rollers were too far apart. By October 8th John Marshall was trying to draw shorts of from twelve to sixteen inches and finding the results very uneven, and

24

the ends of the fibres were falling down between the rollers and going round with them, rather than progressing to the next set of rollers.

Experiment number 17 started on October 11th , and John Marshall seems to have been trying to draw the flax out into a sliver in a similar manner to that used for worsted but the result was still unsatisfactory and he recorded:

" from the above experiment it appears that flax will not spin with rollers the common [usual] way because the fibers will not stick together so much as to hand forward from one roller to another especially at such distances [between rollers] as the length of the fibres requires them to be" and "it will not draw after it has got much if any twine[twist]" [4].

On this disheartening note the first series of experiments ceased and John Marshall did not record any more experiments for seven months.

John Marshall returned to experimenting in June 1789, when he wrote up an experiment he called

"Wetting the yarn by a pair of rollers before the fluted roller before it was twined" [4],

this appears to be an attempt to improve the quality of his line yarns. He introduced a pair of pewter rollers, to replace the damp cloth used in Kendrew & Co.'s method, which caused the sliver to be wet in a loose state, and by making the rollers draw out the sliver, it was made more even and smoother and it spun very well until the bobbin was half full after which the bobbin pulled the sliver too much and broke it.

In the middle of July 1789 John Marshall paid a visit to Darlington to see what progress Kendrew & Co. were making in progressing their machinery and did not gain any satisfaction. While there he tried some experiments both at Kendrew's Darlington Mill and at Porthouse's Mill at Coatham. These again seem to have had some promising aspects but once again yielded disappointing results, with the carding equipment at Coatham refusing to doff (take off) the sliver as it could not be extracted from the hackle teeth on the drawing rollers.

Back home at Scotland Mill he met the same problem in another experiment he dated August 1789 for which he wrote the following conclusion:

"There appears to be no chance of spinning 'line' without having previously obtained a regular disposition of the fibres so that there may be a regular succession of ends and not too many come forward at a time. The card cylinder in the manner we use it has the desired effect as to the disposition of the fibres, but it has a very great fault in producing an infinite number of knots upon the 'line' and likewise breaking the fibres which makes bad yarn and appears to be the chief cause of the great number of breakages in spinning" [4].

So between June 1788 and August 1789 John Marshall and others had conducted an extensive set of experiments with both 'line' and 'tow' work to try and spin finer yarns with 'line' and to spin 'tow' and improve the value of the product they were taking to market. All they could produce from Kendrew & Co.'s machines were coarse linen yarns. They had explored the properties and peculiarities of flax, in comparison to the known attributes of cotton and wool, and the differences that existed between the two ends of the flax spectrum of fibres in 'line' and 'tow', and on occasion came close to finding the way to spin' tow '. They had also explored many variations in machinery for carding, drawing and spinning with different numbers and sizes of rollers, different draws or tensions exerted (by differential roller speeds) on the fibres at different stages within any machine. However they did not appear to be able to identify with any precision why a particular set-up did not work satisfactorily, or what to change to achieve the solution. John Marshall's consolation must surely have been that the problems relating to spinning the unyielding flax fibre seemed to be eluding his competitors as well as him. They continued to spin much coarser yarns than they wished to place in the market.

So in December 1788 John Marshall starts to record a different experiment, moving beyond spinning he started to look at a weaving loom. It is probable that John Marshall was aware that Edmund Cartwright had taken out four patents for power looms between 1785 and 1788 and in 1787 had set up factory with twenty looms in Doncaster, and this was the catalyst that prompted the new experiment. At Scotland Mill John Marshall records that they began to work four looms made by John Jubb (a leading early Leeds machine maker), and found them very unreliable partly due to the

"badness of the workmanship". "Jackson [an operator] *did not on an average weave more than 2 pieces a week of our own spun warps and foreign wefts. Ranson once wove five feet but in general only about three pieces a week."* [4].

John Marshall wrote that little change was made until September 1789

" when we gave over spinning and set Matt Murray to work on a new loom" [4] These words could be taken to imply that Matthew Murray was already working with John Marshall on the spinning experiments and was therefore not available to work on the loom until September, but it is unclear.

But Matthew Murray has re-emerged onto the scene working for John Marshall at Scotland Mill near Leeds, our last factual placing of him being the occasion of the baptism of his daughter Margaret in Stockton on Tees on 22 October 1786 nearly two years before.

It is not known when Matthew Murray joined Marshall Fenton & Co or the circumstances. We have Samuel Smiles' account:

"trade being slack at Stockton, he [Matthew Murray] found it necessary to look for work elsewhere. Leaving his wife behind him, he set out for Leeds with his bundle on his back, and after a long journey on foot, he reached that town with not enough money left in his pocket to pay for a bed at the Bay Horse inn, where he put up. But telling the landlord that he expected work at Marshall's, and seeming to be a respectable young man, the landlord trusted him; and he was so fortunate as to obtain the job which he sought at Mr Marshall's, who was then beginning the manufacture of flax, for which the firm has since become famous" [5].

In a footnote Smiles also states:

" It is possible that Matthew Murray may have obtained some experience of flax machinery in working for Kendrew, which afterwards proved of use to him in Mr Marshall's establishment" [5].

The part about walking to Leeds is probably correct as it was the only way for a person without money to move around, stage coaches were prohibitively expensive, and history tells that another future character in the book, William Murdock walked a lot farther in order to find employment with Matthew Boulton. The possibility of Murray having had involvement with Kendrew and Porthouse has been covered in Chapter 2. Stockton as was seen from the description in the previous chapter was at the time a centre of sail making and if Murray had travelled there to work in such an establishment, or to work in a machine makers shop supplying the local area, he could also well have rubbed shoulders with other machine makers who were familiar with what was happening at Darlington a few miles away. It is therefore most likely that the walk did have specific employment with Marshall in mind.

Despite the fact that there had been no satisfactory outcome as yet to the experiments in terms of a fine enough yarn and machinery to spin it, John Marshall must have remained optimistic because he was already planning the next move. We know from his notebooks that he was familiar with the work of Joshua Wrigley of Manchester, who built his adaptation of Savery engines to pump water up to a cistern to drive water wheels. The water wheel was seen at the time as giving a much smoother drive for textile machinery than steam engines, with less chance of sudden stoppages and consequent damage to machinery and broken yarn. John Marshall noted that Wrigley was of the opinion that only Boulton and Watt crank engines produced a motion sufficiently regular for spinning. On the 25th of February 1789 John Marshall wrote to Boulton and Watt asking various questions about their engines:

"1. Whether you would recommend a crank engine or a water wheel for working machinery that requires an uniform motion. We want ours for spinning & would wish to know if you have any where erected a crank engine for that or any similar purpose and whether you think the motion will be as regular as a water wheel, and whether in your engines the crank is ever liable to turn the wrong way round as we understand it is in engines of the common construction.

2. What difference there will be in the first cost of an engine of your principle from a common one of the same power, and in the first cost of a crank engine, including the large fly wheel, & of one to pump water say 12 yards high (that would turn the same machinery) including the pumps. We want ours to turn three thousand spindles, if you know the power required we should be glad you would inform us what you think the expence would be.
3. What difference there would be in the expence of coals in an engine of yours working machinery by a crank from a common engine pumping water sufficient to work the same machinery.
4. Upon what terms you erect them"[6].

On the 17th of March Boulton and Watt replied as follows:

"We produce circular motions in our Engines, not by means of a crank but by two wheels one of which revolves round the other, which we call the rotative motion, & which we esteem to be more regular than a crank. We have at Nottingham, erected several engines with that motion for spinning cotton & one for the same purpose for Messrs Peels & Ainsworth at Warrington, all which give satisfaction, and are not complained of for turning the contrary way, which if they were liable to could easily be prevented.

We are ignorant of the costs of common Steam Engines, but believe, of the same powers & equally well executed they will not be cheaper than ours. Engines employed to raise water to work water wheels must be much larger than those which perform their work immediately by rotative motions, as about half their power is absorbed by the pump & wheel & they cannot make so many strokes per minute as those which work rotative motions.

It is commonly reckoned that 100 cotton spindles with their carding & co are equal to the power of one horse but if the machinery is very well executed less power may be sufficient. In respect to spindles for spinning wool, we have had no experience, but are now erecting a pretty large engine for that purpose.

Therefore supposing your spindles to be for cotton, or to require the same power, an engine equal to the force of 30 horses working together would be sufficient.

Such an engine, of the construction we call double engines (because the piston is acted upon both in its ascent & descent) would require a cylinder of 28 inches in diam. 6 ft long in the stroke & making 19 strokes a minute, which would burn about 360lbs weight, of our coals here, per hour, but with some coals would require a smaller quantity; if best Newcastle coals not much above 3 bushels an hour. As we have not executed any engines by the piece, and very rarely provide all the materials, we are not able to speak precisely to the cost, but expect it would not exceed £1000, independent of the engine house and mill work, but including the boiler & fly wheel with the first rotative shaft.

If you should otherwise approve of our proposals, we are willing to undertake all the cast iron, hammered iron, brass & copper work of the engine for a fixed sum, at least all that can be conveniently exercised here.

Our terms are, that our employers pay for the materials and erection of the engine, and also pay us for our trouble & the use of our inventions, £5 [.......] annually for each horse power the engine is equal to, which in the present case would be £150 [....] per year, to be paid us during the term of our patent 12 years to come. We furnish all necessary drawings and directions for fitting up the engine, and send a good workman to assist in the erection, who is paid by our employers"[7].

John Marshall was presumably not impressed, as when he moved to Holbeck he initially installed a Savery type arrangement provided by Wrigley. Some detail will be recorded of the four engine purchases from Boulton and Watt in which John Marshall was involved as they show an uncomfortable relationship with which John Marshall was obviously increasingly dissatisfied. As John Marshall's chief mechanic Murray will no doubt have been heavily involved in discussions with his employer on the subject, and here may be the seeds that encouraged Murray to enter the Steam Engine business himself, with John Marshall's active encouragement.

Meanwhile the experiments with the loom carried on until January 1790, with a little further work in June 1790, but while it is recorded in some detail with more than one diagram, a satisfactory loom did not materialise. In this the first major task mentioning Matthew Murray he was unsuccessful, and the challenge of a viable power loom was not met for another quarter of a century and in that time Marshall & Co. continued to use hand looms. However Matthew Murray must have displayed enough drive and knowledge for John Marshall to realise his potential as before the loom was finally abandoned a second series of spinning experiments was commenced in January 1790.

This new series of experiments marked the turning point in John Marshall's fortunes as at last real progress was made, and the experiments led directly to the taking out of a patent in Matthew Murray's name in June 1790.

The first experiment concerned laying out long tow and drawing it and twisting it at the same time. However

"we could not get an even sliver and supposed it was owing to the rollers being at too great a distance from the cylinder for the length of the staple" [4].

They then tried the set-up with short tow which drew worse and less even, sinking to the bottom of the heckle teeth and being difficult to extract, which they thought proved that the uneven sliver was not a result of rollers being too far from the cylinder and John Marshall continues

"Matthew thought it was occasioned by the tunnel conducting it to the rollers being too straight for it to get through" [4].

The machine they were working is shown in a diagram in John Marshall's books thus:

"a.is a cylinder of 10 inches diameter covered with heckle teeth, on its axis is a contrat wheel of 80 cogs worked by a suite of 7 [cogs] on the upright shaft (b). (c) a wheel of 96 cogs worked by a suite of (d) of 12 on the axis of the roller frame. (e) is a fixed wheel in which the wheel (f) runs as the frame turns round and by the suite (g) of 10 cogs working in the wheel (h) of 20 [cogs] moves the fluted roller which is 5 inches in circumference [and] therefore it has a draw of nearly 6 and gives the sliver a twine of 2 ½ inches

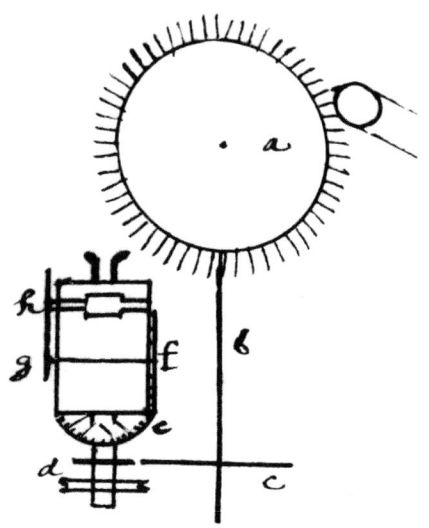

7 Flax spinning experiment January 1790.

They widened the tunnel, moved the rollers half an inch nearer to the cylinder, and altered the draw to 12 and the twist to one in five inches. This gives an idea of the number of variables they were having to contend with, many of which remained as issues in the design of flax machinery for decades. Depending on the fibre lengths would be the distance between the rollers; the differential speed of the rollers between the slow back rollers and the faster front rollers and the diameters of the rollers when balanced would create the correct draw for the type of flax being used; and with the final machine before spinning, the roving frame, the correct twist would need to be put in the yarn, sufficient to enable it to be spun, but if it was too great the slivers would not draw and would tangle and break. Some improvement was seen but not enough so by February 6th the draw and twist were changed again, and further changes made in laying out the sliver, but the sliver began to rope.

It is interesting to note that in this formal age when an employee was classified as a servant, John Marshall had started referring to Matthew Murray simply as Matthew. Compare this to his references earlier in the loom experiment to Jackson and Ranson. While Marshall and Murray were of an age, both being born in 1765 and were obviously working together extremely closely, it indicates that John Marshall had already developed a high regard for Matthew Murray's abilities.

For what John Marshall classed as experiment 2 with long tow, dated February 8th, the diameter of the cylinder was reduced from ten inches to two inches with heckle teeth one and a half inches long, the draw was changed to ten, by altering the gearing of the wheels, and the twist to one in two and half inches. They thought that the cylinder size was about right but the rollers could not pull the fibres out of the heckle teeth. Murray suggested reverting to a previous set-up of giving one motion only ("*as the present plan of giving two motions to the rollers was complicated*") to the rollers and to cut away part of the rollers so that the twist being given by the spindle would pass the roller and begin to work on the fibres between the cylinder and the rollers, before the rollers drew the fibres further

"*which would answer the same end as twining and drawing at the same time*" [4].

By February 11th, experiment 3, they were ready to try the fluted roller with one side cut away, but it did not answer due to the amount of twist in the spindle and the rollers breaking the sliver. So they placed the heckle roller at right angles to the feeding and drawing rollers and the sliver came very well off the heckle roller and drawing and twisting by hand produced a perfect roving, it therefore appeared certain that the principles would answer. John Marshall concludes his account of this first breakthrough experiment with:

"*Anything that would take & keep hold at a certain place would draw it truly because the twine which was in the sliver above that place would run forward as the sliver was drawn out & by that means take hold of the fresh fibres as they came forwards & likewise require less twine from the spindle*" [4].

Still looking for solutions and presumably ways of doffing the fibres from the hackle pins John Marshall recorded on February 20th:

"*Sent Matthew into Cheshire to enquire into their mode of spinning worsted by pincers. He could not get into their mills but was told that the pincers did not let the twine run up from the spindle. We therefore determined to go upon the old plan of drawing & twining at the same time. Made wires with springs to push the sliver out of the teeth of the heckle, by rising up at the point of drawing, which answered the end pretty well*" [4].

In experiment 4 of March 1790 they began to make a four spindle spinning frame with a revolving card sheet on the same principle. The roving seldom broke and when it did it was because it contained irregularities.

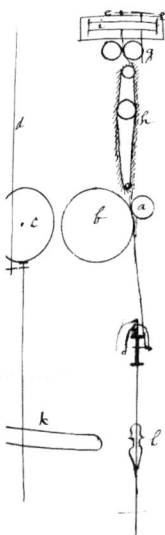

8 Flax Spinning experiment March 1790.

(h) is the revolving card 7 inches long by 1 inch broad moved by a ¾ inch back roller over a ¼ inch front roller

(l) the roving bobbin

(k) the drum 10 inches Sphaft (sic) 2 inches

drawing twist 2 in 1 inch

draw 6 ½

Spinning twist 7 in 1 inch

(a) the 1 ¼ inch diameter fluted roller

(b) the 4 inch diameter wood roller

(c) contrat wheel 72 cogs

(d) upright shaft on the bottom of which is a wheel of 31 worked by a suit of 10 [cogs] on the drum shaft, on top of the shaft(d) is a pulley 10 inches, which works another of 3 inches on top of the frame in which the back rollers, revolving card and roving bobbin all turn round together

(i) is a fixed wheel of 12 cogs round which the wheel (f) of 26 cogs works and by a worm at the other end of its shaft works a wheel of 20 [cogs] on the back roller, which works another wheel of 20 [cogs] on the back roller of the revolving card [4].

Trying for additional improvements, with experiment 5, they added a stationary roving bobbin above the back rollers which gave too much twist above the back rollers. So in experiment 6 they removed the back rollers, but the machine would not spin because the roving came forward irregularly from the bobbin and would not enter the revolving card when it was twisted hard. The solution as recorded by John Marshall was as follows:

"Matthew proposed to draw the sliver through two revolving sheets [of] leather instead of card, because the sliver has so great a tendency to go round with the card especially if any tooth be bent or out of order which it is very liable to be. One great difficulty in spinning [line] seems to be to keep back the short fibres sufficiently without holding the long ones too much, which a card seems to do sufficiently well, but in drawing it through a set of weighted rollers, e.g. 4, a long fibre has four times the weight upon it that a short one has, and a cylinder is worse still because the long fibres are held faster still by the curve and the short ones not at all" [4].

This proposal formed the basis of experiment 7 on April 3rd:

"Tried two sheets of leather instead of the card which seemed to answer the end of keeping back the short fibres without too great a pressure on the long ones equally well as the card; without its fault of taking the sliver round with it and it seems unnecessary to have any card either in the drawing or spinning for cleaning the flax or tow because it ought to be perfectly cleaned before either by the heckle or cards" [4].

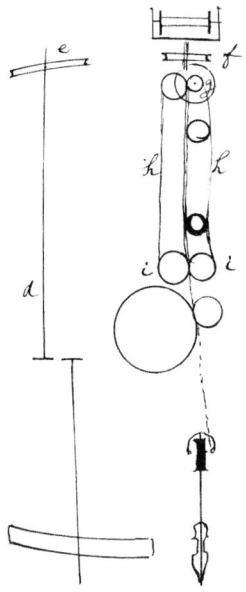

9 Flax spinning experiment April 1790.

The diagram shows that the card has been replaced with leather belts driven by a wheel (g).

Almost immediately they made a new drawing frame with the front rollers and cans static but allowing the back rollers and leather sheets to oscillate through three quarters of a revolution:

"it did not answer at all, at each turn of the back rollers it opened the sliver which came through the front rollers broad & not contacted, then when they had got half way & were at right angles with the front rollers it came through narrow & well contracted"[4].

So on April 5th, not letting the momentum drop, they performed experiment 9:

. *"Made a drawing frame with the delivering rollers working at right angles to the drawing rollers which had the effect of perfectly contracting the sliver without doubling back any of the ends of fibres which any sort of fixed contractors will unavoidably do. Long tow drawn three times over made a perfect roving"[4].*

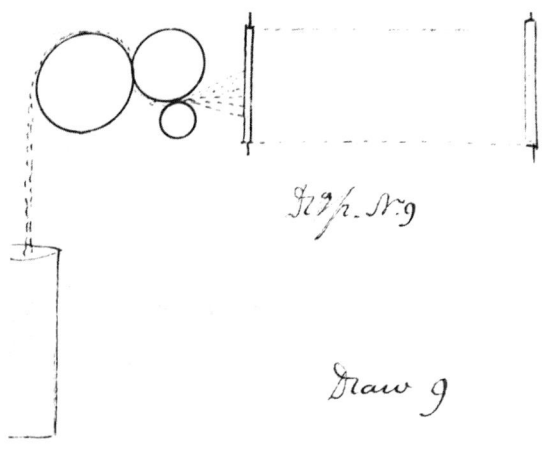

10 Flax spinning experiment (two) April 1790.

On April 6th they refined the process with different draws and twists and also determined to put a wheel on the back roller of each sheet.

They had achieved success and produced a spinning machine, which though not perfect, did the job. As John Marshall summarised the whole episode pithily in his life:

"and worked some machinery on the patent plan of Kendrew and Co. This did not answer, and we tried experiments, and took out a patent for a plan of Matthew Murray's our principal mechanic" [1].

Here Marshall recognised that the technical advances were due to Matthew Murray and the patent that ensued was number 1752 of the 1st June 1790 [8]. The costs were paid by John Marshall or Marshall & Co as they would have been beyond Matthew Murray's pocket at the time. Certainly an account survives in John Marshall's private ledgers for 1793/4 for the cost of a patent, and this was the date of Matthew Murray's second patent, with the charges running out at just over £190.

Murray's patent specification reads as follows:

"The machine is made and used as hereafter described in the two Drawings or Plans added hereto, numbered 1 and 2, and severally marked.

*Number 1 is a drawing frame, in which **a** is a crank on the axis of the front roller **f**, which, by the rod **g**, whose pivot **h** slides in the groove **i**, turns the crank **k**, and gives motion to the axis on which the worms **b,b**, are fixed. The same motion may be equally well communicated by two or more wheels. The worms **b,b**, work a small wheel on the axis of each of the rollers **c,c**. Round each of these rollers, and the rollers **d,d**, revolves a leather strap **e,e**, which must be something longer than the fibres of the material to be spun, and which must be of different breadths according to the quantity of materials that is to pass through them. Those for the first drawing frame are about six inches abroad. The rollers **c,c**, and **d,d**, are pressed together by four springs, which are fixed on two flat bars of iron, one above and the other below, in which the pivots of the rollers **c,c**, and **d,d**, run which are not shewn in the Plan to prevent confusion. The rollers **d,d**, are wrapped round with woolen yarn or any soft substance, that they may not hold the sliver so fast but that it may draw freely from between them. The rollers **c,c**, and **d,d**, are perpendicular, and rollers **f** are horizontal, consequently at right angles to one another. By this position the sliver is converged into a point between the rollers **f**, which must be on a level with the middle of the rollers **d,d**. Motion being given to the front rollers **f** (which are weighted) as usual in spinning mills, the material to be spun is taken in a sliver made by the card comb or hackel, and being introduced between the rollers **c,c**, is carried forwards between the leather straps **e,e**, through the rollers **d,d**, and is drawn out by the front rollers **f** to a proper size.*
The roving and spinning frames, as to their drawing principles, are made as above described, except that in the spinning frame, for greater convenience, the back rollers and leather straps are placed over the front rollers, as represented in the Drawing numbered 2, and standing at right angles with them, and are worked by a crank worms and wheels, as already described in the Drawing frame No 1. The other parts differ in nothing from the common spinning machinery.

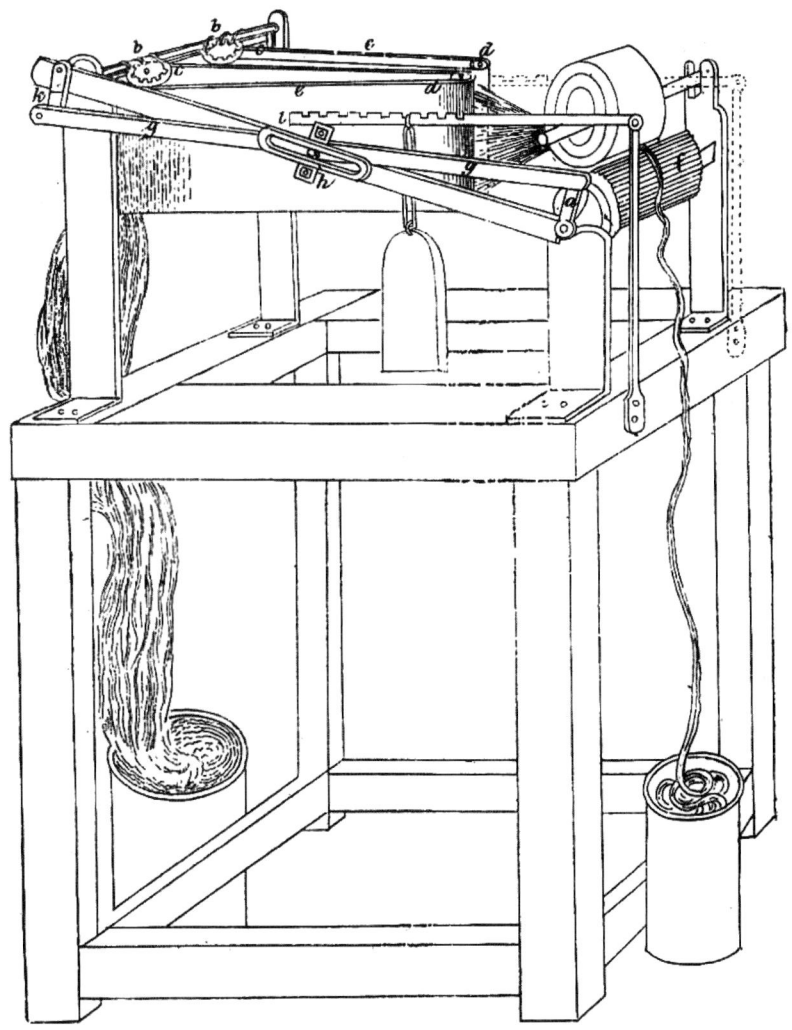

MURRAY'S SPECIFICATION. NO. 1752. 1ST JUNE. 1790.
11 Murray's Drawing Frame Patent no. 1752.

For some materials and particular kinds of yarn the spinning frames may be made as described in the Drawing numbered 2, in which **a** *is a double crank, the same as before described in the Drawing numbered 1, which communicates the motion from the roller* **p** *to the small axis on which the wheels* **b,b,** *are fixed .*

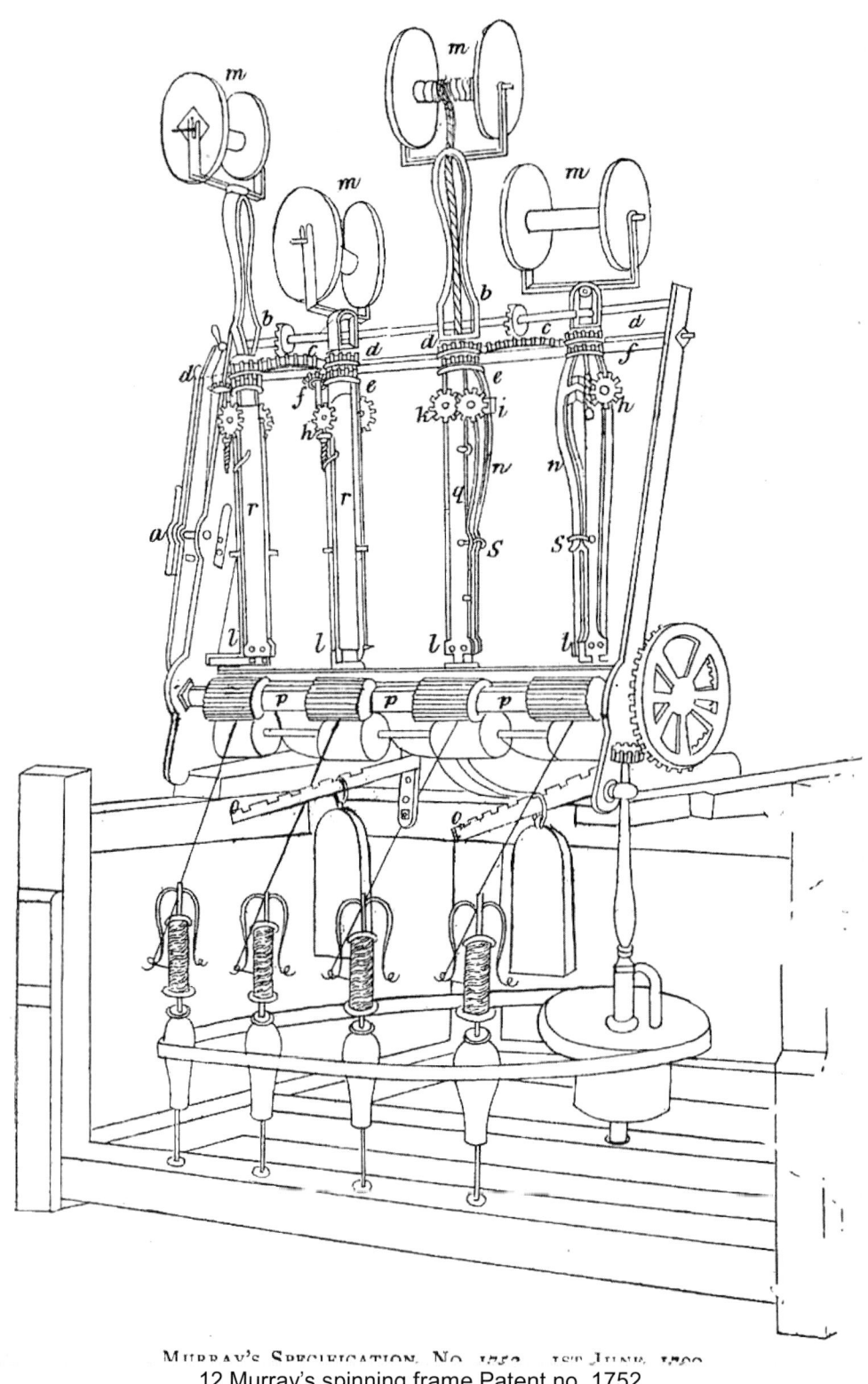

12 Murray's spinning frame Patent no. 1752.

   The wheels **b,b**, work into small wheels (which cannot be shewn in the plan), which are fixed on a stud to the wheels **c,c**. The stud wheels **c,c**, work into the wheels **d,d,d,d**, and turn round the frames **q,q**, in which the back rollers are fixed. **e** is a wheel fixed to the rail **t**, within the wheel **e**; the neck of the frame **q,q**, turns freely round the wheel **e**; the wheel **f** turns as the frame turns round, which, by the small upright shaft and worm **h**, gives motion to a wheel on the end of a small roller. On the *other end of the roller is a wheel **i**, working into another wheel of equal number on another roller **k**.*

*Round each roller revolves a small leather strap **r**, which pass round two small rollers **l**,**l**, which are fixed at the bottom of the frame **q**, and cannot be shewn in the plan. The roving bobbins **m** are supported by forks fixed to the neck of the frame **q** and turn round along with it. **n** is a spring which presses together the back rollers **i** and **k**, the end of it being held down by a loop **s** about the middle of the frame **q**. **o** is a weighted lever which presses together the front rollers **p**. Motion being given to the spindles and roller **p**, as usual in spinning mills, the roving is introduced through the neck of the frame **q**, through which there is a perforation for that purpose between the back rollers **i** and **k**, and, being carried forwards between the leather straps **r** and rollers **l**, is drawn out by the roller **p** to a proper size for the intended thread, and twisted into a thread by the spindle as usual. The back rollers turning round at the same time that the sliver is drawing out, give it one twist in an inch (but may be varied by the number of cogs in the wheelwork above, according to the different materials to be spun), by which means the fibres are completely contracted, and made ready to be twisted into a thread.*

The above described machines may be worked or used by a water mill, horse mill, or any other kind of mill, or by hand.

In witness whereof, I, the said Mathew Murray, have hereunto set my hand and seal, this Twenty-second day of June, in the year of our Lord One thousand seven hundred and ninety"[9].

It is interesting to note that five weeks before the above patent was taken out, Edmund Cartwright had been granted patent no 1747 in which it can be deduced he uses both the drawing of the fibres through cloth or leather and the setting of some rollers at right angles to others. He did not however combine the ideas. But Murray's drawing through leather bands was the innovative breakthrough and A. V Pringle wrote in 1951 that the system gave more complete control of the flax fibres – the first really successful arrangement, but that it can only be used for coarse fabrics (although finer ones than had been spun on Kendrew's machinery) [10]. He also noted that the principle was still in use at that time for cotton spinning. This idea coupled with putting the leather bands at right angles to the drawing rollers gave the first machine, and it was all achieved in about three months from the first breakthrough.

John Marshall had been experimenting since June 1788 to achieve a standard of machine that would grant him commercial success, and his perseverance coupled with Matthew Murray's technical ability at last promised that there was a future in commercially spinning line flax, and if the technical issues for this could be overcome then maybe so could those surrounding the more problematical tow.

The successful results of the experiments in April presumably were the final pieces of the scheme that convinced John Marshall to proceed further with his plans to surrender Adel Mill and move closer to Leeds. On 17th May 1790 John Marshall's accounts record the payment of £600 to William Naylor for land. Rimmer informs us that this was the freehold land in Water lane where Marshall built his first mills in

Holbeck. Mill A, as it was to be known later on, must have been started in the next few months.

Meanwhile in the experiment shops John Marshall and Matthew Murray remained busy. Their first viable machines for processing and spinning line were settled but there remained the outstanding challenge of putting tow to good use. There was still a long way to go. In June 1790 under experiment 13 John Marshall wrote:

"We never yet could get a sliver from the Cards either tolerably even or that would hold together to hand up to the rollers of the drawing frame which was owing to the fibres being very much crossed & intermixed" [4].

They proceeded to make various carding engines to try and improve the preparation of the tow before it was offered up to be spun. The first results were poor as on June 8th it was recorded:

"for as there was no working card it never sunk into the teeth but was carried over all the cylinders in lumps & patches without drawing it properly out" [4].

Various modifications were introduced through June and July without the desired outcome.

In July Mr Dearlove tried an experiment using a wool comb, which produced a very even sliver but the labour costs were prohibitive. And by the end of July they found that the machinery was becoming too complicated and expensive and that they were unsure it could work for tow unless it was prepared in a much better manner. New plans of both drawing and roving frames to accommodate two slivers were experimented with, followed by a spinning frame. By December 1790 they made a new carding engine and this time it would not make an even sliver.

On February 23rd 1791 Matthew and Mary Murray's second daughter was born and baptised Ann on April 10th.

By February 1791(experiment no 31) they were back working with parallel rollers, rather than one being at 90 degrees to the other as in the patent, and the results were good enough that they determined to alter all the machinery drawing, roving and spinning to the new plan. After this no experiments were recorded until September 1791.

The most likely explanation is that everybody was fully occupied in getting Mill A into production. John Marshall was no doubt project managing the whole job and Matthew Murray would have been making and installing machines and presumably the millwork to drive them. Rimmer tells us that the four storey mill was erected by September 1791 and had as previously stated a Savery type engine which pumped the water up into a cistern which then powered a water wheel. An illustration of such a Savery engine is given in Farey's 'Treatise on the Steam Engine '[11]. Farey also supports the memory of Stephen Marshall (one of John Marshall's grandsons) regarding the use of an atmospheric engine to begin with at Mill A [12].

In October 1791 (experiment no 33) they made a new carding engine and *"found no fault with it but that the wheels of the fillet cylinder made a great noise"* [13]. Work continued on Drawing, Roving and Spinning Frames for the remainder of 1791.

By this time with the expenses of the Mill, the Engine and machinery the finances of the concern were becoming stretched and at the end of October John Marshall took a mortgage with James Armitage of Hunslett for £2500 at 4 ¾ per cent interest per annum. This obviously did not fully meet their needs as the next month John Marshall and S Fenton took out another bond with James Armitage for £1500, i.e. £750 each, at an interest rate of 5 per cent. It would also seem that the Mill was not functioning as expected, because by April 1792 John Marshall was in correspondence with Boulton and Watt for one of their 30 Horsepower engines, but it was decided that one less powerful would suffice. The following letter from Boulton and Watt demonstrates that buying an engine from them involved considerable effort on the part of the customer as while Boulton and Watt supplied the main metallic parts, all the framing, boilers, brickwork etc had to be carried out by the customer. Then when everything was ready Boulton and Watt would send an experienced man to assist in the erection and starting of the engine.

"Messrs Marshall Fenton & Co

Soho Birmingham
15 May 1792

Gentlemen
We have your favour of the 11th accepting the 20 horse engine. We accordingly shall proceed with it as fast as possible.
Holes may be left in ye wall for the ends of the pieces G & K, and the pieces put in afterwards and wedged fast. The backframing will, with the cylinder, be steadied to the back wall by iron bolts, which it is not necessary to shew at present.
The sills of timber must stand upon brick or stone work to be built above the line of bottom of engine house. If the ground be good it is no necessary to flag the bottom. The slits in the strap xy are for cutters to tighten up the straps.
The fly wheel may be cast in open sand, if you think it will be handsome enough, as weight is the chief object.
We always use a damper as it is necessary to regulate the fire, and is therefore of advantage.
We think a 2nd engine may be applied in your old engine house, or somewhere in the present reservoir, to combine its effect with those of the first one in the lying shaft at some place, though we cannot say at present which would be the best, not having time duly to consider it; but we think you need not prepare much for it, as by pulling down a piece of a wall or something of that kind, it may be done. Your situation is not such a one, as would admit of the most eligible method of combining two engines.

We are
etc John Southern for B&W·

The premium in a gross sum payable 3 months after the engine is finished will be £530. Annual premium £100 till midsummer 1800.
Your determination which you will adopt will oblige. B&W JS" [14].

The letters in the text refer to the Boulton and Watt drawings. The Engine was a double acting 20 Horsepower machine with a 24 inch cylinder, a 5 foot stroke and a sun and planet motion.

The postscript represents the fee payable to Boulton and Watt for the use of their patent, which at this time, as indicated could be settled by a lump sum, or in six monthly instalments called the annual premium. On top of this Boulton and Watt would invoice for all the materials supplied. Marshall Fenton and Co. opted for the annual premium as is evidenced by an amount shown prepaid to Boulton and Watt in the accounts at April 6th 1793 [15].

While the negotiations for the engine progressed and the foundations and supporting structures for it were started, a new carding engine was being built as experiment no 39 of May 1792.

"We made a Carding Engine for laying out heckled line or shorts which our tow Carding Engines would not do without breaking & injuring the fibres. ...The Engine made an excellent sliver of line perfectly level & free from patches & apparently not at all injured. If the waste be not too great it will certainly be infinitely preferable to any hand spreading. We have not yet had sufficient experience to determine whether or no it will stand instead of the fine heckle" [13].

The diagram accompanying this experiment clearly shows that this was the basic carding engine that would become the subject of Matthew Murray's second patent.

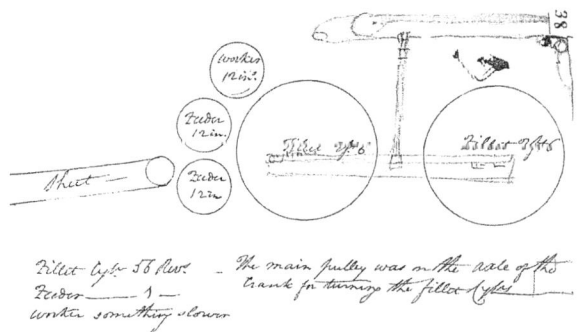

13. carding engine experiment May 1792.

On the 30th May Boulton and Watt sent, along with some more technical details and a correction of a mistake, a draft of the agreement. John Marshall obviously replied to the effect that he did not wish to have any restriction on relocating the engine (a standard Boulton and Watt clause) and for what it was all going to cost him, he wanted any improvements made by Boulton and Watt incorporated at no extra cost. Boulton and Watt offered little movement in their response of the 15th June:

"It always gives us pains to refuse what our friends seem to desire, but our compliance with the alterations you propose relative to the removal of the engine, and concerning the improvements, might be attended with serious inconveniences to us. We have a higher premium for our engines in some parts of Britain than in others, & there are other reasons which have made us limit the power of removal without our consent to some reasonable distance such as ten miles, at the same time it is scarcely probable that we should refuse our consent, unless the engine were to be removed so as to operate to our detriment. Ten miles is the distance we fix, but if you wish to have that distance doubled, we have no objections in your case, but cannot agree to its being unlimited.
We have at present no expectation of making any improvements on the engine, but if we were to make any that were of consequence enough to [need] a patent, it would not be reasonable that we should not be recompensed for the same according to its merit, at least we should not wish [to tye ourselves up] from it; though we have hitherto not made any charge for such things, but contented ourselves with our original premiums. Whatever concessions we make to you, others would claim & in similar circumstances we could not in justice refuse them and in consequence it would cease to be our interest to expend merely to make improvements, we could reap no benefit from "[14].

The agreement as negotiated did grant Marshall Fenton & Dearlove the right to relocate the engine within 20 miles of Holbeck. The agreement was signed, sometime at the end of the year (1792) by John Marshall, James Fenton and Ralph Dearlove. Perhaps Samuel Fenton was away on business or unwell. By 23rd July Boulton and Watt wrote to hope that their man James Murdoch had arrived to erect the engine. James Murdoch was the brother of the better known William Murdoch, but did not enjoy a long career with Boulton and Watt as he was dismissed for misconduct after erecting an engine in Cadiz in 1795.

James Murdoch wrote to John Southern from Stockport on October 8th to advise them among other matters that "Mr Marshall's engine went to work for good Sept 18th". Among the other matters he confirmed that Mr Kendrew's engine had gone to work on the 5th January 1792.

Work was still progressing to improve the flax machinery and in October, in the report of experiment 41, Matthew Murray is again mentioned (drawing not given):

"We began to be more than ever convinced that flax could not be drawn even on the usual plan with rollers, & that it would never be spun to perfection till some other method was found out which would answer for fibres of unequal length. We again recurred to our old idea of drawing by cards. Matthew proposed the plan of making a drawing frame on the same principle as the carding engine, which certainly produces a perfect sliver free from twits or patches & it seems as if all we wanted was to join together a number of those slivers & reduce them to a proper size for spinning without injuring their quality, which can certainly be done by a machine on the same principle as that which first formed those slivers. We therefore made a drawing frame exactly upon the same principle as the carding engine according to the annexed plan, where (a) is a cylinder 2 ft diameter covered with a fillet 1 inch broad with 3 pair

feeding rollers (b)(c) likewise covered in fillet 1 inch broad. Three separate card slivers passed through the rollers (dd) which were made of box & between which & (b)(c) there was a tin contractor. We thought it necessary to put the slivers through separate back rollers that the cylinder might not be too much loaded in one place. The cylinder had a draw of 12 from the back rollers. The sliver was lifted out of the teeth at the front of the Cylinder by wires which were raised up a sufficient height from the surface by an inclined plane, the sliver then passed through the delivery rollers (e)(e)" [13].

The machine produced the required results but they recognised that it was far too complicated to use in production. The next drawing frame made in November "failed completely". It would appear that they were trying to process too much material at once as indicated by the conclusions expressed at the end of December 1792 (experiment 43) after making comparisons with the attributes of wool and cotton:

*"Flax Hemp Tow &c are fibres of unequal lengths, not elastic, & have one bad property which none of the others (*i.e. wool and cotton*) have that they are full of branches & divisible ad infinitum. The inequality of length requires it to be drawn by points (cards) & not by plain surfaces, & its divisibility requires it to be drawn in exceeding small quantities, as it were fibre by fibre, so as that one fibre shall scarcely touch another"* [13].

John Marshall continued to compare flax with cotton and wool through February 1793.

The month of February saw adverse changes in his business, as in that month Revolutionary France declared war on Great Britain. This led to a commercial crisis with share prices collapsing spectacularly. A Select Committee on the State of Commercial Credit was set up and within its reports quoted a Mr Innes, a Director of the Royal Bank of Scotland, as stating that there were:

"several failures, and a very considerable one lately, which is connected with Manufacturers who may ultimately be involved, and where seven of eight hundred persons are now employed" [16].

The committee also said that credit was unobtainable and that manufacturers had begun to discharge their workmen.

On March 23rd 1793 a son was born to Matthew and Mary Murray who was baptised Matthew on 12th of May.

When John Marshall drew up his accounts on 6th April 1793, it was apparent that the partnership of Marshall, Fenton and Dearlove would not survive. The business was running at a loss and they now, in addition to the loans from Marshall's friends and family and Mr Armitage, owed Beckett & Co, the Leeds bankers, £3780/12/5d [15]. The firm also recorded that they owed Matthew Murray £3/3/0d which might indicate a severe cash-flow problem, if they were running up debts with a valued employee. Marshall also records a loss of £500 from selling flax. A very large sum of money had been invested in the new mill, the land, the steam engine and the machinery which included the following:

3 new spinning frames of 48 spindles each,

2 new spinning frames of 64 spindles each,

1 old spinning frame of 48 spindles,

8 old spinning frames of 64 spindles each,

14 Carding engines,

4 Roving Frames,

1 Drawing frame,

1 unfinished roving frame,

60 yards of Hecklers Benches,

28 looms,

24 looms in weavers hands,

and a vast store of machine components no doubt waiting to be erected.

At this point the situation can be summarised by quoting John Marshall:

"When the war broke out in 1793 we had 900 spindles at work. The sudden shock that was then given to mercantile credit brought us to a stand. We had never taken stock since we began our spinning speculation. We were aware of the great sums we had spent, but flattered ourselves with a prospect of immense profits, as soon as the trade was established. So much so that S. Fenton and I had agreed to give up the linen trade to his father, who had lately separated from Mr Wainhouse, and to his brother James.

We had a large stock of flax, the payments for which were becoming due, our book accounts were of small amount, and in the entire stagnation of trade which ensued, we could neither sell our goods nor obtain payments. Our Bankers called on us to reduce our account which we were unable to do and equally so to provide cash for the payment of the workmen. At first my spirits sunk under my situation, and I never shall forget the remorse which I felt, at thinking of the hazardous situation into which I had brought my Mother, who had entrusted me with her all. I was determined however to try to weather the storm, for if the concern had been broken up and Mill and machinery sold, I must have lost all I had, and probably there would not have been enough to pay my creditors. I set about taking stock, and at the valuation we then fixed I found our loss was above £8000 [although the 1793 accounts appear to only show a loss of £3441].

Of this R Dearlove lost what he had advanced, which was about £700. S. Fenton lost £2000 which was all he was worth, and I lost the rest. As they could neither of them be of any further use, I released them from the concern, and took the whole upon myself.

I set my shoulder to the wheel in good earnest. I was at the Mill from six in the morning to nine at night, and minutely attended to every part of the manufactory. I considerably reduced the number of work people, and paid some of my most pressing creditors by the sale of some flax at Hull. Matthew Murray exerted his talents and made some great improvements. The face of affairs began soon to change. I borrowed £1200, from Mr Garforth upon security of my house in Millhill and he was of great use to me by his advice and encouragement.

I have reason to think the breaking out of the war at that time, the most fortunate event that could have happened for me. If we had not received some such check, we should have gone on thoughtlessly with a ruinous expence, which we should probably have been unable to retrieve.

Finding that an increase of capital would be desirable, I entered into partnership with T and B Benyon December 16, 1793, who took half share of the trade, advancing a capital of £9000 which was equal to mine" [1].

In support of John Marshall's assertion of rigorous attendance to the affairs of the mill and Matthew Murray exerting his talents there are to be found a further series of experiments written down between May and September 1793. These appear to have concentrated on the initial process to produce a sliver and the end process of actually spinning. In May (experiment 44) John Marshall wrote

"We tried the Darlington method of spreading line upon a board 14 yards long, it made full as good a sliver as the carding engine & quite as free from patches & seemed to waste & injure the line less, but took more time as a girl cannot spread more than 20 lb per day & a carding engine with 2 spreaders & 1 doffer does 1 cwt.. Tried 20 weighs & 20 slivers from the card Eng. the first weighed 11 lb the last 10 ½ lb so that there is near /20th part waste. Spreading on a board is certainly better for fine line.

We are informed that at Darlington they contract the sliver very little in the front rollers but bring it out nearly the breadth of the sliver going in, & twist it in the cans, they use 2 drawing & 1 roving frame each having a draw of 16 or 18 & the spinning frame the same & their front top rollers are only the breadth of the sliver they produce.

They are said to spin very well. So that it appears from them that the whole art of flax spinning consisted in twisting the slivers which confines the fibres from coming irregularly forward" [13].

The above indicates that Marshall & Co were again in contact with Darlington to see whether they had made any useful progress.

Experiments numbers 45, 46 & 48 which occupied August and September 1793 had to do with the spinning frame and concerned reducing the number of breakages being experienced. By changing the draw from 18 to 36 they reduced the breakages from 25 a day to 11.

They then turned their attention, with experiment 46, to wetting the yarn. This operation was carried out in the Darlington system, but was fairly basic. At first they:

" Tried to wet the yarn on a new plan. We made the front roller of the spinning frame of brass & wet the top roller by a sponge which was fixed on top of it & supplied with water by an Archimedes screw from a trough placed behind. The use of the screw was to bring up a certain quantity of water whether the trough was full or nearly empty, for the yarn requires an exact degree of moisture, too much or too little spoils it. ..... .It had the desired effect of making the yarn smooth & strong & saved above half the breakages, a roving spun wet breaking 12 times a day when a dry one broke 25 times" [13].

Then in September they made further changes as the Archimedes screw was too complex and too expensive. At first they tried to pump water through a small hole onto the roller, but the hole had to be so small that it kept getting clogged. They tried another contrivance which would not fill with water. And then:

"We next made a small trough which was kept always full of water by a communication with a Bird fountain, & fixed a roller with 4 wood bosses running half in the water. The bosses took up a sufficient quantity of water which was taken from them by a sponge & conveyed to the wood roller.
We then took away the sponge, & put a list over each boss of the roller which passed through the water & over another roller fixed above the trough & rubbed against the back of the wood roller. This is the best & simplest way we have tried"[13].

Marshall and Murray had now tried all the components that would end up as part of Matthew Murray's second patent.

It would also appear that John Marshall wanted the improvements rapidly built in new or modified machines in the mill as the following advertisement appeared in the Leeds newspapers at the end of September 1793

To SMITHS

WANTED, THREE or FOUR GOOD

VICE-MEN, and a SPINDLE TURNER

Good Hands will meet with Constant Work, and Wages according to Merit, by

applying to Mr John Marshall, Leeds

None but good Hands need apply; and no Letters will be answered unless Post-paid. It could be suspected that Matthew Murray was involved in the placing of this advertisement as the wording and structure are to be seen in the future advertisements of Murray and Wood and the successor firms.

Although the Boulton and Watt steam engine had been working for over a year, some correspondence at the end of 1793 indicates that John Marshall was not satisfied, and judging from a Boulton and Watt letter of the 19th November 1793, had been withholding payment. In the letter Boulton and Watt acknowledge receipt of two six monthly payments at once. There was an issue over a component, but the main concern appears to be over the conduct of James Murdoch. A large part of the letter is given below as it shows Boulton and Watt on their high horse and not listening to the customer (as John Marshall most probably viewed the matter).

"We will commence by sending you a copy of Mr Lawson's letter, dated 19th Oct, so far as relates to your concern: **"I yesterday saw Mr Marshall who again spoke to me about having his account adjusted, I spoke to Mr Boulton about it when at Matlock, he promised to enquire about it but I suppose he may have forgot. I must own I think Mr M is rather harshly treated, as I do not think he should pay for the faults of Murdock."** *In reply to this our J.S[outhern]. wrote the 22nd to Mr Lawson as under, Mr Boulton having forgotten it which Mr Lawson knew was very probable. "In relation to Mr Marshall's account, we should be glad to know what that gentleman wishes, as I am certain that whatever is reasonable Messrs B & W will readily comply with. Therefore be so good as to desire he will state in what particular he thinks himself overcharged and how much. In relation to the connecting strap,*

Murdock said they did not get a new one, but actually used that which was sent from hence, and when they desired we would charge it as old iron, we informed them, that we could not consent to any such proposition, but would take it back at the full price. If they did not use it, they must have it by them," (and therefore could return it for B&W to make better of it than old iron) "and they cannot desire a fairer offer, than the above was. In relation to Murdock's behaviour, I give it was not such as would do anybody credit, but he certainly was to the description contracted for, viz, an experienced man capable of erecting the engine. If he did not do his duty Mr M should have complained at the time, & if he unfortunately has extorted money from Mr M it is certainly to be lamented, but it cannot be incumbent upon B & W to make it good. B & W will certainly do what is reasonable, but they cannot allow abatements, without being convinced previously that it is just that they should do so. I need not tell you this."

So far Mr Lawson's representation of your complaint, and the reply to him. You will see the former was not specific, and therefore the latter required that it might be made so; but as far as J.S[outhern] is acquainted with the affair, that has not been done. We must therefore beg of you Sir to say, whether you are dissatisfied with the explanation of time or of any other article of charges which was given in our letter to you of the 17th of July last, and if you are for what reasons. Secondly we repeat our offer made the 16th of March last, to take back the connecting strap at the price charged in our account to you. And thirdly that you will be good enough to state by letter the particulars which you think want [additional work---] such observations as you may think will exhibit the reasonableness of your claim. Murdock said that the Millwrights broke the strap because they would not cut room enough for it on the beam, but that he pieced it, please see our letter 16 March.

Upon the whole, then Sir, we beg the favour of your reply as above, and assure you that the credit we attach to your word is by no means brought into competition with Murdock's.; we have latterly had too much reason not too give him almost any credit, and we ought to say in justification of ourselves that his character from Mr Illingworth (where he was before your engine was erected) was very good, and such as could not be objected against. We do not send a scoundrel out, if we know him to be such, but it is not difficulty of belief that we cannot be answerable for [the men's] immorality.

We should not have been so very full upon the subject, but that justice to our character, which we think is in your opinion impeached, demands it: and to shew you that we are not backward to render justice, & have not been in any shape intentionally contiguous, we are now satisfied and hope to make you so.
Our principals being from home, the requisition you make in the latter part of your favour, shall be sent to them & as soon as their will is known"[14].

Unfortunately John Marshall's reply does not seem to have survived but it produced a more amenable and brief response from Boulton and Watt in John Southern's letter of the 9th of December:

"We had yours of the 29th ulto a day or two ago, and on the other side in pursuance of your request in your favour of the 16th do, send you the state of your account after making the allowances on the strap and Murdock's time, both of which we must sit down with the loss with, remaining"[14].

While this exchange was happening work had progressed on the submission of Patent Number 1971 which was dated December 18th 1793, and formally granted on 14th January 1794. As noted under the first patent some of the costs for this patent survive in John Marshall's private accounts including £8/8/0d paid to Craig for drawing plans, £10 to Richardson and Dawson for attesting the specification, and bills(presumably legal) sent to S G Benyon for nearly £180.

In the middle of all this John Marshall states that he formed a new partnership on the 16th December 1793 with Thomas and Benjamin Benyon, merchants from Shrewsbury, who had already had dealings with Marshall Fenton & Co, or at least John Marshall as they appear as creditors for £1000 in John Marshall's accounts for 1792. Whether this sum was due by trade or was a loan is not stated, but as they are listed among the friends and family who had loaned capital to John Marshall, it would seem reasonable to assume it was a business loan.

So the patent was granted and the machinery on which the business was dependant was modified as it would continue to be over the decades either by internally driven ideas or those of the machine makers. Murray & Wood and the successor companies continued to work with Marshall's for many years, to re-machine the mills with the step changes in textile technology. This second advance in spinning was patented as follows:

"Now Know ye that I the said Mathew Murray in compliance with the said proviso Do hereby Declare that the nature of my said Invention and the manner in which the same is to be performed is particularly described and ascertained in the respective plans hereunto annexed and the following description thereof that is to say **Number** *1 is a Carding Engine of which Figure 1 is a perspective view and Figure 2 a ground plan.*
The same Letters refer to the same parts in each. a.a are two Cylinders round each of which is wound a fillet card in spires at about an inch distance from each other. The roller v is covered with plain Cards and Works against the Cylinder A and is doffed by the feeding Rollers c.c . .d,d are two iron standards moveable on the pins e and connected by the rail f. g is an Iron Axis to which are fixed the inner rails h.h. which support the Cylinders a.a . The pivots of the axis g move in the iron standards d.d At each end of the inner rails h.h is a notch i which slides onto the two iron pins k.k fixed to the main framing l. – m is a lever on the axis of which is fixed the curve n which is connected by a strap to the rail f. o,o,o, are fixtures for the lever m. p is a cord which takes hold of a catch that falls into a notch in the curve n. The Engine receives its motion from the Mill by the pulley v on the Axis of which is fixed a Crank w which by the finger x turns the cylinder a. The small pulley y gives motion to the feeders by the pulley z. Motion is given from the feeders to the feeding shut roller, and to the working roller b, by pulleys, and from one feeding roller to the other by the wheels shewn at the ends of their axles.

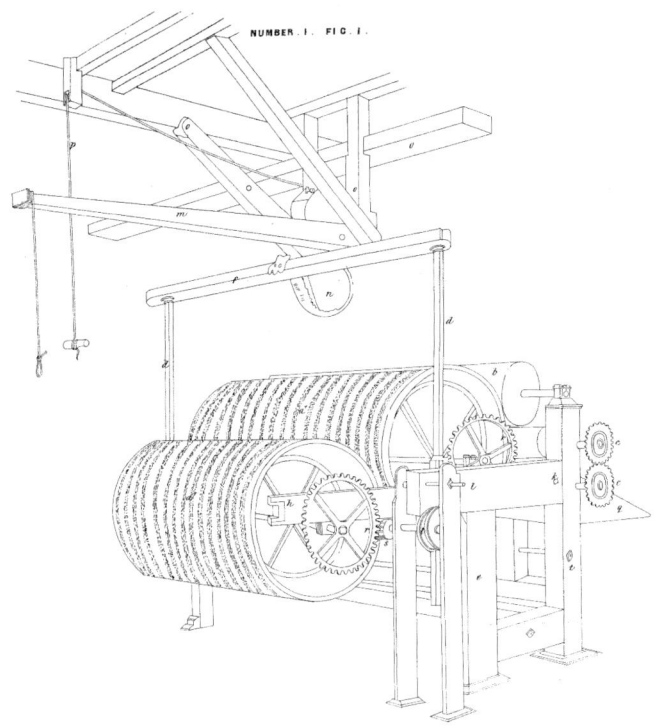

14. Perspective view Murray's carding engine patent no.1971.

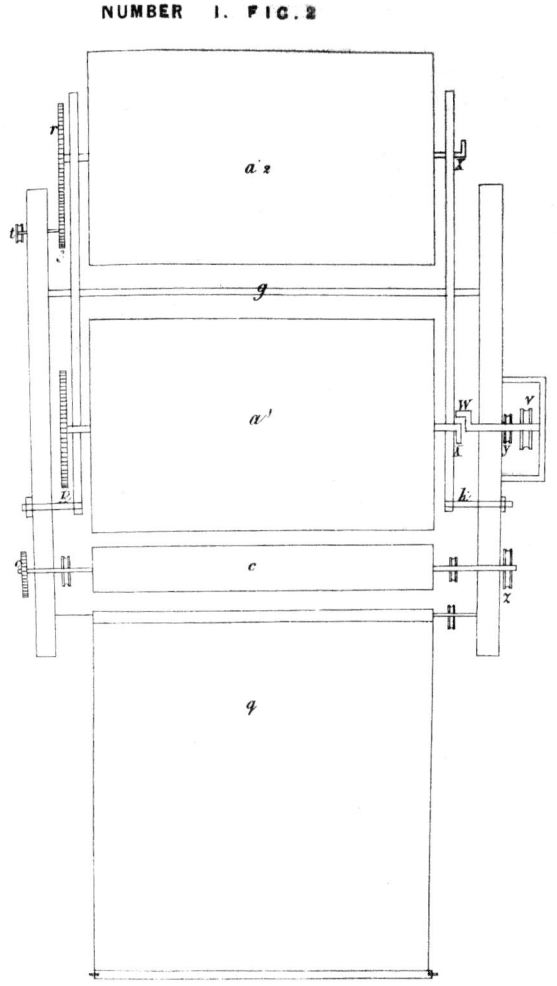

15. Top view Murray's carding engine Patent no.1971.

The engine being put in motion a proper weight of the materiel to be carded is spread upon the feeding sheet q, *and passes through the feeding rollers* c.c *onto the cylinder* a1. *The machine is then stopped and the cord* p *is pulled to discharge the catch, which allows the cylinders on their frame* h.h *to slide forwards of the pins* k.k. *The cylinders are then turned over and change places, and the lever* m *is pulled down, which draws the cylinders and their frame to their former position, in which they are held by the pins* k *in the notches* i. *The cylinder* a2 *is then set to work as before, and the material upon the cylinder* a1 *is taken off from the whole length of the fillet into a can, for which purpose a circular motion is given to the cylinder by the wheel and pinion* r.s. *The pinions* s, *by the pulley* t, *receives its motion from the mill.*
Number *2 is a Spinning Machine of which Figure 1 is a perspective and Figure 2 a side view.*

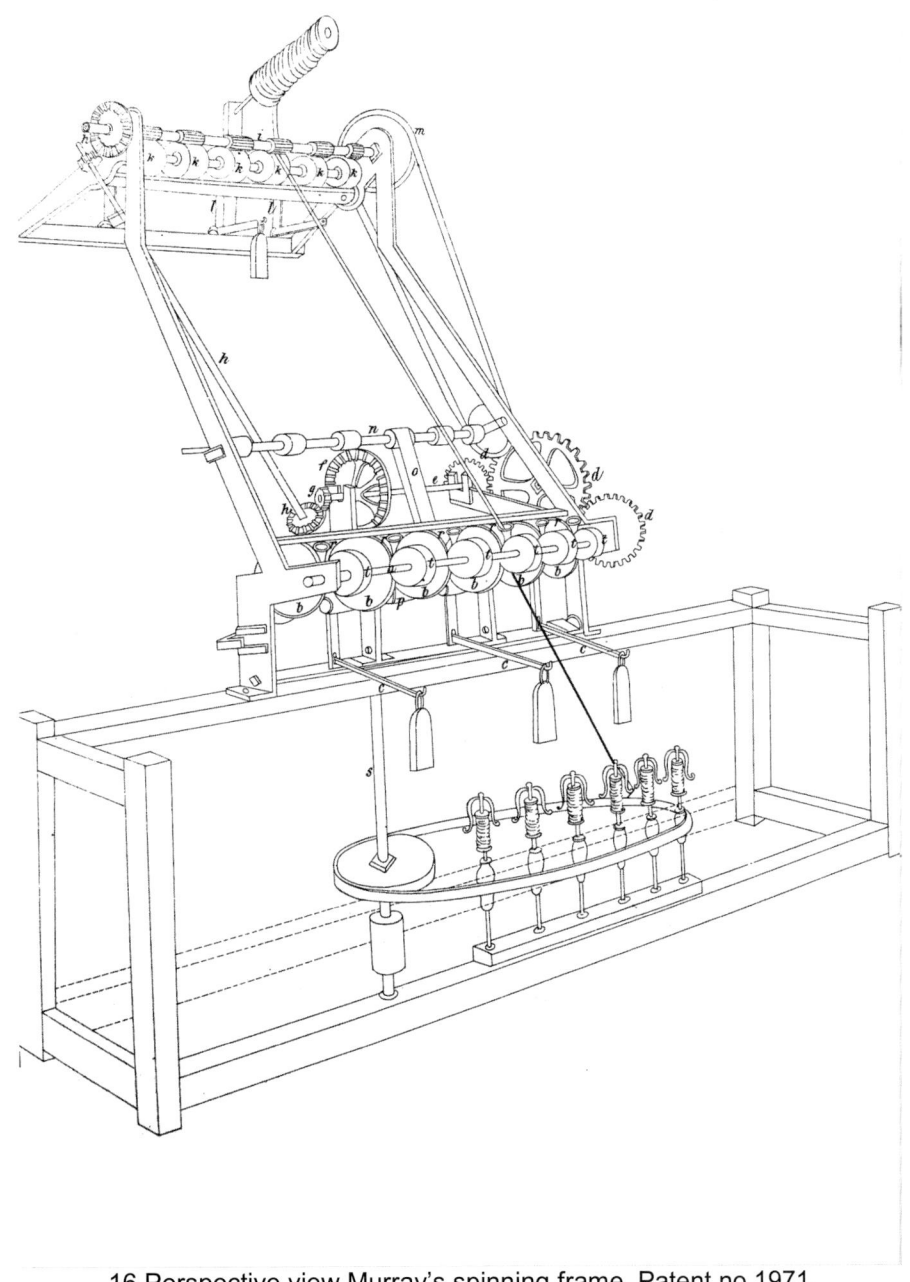

16 Perspective view Murray's spinning frame. Patent no 1971.

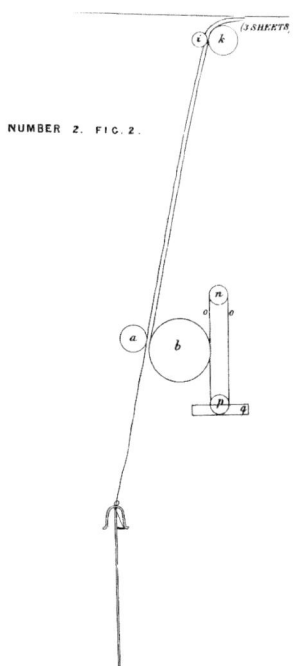

17. Side view Murray's spinning frame. Patent no.1971.

The same Letters refer to the same parts in each. a is an iron roller, upon which are six bosses t made of brass or any other material which corrodes the least with moisture; against these the wood rollers b,b, are pressed by the weighted levers c,c. The three wheels d,d,d, connect the roller a with the back axis e, which by its contrat wheel f receives its motion from a pinion on the top of the list pulley-shaft s. The pinion g on the axis e, by the wheels and diagonal shaft h,h,h, gives motion to the iron roller i, against which the wood rollers k,k, are pressed by the weighted levers l,l. On the end of the roller i is a pulley m which conveys the motion to the wood roller n. The lists o,o, pass round the roller n and lower roller p, and rub against the back of the roller h. The roller p runs in a trough q filled with water or any other liquid. N.B. Some of the lists and weighted levers are not shown in the plan to prevent confusion, motion being given to the list pulley-shaft s and spindles, as is usual in spinning mills. The roving is introduced between the rollers l,k, and passing through the circular guides r is drawn out by the rollers a,b, to a proper size, and is twisted into a thread by the spindle as usual. The list o, revolving by the motion it receives from the roller n, brings up a certain quantity of liquid from the trough q which it deposits on the outer edge of the roller b which wets the untwined substance of the thread in the point of passing between the rollers a and b" [17].

And so the year 1794 started well with a new patent and a new partnership. With new funding available to re-equip the mill with new machines the prospects looked better and a further 4 to 5 acres were purchased in Water Lane and a second mill was started (to be known as Mill B). Mill B had a floor area of 29000 square feet spread over 5 storeys.

The experiments to improve the textile machinery continued to be made with work to the drawing frames and the development of another new carding engine. In April it was time for a little more industrial espionage to see if the issues on spinning short tow could be overcome. John Marshall wrote:

"We sent Matthew into Derbyshire to learn their method of spinning short tow from their carding engines. There are several small mills in the neighbourhood of Matlock. The following calculation & drawing are taken from Miss Willoughby's.... It seems to be a tolerably ready way of spinning coarse yarn from 2 to 3 leas per lb but they cannot manage finer & they cannot spin it without a great deal of twist, which makes it no worse for Candlewick but it would not answer for manufacturing" [13].

Matthew Murray returned with details of the machinery, which were recorded, but the verdict, *"On the whole it is a clumsy expensive machine"* [13], indicated a wasted journey.

On 19th May 1794 John Marshall drew up his accounts for Marshall and Benyons in which he recorded a value of £11 for a lathe in Matthew's room. The accounts also show that the partnership was owed £74/14/4 ¼ d by Matthew Murray, a large sum of money for a man in his position. Among the possible explanations for the amount could be that: Murray made a contribution to the costs of patent number 1971, that he borrowed money for a house or, possibly, that this was a loan connected with Matthew Murray setting up in his own partnership with one David Wood. While the date Matthew Murray left the partnership of Marshall & Benyons is uncertain, sometime in 1794 he was in the partnership of Murray and Wood as according to the accounts by the 5th of January 1795 Marshall & Benyons owed Murray & Wood £285/14/7 ½ d. David Wood had been a blacksmith in Bolton Percy in 1783, and was the son of a blacksmith from Ulleskelf near Tadcaster. He was baptised on November 2nd 1761. It is thought that he had been in Leeds for a few years prior to joining with Matthew Murray, but how or where they met or whether David Wood also worked for Marshall remain to be discovered [18]. However Matthew Murray and David Wood had similar skills, which most probably brought them together in the first place. The January 1795 accounts also show that Matthew Murray's debt had increased to £90/3/6d.

On the 17th June 1794 John Marshall ordered a 28 Horsepower steam engine from Boulton and Watt (27 inch cylinder, 6 foot stroke with sun and planet motion) to provide the power for Mill B. The delivery was scheduled for the New Year.

The last textile machine experiment conducted in 1794 was number 56 in August and involved building some copies of machines seen at Thackwray's in Manchester, which had cost advantages, but also some potential production issues on stoppages, which were probably not an issue: *"We liked it very much"*[13].

During1794 Kendrew and Porthouse attempted to obtain an Act of Parliament to extend their patent, in much the same way that James Watt had. At a meeting of the Corporation of Leeds on the 28th April they had resolved to present a petition to the House of Lords to prevent such a bill passing into law. There had been notices in many local newspapers in May summoning mill-owners and machine makers to meetings to organise opposition. After succeeding in the House of Commons, Kendrew and Porthouse's petition failed to get through the House of Lords and was effectively squashed by the middle of 1794.

This does not seem to have put Kendrew & Co off litigation as they sued John Marshall for non-payment of license fees. The case was heard on the 15th August 1794 at Durham Assizes and subsequently reported briefly in the northern newspapers.

"it was a cause depending between Kendrew, Porthouse, and Backhouse, of Darlington, plaintiffs, and Marshall and Dearlove, defendants, respecting a right which the defendants had set up for the spinning of yarn from flax, by a machine of their pretended invention, erected, and employed in Leeds, in the County of York; and for which, in December last, they obtained a patent, in opposition and violation of the plaintiffs patent machine established at Darlington: but, after a hearing of eight hours, and on inspection of both machines, which were produced in court, it appeared evident, that the principle of the defendants machine were only a piratical imitation of the plaintiffs patent machine, disguised to give it a specious different appearance, and that the pretended right of the defendants was a direct invasion of the rights of the plaintiffs. Upon which the jury immediately gave a verdict for the plaintiffs with £1100 damages" [19].

It is known that Kendrew & Co sued for breach of contract (fees), see below, to which it can only be surmised that Marshall retaliated by stating that he was not using Kendrew & Co's machinery. The judge, Justice Lawrence, presumably then decided to follow the route described in the newspapers. Some additional logic for this train of thought must be that if Kendrew & Co were to sue to overrule the patent they would surely have named the patent holder, and Matthew Murray's name does not appear in any of the reports. A very brief passage in his 'life' Marshall acknowledged the case, and stated that they settled with Kendrew & Co for £500. This may be one of Marshall's memory lapses, which are apparent elsewhere in the 'life' or maybe Kendrew & Co were offered free use of Murray's machines as well as the money.

The impact of the case was not limited to the financial penalty however as Messrs Boulton and Watt had heard that Kendrew & Co had won a case of patent infringement. This resulted in the following exchange of letters. On the 3rd of December 1794 Boulton and Watt wrote to Kendrew & Co (who had recently had one of their steam engines erected) in the following terms:

"having been laid under the necessity of bringing actions against some persons who had infringed upon our Patent, and supposing it probable that the referring to precedents will be necessary upon an approaching trial, we take the liberty of applying to you for a perusal of the papers relating to your late lawsuit etc, [and] to copy such of them as may be of use to us; or if they are not now in your hands to

inform us of the name of your solicitor in London that we may from him acquire the needful information" [20].

John Kendrew replied as follows on the 9th December 1794:

"In answer to yours of the 3rd Inst we should be happy to render you any assistance in our power but we have no papers in hand relative to the trial alluded to. The name of our solicitor in London is James Mainston he lives in Essex Street Strand. It may perhaps not be amiss to inform you our action was brought for a breach of contract, but tho' the attack was made on that ground, the trial turned on the general law of patents. The defendants had very much altered the machine from what it was when put into their hands and had taken out a patent for it: the question therefore to be tried was whether such alterations constituted a new machine or whether the defendants machine tho' very different in appearance was not composed of the parts constituting the original patent machine thus altered to give it a colourable difference. Working models of both machines were produced in court; ours examined and compared part after part by the specification to see that it was the machine for which the patent had been taken which being found to agree we were then considered as invested with all therein contained. The defendant's machine was next examined & on that examination found to be made up of many original parts specified by us. After a hearing of about 8 hours a verdict was given in our favour: the defendant's machine being pronounced nothing more than a disguised copy. Heartily wishing you success. Law Chambre & Wood are our Councel" [21].

With the knowledge of the number of variables that need to be balanced to obtain a satisfactory result, it may be that such a result would not have happened today, and that in these respects the two machines would have been clearly distinguished.

It is very noticeable, that from this point forward, the descriptions applied, in the internal Boulton & Watt correspondence, concerning both Marshall and Murray became at times vituperative and derogatory. Marshall and by implication Murray had been condemned in a Court of Law as patent pirates, and Boulton and Watt knew all about such evil.

Matthew Murray however retained a pride in his accomplishments with respect to the Flax industry and in a letter written in 1859 between two members of the Cornish Loam family of engineers, Uncle to Nephew and both Matthews, the following is recorded as history moving into legend

"Still in Mr Watts days, the engine was a rough made machine and continued so until a Matthew Murrey sprang up, at Leeds, as Fenten, Murrey and Wood. Murrey's genius and workmanship did wonders with the steam engine, in its adaptation to the driving machinery etc, so the detailed parts underwent a thora alteration and improvement by him and through him the Cloth business was interduced into Leeds, he has the first peice made in Leeds, made into a Coat, of which he often swagered" [22].

Notes.
1. John Marshall, sketch of his own life. Marshall Papers MS200 Brotherton Library, Leeds University.
2. Marshalls of Leeds 1788-1886 by W G Rimmer, Cambridge University Press 1960.
3. Marshall Papers MS 200 folio 14, Brotherton Library.

4. Marshall Papers MS200 folio 53, Experiment Book 1, Brotherton Library.
5. Industrial Biography by Samuel Smiles, John Murray 1863.
6. Boulton & Watt Collection, Birmingham Central Library, MS/3147/3/396.
7. Boulton & Watt Collection, Birmingham Central Library, MS/3147/3/87 Letter Book.
8. Marshall Papers MS200 folio 1, Brotherton Library.
9. Patent No. 1752 to Matthew Murray 1790 – Machinery for Spinning Fibrous Materials.
10. The Mechanics of Flax Spinning by A V Pringle, H R Carter Belfast 1951.
11. A Treatise on the Steam Engine by John Farey, Longman Greene & Co 1827, David & Charles 1971, page 123.
12. A Treatise on the Steam Engine by John Farey, Longman Greene & Co 1827, David & Charles 1971, footnote page 444.
13. Marshall Papers MS200 folio 54 Experiment Book 2, Brotherton Library.
14. Boulton & Watt Collection, Birmingham Central Library, MS3147 Foundry Letter Book October 1791 – October 1794.
15. Marshall Papers MS200 folio 2 Accounts, Brotherton Library.
16. Money and Banking in England by P L Cottrell and B L Anderson, David and Charles 1974.
17. Patent No 1971 to Matthew Murray 1793 – Machine for Preparing and Spinning Flax, Hemp, Tow & c.
18. Early Textile Engineers in Leeds, 1780-1850 by Gillian Cookson, The Thoresby Society 1994.
19. The Derby Mercury August 28 1794.
20. Boulton & Watt Collection, Library of Birmingham, MS3147 Letter Book 89, the condition of the letter is not good but the content quoted is clear.
21. Boulton & Watt Collection, Library of Birmingham, Ms3147 Letters 1794. (See index under Kendrew).
22. Letter from Matthew Loam (1794-1875) to his nephew Matthew Loam (1819-1902) helping him prepare for a lecture on the steam engine, Cornish Studies Library, Redruth, Cornwall. Ref t7994044 and Moo01948CC.

Chapter 4
Murray and Wood

Both Matthew Murray and David Wood had fathers who had worked for themselves, one as a whitesmith and one as a blacksmith, and is not surprising therefore that they should wish to follow a similar path. The start was their formation of a partnership sometime in 1794. What the particular catalyst to do it at this point will probably never be known. Trevor Turner speculates that Matthew Murray may not have been happy with the new management partnership, but as Murray and Wood carried out substantial amounts of work for Marshall and Benyons and for both parties after that partnership split, the reasons for Murray and Wood setting up on their own were probably other things. Murray and Wood certainly received a good amount of support from Marshall and Benyons, so the rearrangement by both parties was planned and amicable. John Marshall may have seen the possibilities of reducing his machinery costs long term by having a specialist firm of machine makers giving him priority, and selling the second string machines to other firms. There may also have already been the idea of having a steam engine manufacturer, similarly giving him priority, but also doing business in a different way to Boulton and Watt. The accrual of these benefits to John Marshall, along with satisfying Matthew Murray's desire to progress, all support the business plan.

Murray and Wood were no doubt fully occupied making the machinery for Mill B, as evidenced by the money owed to them on 5th January 1795 (£285/14/7 ¼). While the level of textile machine experiments, recorded by John Marshall, reduced, it is apparent that Matthew Murray was still involved, as he is named in experiment number 58 of May 1795 for *"a drawing frame for line"*[1]. Murray and Wood were operating from premises at Mill Green, Holbeck Lane, Holbeck. This was the address used in the Ledgers of the Kirkstall Forge for early supplies made in 1796[2].

18 Kirkstall Forge Waste Book showing Murray & Wood's address at Mill Green.

In the meantime work progressed with Boulton and Watt on the arrangements for the steam engine for Mill B. This time there were more technical inputs from the customer. In a letter of the 30th July 1794 Marshall and Benyon state that they wish the fly shaft to make 42 revolutions and would there be any objection to *"having the*

Sun & Planet wheels of unequal numbers." The letter goes on to state, in response to a question from Boulton and Watt:

"We mean to make the Boiler here, the fly wheel, rotative shafts, plumber blocks, steam pipe, cold and hot water pumps, for which please send your directions" [3].

John Southern explained in a letter dated 11 August that they did not *"much approve"* [4] of the idea of making the Sun and Planet wheels with different numbers of wheels due to the additional stress imposed on the Sun wheel, but that they would make it that way if there was a benefit to the customer. On the 19th December 1794 Marshall and Benyons informed Boulton and Watt that;

"We wish to have the Governor to stand over the rotative shaft, & receive its motion from the shaft by a pair of wheels; and to be calculated to make the same number of revolutions as the fly shaft. We likewise wish the balls to be made to rise or fall upon the legs of the Governor, so as the speed of the engine may be varied by that means, or by any other means that you may think more eligible" [3].

Boulton and Watt replied to this, among other issues, on the 22nd December

"The governor is made, and intended to go 35 turns per minute; and by having different sized pulleys small variations in the speed can be that way managed, when it is turned by a rope. Query what difference of speed do you wish, what the utmost extent, and how small variation you may want?"

They went on to say that the engine was almost ready but that they had been *"disappointed of the cylinder and air pump"* [5], which presumably at that period were being made at John Wilkinson's works.

It can be imagined that John Marshall and Matthew Murray having conducted so many experiments together were now working, with their experience of the steam engine at Mill A, on obtaining a steam engine that would work Mill B and its machinery in the best manner possible for their requirements. They had obviously decided on some of the less satisfactory aspects and how they would like them improved as shown in their response to Boulton and Watt dated 26th December 1794:

"When the Governor is turned by a rope it answers so imperfectly as scarcely to be of any use, and for that reason we prefer wheels. If it can be made to go 38 turns per minute it will save us the expense of making a pair of wheels as we have some that will do. You need not give yourselves any trouble about making it to vary the speed, as we can do that ourselves. Mr Lawson has been here more than once and seen the situation of our new engine, and we mentioned the above circumstances to him of the Governor etc, which we supposed was the same thing as writing about them to Soho, otherwise we should have writ about them long ago. We will thank you to send an experienced man to fit up the engine, and not more than one" [3].

Boulton and Watt replied on 31st December, they began by addressing another issue relating to the steam pipe and stated that they were sending some new plans of the layout of the boiler and engine, as in discussion with Mr Lawson it appeared the requirements may have been altered. They go on to state that the goods are all ready and will start to be packed tomorrow, and that they will alter the Governor to go 38 turns per minute. They attributed the issues with working the Governor with a rope to "*some peculiarity in your business*" as there were 20 times as many worked by a rope as by wheels. The wheels, they said, were put up chiefly by millwrights who were " *frequently, though not always, fond of much wheel work*"[6]. They had no objection to wheel work where people were happy to incur the expense.

On the subject of wheels and mechanics there is an interesting item in one of John Marshall's notebooks which is dated 1795 and headed 'Theory of Wheels M. M.' which supposes that he and Matthew Murray had been discussing wheels during the year.

"*The teeth of two wheels working together must necessarily rub against one another over so much space as the difference of length of two radii meeting at the centre of action of the two wheels & of two radii meeting at the thickness of a tooth from the centre of action, which is the place where the teeth first begin to act. Consequently the finer the pitch & the less friction there will be upon the teeth.*
The best form of the teeth of wheels is that which is the strongest & at the same admits no tooth to come into contact but that which is in action. The form of the teeth is therefore determined by the relative diameter of the wheels which are to work together.
To find the true form of the teeth of two wheels of equal diameters draw the pitch line at half the depth of the teeth, and setting one leg of your compasses on the pitch line in the middle of one tooth draw the point of the next tooth with the other leg. For wheels of different diameter which are to work together the pitch line of the smaller must be drawn as much nearer the root of the tooth as the relative difference of the diameter, and the pitch line of the larger as much nearer the point of its teeth, and the points of the teeth drawn from the pitch lines as before.
As suppose a wheel 2 ft. diameter working into another 8 ft. diameter the pitch line of the smaller must be ¾ parts of half the depth of the teeth nearer the root, and the larger as much nearer the point.
The less is the depth of teeth & the less is the friction, whilst the wheels are new, but it is found in practice that if the teeth are made very shallow they will soon wear so much as to have more friction than if they had at first been made deeper, because the power being exerted upon a smaller surface of course wears it sooner.
Perhaps the best general rule for the depth of teeth is to make the depth of the acting part ¾ of the pitch."
It goes on to say, although obviously written at a different time:
"*Strength of Wheels*
To find the strength necessary for any given power
Rule The square of the thickness of the tooth multiplied by its breadth will give the number of Horse power that the wheel is adequate to work, if it move at a velocity of its surface of 2 ½ ft. per second of time. If the velocity is greater or less, the power is proportionate.
The best breadth of a tooth is six times its thickness"[7].

On the 2nd April 1795 Boulton and Watt sent Marshall and Benyon a copy of the agreement for the new engine for signature and stated that they had stopped having their royalty premium as an annual payment and now charged it along with the cost of materials as a lump sum to be paid at once or by agreed instalments.

They said they would be glad to hear that the engine parts had arrived [8]. On the 4th of April Marshall and Benyon acknowledged receipt of the engine and requested a good hand to fit it up [3]. Boulton and Watt advised on the 13th April 1795 that the erector would be a George McMurdo, and that they were sending more drawings by coach the next day and requesting an answer on the agreement [9]. This duly came, dated 28th April, with an acknowledgement of the letters and drawings and the statement:

"We object to nothing in the copy of Agreement which you have sent, but the alterations which you have made in the former part of it. In our former agreement you recite the Letters Patent as the foundation of the agreement. In this it appears as if the only consideration on your part were your skill and assistance in constructing the Engine. Your reply on this subject will oblige" [3].

James Watt junior replied (3rd May) as both Boulton and Watt senior were absent, explaining that they were changing their agreements due to the patent term drawing to an end. The agreements would now include lump sums as the annuities on small engines would not recompense Boulton and Watt for their trouble and expense. But he did agree that it would be right to let Marshall and Benyon have the engine upon similar terms to the former one.

However James Watt junior was ill at ease to the extent that he wrote to his father on June 11th:

"Messrs Marshall & Benyon we learn from private information, say publicly they will not sign the Agreement. I shall write to them an expostulatory letter by this post and insist on a positive answer, if it does not come, we have no alternative but to order George (McMurdo) *to leave the premises & apply to the Chancellor for an injunction to prevent them from working the engine. Please advise with Weston* (Boulton and Watt's solicitor) *what can best be done. They had a rough draft of the Agreement sent them before they began to erect the engine and have never objected to any of the essential parts"*[10].

On the same day he wrote again to Marshall and Benyon and in order to give the full flavour of the feelings at Soho the full text is given:

"We have this day been informed of a report of a very injurious nature as we apprehend both to you and to ourselves and as in all cases of doubtful authenticity the surest way to avoid misunderstandings is that of applying immediately to the parties themselves, we shall take that liberty in the present instance.
The report we allude to is that your Mr Marshall has publicly asserted he would not sign the agreement for our second engine.
Being conscious that our conduct towards you entitles us to a different mode of treatment and being ignorant of any motive which would induce you either publicly or privately to break through what was understood to be the leading condition upon which we undertook to erect your engine and conceiving moreover that such an assertion would be extremely detrimental; to your interest, in as much as it would compel us to prevent the working of the engine, we shall attach no credit to the report until you authorise us so to do.
But in the meantime, as our relative positions are by no means pleasant, we must take the liberty of requesting an early answer with the communication of your sentiments and intentions.

We are sorry to trouble you upon so disagreeable a subject, but we conceive it to be of equal importance to your character & to our interest that this matter should be cleared up to mutual satisfaction. In this expectation we remain respectfully"[11].

And indeed the position felt in Leeds was rather prickly as the following missive was despatched to Soho on the 12th June presumably before the above had arrived:

"Please to place the enclosed Bill value Three Hundred pounds to our credit on account. When the Engine is finished & we have your charge for the Man's time we will remit the balance. When our last Engine was made you told us that the payment for the materials was due when the Engine was finished. We some time ago gave Mr Lawson a printed advertisement of Mr Wilkinson's offering bored Steam Cylinders at Bersham at 21/- per Cwt – we shall be obliged to you to explain the reason why you charge them 30/- which we could have cast here at 18/-. As you say that you meant to let us have this Engine upon similar terms to former one, you can have no objection to having a similar agreement. We think there are some parts of the new agreement objectionable, but we have no objection to execute one exactly similar to the former one, & if you please you may fill up one and send it us. Our names are John Marshall, Thomas Benyon & Benjamin Benyon of Leeds in the County of York Flax Spinners" [3].

And on the 16th June John Marshall replied to Boulton and Watt's concerns:

"Yours of the 11th Inst came duly to hand & would have been replied to sooner but that I have been out of town and did not return till today.
We writ to you 12th Inst which will sufficiently explain our sentiments & intentions respecting the agreement. When we gave and you accepted an order for a second Steam Engine without any terms being specified, we considered both you & ourselves as pledged to execute a similar agreement to the former one, & I have never in any company expressed any sentiments or intentions to the contrary. I am for Self & Partners etc" [3].

This crossed with Boulton and Watt's acknowledgement of receipt of the £300 which also stated:

"With respect to the charges of 30/- per cwt for cylinders &c, we have to observe that we have merely recharged Mr Wilkinson's bill to us, without any profit whatever and that we have never had a cylinder cast & [...] at an inferior price. The Advt. you mention was we believe for cylinders of less perfect workmanship and iron of a worse quality than is necessary for our Engines. If any further information is required, we have no objection to your corresponding with Mr Wilkinson himself, with whom we now have no connection" [12].

Then all the discussion closed, at least at the surface level, with the reply to John Marshall's missive of the 16th June:

"We have to acknowledge receipt and return thanks to your Mr Marshall for his favour of the 16th Inst which is perfectly satisfactory.
We always considered you as at liberty to choose between the new and old form of agreement and it is by no means our wish to press the former upon you against your inclination. If at first we had understood your objections in the way they now appear to have been meant, we should have conceded the point at once, more particularly as we do not conceive the variation between the two agreements to be at all material

in the present case. We have ordered them to be engrossed in the old form and they shall be forwarded for your signatures as soon as they are ready" [13].

After some minor further exchanges of information, Boulton and Watt advised by letter dated the 15th July that the Agreements signed by them had been sent by the Sheffield coach. It would have been interesting if there had been further delay, as it was on July 31st 1795 that George Hawks wrote to Boulton and Watt from Gateshead. No doubt due to the cessation of dealings between John Wilkinson and Boulton and Watt; Hawks had recently learned that the engine he wanted from Soho was to take much longer than expected. He had called in at the Low Moor Iron Works near Bradford to enquire whether they could make the necessary parts. As he put the matter in his letter:

"On my way here I called at the furnaces at Low Moor near Bradford where I learned that they were upon the eve of making a cylinder, air pump etc on your principle for a Messrs Murray & Wood, ye former an agent of Messrs Marshall & Benyon of Leeds, who they understand has your leave to erect an engine on your plan" [14].

He went on to say that he understood that an agent of Boulton and Watt had viewed Low Moor's boring machine and been satisfied that they could bore cylinders etc.

On the 5th August John Southern informed Marshall and Benyon that there had been a mistake made in the calculations for the fly wheel and that the rim would need 4 inches added all round.

James Watt junior wrote to his father on September 4th:

" I have also desired Mr Southern to call at Low Moor in Yorkshire and learn what particulars he can respecting the pirate Engine about to be made there for Marshall & Benyon's foreman. Weston thinks that upon the facts stated in Mr Hawks' letter, an Injunction may be obtained both against Marshall & Benyon & against Hird Jarret & Co."[The partners in Low Moor] [15].

It would appear from this letter that Boulton and Watt knew of Matthew Murray already, presumably from James Lawson and the engine erectors, but were unaware that he was setting up on his own. With Boulton and Watt's knowledge of the case brought by Kendrew & Co. and Marshall's recent intractability over signing the agreement, James Watt junior readily assumed that Marshall and Benyon were the true villains in the matter.

It is evident that Murray and Wood were seriously setting up in business at this point, and no doubt beginning to seek and obtain orders from firms other than Marshall and Benyons. There is the evidence of Hawks' letter that they were building a steam engine and in the next couple of months they were recruiting. On the 17th and 24th of August 1795 the following advert appeared in the Leeds Intelligencer:

"Mechanics Wanted
Wanted, a number of White-smiths, Joiners, Wood-turners, and Iron-turners, who will meet with constant employment, by applying to Messrs Murray and Wood, machine-makers, at Holbeck near Leeds August 13th 1795"

This was rapidly followed on 31st August and the 7th & 14th of September by another advert:

"To Mechanics
Wanted, a number of Smiths, viz. Firemen, Vicemen, and Iron turners; also a good Wood turner. The above will meet with good wages, and constant employment, by applying to Messrs Murray and Wood, Machine makers at Holbeck, near Leeds August 25th 1795."

As Murray and Wood did not obtain the land in Water Lane until February 1796 the initial expansion presumably took place at their first premises at Mill Green, Holbeck, although of course they may have already been planning their move, and identified the ground adjacent to Marshall and Benyon's Mill.

On the 21st September 1795 Marshall and Benyon advised Boulton and Watt that George McMurdo had been with them for 16 weeks and had received £14/14/0d, and asking for the balance of their account including the premium for the engine in Mill A. They also advised that they were making the changes to the fly wheel as directed and requesting that the premium date for the new engine be set on the 18th September along with the Mill A engine although they will have used it before then. This to compensate for their additional expense with the flywheel which was not caused by them [3].

Matthew Murray's youngest sister Margaret, who had presumably moved to Leeds to live with her brother on their father's death in 1792 when she was only 14, got married at Leed's Parish Church to James Clayton a whitesmith on 19th October 1795. The marriage was witnessed by Matthew Murray and Robert Bucktrout. A Robert Bucktrout was a signatory to the articles 'for the better regulation and good order of this shop' [16] at Marshall and Benyon's Holbeck dated March 10th 1795.

John Southern replied to Marshall and Benyon's proposal for a rebate on November 14th stating that they expected any cost to be negligible and that it should be remembered that no charge has been made for Mr Lawson's many visits. However he proposed that both engines should be charged from September 1st as opposed to the 18th which he reckoned as a benefit to Marshall and Benyon's of £3/10/0d [17]. This offer was not accepted by Marshall and Benyon who advised that the additional expense would be in the form of the bolts for fixing the extra weight to the fly wheel and drilling the holes and fixing it all up and advise that they will keep an exact account of the cost and would advise Boulton and Watt of the amount [3].

Officially the engine in Mill B started work on the 10th August 1795, spinning commenced in September and by the end of the year they had 1200 spindles in operation [18].

Two agreements covering the purchase of land in Holbeck by Marshall and Benyon and Murray and Wood, dated the 11th and 12th of February were registered on the 19th February 1796 [19].The land named the Leckeys, totalling 2 acres 2 roods and 30 perches, was purchased from the Barstow family by the firms who took half each. Murray and Wood bought the eastern part[19]. It is probable that the funding for the whole purchase was either provided by Marshall and Benyon or guaranteed by them. Some of the part purchased by Marshall and Benyon was immediately offered for sale for building cottages or houses and advertised in the Leeds Intelligencer on February 15th.

On the 13th February 1796 disaster struck Marshall and Benyons as Mill B was totally destroyed by:

"the most dreadful fire we ever remember to have happened in this country" [20].

A description of the event was included in the paper later:

"On Saturday morning last, about seven o'clock, the buildings lately erected for the manufacturing of canvas, linen etc by Messrs Marshall & Benyons, near this town, were discovered to be on fire and notwithstanding immediate assistance was had, it raged with the greatest fury until the afternoon, when their new mill which had been at work about six months and a small building lately erected were almost totally consumed together with a great part of the machinery and other articles within the same; but the fire was fortunately extinguished without doing any damage to their old mill or flax warehouse. We are sorry to add, that by the falling of a wall seven or eight people were killed,(John Marshall records 10 deaths) and about twenty others much bruised and wounded, the greatest part of them were taken to our Infirmary, one of whom is since dead. We are happy to hear that a great part of the property was insured. To the honour of the Gentlemen Volunteers they attended day and night until Monday to guard the property that was unconsumed" [21].

Mill fires were not an uncommon occurrence at this period, due to the fibrous nature of the products being processed, which created a very combustible atmosphere full of fabric dust. The dust could be ignited by sparks or hot machinery parts or drives. Fireproof construction was just beginning but still in its infancy.

James Lawson reported the incident to Soho for on the 17th February John Southern wrote to Lawson acknowledging receipt of three of Lawson's letters one of which had communicated

"the terrible account of the fire of Mr Marshall's Mill, and of the shocking circumstances attending it".

He enclosed papers to Lawson drawn up by Mr Watt after the fire at Albion Mills (a large flour mill in London in which Boulton and Watt had a big financial interest, as well as supplying much of the machinery, which had also suffered a disastrous fire) and also described the methods adopted by Messrs Strutt of Derby, which Messrs Marshall & Co could no doubt view. The system: *"consists of throwing brick arches between beam and beam somewhat thus and tying the beams together by iron bolts to prevent them from yielding to the [   ]"* [22].

19 Southern's sketch of brick arches spanning beams.

As regards the loss of Mill B, Marshall recorded that its value was £10,000 and half was insured, and it was rebuilt within the year. So Murray and Wood would have been very busy replacing all the lost machinery. They put their first advert, of many over the years, for green sand moulders in the Leeds Mercury on February 20th

"To Iron Founders
Wanted Immediately two sober steady and active men, as Green Sand Moulders. Any person answering the above description will meet with constant employ and good wages by applying to Murray & Wood, Holbeck, near Leeds February 17th 1796. None but good workmen, and of good character need apply."

During March 1796 Marshall and Benyon were in correspondence with Boulton and Watt for details on the fitting of a forcing pump onto their 20 HP engine at Mill A along with cisterns and other necessaries for firefighting [23].

In a ledger entry dated Monday 18th April in the surviving books of the Kirkstall Forge (illustrated at the beginning of the chapter) is found the first known reference to Murray and Wood as a customer, when Kirkstall Forge supplied them with with 50 square bars of turning iron weighing just over 20 hundredweight and costing £25/10/3d. This entry also gives the address for Murray and Wood as Mill Green Holbeck.

Royal Exchange Insurance Policy number 151066 survives and indicates that Murray and Wood were operating from the new Water Lane site by May of 1796 as it reads:

"May 13th Matthew Murray and David Wood of Leeds in the County of York Iron Founders.
On a building brick and slate situated in Water Lane in Leeds aforesaid used as an Iron Foundry, Joiners Shop and Model warehouse in which no fire is used for the drying of timber   £400
On utensils of Trade and in trust in the same   £150
£550" [24].

While obviously a setback, the fire at Mill B did not stop the expansion plans of Marshall and Benyon for early in May they were enquiring for a 20 HP engine from

Boulton and Watt, which it becomes clear in the later correspondence was for a new mill to be built near Shrewsbury. Marshall and Benyons required the price to exclude; the boiler, fly wheel, rotative shaft, connecting links, and sun and planet wheels, which they intended to make themselves [3]. Presumably they were to be made by Murray and Wood. On the 9th May Boulton and Watt quoted £750 for the metal materials of a 20 HP engine without the excluded items which they redefined to include more detail. They advised that if they received the order promptly they should be able to deliver the engine by 1 February 1797.

Marshall and Benyon accepted the offer, while adding the connecting rod to the exclusions for £10 to £12. They asked for the payment terms for the £750, in terms of due date, whether instalments were allowed and if they attracted interest. Marshall and Benyon also pointed out that they had returned previous drawings to Boulton and Watt and would require new ones. Boulton and Watt sent a copy of the agreement for Marshall and Benyons approbation on the 24th May advising that the money should be paid 3 months after the goods were delivered; and that due to the large investment being made in their own establishments *"it would be extremely inconvenient to be paid by instalments"*. This was all accepted by Marshall and Benyons on the 10th of June, who advised that for their part the agreement should be in the names of Thomas Benyon, Benjamin Benyon and John Marshall of Leeds and Charles Bage of Shrewsbury. Boulton and Watt sent the contract, signed by themselves, to Marshall and Benyons on 5th July [26].

Murray and Wood declared to the Leeds Community that were in business for all (and therefore not just supplying Marshall and Benyon) by the following advertisement which was placed in the Leeds Mercury on July 9th and the Leeds Intelligencer on July 11th

"MURRAY and WOOD
Desire to inform their Friends and the Public in general, That they have erected and opened a FOUNDRY, in Water-Lane, Leeds, for the Purpose of Casting Iron, viz. Engine Work of all Kinds, Ballance Wheels, Joints, Bosses and Steps. Crank and Octagon Wheels, Grate Bars, Bearers, Frames and Doors. Steam and Injection Boxes, Wheels, Segments, Tumbling Shafts, Plumbers Blocks, Coupling Boxes, and Mill Work in general.
Tapet and Waggon Wheels, Waggon Rails, and Tram Wheels. Rasp Barrels and Paper Rolls Chip Plates Oil Presses and Blocks. Callender Wheels and Plates. Tenter Posts. Press Tops and Bottoms. Press Ovens. Press and Singing Plates. Cotton Spinning Plates. Cotton and Worsted Weights, Carding and Scribbling Engine Rims. Chain Wheels and Strap Pullies. Malt Rollers, Pallisades, Weights, Clock and Sash Weights &c.
Those who please to favour them with their Commands, may depend upon them being well executed on the lowest Terms.

N.B. As they cast twice each Day, any Gentleman may be accommodated with Castings on the shortest Notice, in Cases of Emergency."

At the end of August Marshall and Benyons advised Boulton and Watt that the contract was signed by the Leeds partners, but awaited Mr Bage's signature. They also requested an allowance against the premiums of the Mill B engine which had been stopped for some 6 months after the fire, whilst acknowledging that this was not provided for in the contract (unless the period of stoppage was 12 months or over), but they understood similar concessions had been made to others. Boulton and Watt agreed a period of concession based on the engine starting work again on 11 July 1796, which Marshall and Benyons accepted by letter on the 7th October and remitted the sum due. Boulton and Watt's advice was also sought on the acceptable wear limits for the piston rod of the Mill A engine as ¼ inch had worn off the diameter [3].

Murray and Wood purchased 4 ladles from Kirkstall Forge on the 24th October, and on 31st October 1796 35 Boiler plates (over 27 hundredweight), and 10 bundles of rivet rods which was shortly followed on November 9th by a further purchase of 48 boiler plates (over 33cwt). Unfortunately there is no record of Murray and Wood's customer, but it is further evidence that they had entered the steam engine business alongside machine making and millwork. At this early stage it can be reasoned that the boiler was either for themselves, although if the engine for which Low Moor made the cylinder in 1795 was erected in Mill Green there would have had to be a boiler as well, which presumably could be moved along with the engine, or it may have been a second boiler for this engine. It may have been a spare boiler for Marshall & Benyons or for another unknown customer.

With the delay in the signature agreement over the contract for the Mill B engine during 1795 in mind Boulton and Watt wrote to Mr Bage on the 30th November 1796 saying that Mr Marshall had advised them that the contract was with him for signature and:

"we therefore request you will immediately execute the same and return one to us, with advice by post, as it has been unusually delayed" [27].

It would appear that Mr Bage obliged, as the issue is not raised again.

It is worth noting that at this period Boulton and Watt were in the middle of their litigation with Maberly and Hornblower over patent infringement on the one part and a contention that Watt's patent was not new and the specification insufficient. The matter came to court in December 1796 with the jury deciding in favour of James Watt. However Maberley and Hornblower appealed and the lawsuit continued until January 1799 when Boulton and Watt won the case. However their damages received (as opposed to awarded) probably covered about one fifth or less of their legal costs. Boulton and Watt had successfully applied to the courts for an injunction against Edward Bull (June 1793) to prevent Bull from erecting any more engines. However the case did not finally go to trial until a year later, during which major attacks were made on Watt's specification, which when judgement was made in 1795 the judges were split 50:50. However the injunctions on Bull remained. In 1796 Boulton and Watt had also been dealing with pirate engines made by among others Bateman and Sherrat of Manchester, Sturges and Co, and Low Moor, both near Bradford. Financial settlements had been reached with Bateman and Sherrat and Sturges and Co, whilst Low Moor were found innocent of intent. Sturges and Co for example reached a peace agreement with Boulton and Watt on 27th July 1796 agreeing to pay a £1640 premium on the two engines in their own works. Boulton and Watt got on so well with Mr Smalley of Low Moor that he ended up making at least the boring bar for the Soho Boring machine. This brief background of a very complex matter plus their knowledge of the Kendrew case explains why Boulton and Watt were so suspicious of the actions and motives behind Marshall and Benyons and Murray, but also why they were unwilling to take action in the courts unless they had a cast iron case. Litigation had cost them dear, even if on balance they were ahead on judgements on the basis that they had won some and not absolutely lost any.

The reasons for the new partnership of Benyons Marshall including Bage were extremely relevant to the development of Murray and Wood. Charles Bage was a Shrewsbury wine-merchant with a penchant for science and a correspondent with William Strutt of Derbyshire, who was a pioneer in the early fire proof construction of mills. Bage also knew the Reynolds of the Ketley Ironworks who built an iron aqueduct on the Shrewsbury canal; and Thomas Telford, although probably at a later date. Strutt's work has already been referenced above in the letter from John Southern to James Lawson after the Mill B fire. The Strutts were building a calico mill at Derby and a warehouse at Milford in 1792, just after the fire at the Albion Mills in London joined the long list of mill fires, and the Strutts determined to make their buildings fire resistant. In the buildings cast iron pillars were used to support Baltic fir beams 12 inches square, with brick arches in between to support the floors also of brick. The wooden beams were cased in sheet metal. The roof beams were of wood. These techniques were also applied in building the new cotton mill at Belper which was built between 1793 and 1795. These ideas and methods were communicated to Charles Bage who undertook the design of the new mill for Benyons Marshall and Bage in 1796 at Ditherington just outside Shrewsbury. Charles Bage however took the process one stage further and the new five storey mill introduced cast iron beams and tie-rods as well as columns and a basic cast iron structure for the roof thus eliminating wood apart from the staircases and wooden trusses and battens to support the slates on the roof. While there are one or two other possible contenders, Ditherington, which survives today, is generally held to be the first multi-storey iron framed building in the world. The iron work for the original mill was procured from William Hazeldine's foundry in Shrewsbury. It is interesting to note that the first surviving letter from Charles Bage to William Strutt refers to decisive advice in favour of inverted arch foundations for walls.

So at the dawning of 1797 Ditherington Mill was under construction, while in Leeds building was progressing at Murray and Wood's new factory in Water lane. Both were of fire-proof or fire retardant construction. Ditherington was the more advanced as noted above. However the early buildings of the Murray and Wood factory, which would become known as the Round Foundry, and is roughly contemporary with Ditherington have a mixture of massive oak columns and cast iron columns with massive oak beams in lieu of the sheet metal covered baltic pine. It is probable that Murray and Wood were involved in setting up Ditherington to some extent in terms of textile machinery, the manufacture of the steam engine parts not being supplied by Boulton and Watt and general millwork. They would therefore probably have been aware of the options from talking to Charles Bage and also through discussions with John Marshall. Their budget at this point may not have stretched to cast iron beams, or the stress calculations may not have been clear, but they had certainly learnt the lesson from the burning down of Mill B, and used some of the latest construction technology for their buildings.

In February 1797 Benyons Marshall and Bage asked Boulton and Watt for the delivery date for the engine, and followed up in March with a request for the details needed to erect the boiler and grate which was accompanied by a drawing of the Engine House. At the end of May Boulton and Watt duly sent the drawings for the seating of the boilers and advised that they would probably be in a position to send the engine materials "in the course of the ensuing month". Further queries concerning the grate bars, the height of the boiler, the opening for clearing the ash and the weight of the engine (in order to calculate the size of the team of horses etc. to transport it) followed from Shrewsbury on June 2nd, but by June the 10th Benyons Marshall and Bage were writing to advise their preference for having the engine shipped by water to Stourport. Boulton and Watt replied on the 12th June to the queries that had been posed on the 2nd and advised that if the engine was to be transported by water they should have the parts ready perhaps by the 26th of June, and asked if there was enough water in the River Severn for the boats. In the event the engine parts started arriving by July 2nd, and Joseph Varley was despatched by Boulton and Watt to erect the engine.

While this normal business was taking place James Lawson was in Leeds and wrote to Boulton and Watt on 25th July, and after reporting on their normal customer business included the following:

"I refer you to Mr Southern who I suppose will be with you on Saturday for information as to the Engines Erecting here both by Hurst & M Murray (Mr Marshalls man) who are now erecting several so as to be made double Engines so soon as the patent expires – one erected by Hurst I have seen it has Par[allel] motion & Sun & Planet Wheels & the place for the Air pump & Condenser is contrived so as to be put in without altering anything – in fact the Cylinder & Lower Nozzle is nearly a copy – this is the Engine I before mentioned which Hurst wanted to pay the premium for – I am much afraid this method of theirs may prevent some orders here – and I should think allowing them to pay the premium & make them in their own way would be all or perhaps more than all gain – for I think they would not be able to execute so as to give satisfaction at first which would make others wish to go to the original makers"[28].

This was followed on the 23rd August by a letter from Manchester in which Lawson remarked:

"As you do not wish to allow Hurst or Murray to apply Air pumps &c I judged it the most prudent to keep out of their way & have not seen either since"[28].

Benyons Marshall and Bage wrote to Boulton and Watt on the 11th of September to state that Joseph Varley had erected the engine, conducted himself to their satisfaction and his credit and been advanced £6/6/0d which would be put to Boulton and Watt's account. Ditherington Mill was in business.

The evidence from Kirkstall Forge shows that Murray and Wood were still active as during September 1797 Kirkstall Forge's ledger records that they spent over £150 on boiler plates. The following advertisement demonstrated their intent to actively participate in the steam engine business. It appeared in the Leeds papers in September and October 1797

"TO SMITHS &C
WANTED
TWO FIRE-MEN that can undertake
Heavy Work in the STEAM LINE.
Likewise ONE GOOD WORKMAN, as a GREEN
SAND MOULDER in the Foundry Business; and also
A PERSON who has been accustomed to make IRON
BOILERS for Steam Engines.
None but good Hands need apply, as good Wages and
constant Employment will be given.
For further particulars apply to Murray & Wood
Water Lane Leeds."

James Lawson reported in again on 7th October, this time from Liverpool, to Matthew. R. Boulton (Matthew Boulton's son) and stated:

"I came here on Monday last, having gone round with Mr Lee by Langollen. At Shrewsbury I waited on Mr Bage & found Mr Benyon also, they were very civil to us & promised to remit in a day or two which I suppose they have done before this time. The drawings Mr Bage will not give up without you insist on it , by letter, which I should not think of much consequence as I have no doubt they are already copied – and I have seen some drawings of Murray's – which were copies of the Leeds Engine" [28].

A deed dated the 23rd and 24th of October records the purchase of a piece of ground known as Kays Close in Leeds, adjacent to the Leckeys, by Matthew Murray and David Wood with John Marshall acting as guarantor for them.

Boulton and Watt wrote to Benyons Marshall and Bage on November 3rd 1797, much of the letter concerning payment is lost but the final paragraph can be made out:

"May we at the same time request the favour of you to return the drawings left in your possession by our man" [29].

Shortly after this Boulton and Watt received the well-known letter from John Rennie which he wrote from London on November 8th advising them:

"There is a Mr Murray an Engine Maker at Leeds who is now erecting a 40 horse & sixteen horse steam Engines – this man makes very free with your patents – would it not be well to look sharply after him" [30].

James Watt junior was ill at this time and staying with a Dr Beddoes at Clifton Bristol, however in the light of Lawson's intelligence and Rennie's letter, M R Boulton wrote to him there on November 7th:

"Indeed it begins to be necessary to be prepared against our opponents for the encroachments of Murray at Leeds are rather serious".

He then repeated Rennie's remark about Murray and carried on saying:

"I rather think he is mistaken in the latter assertion it will however be proper to have his motions scrutinised by the vigilance of Lawson & to devise some method of checking his career – His patrons Benyons Marshall & Co have not been very punctual in the discharge of our [ ] & retain our drawings notwithstanding our applications to have then returned – Another competitor has declared himself – the Rev Mr Cartwright….." [31].

On the 17th November M R Boulton instructed James Lawson again to find out what Matthew Murray was doing:

"Your vigilance will be wanted at Leeds to inspect a little into the proceedings of Mr Murray –
Mr Rennie writes that he is erecting a 40 & six 10 Horse Engines & is making free with our patent.

In the latter point we are inclined to think he is mistaken as Mr Rennie perhaps means the patents for parallel motion Rotative Wheels which are expired. It will however be necessary to know how far his encroachments extend & what is to be feared from this competitor. The first point to be ascertained is whether his Engines are really infringements upon the essential principles of our Engine Patent, if, he is found to trespass we shall know how to deal with him, but if, as we have been given to understand he is only preparing to employ them at the expiration of the patent we should not notice his proceedings for the present. We wish you however to get some facts respecting the consumption of his Engines in proportion to their work & of their original cost with any other particulars you think of importance, observing to procure this information silently & especially with the privaty of Murray.

We think these precautions of Moment as otherwise our enquiries might tend to impress our antagonist with exaggerated ideas of his own importance & in the minds of the prejudiced manufacturers raise the reputation of his Engines by our appearing to dread them. We leave it to your ingenuity & better knowledge of the local to devise the means of carrying our wishes into effect persuaded from your former success we entrust them to skilful hands – Should you in the course of your investigations have an opportunity of learning the prices given by Murray for Dry Green & Open Sand Moulding it would be acceptable information as we are proposing to put our Moulders upon Piece Work.

Recollect likewise that you have a standing order to procure if possible one or two more such hands as [   ] endebted by your researches" [32].

This letter is worth bearing in mind as the story of the rivalry between Boulton and Watt and Fenton Murray and Wood unfolds. Lawson is tasked with industrial espionage, some of which could be argued to be legimate, in terms of patent infringement, but also to obtain pricing information, and apparently to obtain skilled workers.

On *t*he same day M R Boulton replied to John Rennie's business matter and added:

"Mr Murray's proceedings are not unobserved & if he is found trespassing you may be assured it will not be with impunity. Thank you equally for your friendly information" [33].

The feud between Soho and Water Lane was now well and truly launched, and in the minds of Soho Murray remains completely linked to Benyons Marshall & Co. James Lawson, who would have known Matthew Murray quite well already from his visits to Mill A and B and his enquiries in Leeds, acknowledged his instructions adding:

"Murray I well know will go as near the wind as he dare" [28].

James Watt senior wrote to his son at Clifton mentioning among other business which included the Cornish patent infringers:

"We hear however that some mischief is brewing again at Leeds at Manchester all are quiet, One Murray is the Leeds operator; we are not sure however if he is doing any more than making Engines which may be altered to ours when the patent is done, Lawson is gone to make enquiries" [34].

James Lawson provided some details on the 21st November and confirmed the Boulton and Watt suspicions that Matthew Murray was making engines to their patent but in such a manner that they could be operated as a Common Engine (atmospheric or non-patent) until the expiration of Boulton and Watt's patent in 1800; if permission was not granted prior to that date to use the patent parts. If so the patent parts could be added to the engine after patent expiry.

"I learnt that Murray is doing
1st a 40 Horse Engine for Mr Fischer (Fisher) *who is a respectable & pretty large manufacturer. This M makes a complete B&W's Engine with one Boiler & all the wood work- for abt £800 (as it is I believe a few pounds more, I can learn exactly how much) – this also includes erecting – The Engine House is built but the Engine is not yet begun putting up – but I understand is partly made.*
2nd a Ten Horse Engine at Hull I believe this is already erected but am not sure.
3rd One for his own manufactory the Engine House is built but the Engine is not yet begun putting up – I saw in passing his works an Air Pump (part of Watt's patent) laying and several parts of Engine Materials.

The plan he is going on with is to erect the Engine same as B&W's and then ask their permission to work – which if they refuse he will set them to work as a Common Engine till the term of the Patent expires. This is common report that I have only heard in conversation without making particular enquiries and I have avoided seeing Murray as he makes no secret of what he is doing and will probably ask questions I am not prepared to answer.
I shall therefore see some of the Engines at Horbury & Birstal etc – till I have got your instructions.
I believe I shall get a list of his prices for different sized Engines tomorrow" [28].

Lawson was as good as his word and wrote again the following day with information he had obtained from Walker a Cotton Spinner. This was probably Ard Walker who did in fact go on to buy an engine from Murray and Wood. Lawson listed the information he had received from Walker:

10 HP, 18 inch cylinder, 4 feet stroke, 27 ½ strokes/minute complete	£300
20 HP, 24 inch cylinder, 5 feet stroke, 23 strokes/minute	£470
40 HP, 33 inch cylinder, 7 feet stroke, 16 strokes/minute	£834.

The estimates were broken down and Lawson gives as the example that the 20HP is made up of £300 Iron materials, Boiler £100 and wood work £70, brickwork was excluded. Lawson added that Mr Gott had seen some of the work in Murray and Wood's shops and judged it well done and very nearly all copies from Boulton and Watt. Mr Walker also told Lawson that Murray and Wood gave unquestionable security. Walker was also reported as saying that perhaps he should wait until some of Murray and Wood's engines had been tried [28].

On 27th November James Watt junior wrote to his father:

"This Mr Murray seems to be ripening apace & to call for the kind hand of all assiduous care. I have therefore introduced him to our friends Weston (who was the

solicitor of Boulton and Watt) *with such queries & suggestions as occurred. Question and answer shall be sent to Soho as soon as received*"[35].

And on the same day Lawson provided even more information, opining that Boulton and Watt would sell no further engines until there had been a trial of one of Murray and Wood's engines:

"*I heard on Saturday that he means to begin at Mr Fischer's Engine with a close top on the Cylinder* (i.e. on Boulton and Watt's plan)*, but from the duplicity & cunning of his employers there is no knowing what they mean to do as I saw the Cylinder for Fischer's Engine which is an exact copy (of Ours) – it is the only part of the Engine yet at the premises they are preparing to put in the condensing cistern.*
The Engine for Hull is gone there some time since and is putting together.....
I cannot conjecture how he means to do Engines at the prices given in the estimates except that Marshall means to get orders to insure future trade, for in the way they are doing the work they must lose much by it. They have built very large Smithies and other workshops – which are just covered in & seem to spare no expense" [28].

James Watt junior's letter to Weston is quite exhaustive, but again the main thrust is against Marshall and Benyon:

"*In spite of the signal example we made at Leeds we are more than menaced with a new attack and of a new Species, from that quarter.I mentioned to you formerly that we expected some foul play from our Correspondent Messrs Marshall & Benyons of that place........ That Engine has been erected by us upon one of the new Agreements and is already in fact paid for, that the whole payment will follow, we have no doubt. It is not in that way in which they mean to annoy us. A man of the name of Murray who is their foreman, has lately erected a foundry at Leeds & begun to make Engines of all sorts in which we have no doubt he is supported by them although they do not appear as partners in the undertaking. This man having had daily opportunities of inspecting our Engines........ He has accordingly entered into contracts with a house at Leeds and with another at Hull to erect for them, for a stated sum, compleat Engines of B&W Construction. To be furnished (as we understand) with Airpumps. Condensers &c but these not to be used, unless the Parties [agree] it, before the Expiration of B&W's Patent. He is also making another Engine for his own Manufactory. .... We are moreover informed that when these Engines are finished, it is Murray's intention to apply to us for a Licence – and if refused, to work them as Common Engines during the remainder of our term – This Gentleman does not skulk in obscurity but openly avows his views & intentions, [in] that he has probably taken his ground, and with the benefit of Messrs Marshall & Benyon's legal experience is determined to maintain it. But as we have seen you pull down greater men than him we do not despair of your finding him vulnerable [   ] proceedings. The following suggestions are submitted to your consideration. Let Lawson employ himself in making it generally understood that B&W will not licence Engines .... May it not be right also for Lawson to see the Gent'n for whom the Engine is intended at Leeds and at Hull & inform them of the above, and moreover that if they are found in a degree to encroach upon us we shall without further ceremony proceed against them at Law. Are they not in the first instance, as well as Murray, liable to an Action for having entered into an Agreement to erect a contraband Engine and might not as a Species of Light Artillery a Bill be filed to compel disclosure of the Agreement.......   Their Engines will of course become subject to injunctions prior to their working....... But supposing they retain those parts about the Engine without using them (which will of course be attended with a diminution of power and increased expenditure of fuel) are they then also liable to an injunction"* [33].

Weston shortly received further input from James Watt senior:

"James has wrote you concerning Mr Murray or rather Marshall & Benyons proceedings which are more formidable' there we know of they are making our Engines at less than half price (we believe to their own loss) with an intention of knocking up our trade & upon the idea that the Engines once made we must grant our licence on the payment of the ancient premium...." [33].

He also asked whether they should contact Murray's customers to advise them they would not grant licenses under the old rules in these cases, whether they should simply remain quiet, or whether they should seek injunctions. He also wrote to his son James and gave his interpretation of part of Lawson's correspondence in that he stated that Marshall was giving security for Murray's performance. While this may have been the case there is no independent verification.

On the first of December, although dated November 31st, M R Boulton wrote to James Lawson and this letter probably contains the most straightforward expression of Boulton and Watt's feelings on the matter:

"The prices you state convince either Murray's folly or some grand plan in[stead]. It is absolutely impossible to furnish the materials with a common manufacturers profit for the amount of his estimates & especially as you say the work is well done. If Mr Marshall thinks to drive us out of the market & ensure himself afterwards a [snug] trade by the common manoeuvre of lowering the prices he will be mistaken. Does he imagine our capital, our activity or our experience & knowledge inferior to Mr Murray's or does he reckon upon his greater perseverance? I trust he will not find us deficient in either point, at least it shall not be without a contest proportionate to the importance of the object that he obtains possession of the field & if he succeeds in depriving us of the advantages we had a right to expect from our extensive & expensive preparations we must turn flax spinners & indemnify ourselves from the reputed great profits of his lucrative trade.
If war is to be declared it shall not be only defensive on our part we shall not content ourselves with repelling the attacks of such an aggressor he must prepare himself for retaliation & the means are perhaps more within our power than he supposes" [36].

By the 5th of December Lawson had been to Hull and was able to inform that the engine in Hull was for a Mr Charles Shipman who was in the business of spinning coarse flax to make sail cloth. The engine was not complete but what Lawson saw was a direct copy *"a B & W's 10 horse Engine"*

The following day James Watt junior forwarded Weston's advice to M R Boulton, which was that Lawson could safely speak to Murray's customers about the situation. It is notable at this point that there is no reference made to David Wood as at this time he is either ignored or unknown to Boulton and Watt. Weston declined to answer the other points until there was fuller information [37]. The covering letter suggests that Lawson on seeing Fisher should emphasise *"the illegality of their agreement with Murray"* and *"the improbability of Murray making a complete engine"* as good as ours, the imprudence of dealing with someone who has never made an engine and is devoid of honesty and principle. Lawson should tell Fisher what happened to the others who infringed James Watt's patent: Hornblower, Bull, Bateman and Sherrat, Sturges and Co and Maberley; and advise Fisher that if they proceeded they would be held as liable as Murray. J Watt junior suggested they ask John Southern to calculate the loss of power and the extra fuel costs that Fisher would face if he operated the engine "atmospherically" [38].

M R Boulton passed his instructions to James Lawson:

"You are to wait upon Mr Ficher..... inform him of your reasons for believing the Engine preparing for his factory by Mr Murray to be an infringement of our Patent. .....to appraise him that he must not expect the Engine to be licensed without payment of a greater compensation than the former annuity premiums...... Such are the leading arguments with which we wish you to accompany the admonitions to Mr Fischer & you may subjoin that in appraising of what you know to be the sentiments of your Employers you have been activated by a mutual regard to his & their interest wishing by such previous declaration to preclude the necessity of unpleasant measures hereafter – You will of course understand that the whole of the foregoing is to pop in conversations & carefully avoid [giving an item under your hand].

It will not be less prudent to make your interview as short as civility will admit ...... & to bear constantly in mind that you are conversing with a colleague of the cunning Mr Marshall. It is not intended that Mr Fischer should be the only person you are to communicate the foregoing sentiments....

Let us know what you have seen or heard at Hull" [39].

Matthew Murray did indeed want a license for the engine he was building for Fisher and Nixon, as shown in the letter Lawson sent to M R Boulton on the 7th December prior to receiving the above instructions:

"I met yesterday with the Clerk of Murray – who said that Murray wanted to see me to know if B&W would accept £200 to permit him to set the Engine for Mr Fischer to work at first as a Patent Engine, I understood that Mr Fischer had offered that sum. I told him I knew nothing of Mr Fischer's Engine & I could not say what B&W would take but he Mr Fischer might write as I was going out of town. I found from him that Murray was gone to Hull so that I shall probably see tomorrow more of what he is doing.
Mr Fischer is a man of respectability and one of largest Manufacturers I believe next to Wormald & Co (as a Manufacturer) (Wormald,Fountaine and Gott, in full). I have been several times in his company but he has never spoke on the subject."

James Lawson wrote to Boulton and Watt again on the 11th of December to say that having received M R Boulton's instructions he had called on Mr Fisher and had seen him that morning. He had passed on to Mr Fisher all the messages required, however Mr Fisher had informed him that the agreement with Mr Murray was for 40 HP when worked as a patent engine but it was to be worked as a common engine for the rest of the patent term. His inducement had been price and

74

Mr Murray was bound to produce an engine equal to a Boulton and Watt one. He wished to see Lawson to ask what Boulton and Watt's terms would be to allow the use of the engine as a patent one from day one. Lawson had replied that all he could say was that it would not be on the normal terms. Mr Fisher then informed Lawson that he thought Boulton and Watt were *"entitled to all they got by it"*. Lawson went on to say he had visited the engine house and had seen the parts, although not the cylinder bottom, and stated:

"I do not think it has been Murray's intention to work as a common engine" [28].

However when he wrote again two days later he had to modify that view:

"Since I last wrote I have been at Murray's shops – but find he has not yet made either the Top or Bottom for the Cylinder for Mr Fischer...... Only I find from Mr Fischer's foreman that he is making preparation for working as a common Engine"[28].

Due to the level of industrial unrest in Leeds Mr Fisher, on the advice of Benjamin Gott a partner in Wormald Fountaine and Gott, the leading woollen manufacturers in Leeds, had ordered some guns from Boulton and Watt and when they sent the invoice on the 19th of December Boulton and Watt explained their position as regards the engine:

"Our Mr Lawson has informed us of his interview with you respecting the Engine preparing for your Manufactory by Mr Murray & upon the supposition that you may have been led to believe by Mr Murray that we should be induced to licence his Engine we thought such previous communication of our [   ] [   ] [   ] to the respectability of your character & likely to avert legal measures which would be [] unpleasant. Whatever may be the views of Mr Murray (and certainly they must excite mistrust) in undertaking to furnish you with an Engine at a price which is very far [   ] to the employer much experience & knowledge on the subject of Engines.
We take the liberty further to remark, understanding that it was in contemplation by Mr Murray to provide your Engine with Condenser & other parts necessary for the application of our principles, that if such a design is persisted in we shall be under the necessity of counteracting it by legal measures in which you must unavoidably be involved. Obviously such a practice if allowed would defeat in a great measure the object of the patent in securing to us exclusively the benefit of our invention.
We shall therefore consider ourselves called upon to oppose in the first instance such a material encroachment upon our rights & which we are induced to hope when you are aware of its detriment & injustice towards us, will not meet with your countenance" [33].

Fisher Nixon and Fisher made an offer to Boulton and Watt on December 23rd 1797, which addressed payment for the muskets and continued:

"In reply to the chief subject of your Letter, we can assure you with much sincerity, it never was our intention to infringe upon your Patent by the Engine Mr Murray is now making for us, as we mean it to be a common Engine, so constructed, as not to admit of a single doubt, of its interfering with your Patent. We shall not require from it more than 17 Horses but from the information we have receiv'd from Mr Lawson, regarding the considerable saving of Coals, that is attached to your Principles, we are from this consideration induced to offer you £400 for your Permission to suffer the present

Engine now making for us to be converted into a patent one, being as we are informed nearly your usual Premium at the rate of 5£ per Horse, as there will remain little more than two years of yr Patent after our Engine is likely to be set to work. If you do not close with this proposal, you may rest assur'd that to the best of our knowledge we will not do anything in this Business that might draw upon us your Censure, or make us liable to a prosecution" [40].

Unfortunately for Fisher and Nixon, Boulton and Watt were not going to accept any such offer and replied on December 26th:

"At the same time that we entertain a just [sense] of the candid & liberal sentiment manifested in your letter we must for the reasons before assigned beg leave to decline the acceptance of your proposal" [33].

In a list of Boulton and Watt engines at Leeds written by Lawson at about this time he noted under the listing of the engines for Marshall & Co:

"20 Horse 1500 flax spindles. 28 Horse 2300 Spindles. [Spins] 12 tons of Flax per week [ ] from information of Murray" [41].

So by the end of 1797 the commercial war had been internally declared by Boulton and Watt on Marshall & Co and Matthew Murray and their spotlight on activities at Leeds would continue for a number of years whilst Boulton and Watt tried to bring Murray to heel by legal means. However Murray continued to avoid infringing their patent by building atmospheric engines designed to be converted to condensing engines after the expiration of the patent in 1800. Murray also, as recorded by Lawson, stated that this was his purpose.

Thomas Butler of Kirkstall Forge recorded in his diary for Wednesday 17th January 1798:

"After breakfast G. B jnr and I went to Leeds. We called upon a great number of our customers with Xmas accounts but we were not very successful. We received about 140£. In the evening we spent a very pleasant hour with Mr Murray at the Elephant and Castle" [42].

Whilst the reference cannot be completely attributed to Matthew Murray it certainly sounds like him.

The known records are then quiet for a couple of months and the next matter to be heard about Murray and Wood was the purchase of more ladles from Kirkstall Forge in March [43] and then at the end of March the main theme started again with a letter from Fisher and Nixon to Boulton and Watt displaying their determination not to infringe the patent:

"Our Engine being nearly finished we are desirous that it should be inspected by your agent Mr Lawson or any other Person you may appoint, prior to its being put to Work. For notwithstanding the strictest attention has been paid to withhold everything, which we conceive would in its operation interfere with your Patent, we may unintentionally have erred, & it will afford us a satisfaction your instructing Mr Lawson to view the Engine. For should it appear there was any actual infringement

on your Patent we may have the opportunity of removing it on its being pointed out to us" [44].

Boulton and Watt replied on April 12th that Lawson would be in touch when he had finished his current undertakings. However it is not until June that the meeting was achieved, interestingly James Watt junior was in Leeds at the time of the meeting, but he did not attend. He wrote to M R Boulton on June 10th:

"Lawson yesterday called on Fischer whom he did not meet with. I propose for him to call again & see the Engine on Monday. I understand the Engine goes well; but burns more coals than Mr Gott's without using more than 15 or 16 horses power. Gott says he would pay up the difference between our price & Murray's to be allowed to have an air pump. [I query] whether upon this footing it might be advisable to permit him, or upon any other footing, say 2/3 of saving instead of 1/3.

Murray is pushing himself everywhere into employment for Mill Work & Engines & he must succeed by dint of impudence, if not by ability. As I see no way of opposing him but by countenancing the establishment of a small foundry here in opposition to him as his local advantages must otherwise deprive us of all chance of competition. The fellow reports that he has still orders for several Engines, but I believe they are all in buckram as no names are mentioned except Marshalls for whom he says he is to erect one here and one at Shrewsbury. He sells his castings, say wheels & green sand at 13/-" [45].

The above letter introduced one of James Watt junior's continuing themes in the contest with Matthew Murray, that of Boulton and Watt setting up a rival foundry in Leeds or thereabouts. The next letter introduces a further strand which becomes very important, namely the difference in quality between the output of the two firms, and here it is set with an admission by James Watt junior of the bad state of Boulton and Watt's engines and later on with the statement that Murray and Wood's engine was reported to be extremely well made.

The letter to M R Boulton is dated June 12th:

"We have not had a call from Taylor of Gomershall as was expected & hear of no new Engines being wanted; indeed trade is not very brisk, and at all events, I fear Mr Murray has cut us out, aided and assisted as he has been by the bad state of our own Engines. – One consolation we have which is that he must be losing money by his trade & that even the Jew Fischer (for he is no better) is biting his finger nails with vexation at having three stories of his Mill unoccupied, owing to the deficiency of power of his Engine. It was examined yesterday by the [Dieurs], Lee, Ewart, Gott & Lawson in the presence of Messrs Fischer & Murray when our friend Peter [Ewart] was very severe in his animadversions & I understand left Fischer & Murray neither so much pleased with each other, nor with performances as before. They think Fischer would go high to purchase the right of using the Air Pump. I have thrown out a loose suggestion that I hardly thought we should licence him for less than £1000 & I did not know whether we should do it for that. I expect the determination of B&W on that point, in case overtures should be made on my return. He uses Rotative Wheels & the Parallel Motion. May examine whether both these patents are expired. I know that one is, but have forgotten which. If there was only a fortnight remained unexpired, I think we should indulge ourselves, in the pleasure of giving the mortification to Murray of taking it down & it might even furnish a good ground of action against that delinquent. The Engine is reported to be extremely well made &

the castings particularly good. It is for the most part a servile imitation, the slight alterations which occur will be pointed out by Lawson. The nozzles & sidepipes are cast in two pieces only. The Cyl[inder] made at Rotherham [probably by Walkers], whether any other parts were made out of his own foundry I have not heard. He did not ask any of the party to see his foundry but I do not learn that there is anything out of the common way about it. I have been thinking that the only way to stimulate Abram [Storey] to proper exertions at the Foundry [Soho] would be to state to him this rival ship of Murray & the injury we have sustained & may still sustain from him & to send him to Leeds to see these castings with his own eyes. There is no other way of making him ashamed of ours, or sensible of the necessity of improvement.

On the other side I send you copy of the prices of Murray & Wood for Millwork. They have offered to undertake a new mill of Mr Gott's at these prices & to guarantee the Millwork at 2 per cent per annum.

Mix't Iron Tumbling & upright shafts }	
including patterns & fixing	*} 14/- per cwt*
D° pillars, headlocks & carriages	*14/- "*
Bevil wheels & spur d° & nuts }	
of all kinds for gearing	*} 16/- "*
Mill upon new construction	*}*
Common Sq. couplings of all }	
Sizes	*} 14/-*
Circular D° & plummer blocks	*16/-*
Bell Metal Bushes	*@16 per lb.*

The whole of the above include patterns drawings & instructions for completing the Mill.

Prices of old Millwork by Murray & Wood:

Iron Shafts	*14/- per cwt.*
Pillars headlocking & carriages	*13/-*
Bevil Wheels & spur D° & mts }	
upon the old construction	*} 14/-*
Common Sq. coupling of all sizes	*13/-*

NB these do not include patterns of any description.

Murray acknowledges he was burning twice the coals at Fischer's that were necessary & was not doing one quarter of the work it was capable of doing with the addition of the air pump. – The Cyl[inder] had a top to it but a passage for the atmosphere to act on the piston. It is probable he uses grease to keep it tight, although our inspectors could not ascertain that point" [46].

M R Boulton replied with a very long and detailed analysis of their position on the 17th June starting with:

"We are fully aware that Murray from his superior workmanship is likely to become a more formidable rival than any of his predecessors in the piratical list & whatever doubts may be entertained in regard to the policy of licensing Fischer's Engine none exist upon the necessity of using every exertion to perfect & improve our work. We are agreed as to the propriety of the stimulus you propose for Abraham, the only query is, if he should go to Leeds to meet you or defer the journey till Murdock comes here when probably his absence will be less felt. There is at this time a great number of preparing orders in the dry sand line which I fear would be much bungled without

his superintendence & on the other hand it would not be altogether prudent to trust him in such dangerous company without a Mentor."

The letter continued with a discussion about how much any premium should be if they were to grant it. In this section he disclosed that Boulton and Watt sought to achieve a profit of 20% on the manufacture of the engine components in addition to their premium, and evaluated that their lost profit from Murray and Wood's engine for Fisher and Nixon was £757/7/0d. One option for a valuation of any premium. The second option he gave consisted in charging the difference in price which he calculated at £886 based on Murray and Wood charging £834. And immediately it is obvious why manufacturers such as Fisher and Nixon and Ard Walker were keen to pursue Murray and Wood's offers when these were less than half the price charged by Boulton and Watt. M R Boulton goes on to analyse the options over whether they should offer a license or not and what the considerations were

1. What would Boulton and Watt lose by not licensing engines erected by Murray and Wood?

2 Would such a policy discourage Murray and Wood and prevent a dangerous rivalry and maintain their patent?

3 If they were to license Fisher Nixon and Fisher's engine what would be the downsides.

If they charged the difference in price as a premium then there can be no reason for customers to buy Murray and Wood's engines except if one party succeeded, possibly through legal action, in obtaining a ruling that such a premium was excessive. This risk has to be evaluated because if real, they would in theory still have made money on engines erected by Murray and Wood. The fear here was that once the situation was known other engine makers would imitate Murray and Wood nationwide; and more steam engine manufacturers would spring up immediately and they would then be in a position to realistically challenge Boulton and Watt from 1800 when their patent expired. He then argued that it would be better for Boulton and Watt to reduce their prices and withhold any license to Murray and Wood or their customers, and that their overall income would not necessarily suffer. It is then pointed out that Murray and Wood were making a loss, and presumably doing this to achieve a hidden objective which would in the end produce profits for them and their patron. The only objective seen was that they intended to make Boulton and Watt share the market. If correct then they should ignore all the pleas which would put pressure on Murray and Wood without *"perpetual hostilities"*. The concerns about being taken to court over excessive premiums were held by Boulton and Watt due to their perception of Marshall's malignant spirit which to them had created the whole issue. M R Boulton reminded his partner that in the past they had settled with all the pirates at their standard premium of £5 per horsepower, and hence the concern of being deemed excessive in law if they attempted punitive premiums. He went so far as to propose that Marshall might encourage Fisher and Nixon to pay a high premium in order to instigate a legal challenge. He concluded by advising that they did nothing, offered no premium and observed what happened. The above seems a very able analysis of the market, where it can be seen that M R Boulton was capable of leaving aside any emotion and coolly examining the issues of the business.

By July M R Boulton was admonishing Lawson that his recent correspondence was entirely silent concerning Matthew Murray. So on the 26th of that month Lawson advised that:

"I have not heard of anything new doing by Murray except a new Engine for Messrs Marshall & Benyon which I only heard of this day as I shall go and see his own engine – as he has beg'd I should . I may probably know more"[47].

There follows some interesting correspondence concerning engines at Wilkinson and Holdworth & Co whose mill was at Hillhouse Bank. Richard Paley, who was a partner with Sturges & Co in the Bowling Ironworks, was also a partner in Wilkinson and Holdworth, and had a 45 HP Bowling Ironworks steam engine installed at Wilkinson and Holdworth circa 1796, which was one of the pirate engines involved in the settlement of Boulton and Watt's action against Sturges & Co. However this had been resolved between the parties and a Boulton and Watt engine had been erected also. In 1798 this engine was either being replaced with a larger one or upgraded, the displaced parts or engine being used by Boulton and Watt to satisfy a requirement from M'Connel & Kennedy Cotton Spinners of Manchester. M R Boulton wrote to Lawson on 9th August 1798:

"We are glad to hear that Holdforth & Co's Engine gives satisfaction & we hope the vigilance & assiduity of Gavin (McMurdo a Boulton and Watt engine erector) will soon remove the other grievances. – Nothing will more effectually co-operate with our refusal to Fisher in checking the insolent boastings of Murray than the good condition of our Engines at Leeds & its neighbourhood. – A present of a couple or more Guineas to Gavin would be a seasonal douceur. It would fortify him against the seductions of Murray & animate his zeal in the reformation of the uncleanly" [48].

The letter carried on to comment on another pirate activity in which Arthur Woolf was erecting a Hornblower and Maberley engine for winding at Newbottle colliery [Tyneside] with:

"Woolfs business must sleep till the Courts open."

On 15th August James Watt junior wrote to his brother Gregory who was based at Soho at the time:

"I have seen Gavin from whom I learn that Murray continues to run down our Engines with all the scurrility of which he is master. He [Gavin] has made a very good job of Holdforth & Co Engine but I fear they have too much connection with Murray to be satisfied with it – I mean with the price of it. Young Holdforth had refused to deliver up the drawings to Gavin, saying they were as much his property as the Engine, but I have given him to understand a different story & insisted upon their delivery this very day. He says they have neither been seen by Murray, nor any copies taken of them: both of which assertions I am inclined to disbelieve" [49].

Any suspicions that Boulton and Watt had that all was not well at Leeds must have been wholly confirmed when on October 3rd Wilkinson Holdworth & Co sent Boulton and Watt a bill for a new rotative fly wheel shaft made by Murray and Wood.

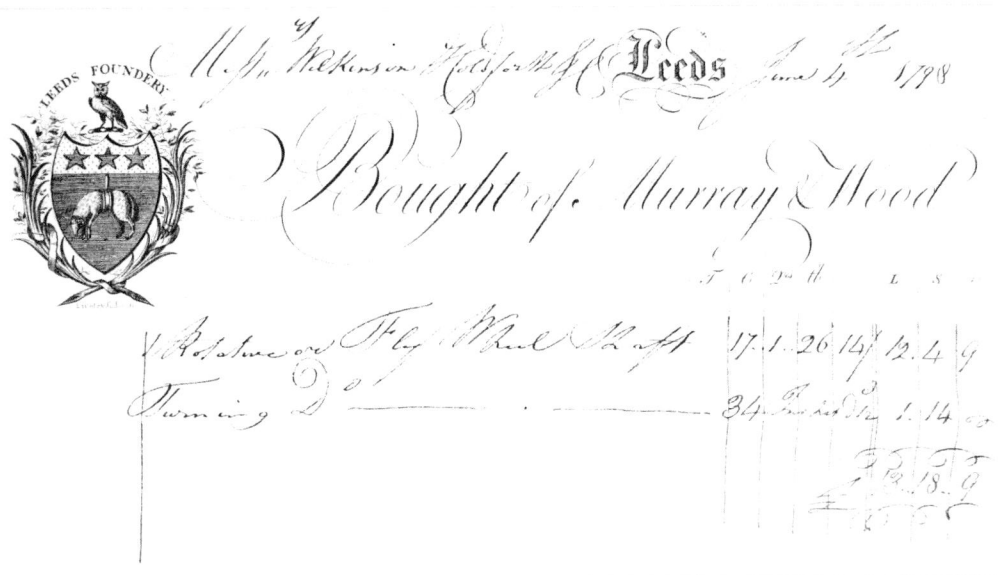

20. Murray and Wood invoice to Wilkinson Holdsworth sent to B & W for reimbursement.

The charge had apparently been agreed by M R Boulton on a recent visit. The letter went on to detail despatches of engine parts (which probably included the above) to M'Connel & Kennedy and also to resolve an issue over Gavin's (McMurdo) time for undertaking the work.

A piece of land within the Leckeys had been sold by Murray and Wood and their wives to John Kirby a maltster of Holbeck on the 25th and 26th of September, and duly registered on the 12th of October [50].

An indication of the progress in building the works in Water Lane is given by Royal Exchange Insurance Policy number 164622 for the Xmas Quarter of 1798 dated November 7th.

"Wm & Richd Hargreaves, Joiners Jno Cave & Jas Longley
Bricklayers All of Leeds in the Co of York
On a Building used as an Iron Foundry situated in Water Lane
in Leeds as Tenants Messrs Murray & Co   *150*
On a building adjacent not communicating used as a Model Warehouse
only   *150*
On Utensils of Trade & in Trust therein   *200*

On a Building [new] used as a Mill for Casting Iron & Machine
Makers Shops over   *400*
On the Steam Engine adjacent & communicating only by the Aperture
thro' the Iron Shaft passes.   *100*
All Bk & Slated   *£1000"*

Whereas there was only one building itemised in the first policy, there were now three buildings and a steam engine. The mortgage or sale and leaseback reflected in the policy was covered by documents dated the 29th and 30th of October and 1st November 1798, and was underwritten by John Marshall, but appears only to have been registered in February 1799 [51]. So while the business of Murray and Wood was progressing it would appear that it could not support the significant building programme, and that land needed to be sold and mortgaged in order to carry on. In December Murray and Wood advertised for people again, this time for Brass founders:

"To Brass-Founders. WANTED IMMEDIATELY A PERSON who perfectly understands the Business of casting Brass, in the Machine Line, by applying at Messrs Murray and Wood's Brass and Iron Foundry, Leeds, will meet with good Wages and constant Employment. N.B. None need apply but good Workmen; nor no Letters answered unless Post-paid.
Leeds, Dec 13th 1798" [52].

Another early engine by Murray was probably put up in 1798, although it may not have been until 1799. John Brookes of Armitage Bridge House recalled in a letter dated March 7th 1878 that his father had Murray put up one of the first steam engines near Huddersfield. It was put up at the Brooke's mill at Honley, where they were located before moving to Armitage Bridge Mill. John Brooke had always remembered that one of Murray's foremen (or erectors?) was a Frenchman named T Debos [53].

The projected visit of Abraham Storey who was in charge of castings at Soho took place at the end of 1798 or early in 1799 and he was accompanied by William Murdock. Abraham Storey had previously worked for the famous iron founder John Wilkinson during the time he made cylinders for Boulton and Watt. For once a description of this visit can be given in Matthew Murray's words before we see how Boulton and Watt viewed it. Writing some years later Matthew Murray described it thus

"Mr Storey, manager of their Foundry, and Wm Murdock, Superintendent of the workmen at Soho, some time back visited our works at Leeds, and from their assuring us of Messrs Boulton, Watt and Co's friendly disposition, were admitted into every part of the manufactory, by Mr Wood and myself; they were even permitted to take patterns and specimens of our workmanship, and we know that upon their return to Soho, many of our improvements were immediately adopted, and the Engines made after that by them, were in part constructed on our plans. Mr Murdock, upon taking his leave of us, expressed a wish, that as they and we were certainly the best

Engine makers in the Kingdom, we should always be upon good terms, and that if ever I should go to Soho, they would be very glad to shew me all their Works" [54].

While this statement was definitely political it is probably realistic; Boulton and Watt's correspondence explained why they sent Storey and Murdock to Water Lane and the approach outlined above was the only kind of approach that would have produced the manufacturing improvements at Soho. Murray and Wood may to a large extent have been ignorant of the bitter feeling they had created at Boulton and Watt but it is unlikely as it was certainly reciprocated by Murray. It is stated more than once in Boulton and Watt's correspondence, that what Murray wanted was a license. He probably did not want a war, but to go about his business. There may have been a naivety about this attitude, after all his old boss John Marshall used to send him off to ascertain the secrets of various pieces of flax machinery, so he had become accustomed to industrial espionage. But Murray and Wood either saw the visit as an olive branch or as an opportunity to gain information about Soho.

So what was the attitude on the other side? M R Boulton wrote to James Watt junior on January 17th 1799 after the visit had taken place, and after some detailed description of other business, stated:

"For more than a week past my time & attention have been almost wholly engrossed by our contest respecting the Poor State of the Foundry.........
Murdock & Abraham are returned from their excursion highly delighted & full of panegyrics upon Murray's excellent work. Abraham is now entirely convinced of his inferiority & what is more of the possibility of amendment & he is now actually making trials of different substances to mix with the sand with the view of giving a better skin to the castings.
We have likewise written to G W Murdock to send a boat load of the sand used by Murray.
It seems his forge work is still better than his Castings & we are trying therefore to infuse a spirit of amendment into this department of the work. – They were admitted into every part of Murray's manufactory & spent two evenings with him & by virtue of plentiful doze of ale succeeded in extracting from him the arcane & mysteries of his superior performance.
The Low Moor Co have undertaken to cast the [Boring] Rod & Smalley has promised to make one in every respect before it is sent away" [55].

He continued on these matters in a further letter dated 1st February containing the following extracts:

"I have to submit to your consideration a proposal for including in our contracts, the erection of the Engine & setting of piece work to our Engineers.
.... Our competitors Bateman & Sherrat, [   ] Murray &c always engage to put the Engine to work on their own [cost] and it is the unanimous opinion of both Lee, Ewart & Lawson that this very circumstance has procured them several customers.
....
I cannot however altogether suppress the pleasure of announcing to you a most promising alteration in the spirit of our forgers & the appearance of their work. I should first inform you that Murray presented Murdock with a specimen of his forge

work, no doubt with a view of exciting his astonishment and perhaps a despair of ever attaining the same perfection, for I must candidly acknowledge it was the most beautiful and perfect piece of work I ever beheld. Fortunately however it has been the means of producing a very different spirit.

The Emulation of our Men has been awakened & they now vie with each other in adopting the tools & improvements suggested by Murdoch. Our first [   ] essays have succeeded completely & we entertain great hopes of turning out the working gear in such a manner as will render filing a superfluous operation.

Some other tools partly in imitation of Murray & partly of the suggestion of Murdock have been introduced with the greatest success into the fitting department. The new method of turning the larger lathes by means of the endless screw as shewn in the drawings of the Boring Mill has been tried upon our Air Pump lathe and found to answer every expectation. .... I wish I could speak with the same exultation of the exertions in the foundry department Abraham's conjurations have not hitherto produced any perceptible alteration in the appearance of the casting & as usual he is much more expert in divining the causes of the failures than the remedies. Everything in its turn has been considered as principally in fault & now the sand is chiefly blamed. I have therefore ordered a boat load of the same sand Murray uses" [56].

There is an implication in these letters that Murray was plied with too much ale and therefore gave away what Storey and Murdock wanted. At this period in his life Murray was a frequenter of the ale house and it may not have been Murray who was the worst for wear. If Murdock and Storey were under the influence it may explain to some extent how much information on the working of Soho passed to Murray This became a major issue to Boulton and Watt later as will be seen and caused a witch hunt among their employees, however the explanation may be simpler than they thought. Murray had after all served an apprenticeship in obtaining information while working for John Marshall. Matthew Loam in the letter referred to in Chapter 3 also described Murray (though whether this was fact or myth is unknown) as:

"an excellent fiddler, and if he heard of anything new, anywhere, he would be off with his fiddle, fiddle to the men, get employ sometimes as striker, other times as smith, get the secret, return, adopt it" [57].

So on the whole Boulton and Watt were probably feeling quite pleased with themselves when another letter from Benyons Marshall and Bage arrived, which may well have been foreseen, for it asked for the terms for operating a Murray and Wood steam engine at Ditherington. Lawson's intelligence of the previous July had been correct. The Boulton and Watt response survives and some part was given in Chapter 1, but unfortunately it is in one of the letter books and is not in the best of condition. The following can be deciphered:

"By your favour forwarded to us from Soho we [observe] that Messrs Murray & Wood of Leeds in consequence of the   [   ] [   ] connection   ]
[   ] which we hitherto considered unfounded rumour [   ] are about to erect for you a 40 Horse Engine so constructed as not to [   ] with our patent and yet be convertable in [   ] [   ] Engine [at pleasure]

in answer to the above & to your subsequent request, we shall only remark that having as you know at a great expense made preparations for supplying the public with Engines constructed to our Principles We see no reason why Messrs Murray & Wood should be made partners to our Parliamentary Privilege. We shall at all times be Ready to execute your orders for Engines of the Above or any other power & with Thanks for your former favours, we
Remain respectfully Gentl[n.58].

And Matthew Murray would not be put down, as Boulton and Watt soon became aware he was filing for a patent on steam engines, within the period of their patent!

James Lawson was immediately put on the spot again by a letter, addressed to him c/o Wormald & Co, from James Watt junior of the 27th March:

"We are just informed by our legal Reporters, that Matt[w] Murray has applied for a patent for 'Certain New Methods and Improvements on the Steam Engine for the purpose of saving Fuel, lessening the expence of erecting Steam Engines and producing a more steady motion therein than by any means at present practiced.' [No 2327 of 1799] It is stopped at the Solicitor General's Office by our Caveat for the present, and we wish this information may enable you to ferret out the nature of his invention and in what manner the same is to be performed. It might be of consequence to us to learn this Gentleman's motions at present, as we are persuaded that he and his employers M & B are playing a deep game. The firm at Shrewsbury have written some artful letters to solicit our consent to Murray's applying our principles to a 40 Horse Engine for them, and do not seem disposed to take a refusal, which however they must digest, or go to Law with all speed" [59].

And the gloves were coming off and a new idea had come to Boulton and Watt as the letter goes on to delineate:

"Mr [M] R Boulton seemed to think that Wood might be detached from Murray and prove a very useful person to supply Vivian's place. We have heard from various quarters that he is much dissatisfied with Murray's proceedings, but whether he might be induced to accept an inferior situation, after having been a master himself, or to engage with us for a term of years is what we have to learn. We should be willing to give a handsome salary, if the man's character is fair and his sobriety & attention could be depended upon. Perhaps you may have some opportunity of getting information upon these points and might sound him personally without making a direct offer. I believe he is acquainted with the Walkers of Wortley, who could procure you an interview" [59].

On the same day James Watt junior informed his father both of the intent to suborn David Wood and of Matthew Murray's upcoming patent.

James Lawson replied to James Watt junior promptly on the 30th March:

"I have this morning yours of the 27 – Before I received yours I was informed of M Murray's having applied for a new Patent & that it was stopped by your Caveat. I have not yet been able to learn further than that it is to be an Engine without a Beam & is by his account to cost little more than half the present price & to save 1/5th of the fuel. I understand he intends going to London next week and is very loud in his complaint against B&W for attempting to prevent his obtaining a Patent.

I am acquainted with his partner Wood but should think that he's too far engaged to leave the concern which is not merely for Engine work but for Mill work in General & also machinery for both flax & cotton.

I have no doubt M&B [Marshall and Benyon] are at the bottom of everything & will take every precaution to prevent losing Wood.

I shall endeavour to learn if there is any disagreement among them. Wood from the little I know of him is a steady and industrious man & has a good character & I find Murray has been more steady of late. I may probably know more tonight & shall inform you.

I find the casting sand is got about Castleford & belongs to T Smith Esq I shall if the weather permits, for the country is covered with Snow, go there tomorrow and will get a boat load sent as you direct. .....

I have just been with Messrs M&B and inclose bill £120 for 1 years premium as I found them both at the counting house but they as usual dry and only observed that the Premium would soon be at an end – and inquired if there were many new Engines going on – They are [busy] with one for their Mill here (in addition to the last Engine of 28 horses) of 20 Horse power – The House is finished and the framing for the Cylinder and Cistern put in" [60].

On the 27th of March James Watt junior wrote to the firm's solicitor Weston concerning Benyons Marshall and Bage's letters asking for a license to use Murray and Wood's engine at Castle Foregate Shrewsbury (Ditherington):

"Murray's proceedings may perhaps be intimately connected with the inclosed correspondence of his masters.
We must request you to dictate an answer to their two last letters (which are not copied in their proper order) such as may have all the effect of a refusal, without being likely to injure us, if compelled to proceed at law, as will most likely be our case. The sooner the better" [58].

Weston must have been very prompt in his reply as on April 1st James Watt junior sent a further letter to Benyons Marshall and Bage:

"We intended our letter of the 22nd Feb to be a full and final answer to your letter referred to in it and considering it in that point of view, we did not conceive any further reply necessary to the repetitions in yours of the 25th of the same month.

To your late favour of the 23rd Ult we also think we cannot reply better than by referring you to our first letter upon the subject, in which we have intended clearly to express our dissent from your proposal. We have neither leisure at present, nor do we feel ourselves at all called upon to [explain] further the motives of our refusal. Whenever we see occasion ample cause shall be assigned.

Your reasons for preferring Mr Murray's services to ours are no doubt satisfactory to yourselves, however ill-founded they may appear to us. We however deem any investigations of that subject here to be quite superfluous.

As the Expiration of the Term of our exclusive privileges, has (like everything else about it) been made the subject of ungenerous Cavil, we shall refer you to the prolonging Act of Parliament itself, which you will find printed in the statutes at large. It received the Royal assent in June 1775" [58].

By the 3rd April Lawson stated that he could learn no more regarding Matthew Murray's patent.

Boulton and Watt were also thinking of patents, and unusually for them recognised William Murdock's input to the firm by proposing to take out a patent in his name. James Watt Junior to Matthew R Boulton:

"We have had some conversation with Murdock about securing his late improvements upon the Steam Engine by Patent, I would propose the title to be "Certain new methods of manufacturing & constructing Steam Engines", and under the following articles might be specified
1st a new method of boring Cylinders and of turning Cast Iron
2nd a new method of constructing the valves of Steam Engines by means of sliding valves connected together &c.
3rd a new method of working the valves of Steam Engines by means of hollow stems &c.
4th a new method of constructing the Cylinders and Steam Cases of Steam Engines by means of conical joints &c.
5th a new Rotative Engine
6th another new Rotative Engine.
These two last Murdock will explain to you. I am not quite confident that some of the above articles may be contestable upon the grounds of prior use, but I think it may be worth our while to run the risk for the sake of the advantages of a Patent. I do not give the above as the best titles for the particular inventions, but merely to call your attention to them. When you have considered the subject and had some conversation with Murdock, I shall be glad to learn your sentiments. Murray's Patent is a stimulus to us to be active, perhaps" [61].

On the 5th April Lawson reported an evening's conversation with Matthew Murray:

"M Murray came & spent the evening with me but tho' we had much general Engine chat I could not gain anything of consequence. He has said that it is an engine without a Flywheel and that the Cylinder is to be Horizontal. He says he is making a Model which he takes to London with him, this he says will be three weeks or a month in getting ready. I found he wished to know the date of your Caveat which only is in force for a certain time, & said if he thought it was taken out long he should <u>wait.</u>
I have however heard from others that he is not so fond of his scheme as he was.
I also found that the Foundry and building are on the ground of Mr Marshall, and that they have already gone so far as to find difficulty in getting Iron which they cannot now do without Mr M's being surety. He said in speaking of Fischer's Engine that they had lost near £ 200 by that job. His price for a 16 Horse is at present £400 i.e. a 16 Horse when altered.
He has several orders – he got one from a neighbour of Mr Gotts tho' Mr G did what he could to prevent it" [62].

On April 10th James Watt Junior updated his father on the situation both as regards the engine for Ditherington and at Leeds.

In May 1799 between the 20th and the 26th Matthew Murray took up William Murdock's invitation to visit Soho. The outcome is summarised in a letter to Lawson from James Watt junior dated 31st May:

"Matt^w Murray was at Birmingham during my absence (according to the Boulton and Watt collections itineries J Watt junior was absent at Wrexham from 20 to 26 May) & visited Ab^m [Storey] at the Foundry. Murdock got himself denied at which he was

much irritated. He was offered to be shewn the Button & other branches of Soho manufactory, but as the Engine Shops were not nominated in the list of sights he declined seeing any of them with some indignation – W Harrison went to visit him at Birm^m & learnt from him that he was acquainted with the new Nozzle at the Mint, who has been the traitor is not known" [63].

As Matthew Murray put it very succinctly in his later statement:

"*I did go to Soho, and was refused admittance into their manufactory of steam engines*" [54].

If Matthew Murray had been in any doubt over how Boulton and Watt felt about him that was no longer the case and it is probably from this time that it is noticeable that Matthew Murray started becoming increasingly hostile to Boulton and Watt.

Boulton and Watt also learnt something new about Matthew Murray that was really to exercise their minds, and that was his ability to hear about and put to use anything new that was useful in his line of business, such as the nozzles at the Mint engine. James Lawson also had suspicions about the relative exchange of information during the Storey/Murdock visit to Murray and Wood. As he stated when he replied to the James Watt junior letter of 31st May:

"*The only person who could have informed M Murray of the new nozzles at the mint must be M McM[urdo]. Or Murdock & Ab[raham Storey] in their conversations with him or else he must have a correspondent at Soho which I think you will not fail to find out*" [64].

In June Lawson was advised that an employee of Boulton and Watt named Parker has absconded and was believed to be in the Leeds area.

On June 18th James Watt junior wrote again to Lawson and added a postscript:

"*can you learn at Leeds whether it is true that Murray has erected, or is erecting any Engines with the nozzles constructed like those at the Mint. When here, he said he had erected two in that manner, but Gavin says he supposes this to have been a lie it is very necessary we should be informed of the truth as we had some intentions of taking out a new Patent for this & some other things*"[65].

On 21st June Murray and Wood purchased 51 bars of turning iron and 10 bundles of rivet rods from Kirkstall Forge [66].

James Lawson reports back on June 25th that he has been unsuccessful in finding Parker the absconder and that:

"*I cannot find that M Murray is doing anything new only boasting great things of his new Patent. I understand from Mr Gott that Murray in giving estimates says so much done as Boulton & Watt work – and so much extra done my way – from his having been at Soho I thought it best not to go near him personally – Mr Lee was with Mr Marshall & saw his Mill in which he says is very good firm work. ….*"[67].

Mr Gott a good friend of Boulton and Watt, but not above employing Murray and Wood when it suited him, wrote to James Lawson on June 26th to advise him that John Marshall had told a Mr Cookson that the Boulton and Watt patent would expire in November.

In June and July Murray and Wood were buying boiler plates of Kirkstall Forge which were being delivered by a carrier named Riley:

"   **28 June 1799** – in ledger
Sold Murray & Wood Leeds
To 15 plates 2 ft 8 In by 20 3/8   8. 3. 4
To 15 plates 2 ft 4 In by 20 3/8}
To   4 plates 3 ft   by 20 3/16}   9. 1. 0
To 15 plates 3 ft 6 In by 20 3/16   5. 3. 3
To 17 plates 4 ft 6 in by 20 ¼   10. 2. 0
To   4 plates 4 ft 6 In by 10 ¼   }
To   4 plates 5 ft 6 In by 15 3/8 }   5. 2.14
To   2 plates 5 ft 6 In by 20 3/16}
*   76*   <u>39. 3. 2</u>   28/- £55-18-3
Per Riley
2 July 1799
Sold Murray & Wood Leeds
To 10 Plates of iron to order   Cwt 3. 3.22   28/-   £ 5-10- 6
per Riley 6 ladles 9 in diam 4 ½ deep   46   6?   £ 1- 3- 0
                                                           £ 6-13- 6*" [66].

At the end of a letter to Weston dated July 9th regarding Hornblower's piracies James Watt junior asked two questions

"It is presumed that Murray must by this time have given in his Specification. Please to procure us a copy – Whose are the caveats that obstruct your progress" [68].

On the same day James Watt junior was informed by Lawson that both Mr Marshall and Mr Sherrat (Manchester engine builder, one of the pirates) were informing everyone that the Boulton and Watt patent ended in November and that when he was asked he had said next summer.

By the 14th of July James Watt junior had a response from Weston and relayed it to M R Boulton:

"Murdock's patent is stopped by a caveat of Matthew Murray's. This is highly unlucky, more particularly as it is essential he should go up to town this very day to explain his Invention to the Solicitor General on Tuesday. I hope he will not be detained, for it will be grievously inconvenient to our progress at the foundry........
Marshall at Leeds & Sherrat at Manchester are very busy in circulating the report of the expiration of our patent in October" [69].

Matthew Murray was playing tit for tat, they had put a caveat on his patent so he responded in kind. What is more he was granted letters patent for his first one concerning steam engines on the 16th July 1799 with Patent no 2327 *"having brought to perfection certain new methods and improvements on the steam engine, for the purposes of saving fuel, lessening the expence of erecting steam engines, and producing a more steady motion therein than by any means at present practised".*

Weston wrote to James Watt junior on July 17th to say:

"The time for Murrays Specification being inrolled is not out. – I understood from him he would call at Soho to shew you the model. I saw it and can assure you it is quite unlike Murdocks" [68].

It is unimaginable that at this time Matthew Murray was going to open himself up to further snubs by presenting himself again at Soho.

By the 8th of August Murray and Wood were once again buying boiler plates from Kirkstall Forge along with more rivet rods.

On the 10th August 1799 a deed of assignment was signed by Matthew Murray and David Wood relating to the Leckeys and the buildings upon it with one James Fenton of Walworth in the county of Surrey Tanner. The financially challenged partnership of Murray and Wood had expanded, and the new partnership of Fenton Murray and Wood was born, which with additions and contractions would at least have these three gentlemen as constituents for the next twenty years.

Notes.

1. Marshall Papers MS200 folio54 Experimental Book 2
2. Kirkstall Forge Tradesman's Waste Book No 9 1796-1799 West Yorkshire Archives (WYAS) Leeds.
3. Boulton and Watt (B&W) Papers MS3147/3/435.
4. B&W Papers Foundry Letter Book MS3147/3/171 page 265.
5. B&W Papers Foundry Letter Book MS3147/3/172 page 14.
6. B&W Papers Foundry Letter Book MS3147/3/172 page 16.
7. Marshall Papers MS200 folio57 General Notebook 1790-1830.
8. B&W Papers Foundry Letter Book MS3147/3/172 page 41.
9. B&W Papers Foundry Letter Book MS3147/3/172 page 45.
10. B&W Papers MS3147/3/45/27.
11 B&W Papers Letter Book MS3147/3/09 page188.
12. B&W Papers Letter Book MS3147/3/89 page195.
13. B&W Papers Letter Book MS3147/3/89 page198.
14. B&W Papers MS3147/3/412/26.
15. B&W Papers MS3147/3/45/30.
16. Marshall Papers MS200 folio43.
17. B&W Papers Foundry Letter Book MS3147/3/172 page 172.
18. Marshalls of Leeds 1788-1886, W G Rimmer page 45.
19. WYAS Wakefield vol. DS page 614/5.
20. Leeds Intelligencer 15 February 1796.
21. Leeds Intelligencer 20 February 1796.
22. B&W Papers Foundry Letter Book MS3147/3/172.
23. B&W Papers MS3147/3/281/7, MS3147/3/435/16, MS3147/3/172.
24. Royal Exchange Fire Policy No 151066 vol 32, London Metropolitan Archives.
25. B&W Papers Foundry Letter Book MS3147/3/172 page 144.
26. B&W Papers MS3147/3/435, MS3147/3/172.
27. B&W Papers Foundry Letter Book MS3147/3/172 page 220.
28. B&W Papers MS3147/3/272.

29. B&W Papers Letter Book MS3147/3/272.
30. B&W Papers MS3147/3/298/15.
31. B&W Papers MS3147/3/35/11.
32. B&W Papers MS3147/3/282/60.
33. B&W Papers Letter Book MS3147/3/92.
34. B&W Papers MS3219/6/1/131 or MS3219/4/118/49.
35. B&W Papers MS3147/3/46/88.
36. B&W Papers MS3147/3/282/64.
37. B&W Papers MS3147/3/46/90a.
38. B&W Papers MS3147/3/46/90.
39. B&W Papers MS3147/3/282/65.
40. B&W Papers MS3147/3/410/19.
41. B&W Papers MS3147/3/548/9.
42. The Diary of Thomas Butler of Kirkstall Forge Yorkshire 1796-1799. Privately printed at The Chiswick Press, 1906. Page 194.
43. Kirkstall Forge Ledger 1787-90. WYAS Leeds.
44. B&W Papers MS3147/3/410/20.
45. B&W Papers MS3147/3/47/13.
46. B&W Papers MS3147/3/47/14.
47. B&W Papers MS3147/3/272/32.
48. B&W Papers MS3147/3/283.
49. B&W Papers MS3147/3/47/28.
50. WYAS Wakefield vol. EA page 359.
51. WYAS Wakefield vol. EB page 383/4.
52. Leeds Mercury 15 December 1798.
53. The Huddersfield Chronicle 15 June 1878.
54. Matthew Murray's public answer to Boulton and Watt after they placed adverts in the newspapers stating that his patent of 1801 had been overturned. Published in selected newspapers, see Leeds Mercury 23rd July 1803.
55. B&W Papers MS3147/3/37/3
56. B&W Papers MS3147/3/37/9.
57. Letter from Matthew Loam (1794-1875) to his nephew Matthew Loam (1819-1902) helping him prepare for a lecture on the steam engine, Cornish Studies Library, Redruth, Cornwall. Ref t7994044 and Mooo1948CC.
58. B&W Papers Letter Book MS3147/3/93.
59. B&W Papers MS3147/3/284/30.
60. B&W Papers MS3147/3/274/22.
61. B&W Papers MS3147/3/48/27.
62. B&W Papers MS3147/3/274/24.
63. B&W Papers Letter Book MS3147/3/94.
64. B&W Papers MS3147/3/274/28.
65. B&W Papers MS3147/3/284/41.
66. Kirkstall Forge Tradesman's Waste Book No 11 1799-1800 KF4/8, West Yorkshire Archives (WYAS) Leeds.
67. B&W Papers MS3147/3/274/32.
68. B&W Papers MS3147/3/48/58 annex.
69. B&W Papers MS3147/3/48/57.

Chapter 5

Fenton Murray and Wood – the feud years with Boulton and Watt.

So in August 1799 James Fenton, who along with his elder brother Samuel Fenton, had been in partnership with John Marshall, Mrs Marshall and Ralph Dearlove, joined the partnership of Matthew Murray and David Wood at Water Lane. Their father another Samuel Fenton had been a partner in Marshall Fenton & Co Linen Merchants with Jeremiah Marshall, John Marshall's father. James and Samuel junior had another brother William who was an attorney, and many of Marshall & Co.'s legal papers referenced a William Marshall as their lawyer, no doubt one and the same. William was also eventually to inherit James' share in Fenton Murray & Co, when James, who never married, died. So there were strong links between the Marshall's and the Fenton's and James joining Murray and Wood would have been an extension of the ties between the two organisations. A brother of Samuel senior, Philip Ibbetson Marshall was a merchant in Riga, where much of England's flax originated. He retired to Hampstead, London where he purchased a property named the "Clock House" which survives today as Fenton House, part of the National Trust.

The new partnership was soon announced in the Leeds Newspapers by the following advert which appeared in August 1799:

*"To YOUNG MEN
WANTED, A Number of Young Men, from Sixteen to Eighteen Years of Age, that would engage for a Term of Years to work in a Steam Engine Manufactory; where they may learn a valuable Business, and meet with constant Employment by applying to Mess. Fenton, Murray & Wood, Iron-Founders and Engine-Manufacturers, Leeds"* [1].

The shortage of skilled workmen for the mechanics and iron founding trades was becoming acute with many firms recruiting at long distances, from any town where the skills might be found. This situation continued for some time with Portsmouth Dockyard for example recruiting in the early 1800s from as far as Birmingham [2].

Fenton Murray and Wood were busy as shown by purchases in August and September of rivet rods, iron and boiler plates from Kirkstall Forge, the costs of which were offset to a small degree by the sale of their scrap back to Kirkstall Forge [3].

On the 9th September Fenton Murray and Wood wrote to Boulton and Watt concerning John Parker the absconder from Soho and received a reply from Gregory Watt (James Watt Senior's other son) dated 11th September 1799:

"We this morning received your obliging favour of the 9th instant and beg leave to offer you our acknowledgements for the polite attention you have paid to our Advertisement. John Parker who absconded from our Service was about 5 feet six inches high with black hair – a pale complexion and much pitted with the small pox – He is about 28 years of age.

We have directed our agent Mr Lawson to wait upon you and if Parker is identified by him we have given him directions to regulate his further proceedings" [4].

At the same time Gregory Watt chased James Lawson [5] advising that he needed to visit Fenton Murray and Wood on this matter and on the 13th he wrote to Lawson:

"From the time you left Leeds I apprehend you cannot have received my letter announcing to you that Parker was at Leeds with Murray and that Murray had written to inform us of his being there. You have I find by your letter of the 11th rec'd this information there but scrupled to get Parker apprehended because you could not identify him. I think this difficulty may be very easily overcome for if he answers the description given in the advertisement w[h.] is not a little remarkable he may be arrested on suspicion and if you cannot get him sent here without farther evidence he may at least be detained until we can send someone to identify him – this for your government when you return to Leeds" [6].

Matthew Murray had in the meantime lifted his caveat on Murdock's patent, or it had time expired as Gregory Watt wrote to M R Boulton at Truro to inform him that:

"Murdock's patent has at length made its way thro' all the caveats and passed the great seal. This event has removed the embargo from the two Horse Engine made for Mr Symonds which we shall send off immediately" [7].

Mr Symonds was the owner of a brewery in Reading Berkshire, and the engine incorporated some of the improvements included in Murdock's patent, which if it had been delivered before the patent was granted would have given grounds for invalidating the patent. By lodging his caveat on Murdocks's patent Matthew Murray had caused delay in the delivery of this engine.

Lawson wrote back to Gregory Watt from Kendal on September 17th:

"I yesterday rec'd yours of the 11th before leaving Leeds. I saw Murray who informed me of his having wrote to B&W about Parker who tho not known to me personally I have no doubt is the delinquent in question as said to Murray he had wrought at Soho.
I desired Murray to say nothing to him or any one as I supposed you would send a Warrant & have him taken up. For I find from one of the Leeds Justices that it was necessary first to have a warrant from Soho & then any other justice could back it on his being identified . ...." [8].

By the 23rd Lawson was back in Leeds and informed Gregory Watt of the legal process required to recover Parker:

"I have made enquiry about taking up J Parker but find it can only be done by getting a Warrant with you, which may be sent by any person who can identify the man, this person must see the Warrant signed which on his making oath to, will be backed here, and then he (the man sent from Soho) may be made [a] constable to take him home.
This I find from Mr Nicholson as well as Mr Markland one of the Justices is the only method by which he can be taken up.

I have been with Murray who will take no notice till you send for him. If you send immediately the person will find me & I will see to the rest" [9].

On the 27th September Gregory Watt sent three letters to Leeds. The first was to Lawson which, among other business, advised that the warrant was on its way:

"We shall despatch Mr Hurt this evening to Leeds with a Warrant against Parker. He has seen it signed & has been sworn in as constable for Staffordshire. You will please to procure the assistance of a Leeds constable to take up Parker & to assist in escorting him here. It will be proper to make a previous Agreement with such constable, for what we are to pay him over & above his expenses. In short as you seem au fait in the business we trust you will give necessary advice and assistance" [4].

The second letter was to Fenton Murray and Wood and to the same purport:

"The bearer Mr Hurt, our clerk, is furnished with a Warrant against John Parker, of whom you had the goodness to give us information as being now in your employ.
We shall esteem ourselves obliged by any further assistance you may be able to give us towards his apprehension and remain respectfully" [4].

The third letter was to Boulton and Watt's good friend Mr Gott:

"We have dispatched the bearer Mr Hurt to apprehend one of our articled Servants of the name of Parker, who has absconded for some time & is at present with Messrs Fenton Murray & Wood of your town.
In case Mr Lawson should be absent or any difficulties should occur, we beg to request your friendly advice and assistance" [4].

The above exchange is of interest for two reasons. Firstly it is the only known surviving formal exchange of letters between Fenton Murray and Wood and Boulton and Watt and as is clear was initiated by Fenton Murray and Wood to advise Boulton and Watt that their absconder was at Water Lane. Secondly it emphasises the extent firms were prepared to go to recover skilled employees who were indentured. In those days an employee under indenture was bound at law to serve the term stated. It is to be compared with the incident that shall be described at the close of 1799.

During October and November Fenton Murray and Wood bought fly iron, ladles and bars from Kirkstall forge [3]. Presumably the engines for Marshall & Co for Leeds and Ditherington were in place, but it is unclear when they started operation and they had initially to operate without condensers and air pumps to avoid contravention of Watt's patent. John Marshall, having openly asked for a license, would not give Boulton and Watt the satisfaction of proceeding against him at law. Fenton Murray and Wood were now building two other steam engines one of which is probably the one previously referred to by Lawson in his letter of 5 April 1799 to Boulton and Watt, details of which were contained in another of Lawson's letters, written from Manchester on the 28th November 1799:

"I inclose you a sketch of G & J Wright's Engine – which has been made by Messrs Sturges & Co & M Murray and is in its present state one of the worst I have seen, tho' they have had every Eng^{r.} in the neighbourhood to assist them.

I was with Mr Gott at M Murray's shops (about some mill work he is doing for them) and found they were at work on Air Pump, Buckets, of diff^t sizes, and they report at Leeds that Mr Marshall means to put one to his new Engine at Leeds immediately. It is also said the materials are sent to alter the Engine at Shrewsbury, and also to Hull. Gavin [McMurdo] will I have no doubt keep a good look out at Leeds & will inform you.

I was glad to find that M Murray is setting an Engine at work (at Mr Closes the Dyer whose premises join Mr Gott's Mill) that his wheels broke & that they have been altering and adjusting ever since, & that Mr Close already regrets his not taking Mr Gott's advice in having an 8 horse from you, which he already finds would have cost him less" [10].

So even if some of the new engines were not putting up as well as hoped, it is clear that Fenton Murray and Wood were assembling all the parts to convert their customer's engines to condensing ones as soon as Watt's patent expired. Further purchases were made from Kirkstall Forge, possibly in connection with the above engines.

The next item of record should be considered in the light of the Parker case as this time there is evidence of workmen being enticed away from Boulton and Watt to go and work for Fenton Murray and Wood. It comes in the form of an affidavit sworn before a Justice of the peace in Staffordshire by William Harrison. William Harrison was one of the senior workmen at Soho and on Matthew Murray's abortive visit there it had been Harrison who went to see Murray in the evening and discovered that Murray knew about the valves on the Mint Engine.

"County of Stafford
The Information and complaint of William Harrison of Smethwick in the parish of Harborne in the County of Stafford Engineer.
Taken upon oath before Nathaniel Goodrey Clarke Esquire, one of His Majesty's Justices of the Peace for the said County of Stafford this seventeenth day of December 1799
Who saith that he is a workman at Soho Foundry in Smethwick aforesaid carried on by Messrs Boulton & Watt and that on Sunday the fifteenth day of December instant James Brennan came to this informant at the said Soho Foundry and offered him more wages than he received from Messrs Boulton and Watt if he would go and work for Messrs Fenton Murray and Wood of Leeds and which this informant afterwards refused to accept.
Taken sworn and acknowledged the Day
and year above written, before   *sig William Harrison*
sig. N G Clarke" [11].

Clarification of the issue came in a letter Boulton and Watt wrote to Mr Gott on the 19th of December. Here it is made clear that Brennan is actually trying to persuade workmen who do not have Indentures or Articles to leave Soho. So again what was happening was working up to the point of the law (as with the steam engines), where an absconder was articled Fenton Murray and Wood volunteered their return, if they were not articled they were fair game. It is worth comparing Fenton Murray and Wood's attitude with the implied similar policy in the letter from M R Boulton to Lawson of 5 November 1797, when Lawson was reminded that he was supposed to be procuring two more workmen (see Chapter 4). Boulton and Watt realised that it was up to them to keep their men after their contracts expired as shown in the next part of the letter; which relates to Gavin McMurdo one of their erectors, where again the issue is that Boulton and Watt fear that at the end of his current contract he will, as legally entitled to, go and sign up with Fenton Murray and Wood.

"From some hints that have been conveyed to us in consequence of these transactions we have great reason to apprehend that Murray has availed himself of the frequent sojourn of our man Gavin at Leeds, to tamper with him and induced him to enter into his service on the expiration of his engagement with us. This luckily does not happen for a 12 months to come, and we perhaps are yet in time to frustrate this attempt also which we are determined to do even at the expense of a considerable sacrifice and trust that with the assistance of your good offices we shall succeed. Gavin's present wages from us are 20 shillings per week and he continues upon these terms until January 1801 but if he enters into a fresh engagement for 5 years from this time we shall be willing to give him during the first half of the term 25 shillings per week when from home and 26 shillings per week during the remaining half of the term when from home, but when at home (which we hope will seldom happen) 21 shillings per week as we cannot then make more of his time" [4].

Mr Gott was asked to procure Gavin's assent and signature to the new articles on behalf of Boulton and Watt. The letter concluded:

"You will be fully sensible of the necessity of managing this affair with the utmost discretion and despatch, so as totally to preclude any intercourse with Murray, which would certainly prove fatal to our intentions as his offers are regulated by no other rule than that of injuring us. We rely upon your friendship to excuse this trouble, and remain with the utmost regard" [4].

Then on the 22nd of December Lawson wrote from Edinburgh:

"I found that M Murray rounds all his bolts in suages at his own shops.
I forgot to mention that Crowden is now with him, Bateman & Sherrat having turned him off" [12].

Crowden was an ex-employee of Boulton and Watt, who had previously worked for the 'pirate' engine makers Bateman and Sherrat in Manchester.

Mr Gott advised by letter dated the 22nd and 23rd that he had procured Gavin McMurdo's signature to a new agreement. In Boulton and Watt's acknowledgement they advised Mr Gott that they had now signed all their men up to articles with the exception of the smiths and that the smith's wages were equivalent to those paid by Murray. They also re-raised the subject of going into partnership in a flax mill, the business to include Mr Lee of Philips and Lee of Manchester, another friend and user of their engines.

At the end of December Fenton Murray and Wood bought 36 boiler plates from Kirkstall forge with 12 bundles of rivet rods [3].

And so by the end of 1799, while the steam engine issues could be expected to dissipate with the expiry of Boulton and Watt's patent, the relationship between the two companies had rapidly deteriorated over the issue of enticing personnel from one to the other. This item was to become the major grievance felt by Boulton and Watt, and the tone of their internal correspondence continued to seek legal sanctions against Murray, even though they had signally failed to find any proof of either Fenton Murray and Wood or Marshall and Benyons infringing Watt's patent.

The New Year (1800) saw further moves in regard to setting up a flax mill to repay Marshall and Co for entering the steam engine business through Fenton Murray and Wood, as the position was viewed by Boulton and Watt. Mr Lee wrote to Boulton and Watt on January 8th:

"In consequence of Marshall calling and asking to see our Mill I placed myself in a situation that he could not decline introducing me into his, and I found him upon the whole as communicative as I expected. I have now the same claim upon Bage who paid me a similar visit a few weeks since, which I shall certainly return the earliest opportunity. The last time I was at Leeds Murray (unask'd) took me through his workshops and shew'd me the Machinery he was preparing for Marshall; I availed myself of a pretended difference of opinion about the form of wheels to desire him to call upon me afterwards and I then learn'd that he was under Articles not to make Machinery for any other person, but he added slyly there was no restraint against giving Instructions or Drawings. From that moment I saw his palm was open to Corruption. This you will keep a profound secret till it can be render'd serviceable to us" [13].

Business at Water lane continued with metal purchases from Kirkstall forge [3], and for the first time an appearance in a directory, that of Binns and Brown's 1800 Directory of Leeds

"Fenton, Murray and Wood, brass and iron founders Water Lane" [14].

The mortgaging of the buildings for the foundry at Water lane are detailed in the Royal Exchange Insurance Policies no's 172611 and 172612,dated Lady Day, which also lets us know that the boring mill was now up and running

"William & Richard Hargreaves, Joiners & John Cave & Jas Longley Bricklayers all of the Parish of Leeds in the Co of York.

On Two Buildings adj[acant] & Communicating sit[uate] in Water Lane in Leeds af[oresaid] used as Model Warehouses only	*400*
On a Building [new] used as a Mill for Boreing Iron with Machine Makers Shops over it and on the Steam Engine House adj[oinin]g & comm[unicating] therewith only by an aperture in the Wall to Admit the Shaft	*1000*
All Brick & Slated & in the Occupation of Messrs Fenton Murray & Wood only."	

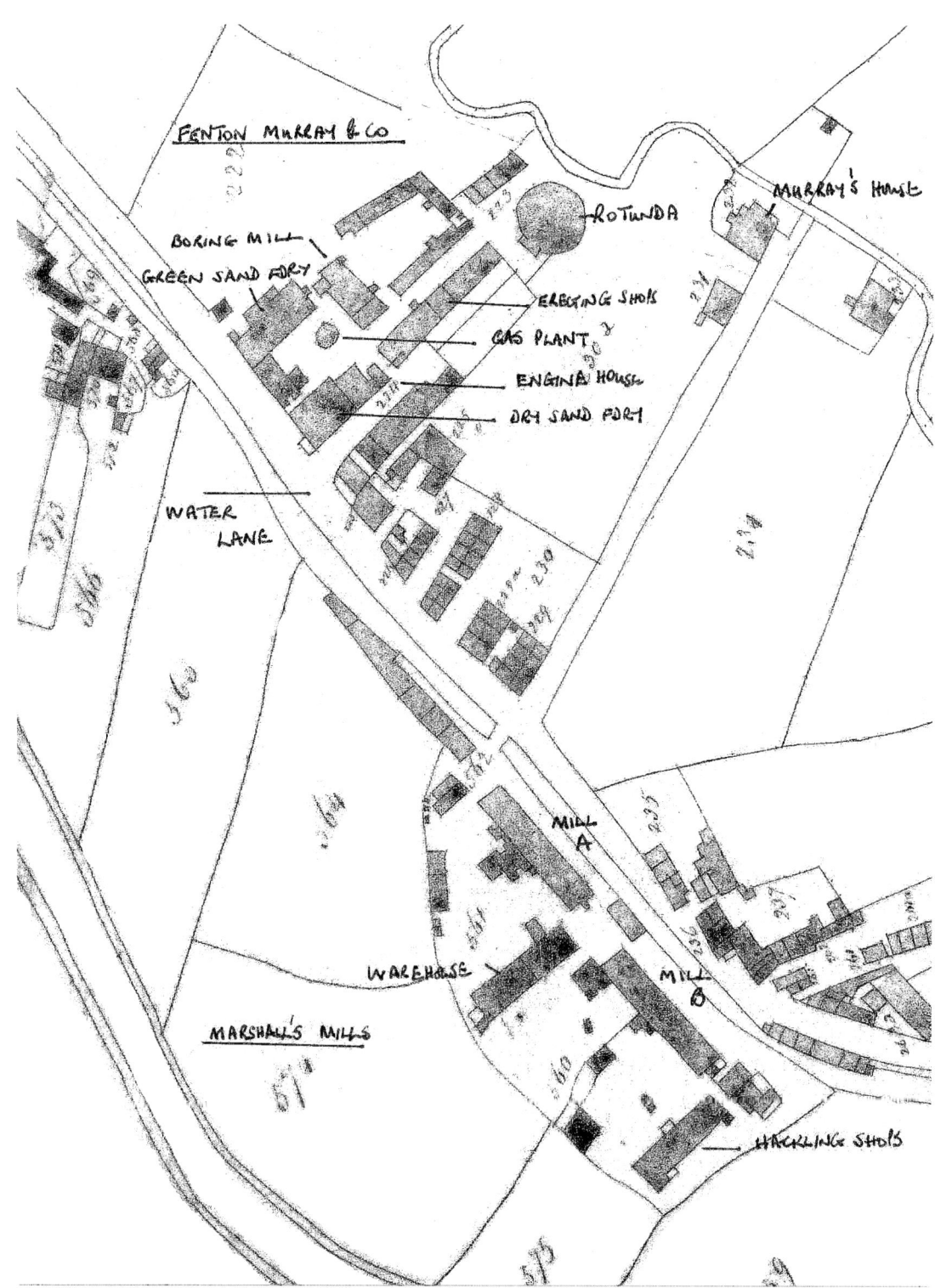

21 Later tithe map extract showing the Round Foundry and early Marshall Mills.

"*Fenton, Matthew Murray & David Wood of Leeds in the Co of York Iron Founders &*
Machine Makers
On Utensils and Stock in Trade in their two Buildings
Adj[oin]*ing & comm*[unicatin]*g used as Model Wareh*[ouse] *only situate*
in Water Lane in Leeds af[oresai]*d* 600
On their Steam Engine near comm[unicatin]*g with their adj*[oin]*g*
Boreing Mill only by the Aperture in the Wall through which the
shaft passes. 150
All Brick & Slated £ 750" [15].

By May there was an indication that the company was beginning to become part of the community and stepping up to the responsibilities of a medium size employer. The local papers reported on the 5th of May within a complete list of subscribers:

"HOLBECK
A LIST of the SUBSCRIPTIONS at HOLBECK, in the Parish of Leeds, which have lately been distributed in Soup to the Poor of that place

John Marshall	*20-10-0*
Thomas Benyon	*10-10-0*
Fenton Murray & Wood	*6-06-0"* [16].

Shortly thereafter on the 7th May Fenton Murray and Wood bought boiler plates and rivet rods from Kirkstall forge again, however they were now marked for delivery to James Beaumont of Gildersome. James Beaumont was a boiler maker, who was also used by Kirkstall forge, and so either Fenton, Murray and Wood had lost their boiler maker or were too busy with engines and other work to make the boilers themselves. They appear to have continued to employ James Beaumont until October 1800, with a large number of purchases recorded over this period as follows:

"Wednesday 7 May 1800
Sold Fenton Murray & Wood Leeds

To 12 Bdls Rivet Rods	*Cwt 6.".*"	*30/-*	*£9-11-0*	
44 plates 7/16 thick	*32.2." }*			
p Riley to James Beaumonts Gildersome	*} 28/-*	*£61-11-9*		
24 plates 3/16 thick	*19.3.24 }*			
3 do ordered afterwards	*1.2.3 }*			
p to Jas Beaumont Gildersome			*£70-11-9*	

~~*Saturday 17 May 1800*~~
~~*Sold Fenton Murray & Wood Leeds*~~
~~*To 26 Boiler plates to order*~~ ~~*Cwt 11.1.27 ended see folio 419*~~
~~*p J Blackburn to Gildersome*~~
[crossing out per original]
Friday 30 May 1800
Sold Fenton Mury Wood & Co Leeds

14 plates of iron	*Cwt 4.3.1*	
4 bund Rivit Rods	*2*	*£9-9-3*

pr James Beaumont Cash
            30
Monday June 23 1800
Sold Fenton Murray & Wood Leeds

To 24 plates	*2..6 by 20*	*Cwt 11.1."*	*28/-*	*£15-15-0*
	3..1 by 4			
per do		*15:15:0*		

Friday June 27th 1800
Sold Fenton Murray & Wood Leeds

To 14 Bdls Rod Iron best ¾	*Cwt 10.".*"	*£15-0-0*

Pr Blackburn to Gildersome
July 3 1800
Sold Fenton Murray & Wood Leeds

To 2 Boiler plates	Cwt 1.1.22	28/-	£2-7-6
4 plates which came to Clipp'd			
per Riley over again			
30 Bdls of Rivet Rods	15.".''	30/-	£22-10-0
			£24-17-6

July 12 1800
Sold Fenton Murray & Wood Leeds

4 plates 3 foot 8 in by 16 in }			
18 do 3 ft 4 in by 17 in}	Cwt 2.2.23	28/-	£33-3-9
11 do 2 ft 9 in by 18 in }			

To J Beaumont Gildersome
J Clark
July 15 1800
Sold Fenton Murray & Wood Leeds

5 plates 3 ft 4 in by 17 in }			
4 do 5 ft 9 in by 9 in }	Cwt 33.".9	28/-	£46-6-3
45 do 2 ft 10 in by 17 in }			

to J Beaumont Gildersome pr Riley.
Wednesday 23 July 1800
Sold Fenton Murray & Wood Leeds
To Boiler plates viz

2 plates 3 ft 9 in by 9in ¼ }			
4 plates 2 ft 2 in by 9 in 3/8 }	Cwt 5.2.13		
4 plates 5 ft 9 in by 12 in 7/16 }			
26 plates 3 ft 9 in by 17 in ¼	13.3.19		
43 plates 2 ft 2 in by 17 in 3/8	22.2.17		
per Riley	42.".21	28/-	£59-1-3

to James Beaumont Gildersome.
Wednesday 30 July 1800
Sold Fenton Murray & Wood

To 18 Boiler plates to compleat the Boiler	Cwt 8.3.21	28/-	£12-10-3
8 Bdls Rivet Rods	Cwt 4	30/-	£6
			£18-10-3

Pr Riley James Beaumont Gildersome
Friday 1st August
Sold Fenton Murray & Wood Leeds

To 23 Boiler plates 2 ft 7 in by 18 in]			
13 do 4 ft 2 in by 20 ½ in }Cwt 21.2.18	28/-	£30-6-6

per do.
27th August 1800
Sold Fenton Murray & Wood Leeds

To 4 boiler plates	Cwt 2.3.9	28/-	£3-19-3

pr their cart   old boiler dressed
          56.3.26
Friday 12 September 1800
Sold Fenton Murray & Wood Leeds

To 5 plates cut to particular order	Cwt 2.2.21	35/-	£4-14-1
15 do 4 ft 4 in 13 ½ by 11- 9.2.14 }			
14 do 4 ft 4 in 3 ½ 2.2.3 }	12.".17	28/-	£17-0-6

for the bleaching pan
also for the large boiler

8 narrow plates 2 ft 6 in and 3 [t] by 9 ½}			
24 plates 2 ft 6 in by 18 ------ }	22.2.12	28/-	£31-13-0
8 plates 3 ft by 18 }			
pr Riley			£53-7-7

September 15 1800
Sold Fenton Murray & Wood Leeds
To 24 plates 2ft 6 in by 18in 7/16   Cwt 13.1.0}
   23   do   3ft   by 18in 7/16   15.3.3}   28/-   £40-12-9
pr do.
September 22 1800
Sold Fenton Murray & Wood Leeds
To 12 Boiler plates to compleat order   Cwt 7.2.1   28/-   £10-10-3
   2 Bdls Rivet Rods   1   30/-   £1-10-0
Pr Riley to Jno Beecroft to be forw'd to   £12-0-3
       Jas Beaumonts Gildersome
October 7 1800
Sold Fenton Murray & Wood Leeds Foundry
To 20 plates 2ft 6 in by 18   Cwt 12.2.25
   16   do   3ft   by 18   11.2.11
                       24.1.8   28/-   £34-5-0
    10 bdls of Rivet Rods   5   30/-   £ 7-10-0
p M Reyley   £41-11-0
October 8 1800
Sold Fenton Murray & Wood Leeds Foundry
To 45 plates   Cwt 31.3.11   28/-   £44-11-9
p Matthew Reyley
October 9 1800
Sold Fenton Murray & Wood Leeds Foundry
36 plates 2 foot 6 by 18   Cwt 21.3.26
 6 plates 3 foot   by 18   4.1.21
                       26.1.19   28/-   £36-19-9

p Matthew
October 10 1800
Sold Fenton Murray & Wood Leeds
To 44 plates 2 ft 2 in by 17 in   Cwt 20.3.17
   15 plates 3 ft 4 in by 17 in   10.2.5
    9 plates 3 ft 4 in by 17 in }   6.1.23
    9 plates 3 ft 8 in by 16 in }
                       37.3.17   28/-   £53-1-3
October 11th 1800
Sold Fenton Murray & Wood Leeds Foundry
To 17 plates 3 ft 4 in by 2 ft 9 in
       by 17 in   Cwt 10.3.14
   22 do 2 ft 10 in by 17   13.".14
                       24.".".   28/-   £33-12-11

P M Reyley they inform that the above
          was short wt 0.1.13
October 14 1800
Sold Fenton Murray & Wood Leeds
To 46 plates 2 ft 10 in   Cwt 22.2.5
   8   do 5 ft 9 in   5.1.25
                       28.".2   28/-   £39-4-6

Wednesday October 22 1800
Sold Murray & Wood Leeds
27 Boiler plates which compleat
       the whole order   Cwt 13.3.7   28/-   £19-6-9
per do
Tuesday October 28th 1800
Sold Fenton Murray & Wood Leeds

To 80 Boiler plates which compleat   *Cwt 41.1.16*   *28/-*   *£57-19-0*
their order for Gildersome
Pr Riley to James Beaumont
              4 bdls Rivet Rods   *2*   *30/-*   *£3-0-0*
per do   *£60-19-0"* [17].

While the above is rather repetitive it clearly demonstrates that Fenton Murray and Wood were very busy and while it is unclear how many boilers were being made during this six month period, the records show entries for orders being completed 4 times. The customers are uncertain, although a further clue may lie in the following advert from the Leeds Papers which appeared in October and November 1800

"LEEDS
To be SOLD by Private Contract
TWO WATER WHEELS, Twenty Feet Diameter and Ten Feet wide, HEAD STOCKS, PIT WHEELS, WHALLEYS, &c –Also a single powered PATENT ENGINE, of Thirty-six Inches Cylinder.
For further Particulars enquire of Fenton Murray and Wood" [18].

This implies that one of the installations was probably to replace a power source similar to the first one erected for Marshall & Co, a Wrigley type of Newcomen engine pumping water back up to top reservoirs to maintain a flow over water wheels. It is possible that this was at the firm of Coupland Wilkinson and Coupland, Cotton Spinners, who were known to operate such a system but are then recorded as having a 40 HP Fenton Murray and Wood engine by 1802 [19]. Another candidate was Ard Walker's Waterloo Mill which also replaced a waterwheel with a Fenton Murray and Wood steam engine around this time.

In November and December the firm were again advertising for workmen in the Leeds papers:

"To BRAZIERS, &C, WANTED, THREE GREENSAND MOULDERS, ONE BRASS FOUNDER, ONE BRAZIER, and ONE AIR FURNACE MAN, None need apply but good Workmen
Enquire of Fenton, Murray and Wood, Leeds" [20].

Roughly at the same time they were also announcing that they had been appointed agents for Mr Roberton for making Mr Roberton's patent furnaces [21].

On December 29th an advert for the sale of land by auction on January 28th 1801 appeared in the Leeds papers, the most relevant part relating to the Shoulder of Mutton close:

"LOT1V A FIELD adjoining upon Water Lane aforesaid, called the Shoulder of Mutton Close, in the occupation of Messrs Fenton, Murray and Wood, containing by Admeasurement One Acre, Two roods and Thirty-four perches
All the lots are particularly eligible for the Erection of Manufactories, being situated in a populous, improving and manufacturing Part of the Town, near the Navigation, convenient for coal, well supplied with a constant stream of water, and containing

very valuable Beds of Clay (especially Lots 11, 111 and 1V) …may be entered to immediately.
The Mode of Payment will be rendered easy to Purchasers; for other particulars apply to Mr Marshall, at the Compting –House of Messrs Marshall and Benyons; or at Mr Fenton's Office in Leeds, where plans of the Estates may be seen" [22].

As 1801 dawned business appeared to be moving along, there were rivet rods bought in January, a subscription for the relief of the Holbeck poor of 6 guineas at the beginning of April [23], and also in that month 56 boiler plates in two deliveries the second marked as completing the order. There is no reference to these plates being delivered to James Beaumont [24].

All this activity did not go unnoticed by Soho and Gregory Watt, whose turn it was to travel, wrote to his brother on the 28th April:

"We reached Bradford before dinner on Sunday & found Dawson at Mr Jarrets – we drank tea there & spent the whole of Monday at Low Moor. Smalley has been very ill but is recovering. There is no chance whatever of a situation there for the coalfield of good coal is very confined and all appropriated.
We left Bradford early this morning reached Leeds & visited Mr Gotts works & dined with him. We got payment from Marshall who was very civil.
……….. Murray and the Lowmoor Co are both making as many engines as they can undertake" [25].

The last comment certainly seems to be substantiated by the Kirkstall Forge records which show that between May 5th and July 22nd they supplied Fenton Murray and Wood with no less than 353 boiler plates, which was probably enough for 3 to 6 boilers, depending on the size of steam engine they had to supply [24].

On the 4th August 1801 patent number 2531 was granted to Matthew Murray:

"having brought to perfection certain new methods and improvements of constructing the air pump and sundry other parts belonging to a steam engine, by which considerable saving will be made in the consumption of fuel and an increased power obtained" [26].

August also saw another large purchase of boiler plates, amounting to 165 plates or well over 4 tons [24]. Further evidence that Fenton Murray and Wood were taking advantage of the expiration of Boulton and Watt's patent is provided by a letter of the 10th September from M R Boulton to James Watt junior which included:

"I am not aware of any obstacle to prevent your intended visit to Newcastle where you may perhaps pick up some orders before they are seized upon by Murray who according to Dudgeon's report is endeavouring to rival us in that quarter & has just got to work a new Engine with his patent apparatus which excites much attention" [27].

It was hardly surprising that Matthew Murray was selling engines in Newcastle upon Tyne the city of his childhood.

In September over 77 hundredweight of boiler plates were sold to Fenton Murray and Wood for delivery to James Beaumont, who reported short weight in the delivery [24].

James Watt junior did indeed proceed with his journey to Newcastle and on October 3rd sent an update to M R Boulton which provides more detail of Murray's sales successes on Tyneside but also an account of Murray celebrating success. The description was probably largely true as Matthew Murray certainly was fond of his ale, there are too many references to ignore, and no doubt he strayed, especially considering the amount of time he spent travelling to gain orders, but one might imagine that there may be some exaggeration for effect from a writer who of all the Boulton and Watt clan, seems to have taken Matthew Murray's successes personally.

"The Engine which Murray has erected here is for a Corn mill & has a 25 inch Cyl which he calls a 20 horse power. It has only been slightly loaded as yet, but is said not to be very tight, altho' external decorations have not been spared.

W King & his partners at Yarrow Colliery who had estimates from us sometime back, have given Murray an order for a 52 inch d[ou]ble, to work 2 sets of pumps 12 inch diam. 120 fathoms deep. He contracts to furnish all the materials exclusive of boiler for £1600 & to complete the Engine in 9 months. He asserts that his Engines are much more powerful than ours of the same size in consequence of his improved construction of the Air Pump and he intends in this case to have no diagonal rod, but to work both sets of pumps from the outer end. It is by no means unlikely that he may get himself into a scrape by going out of his depth.

Murray gave out that the Devon Company had also applied to him & that he was to have their order. I have called upon old Hawks but have not yet seen him.

Murray & his partners were to receive £460 for the portion of the materials which they furnished to the Corn mill Engine and upon its being completed £200 were paid into Murray' s hands; immediately on receipt of which he disappeared and no tidings were had of him for 5 or 6 days, when the hue and cry being raised he was detected in a low bawdy house with three whores, one upon each knee & a third supplying him with his favourite beverage. It seems however he had been economical, for he had still £193 remaining of the cash he had received. His wife and daughter came post from Leeds to fetch him & altogether there has been the Divil to pay.

......

Fishwick has bought a second hand Engine of Murrays & is very angry at having been refused information from Soho as to the construction of a forge, which he says we promised him. They have been in blast about 3 weeks. I am to go there on Wednesday" [28].

Fishwick and Hawks had traditionally purchased their engines from Boulton and Watt, indeed it had been George Hawks from the latter company that had reported Murray's engine building to Boulton and Watt in 1795. So the second report on October 10th from James Watt junior was not good news for Soho:

"Hawks & Co plead guilty to having ordered The Devon Engine from Murray. Time they say, was the only inducement. – It is the general opinion of the viewers here that

the Engine making by Murray for Jarrow Colliery will not be sufficient to drain the water. Doubts are also expressed of the security of the Parties" [29].

It is extremely difficult to track the history of colliery steam engines, but there may well be a reference to this engine as still working in 1845 [30].

So within a short space of the Boulton and Watt patent lapsing Matthew Murray and his partners were proving that they were as much of a threat as Soho had feared, Boulton and Watt engine sales in and around Leeds must have dropped considerably and now they were losing customers in Newcastle, and the Leeds firm were entering the colliery steam engine business as well as providing engines for manufactories.

In February of 1802 Fenton Murray and Wood are once more recorded in the books of the Kirkstall Forge purchasing boiler plates:

Friday February 5th 1802
"Sold Fenton Murray & Wood Leeds
To 26 plates to No 1 Model   Cwt 19.1.6
*   27   do   2   do   13.2.22*
per do   33.".."   28/-   £46-4-0
Friday 12th February 1802
Sold Fenton Murray & Wood Leeds
To 21 Boiler plates   Cwt 9.11.11   28/-   £12-14-9
per Riley to compleat the order" [24].

The firm was also keeping up with the latest advances in steam engines as can be seen from a February advertisement in the Leeds Papers:

"Smoke Burning Furnace
The patentee, John Roberton, from Glasgow, begs leave to inform the public
that they may be served with his furnaces for the burning of smoke on terms which
may be known on application to his Agents Messrs Jarrat, Dawson and Hardy at the
Low Moor Iron Works near Bradford, Messrs Fenton Murray and Wood or Messrs
Pryor and Warwick Leeds" [31].

As stated earlier Fenton Murray and Wood had advertised this as a fact in November 1800, but advertisements placed by Mr Roberton in September 1801 had only nominated Low Moor and Pryor & Warwick as agents in the Leeds district. So it would appear that an agreement had eventually been reached by February 1802.

In June the firm were once more the focus of attention from the team at Soho. The catalyst was the departure of a further two of the Soho workmen to employment with Fenton Murray and Wood, namely one Halligan and John Hughes, whose father Joseph was still in employ at Soho. The correspondence, which occurred between the 12th and the 25th of June, has been well reported elsewhere, but as it contains some of the earliest descriptions of Fenton Murray and Wood's establishment and demonstrates the position of the two rival firms it will be summarised here. Before

James Watt junior and William Murdock left for Leeds, to find out how many of his former employees were there and who was recruiting them, he set James Lawson a task:

"Murray Wood & Co are erecting an Engine at Jarrow a colliery in the vicinity of Hebburn which it may not be amiss to see. King [ ] of Hebburn is a partner & principal director. They would not agree to the time we asked & we were not otherwise very anxious about this order, on account [ ] opinion of the parties. Hackworth does the Iron Work & can tell you all about them" [32].

James Watt junior arrived in Leeds on the 12th of June and immediately enrolled the assistance of Benjamin Gott. The letters in question [33] were written by James Watt junior to M R Boulton and include the latter's one surviving response.

It would appear that Fenton Murray and Wood at this time employed a former Soho workman Dixon as foremen or charge hand in their Dry-sand Foundry. In the letters James Watt junior noted that Dixon asked Joseph Hughes to collect some wages from Boulton and Watt that Dixon claimed were still due to him. Dixon had been visiting the Soho Foundry, or at least the area, ostensibly on one occasion seeking absconded Fenton Murray and Wood apprentices. On these visits he was recruiting for the factory in Water Lane. It is not clear whether this was at the specific instigation of Fenton Murray and Wood, or whether they, as fairly common at this age, paid a bonus to any employee who introduced skilled workmen into their employ.

However it should be said that the evidence is that this appears as before to be legal and related to the employment of Soho workmen whose contracts had expired, and who had declined to sign new agreements. It would also appear that Halligan fell into this category, and John Hughes' father Joseph was still working at Soho because he was waiting for his contact to finish before joining his son in Leeds. The situation with John Hughes is not absolutely clear, but if he was still under contract to Boulton and Watt they would not have needed evidence from Leeds to have him arrested.

One of Benjamin Gott's millwrights called Pritchard, who was familiar with Fenton Murray and Wood's factory, called there and from a description identified Halligan, engaged him in conversation and found where he was living. He and Murdock visited Mrs Halligan and found her to be extremely unhappy at the move and very keen to return to Birmingham. She believed Halligan would do so for her sake and asked if he could be taken back, although prior to her declaring her unhappiness he appears to have been prepared to carry on in his new job. His wages were 22/- a week and their annual house rent was £7 probably with taxes on top. John Hughes was lodging with them, but Mrs Halligan had to admit that on the subject of why he left Soho and how he was getting on in Leeds he was very tight

106

lipped. Halligan and Pritchard were entertained that evening by William Murdock to supper and a drinking bout, and Halligan stated that he had not signed articles although was under pressure to do so and was free therefore to return to Soho and asked if he could. John Hughes had had two letters from his father, but had only disclosed that Dixon had been at Soho. Halligan had an interview with James Watt junior to ask to return to Soho and was informed that due 1. to his wife's ill health, 2.his former good conduct, 3, the fact that he did not appear to be "seducing others" this should be possible but he would have to stay at Fenton Murray and Woods as a spy, until told he could return. Halligan was offered a guinea a week and a loan to repay money he owed Dixon.

Mrs Halligan, after hearing the good news admitted James Watt junior into John Hughes room where James Watt junior succeeded in unlocking John Hughes trunk and discovering drawings of Soho machinery and engines with dimensions, but "*nothing but what Murray already knows or soon may know*", but no letters incriminating those who were encouraging workmen to leave Soho. Mrs Halligan advised that Hughes was under articles as yet and that he carried the letters in his pocket. In order to obtain Joseph Hughes letters to his son and any incriminating evidence (of seducing Soho workmen to leave) therein, the same legal process that had been used to get Parker escorted back to Birmingham by a constable nearly three years previously, was instigated again at Birmingham. Mr and Mrs Halligan had told James Watt junior that Joseph Hughes had been the major instigator in their leaving Soho.

Halligan duly signed his articles, and agreed to request Fenton Murray and Wood to transfer him from the Dry-sand foundry under Dixon to work in the Greensand foundry. Halligan also undertook to read John Hughes' letters while he was out one evening and duly reported that there was one from his father of no import and one from Forster (a Soho employee) saying he was happy at Soho and would not be leaving. John Hughes was duly apprehended and bundled off to Birmingham and advice sent to M R Boulton to interview Forster, before John Hughes arrived, to see if he could provide evidence of Joseph Hughes seducing workmen to leave. The upshot of it all was that Joseph Hughes was examined in Birmingham, his colleagues decided that he was trying to sell them for a fee and were not happy with him. John Hughes painted an unpleasant view of Leeds and so in that respect Boulton and Watt's efforts were successful in preventing more workmen heading north. Boulton and Watt however failed to pin any actionable offence onto Fenton Murray and Wood. Boulton and Watt also learnt that a W

Williams was stirring up dissent within the foundry and that Fenton Murray and Wood's new recruiter was called Hansby.

Immediately after an absconder from Fenton Murray and Wood was returned from London by a constable, Halligan told David Wood that Hughes had been carried off from his house the previous night. David Wood's reaction was to swear that:

"that it was nought but a scheme of Boulton & Watt's and that he had been sent there by them as a spy. There was nothing but trouble with their men and he believed the best way would be to have no more to do with them".

James Watt junior also took the opportunity of his visit to further a Soho plan to establish a rival establishment to Fenton Murray and Wood in Leeds, by searching for some suitable premises. He did check with base to make sure that he still had support to do this and was advised by MR Boulton:

"I can say the parties here are not disinclined to your speculations if the land can be had at anything like the market price. – It may perhaps admit of a question whether it will be more eligible to erect a new foundry on those premises or to form a connection with some of the foundries whose establishments are already in activity you will of course with this view make the necessary enquiries as to their character & other requisites".

But James Watt junior concluded this was not an approach that would work;

"The only effectual way to harass them, seems to be by destroying the basis of their ill got fame and setting up a competition which will diminish their orders. All our friends are of opinion that there is sufficient opening here for another works and it does not appear that any of the present ones would be eligible connections; they are men without character & without means".

James Watt junior mentions a field adjoining Murray's premises, some land near the canal and a Malt House which projected into the factory. He was very keen on the latter but the woman who had it would not sell at any price. The canal area was too expensive anyway, so instructions were left with a solicitor Mr Upton of Nicholson and Upton to deal for the field with Mr Garforth. This land cost more than expected as well, but a sale was duly registered on the 24th November 1802 [34] with the land passing to Mr Thomas Everard Upton who held it on behalf of Boulton and Watt for the next 25 years.

The land was only registered in Boulton and Watt's name after Matthew Murray's death. Ironically part of this land today contains a building called Matthew Murray House and on its wall in Water Lane is the Matthew Murray memorial organised by E. Kilburn-Scott.

James Watt junior also imparted information on the partners to MR Boulton; of James Fenton, collating the data from all the letters sent on that visit, he said:

"Fenton is, as we formerly heard, a man of some property & was Marshall's partner, who provided him with a share in this business to get him out of his own. Mr Gott

speaks well of him & thinks he would be ashamed of his partners proceedings although he may not have the power to control him. Fenton keeps the Cash & Books. Fenton is in London & not well.
Fenton's character is not very respectable. His brother (William) is an Attorney well skilled in all the nefarious practices of his profession.
Fenton is generally busied in the Counting House & probably keeps the ledger & writes the letters. The general result of my enquiries respecting Fenton, is, that he would do anything for money."

What could be termed mixed reporting. Of David Wood he reported little:

"Wood is the steady man of business who directs the works. Wood & Murray when at home, attend chiefly to the fitters. I hear nothing to the prejudice of Wood."

Boulton and Watt had already gained intelligence on David Wood when they had instructed James Lawson to approach him about a job with them in order to break his partnership with Matthew Murray early on. And although they already had plenty of opinions on Matthew Murray he added:

"Murray solicits orders superintends the erection of Engines &c. He(Gott) passes him for a great scoundrel but a very able mechanic & Gott says he has got great credit from his last patent, no one doubting that the inventions are exclusively his own. Murray at Newcastle".

Also in this run of correspondence between the 12th and 19th June 1802 James Watt junior describes, mainly through the eyes of Halligan, the factory and the work of Fenton Murray and Wood.

Fenton Murray and Wood were believed to be making a great deal of money and these profits were funding new buildings at the works and a superb house for Matthew Murray. The chief of the new buildings, the rotunda, from which the works were to get their local name of the Round Foundry, was described as follows:

"Fame has not outdone the magnitude of Murrays new Edifice. It is a rotundo of about 100 feet in diameter with a magnificent Entrance. The Engine is to stand in the middle and the lower rooms to serve as deposits for Engines and other finished goods. The Upper rooms to be for fitting. It is an excellent building & will not look amiss. It is up to the top of the 2nd story.

22 James Watt Junior's sketch of the rotunda.

I will make a better sketch of it before I leave Leeds than this, but am now much pressed for time".

However James Watt Junior's information was not accurate, for when the building was insured it was certified that that it contained no steam engine nor was it connected to one. It was however steam heated.

The descriptions of the factory and works are given in James Watt junior's words, but are reordered to collect all comments on one topic together:

"The Dry Sand Foundry is about 20 yards long & 12 wide with two Airfurnaces & 3 stoves, one 20 feet by 13 wide for loam, another 17 feet by 13 for boxes and a third 17 feet by 9 for cores. Says he (Halligan) is under no Engagement, tho' most of the other men are, and he has been much pressed to enter into articles. (Halligan) Is employed solely in the dry Sand, which he says they do worse than we do & confirms Dixon's report that the nozzles are all made in dry sand. The moulder is dead who made them in green sand & they have no workman now who can do them. The dry sand is done by Dixon with 3 men & a lad. The dry sand work is very badly done & much patched. (Did it escape James Watt junior's notice that 50 % of the men employed in dry sand were ex Soho?)*

The Green Sand Foundry is on the opposite side of the yard about 15 yards distant & is nearly of the same Dimensions, with 2 Air furnaces & a cupola, but no stove, the cores being dried in those of the dry sand. In the Green sand there are 4 men moulders & six lads. The green sand & loam are both capital. Says he does not think much of their moulders although he is sensible of the great perfection of their green sand works which he attributes entirely to the sand. I have seen the place where the greensand is got which is exactly as described by Lawson. I wish you would write to Mr Smith of Castleford (J Wilson's uncle) to order 40 to 50 Ton or say a barge load, which will be equal to two boats. Says it is exactly the same as we had at the foundry, but that we spoiled it, by mixing too much Coal dust with it. The bottom of their green sand foundry (which is a distinct building from the dry sand one) consists for about 4 feet deep of this sand, which they work over & over again & have never renewed since he has been here (13 weeks). When the mould is formed, a little Coal dust is strewed over it, which is all that is used & he supposes this to be consumed by the Iron so that little or none remains mixed with the sand & though its colour is somewhat affected by frequent use, it is not so much changed as might be supposed. Connecting rods, Shafts & Wheels of every description are made solely in green sand, but no pipes or hollow goods. Boxes, where any are used, are of Wood & patched together for the occasion. Cylinder bottoms, tops & pistons are all made in green sand, until they come to be very large.

The Cylinders above 20 Horse, are done in loam by the piece, by a very good moulder of the name of Joseph Brooke, who has under him three men & a lad. The loam men chip their own goods
There are 5 chippies & 2 air furnace men and one Cupola tender, in all for both foundries. They melt a great deal from the Cupola, which is about 8 feet high & 20 Inches diameter & in which they do not use one third of the Coaks which we do to the same quantity of Iron. Believes the Coaks come from near Low Moor. Says the Iron runs much finer from the Cupola than ours does & about 6 Cwt per hour. The worst burnt stuff they can get comes out excellent. The Beeston coal which is used by Murray & Co contains a good deal of pyrites. I believe their coak comes from Bradford & has a reputation of being excellent.

There are 8 smiths, with a similar number of strikers. Of the fitters he (Halligan) knew little & has seldom been in their shops he thinks that on both foundries they cast upon an average 3 to 4 ton per Day, which I think is as much as we do, (of this 50 cwt is green sand).

They had a terrible blow up a short time ago in making a large cylinder.

His cutter block is pushed forward upon the boring rod by an endless screw, which, or some similar contrivance we must adopt, both to guard against the negligence of the borer & to save part of his wages. (This despite the fact that Murdock had fitted his "endless screw" to the borer at Soho.)
Says they have sent off about 2 Engines since he has been here & have 4 in hand. Employ their pattern makers to erect the Engines. Owen (Samuel Owen another ex Boulton and Watt employee) *for one. The Engines which Murray has now in hand are small ones 10 to 20 Horse but a blowing or regulating cylinder has been made within a short time, for a person at Newcastle upon Tyne, whose name he thinks is Fishwick. We have seen the Engines erected by him here, which are certainly neat, but easily imitable in so far as they are worth imitation.*

He (or Fenton Murray and Wood) *employs altogether about 160 Men the greatest part of whom are engaged. Many are Apprentices. He has much trouble to keep them & has been left by several. The air furnace men have a guinea per week wages, besides house rent & fire. The chippies have 18/- per week – There are 4 book keepers or clerks. The cash keeper & his assistant, who I suppose keep the waste book & journal. A clerk who overlooks both foundries in the capacity of a storer. A Time keeper who also assists in paying the men & regulates the police of the yard.*

The men are paid upon the Saturday about 11 o' clock by the clerks who go into each shop, by which means their wives are enabled to go to market early. This regulation is much approved of by the men & is worthy of imitation for several reasons which I shall state when I see you.

There is little chance of getting more than one or two of Murray' s men, as all the good hands are engaged & at high wages. The Green Sand Moulder (Davey) whom Murdock saw incognito, was willing to travel south, but would not engage & would not work for any master under 27/- a week. I thought at present it might not be worthwhile to push the matter further.

We shall have drawings & complete information respecting his machinery. We saw at Rochdale an Engine of Bateman & Sherrat' s newly erected & working very ill & one of Murray's which was just set up to work with the tapit wheels and usual finery. The Cyl was 21 inches Diam & they call it a 15 Horse power, it burns much coal which is a general complaint of every one of Murray's Engines & cost altogether about £1100. The owners are dissatisfied with it, as indeed they seem to be in every instance where we have gained access. It will be necessary to give the new finish to the Iron Work of Mr Taylor's Engine, though he seems fully aware that the saving of coal is the essential object" [33].

There was some discussion between the Soho men over whether James Watt junior should approach the partners in Fenton Murray and Wood, but James Watt junior was clearly not going to do this. M R Boulton when he suggested such an approach aired the subject of the other action that Boulton and Watt were considering, that of issuing a writ of scire facias on Matthew Murray in relation to his patent 2531 of August 1801. James Watt senior had first come across this process when he was corresponding with James Arkwright. It had been a writ of scire facias that had started the proceedings that deprived Arkwright of his patents. To get behind the Latin, it was a very simple legal writ that at this time was used to repeal Letters

Patent granted by the Monarch. Any person who thought a patent was invalid based on false information or the existence of a prior invention could ask the royal Court of Chancery to request the presence of the patent holder to justify the patent. Boulton and Watt were convinced that Matthew Murray had used information on inventions and methods that he had surreptitiously obtained from Soho. They were keen to bring such a writ against Matthew Murray but it was taking time to get their solicitor Weston on board.

The description above shows us a very sizeable enterprise employing 160 men and in a period of 13 weeks delivering 2 steam engines and having 4 more in hand.

On the 28th June 1802 the King was pleased to grant a further patent, number 2632 to Matthew Murray of Leeds, Engineer for:

"New combined steam engines for producing a circular power, and for certain machinery thereunto belonging, applicable to the drawing of coles, ores, and all other minerals from mines, and for spinning cotton, flax, tow, and wool, or any purpose requiring circular power" [26].

This patent included Matthew Murray's famous 'D' valve and his hypercycloidal steam engine, which worked without a beam using some known mathematical principles.

On the 1st July 1802 James Watt Junior, back in Soho, wrote to Benjamin Gott

"Lawson came from Newcastle to Leeds with Murray in the Coach a few days ago when the latter declared his determined hostility to us & his resolution to have more of our men. He was not at that time at all apprized of our recent manoeuvres" [32].

In August John Marshall re-entered the land market and both Matthew Murray and David Wood were involved. It would appear that their purchases were personal, probably land for their houses, as at this point both appear in John Marshall's ledgers as owing him sums of money. The amounts owing by David Wood were probably the ones seen by Trevor Turner which caused him to attribute a particular friendship between John Marshall and David Wood. However amounts also appear for Murray, and Wood seems to have repaid his loan in instalments, whilst Murray seems to have let his debt lie.

"August 5th 1802
Memorandum that Mr Marshall hath agreed to purchase the within land (which including two pieces of Ground sold to Matthew Murray & David Wood is supposed to contain Ten Acres) for Seven Thousand pounds, Nevertheless it is agreed that the money to be paid for the said 2 pieces of Ground shall be considered as a debt due to the partnership, and that Mr Marshall shall pay and be charged with the remainder of the ten acres after the same rate for which he was to pay for the whole. But if the whole (including the land sold to Murray & Wood) shall exceed Ten Acres then Mr

Marshall is to pay Messrs Benyon at the rate of Nine Hundred pounds an acre for the surplus above Ten Acres. Leeds August 5th 1802

Witness	*Thos Bolland*	*John Marshall*
	William Lugdin	*Thos Benyon*
		Ben Benyon" [35].

Some clarification is seen in John Marshall's accounts for 1804:

"Stock in Trade of Marshall & Benyons 2 July 1804

		£ s. d
Land to J Marshall 9 acres 2 roods 81/2 yards £700		*6687 – 3 - 9*
	34 yards £900	*191 – 5 - 0*
D^o to Mathew Murray	*637 yards @ 4/-.*	*127 - 8 - 0*
D^o to David Wood	*1537 yards @ 4/-*	*306 –8 -0"* [35].

The land was to the south of Water Lane, Mill A and B being to the north, and was that upon which Mill C and D would eventually be built. As well as selling some to David Wood and Matthew Murray, it was probably some of this land that John Marshall rented to Fenton Murray and Wood.

On the 14th August 1802 the Leeds papers show Fenton Murray and Wood recruiting again:

"TO FOUNDRY MEN WANTED, TWO GREEN SAND MOULDERS, who have been accustomed to Engine and Machinery Work.
None need apply but good Hands, as they will have good Wages and constant Employment.
Apply to Fenton, Murray and Wood, Leeds" [36].

However Boulton and Watt's determination to undermine Matthew Murray had found a direction and had settled out on a course invoking legal action, as James Watt junior advised Benjamin Gott on the 17th August:

"Our Lawyers are of opinion that there is not the smallest doubt of your townsman's (Matthew Murray) *Patent being set aside with disgrace, but nothing can be done until the courts at Westminster recommence their proceedings"*[32].

Fenton Murray and Wood's advert in the Leed's papers cannot have produced the number of green sand moulders needed, for in September the following advert appeared in the Newcastle papers:

"TO FOUNDERY MEN WANTED, Two Green Sand Moulders; One
Air Furnace Man; and One Man who has been accustomed to work a Cupola.
None but sober, steady, good Workmen need apply, as they will receive good Wages, kind Treatment, and constant Employment, at Fenton, Murray, and Wood's Steam Engine Manufactory, Leeds, Yorkshire, to whom Application may be made personally, or by Letter" [37].

In Soho preparations continued to prepare an action against Matthew Murray's patent of 1801 and in November of 1802 James Watt junior was in London meeting the lawyers and sent the following intelligence to M R Boulton;

"I have had one conference with Weston respecting Murray's Patent & he is to set to upon the case, as soon as we can be supplied with the following drawings:

1. *A drawing of the nozzles at the Mint Engine upon a scale sufficiently large to show the parts distinctly.*
2. *D° of the nozzles of Mr Symonds Engine.*
3. *D° of the old nozzles, to point out the variation*
4. *D° of the executive circle & its application to the Engine at Soho Foundry.*
5. *D° of other applications of the principle of opening the valves by the circular motion derived from the Rotative shaft, which have been devised or executed by us.*
6. *There should also be a complete drawing of an Engine with the old nozzles & working gear & of one constructed in the new manner.*

When these are all ready, it is proposed to lay the case before Holroyd & to give general retainers to Erskine & Garrow (learned counsel).

If this finds Lawson at Soho, it will be proper to learn from him whether he has obtained any information from Gavin (McMurdo) respecting the seduction employed by Murray & also respecting the conversation which the latter held in his presence & that of W Harrison about the hollow spindles & his means of obtaining intelligence of our proceedings. Please also to enquire whether Lawson has obtained any information from Crowther (Crowden)" [38].

As a consequence when M R Boulton sent his next memorandum of tasks to

James Lawson it included the following:

"1.Examine Gavin McMurdo as to his recollection of the conversation which passed in his presence between Murray & Will'm Harrison when the former boasted of the means which he possesses of obtaining intelligence of our proceedings & that he was apprised of the Socket Valves which were making by us indicating with his finger in what manner the spindle of one passed thro' the other.
Learn also from Gavin whether he is acquainted with any of the means employed by Murray to obtain a knowledge of whatever's doing in our Manufactory.
2. Crowden who formerly worked with B&W is suspected of having acted as his spy in our works - Mr Lawson is therefore requested to use his talents in seducing Crowden to confess the fact & to procure from him such other information as may be [produced] in proof of Murray's surreptitious practices" [32].

On the 1st December James Watt junior wrote to Matthew R Boulton:

"I have your favour of the 29th & shall attend to its contents. The parcel also with the drawings is come to hand and I have found no difficulty in making Weston understand the case from them. He suggests however that models will be much better suited to the apprehension of counsel and Murdock may have a commission given him to prepare them upon a small scale, say a split model of the old nozzles & working gear & another of the new ones with the Excentric. I find that from Weston's numerous engagements, it will take some time to put this matter into train & propose remaining here until the middle of next week" [39].

A further letter followed on the 4th:

"We attend a Consultation today with Holroyd on Murray's patent. My father has deferred his departure until Monday" [40].

This was followed up on the 6th:

"*The split models will not be wanted before the end of January. Holroyd is already sufficiently primed to prepare the application for the Scire facias, which cannot be done before that period*" [41].

Matters were certainly not all going the way of Boulton and Watt as James Watt junior received some annoying intelligence and wrote of the matter to his brother Gregory on the 7th:

"*We sometime ago (about two or three months) sent an estimate of a 40 Horse Engine to a Mr Swann in Goodman's fields London. I am told that he has since ordered one of that size from Murray Fenton & Wood in consequence of their Estimate being £400 less than ours. I shall thank you to let me have a copy of our Letter*" [42].

James Lawson duly started sending in the intelligence requested by M R Boulton, on the 12th of December he wrote from Manchester:

"*I have not yet been able to see Crowden I however suppose Gavin will see him this evening – all that Gavin knows of Murray knowing of the valves "is Murray's saying he knew all the new schemes at Soho – and particularly the new valves", describing them by a Motion of the hand often used for other purposes my modesty prevents me saying more.*
Gavin however says that at the time W M[urdock] & Murray were drunk together with himself and company, every sort of scheme was drawn with Chalk on the floor, whether this was among them he is not certain. He thinks Crowden can know little of the matter. He shall however be try'd. Gavin thinks that Mr Murdock (Not William) now at Leeds , [ ] Hunslett with a Mr Pullen a Boiler maker and sort of Engr knows most of the matter. From the acquaintance I have with him I think him not fit to be trusted. However should you think inquiry necessary I dare say Mr Gott could easily get it done" [43].

And on the 13th he reported the outcome of the meeting with Crowden, although only part of this letter survives:

"*I wrote to you yesterday, I afterwards saw Gavin, who at last found out Crowden. I cautioned Gavin to sound him gently, and he & Gemmel went with him to have a pint together, when after much conversation about Soho &c, Crowden confessed that he had told Murray about the Socket Valves of the Mint Engine, soon after he came to Leeds. Gavin found that he had been in Mint Engine once with Wm Harrison. He denies having given any information about the Mint which was suspected & Gavin observes he knows so little that he could not give much. There has however some things occurred in this conversation that surprises me. Gavin says Crowden seems to know a great many things, going on at Soho, such as the kind of work, & wages the different hands get, so that he must still have some correspondence there. I have not yet seen Crowden but expect to see him tomorrow morning.*
I have so far only told Gavin that the information I wanted was merely for B&W to show that Murray was not the inventor of these things.
I shall now wish to know whether any thing further should be done by affidavit or otherwise. Crowden made this confession before Gavin & Gemmel, and I have no doubt will be ready to give me the same.
He is I understand leaving Bateman & Sherrat at Xmas, and I did not choose to inquire for him at their works – and Gavin only told him I was in town – so that he wants to see me to help him to some job. I find from Gavin he would gladly return to

Soho – that however is out of the question from what I know of him – I shall therefore merely get what I can in conversation" [44].

It is noticeable from the number of references to Matthew Murray and Newcastle, that he was making determined efforts to make inroads into the steam engine market there, with some success. The most notable success he had had to date had been winning the order for the pumping engine at the new Jarrow Colliery. It is not surprising therefore to find that the firm was advertising in Newcastle upon Tyne:

"Patent Steam-Engine Manufactory, Leeds FENTON, MURRAY, and WOOD, INFORM their Friends and the Public in general, that they have made several valuable Improvements in Steam Engines, which lessen the Probability of Accidents, and greatly add to their Durability and Perfection in working; and having considerably increased their Establishment, they have it now in their Power to execute Steam-Engines of all Sizes, in a reasonable Time, with the most substantial and compleat Workmanship, and on reasonable Terms" [45].

Additionally the colliery records for 1802 contain the earliest surviving Fenton Murray and Wood price list:

"Powers	Price	Boiler included	
1 Horse	£100	D° }	
2 D°	£175	D° }	Portable Combined Engines
4 D°	£250	D° }	Ready Money
6 D°	£400	D° }	

		Supposed price of Boilers not included
10 D°	£360	£69 }	
15 D°	£483	95 }	
20 D°	£600	126 }	
25 D°	715	148 }	
30 D°	830	177 }	
35 D°	938	217 }	
40 D°	1045	257 }	
45 D°	1150	274 }	Patent Beam
50 D°	1247	296 }	Engines
55 D°	1343	325 }	
60 D°	1437	354 }	
65 D°	1537	394 }	
70 D°	1633	434 }	
75 D°	1795	474 }	
80 D°	1817	514 }	
85 D°	1914	531 }	
90 D°	2012	548 }	
95 D°	2121	570 }	
100 D°	2231	592 }" [46].	

The list shows that they were already offering engines of up to 100 horse power.

The Jarrow engine and Matthew Murray's activities in Newcastle were clearly on Boulton and Watt's agenda in January 1803 and they wrote to James Lawson:

"We should like to have some further report of Yarrow Engine, which it may be as well for you to see & report progress. We wish you also to ascertain whilst at Newcastle whether it is true that Murray has discontinued the use of his new air pump, in the late Engines he has erected there & whether he has reduced it to the common form in those to which he had previously applied it. This may be necessary as evidence upon the approaching trial, in which we mean to prove it useless. Dudgeon can give you ample information upon this hand & perhaps you may be able to see some of them, and learn how they behaved" [47].

The air pump referred to was an important part of Matthew Murray's patent 2531 and its failure to work was to feature strongly in Boulton and Watt's case against Matthew Murray.

On Friday 24th of January 1803 the Leeds papers remind us that business continued as usual and that Fenton Murray & Co contributed 15 Guineas to the House of Recovery in Leeds [48].

This date was of far more significance to Fenton Murray and Wood as it opened up the opportunities for the firm. A gentleman called Simon Goodrich, who was the Chief Mechanist in the Office of the Inspector General of the Navy, kept a journal and many papers. At the time the Royal Navy was considering setting up a steam powered ropery at or near London in an existing dockyard. As part of the research in January 1803 General Samuel Bentham, the Inspector General, his wife, Marc Brunel and Simon Goodrich set off to visit Mr Grimshaw's ropery in the North of England among other industrial concerns. As was quite usual the opportunity was taken to visit suppliers and other establishments of interest on the way. On the 24th January the group, less Marc Brunel who had diverted, were in Leeds, having visited Mr Whitmore's manufactory in New Hall Street Birmingham (a supplier of steam engines), Soho and the Strutt Mills at Belper. They started their stay in Leeds by visiting Mr Gott's establishment and Mr Gott then accompanied them to introduce them to a Mr More at his Flax Spinning Mill and Gen Bentham and Simon Goodrich went on to visit Mr Drabble in Water Lane. While they were at Mr Drabbles, Mr Gott's clerk called to deliver a letter of introduction from Mr Gott to Mr Marshall and to Messrs Murray and Wood. Gen. Bentham and Simon Goodrich proceeded to Mr Marshal's counting House but were refused a visit to his manufactory for spinning flax and weaving sail cloth:

"but were introduced by his clerk to Mr Murray who shew us all over his very extensive foundry and Steam Engine Manufactory.

The first thing we saw was a Steam Engine for driving a boreing mill

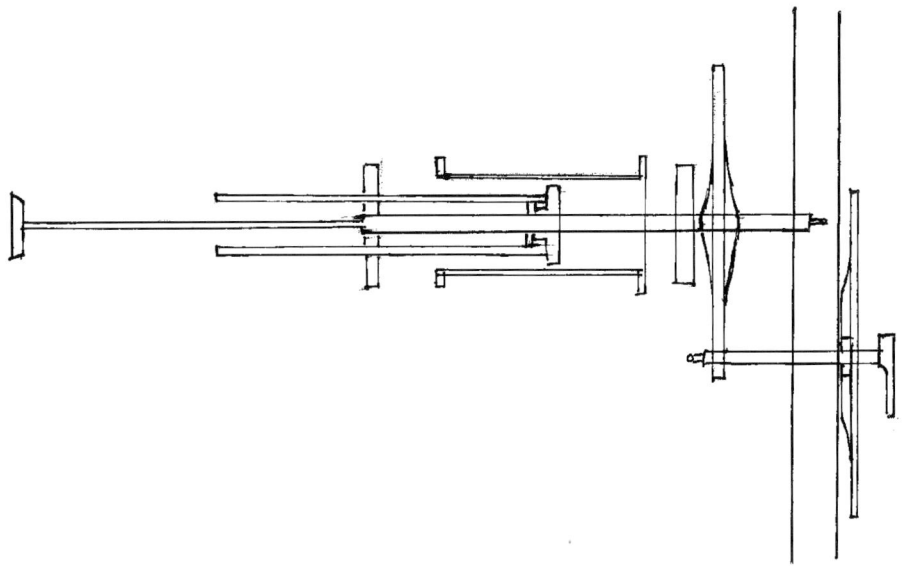

23 Author's representation of Goodrich's sketch of Murray's Boring Machine

with a planet wheel upon a crank revolving within a fixed wheel of twice its diameter producing at once the rotative motion for the fly Wheel on the crank shaft and the parallel motion for the piston rod the upper end of which was connected to a stud projecting from the side of the planet wheel. It is necessary that the center of this stud should be upon the Pitch line of the Planet Wheel. Mr Murray has a Patent for the application of this Motion [the Hypercycloidal] *but proposes not to continue the use of it upon account of its being too complicated and delicate to keep in good order. Mr Bunce made a model exhibiting this motion exactly about 6 years ago he said that he took the idea from a mathematical book which stated that if a circle revolves on the inside of another fixed circle, of twice its diameter every point of the moving circle will traverse in right lines backwards and forwards thro' the centre of the fixed circle.*

Walked thro' several Workshops and saw a variety of tools and expensive engines for doing the work. A cutting Engine for cutting wheels from the smallest size to the diameter of 4 feet with jointed levers so as to cut straight down thro' the wheel or bevelled. Steel, Wrought Iron and cast iron wheels are cut by this Engine. Mr Murray will furnish one of these Engines for 60£. I do not know whether cutters are included. Saw an Engine for fluting rollers for the spinning machines which was completely worked by machinery a boy stood by and had nothing to do but to look on and to take out or put in the rollers when necessary.

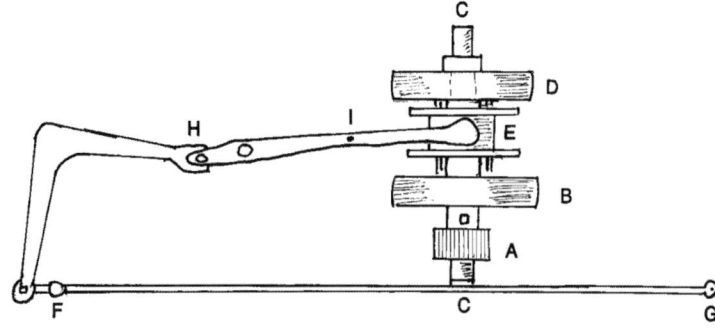

24 Author's representation of Goodrich's sketch of the roller fluting machine

118

*The roller to be fluted was placed in a frame between two centres. This frame was moved to and fro by a rack and pinion – The frame slid upon two triangular barrs. The cutter was fixed in a bridge above the roller and made the groove as the roller passed under it. The roller having passed beyond the cutter the motion of the Pinion **A** was reversed by the action of the frame itself and the frame returned. After the cutter was clear of the groove the roller was moved upon its centres to form another groove also by the action of the machine. The sketch above is designed to show how frame always reversed its motion when it had proceeded far enough either way.*
***A** is the pinion which worked into the rack of the moving frame – **B** is a Pulley which can turn round upon the shaft **C.C** without revolving the shaft. – **D** is another pulley of the same kind; these pullies receive contrary motions by two straps from the same drum one strap being crossed the other not – **E** is a sliding box which cannot revolve without the shaft **C.C** having a fork on each side to engage into either of the pullies **B D** and thereby oblige the shaft **C.C** and pinion **A** to move the same way as the pulley that the box **E** is engaged with – Suppose the frame to be moving from **F** towards **G** by the action of the pulley **D** and that it pushes against **G** by a spring fixed at the end of the frame till the spring is set up sufficiently as by its recoil to move the rod **G F** , the crooked lever **F H** and the forked lever **H I** which last shifts the sliding box **E** from the pulley **D**, engages it with **B**, this causes the frame to return from **G** towards **F** till it pushes against **F** in the same manner and shifts the sliding box **E** from **B** to **D** which causes the frame to move back again and so on continually.*
Returned highly pleased with what we had seen at Mr Murray's to Groves Hotel and dined" [49].

The detailed description of the machinery was typical of Simon Goodrich. At the time Gen. Bentham and Goodrich were heavily involved in managing the establishment of the Block Mill factory in Portsmouth Dockyard which has been called the first mechanised production line in the world. The machinery was designed by Marc Brunel and made by Henry Maudslay. Bentham and Goodrich would have been impressed to find a factory in Leeds with much mechanisation and a nearly automatic machine for fluting rollers. For the future it was the beginning of a relationship, even friendship, between Matthew Murray and Simon Goodrich which was to last the rest of Matthew Murray's life, and Simon Goodrich knew all the great mechanics and engineers of his day.

M.R. Boulton writing from Manchester to James Watt junior on March 20th 1803 described inspecting a new Fenton Murray and Wood engine near Bolton:

"Southern & myself inspected the new Engine erected by Murray – a close imitation as to the form of the principal parts & apparently very well made – the Cylinder lid turned and the wrought iron work in general highly finished – The valves are worked by a shaft place below the floor, similar to our present working gear shaft, which has a rotary motion communicated by bevel gear from the Rotative shaft. He uses tappets which work two stiff rods with brackets in front of the nozzles nearly upon the plan suggested by Murdock – upon the whole the motion of the Engine is very smooth. – The house was decorated in great style & the portrait of Murray or the owner of the Engine painted on the Cistern of the Hot Water Pump" [50].

By June 1st M R Boulton was in London and wrote to James Watt junior:

"I am just returned from the Petty Bay Office after executing conjointly with Goodward & Morley a recognizance for payment of the costs & co which may be incurred in the action against Murray. The scire facias will be sent off to Yorkshire tonight" [51].

Boulton and Watt were on the cusp of issuing the legal challenge to Matthew Murray's patent number 2531 dated 11th August 1801.

Notes:

1. Leeds Mercury 24 August 1799.
2. Goodrich Papers Journals and Notebooks Science Museum Library Wroughton.
3. Kirkstall Forge Tradesman's Waste Book No 11 1799-1800 KF4/8, West Yorkshire Archives (WYAS) Leeds.
4. Boulton and Watt Papers Letter Book MS3147/3/94.
5. Boulton and Watt Papers MS3147/3/284/61.
6. Boulton and Watt Papers MS3147/3/284/62.
7. Boulton and Watt Papers MS3147/3/76/31.
8. Boulton and Watt Papers MS3147/3/274/38.
9. Boulton and Watt Papers MS3147/3/274.
10. Boulton and Watt Papers MS3147/3/274/50.
11. Boulton and Watt Papers.
12. Boulton and Watt Papers MS3147/3/274/56.
13. Boulton and Watt Papers MS3219/6/2/L/110.
14. Binns and Brown Directory of Leeds, Leeds Central Library.
15. Royal Exchange Policies No's 172611/2, MS7253 vol. 37. London Metropolitan Archives. The Leeds Intelligencer 5 May 1800.
16. The Leeds Intelligencer 5 May 1800.
17. Kirkstall Forge Tradesman's Waste Books KF 4/8 and KF 4/15, West Yorkshire Archives (WYAS) Leeds.
18. The Leeds Intelligencer 13 October 1800.
19. A Tour through the Northern Counties of England and the Borders of Scotland, Rev'd Richard Warner, Robinson, London 1802, page 240.
20. The Leeds Intelligencer 3 November 1800.
21. The Leeds Intelligencer 17 November 1800, with reference to patent no 2437 of 13 August 1800 awarded to John Roberton.
22. The Leeds Intelligencer 29 December 1800.
23. The Leeds Intelligencer 6 April 1801.
24. Kirkstall Forge Tradesman's Waste Book KF 4/15, West Yorkshire Archives (WYAS) Leeds.
25. Boulton and Watt Papers MS3147/3/78/2.
26. See Appendix 1.
27. Boulton and Watt Papers MS3147/3/39/19.
28. Boulton and Watt Papers MS3147/3/50/20.
29. Boulton and Watt Papers MS3147/3/50/21.
30. Archives reference ZD36 in The North of England Institute of Mining and Mechanical Engineers, Newcastle upon Tyne.
31. The Leeds Intelligencer 22 February 1802, for the advert which excludes Fenton Murray and Wood see Leeds Intelligencer 28 September 1801.
32. Boulton and Watt Papers Letter Book MS3147/3/98.
33. Boulton and Watt Papers
    MS3147/3/51/6 J Watt Jnr to MRB 12 June 1802.
    MS3147/3/51/7 J Watt Jnr to MRB 14 June 1802
    MS3147/3/51/8 J Watt Jnr to MRB 15 June 1802
    MS3147/3/40/8 MRB to J Watt Jnr 16 June 1802
    MS3147/3/51/9 J Watt Jnr to MRB 17 June 1802
    MS3147/3/51/10 J Watt Jnr to MRB 18 June 1802
    MS3147/3/51/11 J Watt Jnr to MRB 19 June 1802
    MS3147/3/51/12 J Watt Jnr to MRB 19 June 1802.

34. Land Registers Vol. EN page 139, WYAS Wakefield.
35. Marshall Papers MS200 folio 2.
36. The Leeds Mercury 14 August 1802.
37. The Newcastle Courant 4 September 1802.
38. Boulton and Watt Papers MS3147/3/51/16.
39. Boulton and Watt Papers MS3147/3/51/22.
40. Boulton and Watt Papers MS3147/3/51/24.
41. Boulton and Watt Papers MS3147/3/51/25.
42. Boulton and Watt Papers MS3147/3/51/26.
43. Boulton and Watt Papers MS3147/3/276/46.
44. Boulton and Watt Papers MS3147/3/276/47, and MS3147/2/61/5.
45. The Newcastle Courant 18 December 1802.
46. John Papers, John 5 page 54; Watson Papers 2/8 page 339. The North of England Institute of Mining and Mechanical Engineers [NEIMME].
47. Boulton and Watt Papers Letter Book MS3147/3/99.
48. The Leeds Intelligencer 24 January 1803.
49. Goodrich Journals Vol. E3 'Memoranda made Journey to the North in 1803 to visit Mr Grimshaw's Ropery and to the West of England'. Science Museum Library.
50. Boulton and Watt Papers MS3147/3/41/1.
51. Boulton and Watt Papers MS3147/3/41/3.

Chapter 6.

Scire Facias, the King v Murray

For some time Boulton and Watt had been preparing their case against Matthew Murray and his patent with all the rigour that they had learnt to apply in their previous litigation against Bull, Hornblower, Bowling Iron works, Bateman and Sherrat etc. They had been incensed by Matthew Murray's activities since 1795 and now they believed that they could at least prove that his patent of 1801 was largely based on information obtained from Boulton and Watt employees and secondly that his new air pump did not work.

With the cost of taking out patents at this period it is worth reflecting on the following points:

a. Matthew Murray was well aware that he was in a commercial war with Boulton and Watt (it is worth pondering what the real conversations held between Matthew Murray and James Lawson must have been, it is possible to detect that they were actually on quite friendly personal terms), and whatever ones opinion of Matthew Murray's character he was certainly not stupid – so why would he patent ideas that he had filched from Boulton and Watt, when he would have known that it would probably play into their hands? Use the ideas if he could make them work, but why patent them? The success of Kendrew in challenging his textile patent would also be in his mind.

b. He had jumped too fast with this patent anyway, because by the time he received the writ of scire facias he would have known that his air pump as contained within the patent, did not work. He was busy retrofitting standard air pumps to all the engines supplied with the patent pump.

The amount of work done by Boulton and Watt to bring Matthew Murray down is evident in the surviving records of the firm, for which there survives a separate section on Murray, in their legal filing section. The files show draft after draft of their position, and these can only be summarised here.

So on the 8th June 1803 James Watt junior sent a letter to Benjamin Gott in Leeds about the scire facias and the issues in the Marshall partnership [1]:

"Our Lawyers have at length got the Scire facias against Matthew Murray's Patent, & I hope we shall be able to bring on the Trial early in July. I should be glad to know how Matthew likes it, and if anything transpires from the enemy's camp shall be obliged to you to inform us.

Mr Bage called about a fortnight ago for estimates for 20 & 56 Horse Engines which we gave him, but although there seems to be a strong schism between him &

Marshall, I suspect he cannot prepare his Machinery without the aid of Matthew, & the latter may not choose to execute it without he has the Engines also" [2].

On the 10th of June he wrote to James Lawson and in among the other business chased him on a matter pertinent to the upcoming trial:

"we are awaiting the result of your negotiation with Crowden with some anxiety as it is necessary we should [loose] no time in arranging the evidence and preparing our briefs"[2].

The 13th of June found him writing to M R Boulton:

"Murdock's model of his own & Murray's inventions is finished & shall be dispatched by the first wagon, with a proper explanation for dissecting & putting it together. I expect you will also want a model of an Engine upon the old construction (to show the plug working gear & nozzles) but perhaps the working model of our own Engine which was prepared for Bulls trial may serve & will save Murdock a week's work.

That model not being anywhere to be found hereabouts (It is not at the Mint, where we can only discover Bulls) is supposed to be either in London Street or Salisbury Square. I remember it being packed up, but am not certain as to its having been sent off. You will please to write to me upon this subject" [3].

Then on the 18th of June came a reply from Lawson to James Watt junior closing out his previous evidence on Crowden from December 1802:

"I have at length seen Crowden, his evidence which I got by degrees is nearly as under – That after leaving Soho he came to Messrs Bateman & Sherrat where he was about 6 weeks – he then left them and went to Leeds – Soon after this the conversation he had with Murray took place, on a Sunday at the Cross Keys a Public House in Holbeck.

He says that Murray asked him if he had seen the new nozzles at Soho, or at the Mint – which he told him he had – but would not say further than that they worked the valves with rods through a stuffing box – that he asked him how the valves worked inside, which he said he could not explain, as he had never seen more than the outside. He cannot fix the time further than he is sure it was before Xmas 1799" [4].

James Watt junior wanted the issue to be absolutely clear and wrote back to Lawson on the 20th:

"The information from Crowden seems sufficient to prove that Murray knew of our nozzles before he took out his Patent.....and as the writer goes to London tomorrow he will write you from there what more is necessary. In the meantime it will be proper for Gavin to refresh his memory as to what passed in the conversation with Murray at Birmingham which I find (by reference to our letters) took place between 24th & 31st May 1799. It will be necessary he should be accurate in his recollection of the date [as] well as of his conversation. You will address to me anything that occurs upon this subject at No 13 London Street. (Boulton and Watt's London Office).
Mr Lee is also able to speak to the time when he saw the socket nozzles at Soho & the Eccentric Circle at the Foundry; also when the construction was applied to his condenser to make it discharge the Air separately, though we do not attach much importance to this circumstance.
I also wish to know what <u>you</u> are able to prove respecting Murray's having discontinued the use of his improved airpump and now adopting the other generally....... from your own observations and not from hearsay. Plan also to give your ideas as to who is the most likely person to furnish such information.

I wish you would again examine Crowden whether he did not see Murray upon his visit in May 1799; I have a strong suspicion he either received his information <u>at that time</u> from Crowden or came over in consequence of prior information from him or someone else. You will recollect he then told Wm Harrison & Gavin that he knew of the new nozzles we were making and of all we were doing at Soho, or had a spy who informed him of all our proceedings, or words to that effect. This you will observe was at least 6 months prior to his conversation with Crowden reported in your letter.
It was also about this time May 1799 that the Mint Engine got to [again] work which perhaps you may be able to prove.
The first Excentric Circles made of wood were [for] the little Engines in Sept 1800, and the present one was applied to the Foundry Engine in Dec of that year, the small Boiler was set with the funnel fireplace in Sept 1800. Murray 's drawing seems a copy of this" [2].

On the 23rd James Watt junior wrote to M R Boulton on the subject of the models needed to demonstrate to the court the workings of the various parts:

"I have been employed this morning in the very dirty job of unpacking and overhauling the old models, but upon a very ample consideration of the subject with Weston, it appears very desirable if not indispensable that a new one shall be made to exhibit the state of the double engine prior to Murdock's improvements .

All the models here, both the split and the working one, are of single Engines, and the demonstration of the differences would not a little perplex the subject, which would be further increased by the necessity of explaining (upon the working model) the difference between a pumping and a Rotative Engine. These difficulties are certainly of no moment, if we had to explain the subject to a mechanician, but they will embarrass, and may mislead our Counsel and the Jury.

I hope Mr Murdock will set about the new one immediately and let us have it here in the course of a week. I shall send down Maberley's & Bulls Models by tomorrows Coach or wagon as the valves may be made to serve in the new model. I recommend that in the Beam & Flywheel & Cylinder it should be an exact counterpart of the one last made, that the only difference may be in the new improvements of working Gear & nozzles.

I see there will be some difficulty in making the Model a working one & split one at the same time, but unless some scheme is adopted to show the ratchet work of the valves, the value of the improvement may not be so easily understood.

I would have sent down the working model also, but I do not think any part of it could be brought into service upon this occasion it being upon a larger scale than the other" [5].

And Lawson wrote on the 23rd June from Manchester:

"I yesterday received yours of the 20th ult and last night went to Gavin, after a good deal of search he found from some memorandums that the conversation with Murray in Birm[ingham] was on Monday 27th or 28th May 1799, at the George Inn Digbeth. He says that nothing more passed than what he has so often mentioned, i.e. that Murray said he knew what secrets were going on at Soho, and describing the Motion of the Stuffing Box valves with his finger &c. From my conversation with Crowder I am pretty sure nothing more can be got from him & Gavin also thinks that he knows nothing more then and does not think that Crowden either heard of or from Murray when at Birm[ingham].

My own evidence cannot go to anything material further than the time of seeing the Mint Engine & the Little Engine at Foundry, with the eccentric, of which I suppose you have sufficient proof.

With respect to the new Airpumps, I have never seen one. The Engine at the end of Newcastle Bridge (On the Gateshead side) had one which was taken away & I saw it working with a common one when at Newcastle last Febr[uary]. *Full evidence may be had from Dudgeon who saw it before & with me afterwards.*

The Person however who could give the fullest evidence on this point is W Murdock (Not the William Murdock who worked for B & W) *who formerly was with Mr Wormald & Co and who is I believe at present either at Wakefield or Hunslett (who you will recollect I expected to have seen last Aug*[us]*t at Blenavon). I find from Gavin he has not behaved so bad as we supposed having returned the money advanced by the Co*[mpany] *it was on acc*[oun]*t of his family he could not go there, since which he has been working for a Mr Pullen of Hunslett. From Gavin's acc*[oun]*t of him I think he is better acquainted with Murrays transactions than any other person not in his employ & that he would not be unwilling to inform of all he knows, as I am well acquainted with him I shall find him out should you think it necessary. I know nothing against him further than being fond of company & drink & more talk than work. Mr Lee & myself can both speak to the time the exp*[erimen]*t made on the air pump on his engine, I have not my memorandums here, I suppose it was in 96 or 97 as Mr Lee thinks it was early in 1800 he [saw] the Mint E*[ngine] *and eccentric at Foundry*"[6].

Matthew Murray's solicitor forwarded an offer to Boulton and Watt dated the 22nd June:

" *New Inn 22 June 1803*

*Gent^n^*

Murray Patent

Solely with a view of avoiding the Expence and trouble of bringing up Witnesses from remote parts on a trial with Messrs Bolton & Watts as to the validity of Murrays Patent, we have to propose on the part of Mr Murray to grant Messrs Bolton & Co full licence to make use of his inventions & improvements on the Steam Engine, provided the present proceedings are discontinued, and that he will not afterwards grant the same licence to any other person.

We are

Sirs

*Your very Ob^t^ Serv^ts^*
Bleasdale & Alexander

Messrs Westons
Solicitors Fenchurch Street" [7].

This was duly forwarded by James Watt junior to M R Boulton on the 25th:

"*Annexed is a copy of the very liberal offer now transmitted in writing by Murray's Agent, to which we have not yet resolved what answer to send; having been much occupied in the preparation of the Briefs*"[8].

It is again difficult to read into this riposte what was going through Matthew Murray's mind, if he had knowingly used Boulton and Watt's schemes as they supposed, then the conclusion might have to be drawn that this time he was confident that what he had patented was distinguishable at law from Murdock's valves. Murray was also probably aware that the main other component of the patent, the air pump, was known by Boulton and Watt not to work. At a simpler level, he may merely have been trying to assess how strong Boulton and Watt thought their case to be.

M R Boulton was not impressed and he compared Matthew Murray with Napoleon:

"*There seems to be a singular coincidence between Murray's magnanimity & that of the Chief Consul (Napoleon) in giving Hanover & co the Prince of Orange. Doubtless the most proper reward for the allies of such a depredator (Napoleon) is a participation in the spoils of his plundered neighbours, we are not however quite reduced to the abject state of receiving with gratitude such a humiliating offer & I trust you & Mr Weston will give the agents of this piratical Chief fully to understand that our connivance at this piracy is not to be obtained by such a shallow artifice. We do not ask for the use of any invention, we have never been molested in the use of any never having employed the inventions of others. We ask Mr Murray to retract the assertion that he is the author of the invention for which he claimed the protection of the Crown & that consequently he is not entitled to any exclusive privilege for the use of them. Let him make the public recantation & we will agree to discontinue the proceedings*" [9].*

And on the 27th June James Watt junior sent M R Boulton advice that the case had been submitted to Counsel:

"*The Case & Proofs (which constitute our Brief) are made out and delivered to our Counsel, Erskine, Gibbs, Garrow and Holroyd. A copy shall be forwarded to Soho for your remarks and those of my father, which may be given in a supplemental paper if it should appear that anything material has been remitted.*
        There is an error in Murray's Specification in quoting the date of the Patent he calls it of the 4th August, whereas the Patent is really dated the 11th August. This error is also introduced into our Scire Facias, our Counsel having inadvertently consulted the Specification instead of the Patent. This might have been a fatal error in Murray, if we had been properly aware of it, and it had been desirable to overturn his Patent by Dry Law; but as the error now exists on both sides, it is supposed it will not materially affect the Pleadings. Should we however obtain a verdict against him, it will be against the Patent of the 4th August, which does not exist, and he may oblige us to recommence our proceedings against the Patent of the 11th August. But as he could have no better chance upon that ground, it may be presumed that he will acquiesce in the decision without incurring the expense of a second trial.
        I reply to the proposal made by his lawyers, Westons have answered merely that it was declined by their clients, it being thought better to assign no reason.
......
        Lawson in a letter rec'd from him this morning, suggests that Wm Murdock formerly engineer at Gotts may furnish important evidence upon the object of Murrays piracies with the view to obtaining such, I shall probably see him before I return hither" [10].*

Various drafts of the case against Murray exist in the Boulton and Watt papers, the first utterings were dated 1801, the closest to a final version appears to have been drafted around the time of May 1803. The following description of Matthew Murray is contained within the 'Case':

"*The Defendant Murray is a mechanic of some ingenuity, but of low, dissolute habits much given to drinking with his own and other people's workmen and by no means nice about the means he employs to gain any object he has in view. This has enabled him to introduce himself into the company of Boulton Watt & Co's workmen and to obtain from them a knowledge of such improvements as were carrying on at Soho*" [11].*

126

The following, again from a document drafted many times gives the fundamental issues raised by Boulton and Watt, but it is only their side of the story, as no trace of Murray's position appears to have survived:

"Observations upon the Patent of the 11th Aug 1801 granted to Matthew Murray of Leeds and on the Specification thereof inrolled in Chancery.

The invention for which the Patent is granted, is described in the Title as an improved Method of constructing the Air Pump and sundry other parts belonging to a Steam Engine, by which a considerable saving will be made in the consumption of fuel, and an increased power obtained.

This Title is not correct in as much as the Methods or Inventions set forth in the Specification are incapable of attaining the ends proposed. No saving of fuel is produced and no increase of power obtained by the separate or conjoint application of any or all of these Methods to the Steam Engine, when compared with such as have been in common use for years in Boulton & Watts Engines.

The Inventions described in the Specification are six in number:
1st. First, an alteration in the construction of the Air Pump which is thus described "In respect to the principles of the New Air Pump I cause the air to be discharged from the Air Pump without its having to make any effort in the opening of valves or passing through a Body of Water, and in causing the Air & Water to be discharged separately and different ways; this is effected by discharging the Air alone by one Bucket and the Water alone by another, or by an Eduction Pipe of twenty eight feet in length."

Upon this we observe that no advantage has been derived in practice either as to the saving of fuel or increase of power, but rather the reverse, and that from its complicated and defective construction it has been found so difficult to keep in order that the Patentee has for himself abandoned the use of it. The principle also is not entirely new, the same having been partially practised by Boulton & Watt several years ago and a Patent was taken out by a Mr Jonathan Hornblower in 1781 for the variety with the long Eduction Pipe.

2nd. The second invention is intitled and described as follows "an improved Method of packing the Cylinder lid, Stuffing boxes for valves &c. I cause the removable parts of each to come in immediate contact with one another this is effected by placing the necessary packing on the upper side of the Cylinder lid (instead of betwixt the Cylinder & Ld as formerly) which prevents the Piston rod receiving friction from any oblique pressure by the Cylinder Lid being screwed down more upon one side than the other, which is unavoidably the case in all other methods of packing Lids, valves Stuffing boxes and other similar things where great accuracy is required."

Upon this it need only be remarked that the inconvenience stated has not been observed to take place in Cylinder Lids constructed in Mr Watts usual manner and that the method described can only be considered as an expensive alteration not possessing the distinctive novelty and utility necessary to constitute a claim to an exclusive privilege. (i.e a patent).

3rd. The third invention is stated to be " An improvement in the valves and Working Gear, the principle of which consists in the advantageous form in which they are made, and in the two uppermost valves being inverted, and the communication by the Valve rods passing through reservoirs of oil or liquid Matter, which will prevent the Air getting into the Engine to impair the Vacuum; the principle of the Cylinder lid is here applied in fixing the Stuffing boxes for the Valves and Valve box lids which is of the utmost importance to their operating well as the oblique pressure (which cannot be prevented in the old method) throws the valves out of their seats and renders them proportionally imperfect .

Upon this we remark that there is no novelty in the form of the valves they being round as usual and that those delineated in the drawing which accompanies the Specification are not inverted as described, although their stems (which he calls the valve rods) are: but in this inversion of the stems there is no novelty, the same having been practised for years by Boulton & Watt.

The like observation applies to the supposed invention of the Valve Rod passing through reservoirs of oil, this having been constantly done in the piston rods of all Boulton & Watt 's Engines and in the valve rods of their Engine at Chelsea for the last 15 years.

There is also no novelty in the Rod of one valve passing through the rod of the other, which he has represented in the drawing annexed to the Specification, but not described, and which is the Characteristic part of this construction. The merit of that Invention belongs to Mr Murdock an Agent of Boulton Watt & Co and was first practised by them in an Engine to work Mr Boulton's Mint in November 1798 and set to work early in the ensuing year and again in an Engine erected by them for Mr Symonds of Reading in September 1799. The whole construction of the Valves & Nozzles described and drawn by Murray is very nearly an exact representation of those made for Mr Symonds and proof will be given that the plan was communicated to Murray by a Workman of Boulton Watt & Co as likewise that Murray himself declared in May1799 that he knew they were then making valves upon this construction and had a spy in their Works who gave him intelligence of all they were doing.

4ᵗʰ.The fourth invention is specified to be "A new principle of opening and shutting the Valves by a circular Motion (instead of the Plugs and Hand Gear now in use) which prevents the Engine making a false stroke, or turning the wrong way round, by which means a Steam Engine is rendered as steady in its motion as a Water Wheel.

This principle of opening and shutting the Valves by a circular Motion is also presumed to be stolen from Boulton Watt & Co by the means of the same Agent. It was applied by Mr Murdock to the Engine which gives Motion to their Works at Soho Foundry upon the 9ᵗʰ December1800, where it has been used ever since and in a much better method than described by this Patentee which has probably been varied to conceal the theft. Boulton Watt & Co have since the above period been in the practice of Manufacturing Engines with this principle applied in various forms.

5ᵗʰ. The fifth Invention purports to be " A new method of connecting the Piston rod to the Parallel Motion as described in fig 10" and afterwards more particularly in the Specification.

This is an alteration of a most trivial nature being only Mr Watt's invention of a cone at the lower end of the Piston rod, applied to the upper end also; and it is proper to observe that the patentee has not described how he gets the rod with this double cone upon it through the hole of the piston. It is not conceived that this Method possesses any advantage over the common ones, or could be the subject of a patent, even if it were so described as to make the same practicable.

6ᵗʰ.The sixth and last Invention is intitled "An improved Method of constructing fire places for consuming the smoak arising from Steam Engines and other fires".

The form of the fire place is partly copied from Mr Watts patent for a similar Invention in 1785 which has since been described in various publications and practised by Boulton Watt & Co with some variations in various instances. It has also been the subject of another patent by Mr Roberton of Glasgow. In Murray's drawing annexed to this Specification is an exact representation of the general construction of a fire place erected at Soho Foundry in Sept 1800 and which has been in actual use there ever since. It may reasonably be presumed that a knowledge of it was obtained by the same means, as of the Valves and Circular Motion. The charging Box, its Rack, and Pinion and the hinging of the Grate are insignificant additions to the former construction, the utility of either of which is very questionable.

It may generally be observed of this Patent, that whatever is useful in it is not new; and whatever is new is not useful.

The 1st, 2nd, 5th & 6th pretended Improvements are in fact none at all, and the 3rd & 4th were previously practised by Boulton Watt & Co but not considered by them to possess enough of novelty to make them the Subject of Patents.

The application for the Scire facias is made with the view of restoring the credit of the latter improvements (the 3rd & 4th) to those to whom it justly belongs and making known the means by which those improvements have fallen into the hands of this pretended Inventor" [12].

At this point, the end of June, Boulton and Watt were planning to call six witnesses:

John Southern who ran the drawing office and was a senior manager at Soho.

John Rennie the eminent engineer who had formerly worked for Boulton and Watt.

James Lawson, the Boulton and Watt agent, who, as has been seen, had been at the sharp end of looking into Matthew Murray's dealings.

George Lee a mill owner from Manchester, who was a customer for steam engines and a friend of Boulton and Watt.

Gavin McMurdo, who had worked for Boulton and Watt in Leeds and knew Matthew Murray, to the extent that Boulton and Watt had been convinced he would go and work for Murray in 1799.

Francis Crowden an ex-employee of Boulton and Watt, whose position has been explained in James Lawson's letters above.

It is interesting that although one of the patents that Boulton and Watt claim Matthew Murray infringed was that of William Murdock (number 2340 of 1799), it does not appear that Murdock was considered to be fielded as a witness or made a party to the action. That Murdock was considered too lowly in the organisation cannot be as he was far more senior to Gavin McMurdo and Crowden. Perhaps James Lawson's letter contemplating what had really passed between William Murdock and Matthew Murray on that well-known visit to Leeds caused Boulton and Watt to either wonder at his reliability in this instance, or possibly William Murdock had declined to appear. It has already been noted that on the 14th of July 1799 James Watt junior had stated in a letter (See Chapter 5) to M R Boulton that Matthew Murray had stopped Murdock's patent by a caveat, and this was the patent Matthew Murray was now accused of copying.

While meeting the party line John Southern in a letter written on the 28th June 1803 is somewhat more cautious than his employers and this was the man called as a witness on 11 out of 17 points;

"I had yours of the 25th.

Theoretically I think there is no objection to Murray's air pump, or we should not have put into effect an apparatus for accomplishing the best part of his object, which was done at Mr Lees in 97, but it did not appear to be attended with any desirable advantage & Mr Lee laid it aside.

Practically there is great difficulty in making air-tight the air bucket in Murray's pump, for as he professes not to take out water by that pump, there will not any lodge on the bucket to keep it tight, as in the case of Mr Watts method, consequently however tight the bucket may be when new packed, it is well known that it will soon be leaky enough by use to permit considerable quantities of air past it. When that happens there is no other means of extricating it, but by the water-pump; but as that is made large enough only for the extraction of the water (so I understand the specification though it is not expressed) the air accumulating will impede & ultimately stop the engine. Perhaps under those circumstances they would inject less water by way of making the water pump act in some degree as an air pump – but what they would thus gain would be very trifling, and the increase of heat of the water would lessen the power of the engine very materially. If they should discover the cause of the bad performance they would probably increase the size of the water pump and finding advantage from the first increase, would make it finally as large as the air pump in Mr Watts engine, & let their pretended air pump lie idle.

He says in the first part of his specification "or by an eduction pipe of <u>28 feet long</u>", when he comes to the particular description of this he says "an eduction pipe 30 feet long or upwards". 28 feet would command but a sorry vacuum.

I think the mode of packing the cylinder lid has merit, and though it is found that the common method <u>with care</u> does extremely well, and in fact needs not be mended, yet the other is new as far as I know, but will not do without care anymore than the old method. For in his method there is some danger of air leaks, which in the old method does not occur, I mean that there is a cause of apprehension on this head of his which does not occur in the old method.

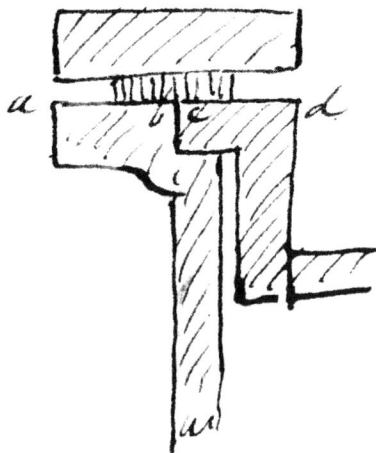

25. Drawing of Murray's cylinder packing by J Southern.

The latter is a single joint, has a double one to keep tight. Thus admitting the groove in the cylinder to lie either not so deep or deeper than the lid is thick, the surfaces ab, and cd, would not coincide and there would be danger that air could leak in, either from one side or other of the packing. I apprehend it would be very difficult to turn them of the same dimensions. The idea is however ingenious, and let the devil have his due.

(Generally I concur with your observations upon his patent, but there are some parts which in a court of justice I should not be willing to say. Not expecting, when I read them, that I was likely to be called to give evidence in the matter, I did not particularly notify these parts, and cannot now specify what they are.)

He states his [3]rd improvement to "consist in the advantageous <u>form</u> in which his valves and working gear are made. (There is nothing new in the <u>form</u> of the valves. There is in the working gear). "In the two uppermost valves being inverted" (They are not inverted, their stems only are inverted, which is not new) .Oil or other fluid if not 'liquid' matter has long been used.

I do not see why you change the term nozzle for steam box, and if there be no sufficient reason, there appears to me to be a weakness in adopting new terms not more significant. I shall expect to see <u>vacuum box</u> substituted for <u>Steam Box</u>. I (reserving old names) say that the improvement of the socket valve consists first in the stem of valve working through that of the other which fits it, that thereby but one external opening into each set of nozzles becomes necessary, and the air leaks are consequently diminished. The fairer action of the valves, is not attributable to this improvement, but to their being pulled open by their own spindle directly through their stuffing box – a practice known & in use many years – I know not what else can be said of them in the style of improvements.

It appears difficult to me to say what is an improvement in the circular mode of opening the valves, unless it be simplifying appearances and rendering certain 'the opening & shutting of the valves in due time. An apparatus was made & I believe applied at the Albion Mill for the "<u>regulation</u>" of the <u>times</u> of opening the valves by a circular motion. This was done by disengaging the catches by the rotative shaft, though the valves were opened by the means of weights instead of the direct force of the engine – which may be added to the number of improvements of the new method. I will attend to the dates in the books, but it may be necessary to have proofs of times from those that made the memorandums in the books" [13].

The other W Murdock was eventually interviewed probably by James Watt junior, who was in Leeds, and supplied the following list of 'Murray Engines' dated the 3rd July 1803:

"*Sam'l Owen* (a former Boulton and Watt employee working as an engine erector for Fenton Murray and Wood) – *at Milton Mill near Huddersfield putting up a five Horse [power].*
John Appleston. (Heptonstall). *Doncaster a partner in the flax mill - Fishergate No1 a pump changed.*
Carrs Dyers Swinegate Leeds was the 2nd & is still working & is the first with socket nozzles erected abt x Mar 1801. 10 Horse said to be = 6 only.
Also Wilson's Buckram makers –Water Lane Holbeck a10 Horse [com[n]* at]*
Brend was sent over by Murray & here men – abt a month after Mr Yetts Mill was burnt" [14].

Sometime in the early part of July, the letter is not dated, Boulton and Watt had an intimation of Matthew Murray's policy from their friend Benjamin Gott:

"*Pritchard has seen two of M[urray]'s men they both were of opinion M[urray] would not bring the matter to issue but each separately declared they had been kept in the dark during the process of litigation & that they only guessed at what was going on – but still they appeared confident he would not stand trial – it was well known you was here yesterday to the men & also to M[urray] -- for when P[ritchard] -- was beg'ing to sound Wood on the subject M[urray] -- came up – asked P[ritchard]-- if he had seen Mr Watt Jnr yesterday for that he was at such an Inn in Leeds & he prevented the chance of further enquiry or conversation. As his social disposition may perhaps have drawn him away this afternoon I have sent P[ritchard] again with enquiries respecting a boiler &c ye result shall be added to this – I should have added that P[ritchard] saw two men who had formerly worked at Soho & he could not find from them any preparations for trial were going on – however I would not be unguarded by such chance information. ¼ past [8] o'clock Pritchard is again returned after an unsuccessful attempt at conversation with Wood who was engaged with Mr Fenton – but he saw their principal sand moulder & their engineer – both of whom said they were of opinion always that the new valves & central motion for hand gear originated at Soho & they did not think M[urray] would contest the point. He also saw a man who formerly was in your employ & who expressed great regret at not having seen you – he said he would have spent or walked a day to see you – I think P[ritchard] may extract from this man how yr information comes from Soho- you shall hear from me again – in the mean time I think so inexperienced a man as I am in such matters need not caution you not to be off your guard so near ye day of battle – but rather have your witnesses at double expenses & inconvenience than have anything to claim – You are not generals who have never been defeated & as everybody must know you have right on your side – be sure you impress that by ye good evidence as you can produce on 7 July"* [15].

The second item on the list of engines and work from the other W Murdock, probably led to Boulton and Watt gaining a further witness as is explained in James Watt junior's letter to M R Boulton dated 4th July 1803:

"Finding I had two spare days upon my hands before McCandlish could receive his answer, I determined to reconnoitre the enemys proceedings and have accordingly been at Leeds, where I learnt that Murray was just returned after 6 weeks absence in London; but I could not ascertain whether he was determined to stand trial or not. I found that one of his Air Pumps had been erected for a Mr Applestan (corrected to Heptonstall) of Doncaster and he not finding it to answer had obliged Murray to pull it down & substitute one of ours. I have sent Upton (the solicitor in Leeds) over to see him with a Subpoena to serve upon him in case he finds him likely to prove a useful witness...... which would render it impossible to be in town (London) by Thursday, on which day I see from Weston's letter to you, it is probable the trial may come in. I have no letter from Weston, although it was positively fixed he should inform me of the precise day of trial and I see he has committed a blunder, or at least acted contrary to what was settled, in including the witnesses here in the same Subpoena with Southern. We had agreed that I should serve the witnesses here (& for that purpose I took the Subpoenas with me) and that Southern & Dudgeon should be included in one Subpoena to be forwarded by you to Newcastle as soon as you had done the needful with it at Soho. I much fear that other blunders will arise and am now reduced to the alternative of leaving this business here unfinished or risking the loss of the trial. I feared from the beginning that something of this sort would happen, but I could not calculate upon Weston's neglect in writing. As it is, I have determined to proceed to London with the Witnesses, so as to be there before the Courts open on Thursday and to make the best arrangements I can with ...." [16].

Upton immediately acted and presumably instructed a local solicitor in Doncaster named Standish who appears to have actually taken the statement of John Heptonstall which survives dated July 4th, along with Mr Standish's letter to Upton and Upton's letter advising that it has been arranged to send Mr Heptonstall to London by coach on the 6th July:

"Witness Statement of Mr John Heptonstall of Fishergate.
The reason I suppose of Mr Murray's double Vacuum Patent Air Pump not answering was a part from the Water that he made use on to condense the Steam and a part from the resistance of the air. He took the Water out of the Condenser and Compress'd it down a pipe of 12 ins Diameter by a Piston & Valve fixed with a rod to the bottom of the Air Pump Piston – it was continually out of repair so that we was obliged to pull it up again – The Air pump nock'd & drove every valve about it to

132

Pieces, which I suppose was from the air being compress'd up so small a pipe from the underside of the Piston on to the top side to be delivered out of a valve at the top of the Air Pump it would be a very good thing if it could be made to answ'r. Water never will be compressed- the Present Air Pump acts very well which is principally upon Bolton & Watts plan it lifts the water out of the Mouth at the upper end of the Air Pump by a large round valve fixed on the top of the Piston – it only forms a single vacuum.

July 4th 1803.

Dear Sirs   Doncaster July 4th 1803
*   I enclose you a copy of the report I have procured from Mr John Heppenstall of Fishergate he wishes me to reserve his own paper that he may refer to it, if you need further information from him. Your present want of prompt answer allowed him no sufficient time to make Drafts of the machinery but which he says he shall be ready to do if requested and to supply all information in his power. I told him he might perhaps be further applied to and that I should have your instructions at any rate to compensate him for his present and any further trouble.*
Signed   H. Standish.

Dear Sir
*   Before I could receive your letter I was under ye necessity of forwarding the Subpoena to Doncaster by the post to our Friend there who will serve it early in the morning & most probably he will be sent off tomorrow by the Rockingham Coach from Leeds or the Highflyer from York both which pass thro' Doncaster ab't 12 o'clock at noon & arrive in London the following day ab't 3 in the afternoon, the former goes to the Bull & Mouth. I have sent you on ye other side a copy of ye report from ye witness with my friends letter. The Evidence seems precisely to meet the point & I am happy that we have so far succeeded in our application & heartily wish you success. If you had desired the Wit[ness] not to be Subpoenaed I sh'd have sent off an Express this Evening countermand his attendance. I sent the Gent 10£ note on account of his expenses & gave him your address in London Street & have no doubt but he will apply to you immediately.*
It may however be proper to have a person at each Coach Office to conduct ye Wit[ness] to Westons if they wish to see him immediately.
I am Dr Sir   Yours faithfully   Thos Everard Upton Leeds 5 July 1803" [17].

James Watt junior's frustration with the whole way in which matters were proceeding was no doubt increased by Thos. Upton's next letter of July 7th which advised that they never sent for their letters in the evening or they would have had his letter the previous day. However Thos. Upton advised that they had subpoenaed Dixon (an ex Boulton and Watt employee at Fenton Murray and Woods) and he would be on the coach that evening and had been directed to go straight to Weston's. He received 2 guineas expenses and 3 guineas for the coach. When James Watt junior arrived in London he was advised that the trial had been deferred until Monday.

On the 8th July John Heptonstall supported his witness statement by making a drawing of Matthew Murray's air pump, which correlates exceedingly well with the drawing in the patent.

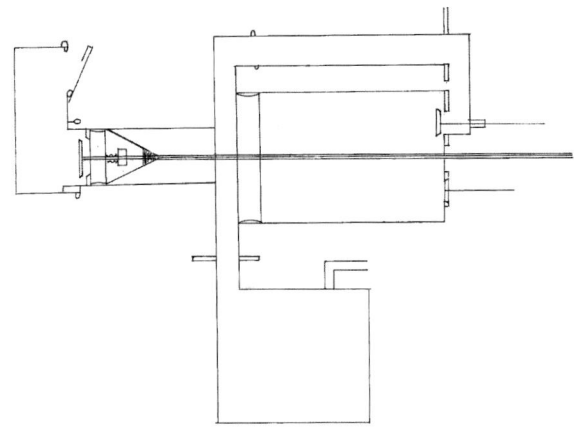

26. Heptonstalls drawing of Murray's Air Pump (redrawn). [18].

The same day the proceedings came to a head as James Watt junior advised M R Boulton:

"After I wrote to you yesterday, we received intelligence that the Court had determined to bring on the trial on Saturday, and that it would be the first cause upon that day. Weston & myself immediately went in search of Murray's Solicitors Bleasdale & Alexander to strike out the Jury and make some agreement as to Evidence. Not finding them at home, we waited upon one of the Partners early this morning & learnt that he had received no instructions from his client to enable him to prepare a Case for Counsel and did not know whether he was in town or not; that he had only seen him twice during six weeks he had been there & he was drunk both times. He added that if he did not receive instructions to proceed by the morning post, he should take out a summons to abandon the Plea & promised to make up his mind by 12 o'clock and would then save us further trouble, if we would call. In the mean time we struck out the Jury, that no time might be lost in issuing the summons for them if necessary. At 12 o'clock we did not find our Gentleman at his appointment, but after some search through the Inns of Court, we met with him & he said he had received his Clients orders to give up the contest and immediately drew up a note purporting that he agreed to let judgement go against the Letters Patent upon the Scire Facias. So that they may be considered as set aside, and the affair ingloriously terminated" [19].

So Matthew Murray had decided not to fight. Boulton and Watt were deprived of their day in court, but had the satisfaction of overturning Matthew Murray's patent. Benjamin Gott had advised correctly. As stated above Matthew Murray knew that his air pump did not work, having already had to change one in Newcastle, and then the one on John Heptonstall's engine, he had already stopped using his stated method for constructing fire places to reduce smoke and within six months or so of the patent had agreed a license to produce the system of John Roberton of Glasgow. If the other items were improved Soho valves, he may have thought it would be difficult to distinguish them in a Court of Law, he had had the experience with John Marshall of trying to defend their flax spinning apparatus against John Kendrew in 1794, when there was strong evidence that they had immeasurably improved and changed the machinery. Matthew Murray was no doubt also under intense pressure from James Fenton and David Wood to concentrate on the business, and as the action had been brought against him to pay any defence costs from his money and not the firms.

Boulton and Watt had won their point but they had not been able to humiliate Matthew Murray in court and James Watt junior's letter continued:

"J Weston thinks they are not aware of the error in our pleading, & in their own specification, so that we shall keep our council upon the point. We shall draw up a paragraph for all the papers, stating the repeal of the Letters patent and the Evidence we should have produced. This shall be my business tomorrow, today being devoted to the celebration of our victory, such as it is.

All the Witnesses are arrived, Dudgeon [included] & have given excellent evidence and the Doncaster Gentleman Mr Heptonstall is still better. He seems to relish the business very well and I expect it will end in his giving us an order for an Engine. When I say that all the witnesses are arrived, I should except Dixon the moulder who has not made his appearance, nor have I heard anything upon the subject from Nicholson & Upton" [19].

On the 9th Of July James Watt junior again wrote of their plans:

"I could not meet with Weston so as to get our advertisement prepared to send you by this night's post, but hope to get it into the Morning Chronicle, Times, Star, Courier, & Traveller of Monday. I shall also take care to have copies sent off for the Manchester, Leeds, Sheffield, Glasgow & Newcastle Papers. You will of course take care of the Birmingham & any others that appear to you necessary.

I have sent off Gavin & Crowden by tonight's Coach. Dixon arrived this Morning, but without examining him, I told him that as his master had declined Trial, we should reserve the important evidence he had it in his power to give respecting his own seduction & that of our other men for another opportunity" [20].

M R Boulton suggested that:

"In addition to the newspaper paragraphs it would be useful to get your narrative inserted in the Monthly & Philosophical Magazine & Repertory as these publications are more likely to fall into the hands of the mechanical & ingenious part of the public than the newspapers, especially among foreigners.

It was always to be apprehended that Murray would cede the field without exposing himself in the conflict & no doubt he will now endeavour to persuade his admirers that he was deterred from proceeding with the suite by the length of our purse. In other respects I don't think your Victory is the less glorious from the circumstances of its being a bloodless one. To defeat the enemy without sacrificing the life of a Roman Citizen was I believe always considered an additional claim to the honour of a triumph" [21].

However even here matters were not to go Boulton and Watt's way for some of the Newspapers objected to the wording and others to the length of the original advertisement. The Morning Chronicle [a London Daily paper] declined them completely on the grounds it only published Law Reports written by its own staff. They had to reconsider the wording and even this reduced version was cut by most of the papers [22].

The wording of the Boulton and Watt advert, which as indicated by the response from the Morning Chronicle, appeared to be a law report, was to be as follows:

"The King against Murray

This was a cause instituted by his Majesty's Attorney General to repeal by writ of Scire Facias, a Patent granted in August 1801 to the Defendant Matthew Murray, an Engine maker at Leeds and a Partner in the house of Fenton Murray & Wood of that place, "for an improved method of constructing the Airpump and sundry other parts belonging to a Steam Engine by which (as he alleged) a considerable saving would be made in the consumption of fuel and an increased power obtained".

The Prosecution was carried on at the instance of Messrs Boulton Watt & Co of Soho, whose principal view in the Repeal of the Patent was to expose the Defendants Conduct, in obtaining by the seduction of their Workmen a knowledge of the Improvements in the modes of constructing and manufacturing Steam Engines from time to time invented and carried into effect at their works and founding a Patent thereon, thus claiming the merit and assuming to himself an exclusive right to the reward due to the labour and ingenuity of others.

Messrs Boulton Watt & Co were prepared to shew by the most respectable testimony that those Articles of the Defendants Specification which had a claim to merit and were apparently new, originated and were practised at their works for some considerable time before the date of his Patent, and that he had surreptitiously become acquainted therewith. They were equally prepared to prove that the other articles of the Defendants Specification were in themselves impracticable, or of too trivial a nature to become the objects of a Patent, and in short to show that whatever was useful in this Patent was not new, and that whatever was new, was not useful.

The Cause was appointed for Trial before Lord Ellenborough and a special Jury of the County of Middlesex on Saturday the 9th of July; but on the preceding day, the Defendant withdrew his plea and consented that Judgement should be entered against him, which has accordingly been done, and the Patent thereby cancelled and repealed" [23].

And to all this Matthew Murray unrepentant issued his well-known response which first appeared in the Leeds newspapers dated 20th July 1803:

"PATENT STEAM-ENGINE MANUFACTORY
Leeds July 20th 1803
I feel myself called upon to vindicate my Character as an Engineer, against a foul Insinuation in a Paragraph inserted in the Newspapers of last week, I suppose by Messrs Boulton, Watt, and Co; they assert, "That every Improvement which was really new and useful, and deserving a Patent, in the one which I obtained in the year 1801, was invented and practised at their Works, and that I surreptitiously obtained a Knowledge thereof from some of their Workmen". I do positively deny, that I ever got the least Hint of the Improvements in Question from any one; indeed a little Observation is sufficient to refute their assertion; If I knew that these Improvements had been invented and practised at Soho, I must have been deficient in Common Sense, as well as Honesty, to attempt to obtain a Patent for what I knew I could not hold. Had they used my Inventions in the manner described in that patent, prior to the date thereof, they certainly would have practised them in the Engines they made before that Period, or taken out a Patent for the Improvements themselves. The Reason of my not defending the Patent was, not from any Fear of losing the Trial as they seem to insinuate, but that I did not think proper to defend it with such Expence as I should most probably have incurred , in contending with such rich and powerful Opponents as Messrs Boulton, Watt, and Co.

136

But had I been guilty of obtaining a Knowledge of their Improvements, if they had any, (but I do not believe they have made any worth Notice since Mr Watt, Senior, retired from the management) it would only have been a return in kind. Mr Storey, manager of their Foundry, and Wm Murdock, Superintendant of the Workmen at Soho, some Time back visited our Works at Leeds, and from their assuring us of Messrs Boulton, Watt, and Co.'s friendly disposition, were admitted into every Part of the Manufactory, by Mr Wood and myself; they were even permitted to take Patterns and Specimens of our Workmanship, and we know that upon their return to Soho, many of our Improvements were immediately adopted, and the Engines made by them after that time, were in part constructed on our Plans. Mr Murdock upon taking his leave of us, expressed a Wish, that as they and we were certainly the best Engine Makers in the Kingdom, we should always be upon good Terms, and that if ever I should go to Soho, they would be very glad to shew me all their Works. I did go to Soho, and was refused admittance into their Manufactory of Steam Engines.

But the World, I believe, cares very little about Messrs Boulton, Watt, and Co. stealing my Inventions, or my stealing theirs; what People want of us, are good Engines. and I am confident that I can make good Ones; and as they hint that no one can do that but themselves, I am willing to end this dispute, for the good of the Public, in a similar mode to one they proposed to Mr Hornblower, with whom they had a dispute some years ago, when Mr Wilson, their Agent in Cornwall, gave him a Challenge for £1000 that Messrs Boulton AND Watt, would produce an Engine superior to that Mr Hornblower had erected at Tincroft Mine, this Challenge Mr H. did not accept.

Now I offer (by way of Trial and Proof of Ingenuity and Workmanship) to make an Engine of one Horse Power, against any one of the same Power, to be contrived and made by Messrs Boulton, Watt, and Co, and I do further offer to deposit in the hands of any Banker in London, if they will do the same, One Hundred Guineas; to become the Property of the Party whose Engine is declared to be most perfect and useful, by twelve Practical Engine-makers, six to be chosen by Boulton, Watt, and Co. and six by me,

MATTHEW MURRAY".

The exchange of vitriol and challenge was a lot wider than has been assumed and so far the following newspapers and journals have been identified; in many cases the Boulton and Watt advert is abbreviated and the words (but not the meaning) vary in some of Matthew Murray's responses:

Newspaper	B&W advert	MM response
London Papers		
The Courier	13th July	27th July
The Star	13th July	25th July
The Times	13th July	none
Provincial Papers		
Newcastle Courant	30th July	23rd July
Manchester Mercury	19th July	26th July
Leeds Mercury	16th July	23rd July
Aris's Birm'm Gazette	12th July	8th Aug
Leeds Intelligencer	18th July	25th July

The Repertory of Arts, Manufacturers, and Agriculture Volume 111 (3) Second Series 1803 carries a reduced version of the B & W advertisement.

Relations between the two firms never recovered and whilst it never sank to such a low position as it reached in 1803, they kept an eye on each other's activities, although not with the intensity of the last few years. Looking back it can only be said that neither side behaved well, but the act which seems to have hardened the war was the visit of Storey and Murdock to Murray and Wood. Matthew Murray never forgave and never forgot and it has to be said he probably considered Boulton and Watt fair game after that. As James Lawson reported to Boulton and Watt after sharing a coach with Matthew Murray from Newcastle to Leeds in June 1802, Matthew Murray "had declared his determined hostility to us"

Needless to say, although there were some remarks passed by M R Boulton as to how to respond to Matthew Murray, Boulton and Watt never responded to the challenge laid down by Matthew Murray.

Notes

1. Boulton and Watt Papers Letter Book MS3147/3/99.
2. John Marshall, the Benyon brothers and Charles Bage had agreed not to renew their partnership at the expiry of the term of the agreement and were in the process of negotiating a split in the partnership assets. Marshall purchased the Shrewsbury mill (Ditherington), while the Benyons with Bage were to build new mills in both Shrewsbury and Leeds.
3. Boulton and Watt Papers MS3147/3/52/11.
4. The original of this letter does not appear to have survived, however the extract quoted is preserved in trial evidence at Boulton and Watt Papers MS3147/6/61/5.
5. Boulton and Watt Papers MS3147/3/52/14.
6. Boulton and Watt Papers MS3147/2/61/6.
7. Boulton and Watt Papers MS3147/3/52/15a.
8. Boulton and Watt Papers MS3147/3/52/15.
9. Boulton and Watt Papers MS3147/3/41/6.
10. Boulton and Watt Papers MS3147/3/52/17.
11. Boulton and Watt Papers MS3147/2/61/7.
12. Boulton and Watt Papers MS3147/2/61/4.
13. Boulton and Watt Papers MS3147/2/61/8.
14. Boulton and Watt Papers MS3147/2/61/10.
15. Boulton and Watt Papers MS3147/2/61/11.
16. Boulton and Watt Papers MS3147/3/52/21.
17. Boulton and Watt Papers MS3147/2/61/13, covers all three documents.
18. Boulton and Watt Papers MS3147/2/61/14.
19. Boulton and Watt Papers MS3147/3/52/23.
20. Boulton and Watt Papers MS3147/3/52/24.
21. Boulton and Watt Papers MS3147/3/41/9.
22. Boulton and Watt Papers MS3147/3/52/25a.
23. Boulton and Watt Papers MS3147/3/52/25b.

Chapter 7

Business Progresses

While Boulton and Watt and Fenton Murray and Wood were exchanging words in the press, the latter also placed an advert in the Newcastle Courant on the 16th July 1803:

"PATENT STEAM ENGINE MANUFACTORY, LEEDS, YORKSHIRE
FENTON, MURRAY, and WOOD, beg leave to inform their Friends, and the Public in general, that they have on Hand, a number of portable Steam Engines, from one Horse Power to six, fixed on Iron Frames, ready for Sale, which are applicable to the various Purposes of Breweries, Distilleries, Wood and Iron Turning, Sawing Marble, Pumping Water, Driving Piles, Threshing Machines, &c, &c The above are of the compleatest Workmanship, and fitted up in the neatest Manner, and will be sold on reasonable Terms.
N.B. Application may be made to Mr Stainburn, their Agent, at the Indian Kings, Quayside, Newcastle. July 13, 1803".

The following month the partnership entered into a new mortgage arrangement with Becketts their Leeds bankers, for part of the Shoulder of Mutton Close.

In September Gregory Watt visited the Tyne Iron Works to whom Boulton and Watt had recently delivered a 16 H.P engine and from whom they hoped for a further order. He described to his brother the other engines working on the site, which included a second hand 10 H.P. Murray engine, which had been purchased at Hull. It was used for turning, and there was an intention to set up a boring mill and use this engine to work it. In another letter that month he reported the winning of the coal at Jarrow and the celebrations that followed the opening ceremonies:

"The only recent novelty has been the winning of the coal at Jarrow below Hebburn. This was affected for less than half the estimated costs. Mr Temple the proprietor was so elated that he invited the whole country to a grand carousel, by advertisement & proclamation. About 1500 accepted his invitation and a most extraordinary excess of uproar and drunkeness ensued. The water is drawn by a double 63 of Murrays, who [appears] to have been sent for to officiate as high priest of these bacchanalian orgies"[1].

The longer report in Sykes [2] records a noteworthy day with schools being opened, great celebrations and nearly 1000 people sitting down to dinner after which there were a great number of loyal and applicable toasts, a description at great variance with the sour note struck by Gregory Watt. The important point was that the business Murray had won whilst in competition with Boulton and Watt at Jarrow had been completed and the engine was working in the Northumberland and Durham Coalfield.

The business of the partnership was developing so well that they were looking for additional capital. An advertisement appeared in The Morning Post (a

London paper) for a partner with a capital of £5000 to invest in 'A Steam Engine Manufactory and Iron Foundry' in the North of England [3]. Particulars could be had on application to Sage, Rawdon and Jennings at No 19 Cheapside. Sage Rawdon and Jennings were London Linen Merchants but were also in partnership with one James Dickenson, trading as James Dickenson and Co, as weavers, warehousemen, and Moroccan leather manufacturers [4]. James Fenton had given his occupation, on the land deed when he joined Fenton Murray and Wood, as a tanner from Walworth (now a suburb of London near Lambeth) in Surrey. It therefore is feasible that he had traded with the James Dickenson and Co as a supplier of leather, and had got to know the partners well. Sage Rawdon and Jennings subsequently held the mortgage on the Fenton Murray and Wood's premises and assets.

In October Fenton Murray & Wood were mortgaging land, which they had purchased from John Marshall, to Joshua Shaw, another Leeds Iron Founder [5].

One of the districts where the business was expanding was in the Northern Coalfields, and it is evident that they were beginning to deal with John Watson the viewer at Willington Colliery, who was subsequently to introduce the Murray – Blenkinsop locomotives to Tyneside. In the Watson papers there is a letter written by John Watson, dated 15th November 1803, although unfortunately the name of the correspondent was not on the surviving copy:

"I do certainly recommend those engines generally adopted at present in our Coal Works which are made by Messrs Fenton Murray & Wood Leeds or those of Messrs Bolton Watt & Co Soho London, the principles upon which they work are nearly the same, altho they will appear to you rather high (expensive)" [6].

Watson supplied his correspondent with prices for two engines to be put on board ship. The first was to raise a column of water 48 fathoms 11 inches diameter and to work 14 strokes per minute each stroke being 7 foot 9 inches for which Fenton Murray & Wood's charge for the metal parts, steam pipes, boiler feed and connections for the spears and beams would be £850 as against a charge of £1019 by Boulton & Watt.

The second engine, which was either for driving machinery or hauling trucks of coal or minerals (winding), or to be used for pumping where it would have been required to raise a column of water 23 fathoms and 18 inches diameter and 26 fathoms at 17 inches diameter with a 7 foot stroke going at 12 strokes per minute for which the metallic engine parts from Fenton Murray & Wood were £1817 (Boulton & Watt £1800). It is tempting to identify Watson's correspondent as Baron Edelcrantz of Sweden who was in England at the time, and ended up purchasing engines from Fenton Murray & Wood for shipping back to Sweden, but it is speculation.

In December the firm is listed by the Leeds Intelligencer as contributing £30 to the Leeds Volunteer Infantry [7].

Sometime in the latter part of 1803, the required addition was made to the partnership. His name was William Lister, and an indenture dated 31 December 1803 / 1 January 1804 shows him taking up a share in the assets of the partnership. It is considered that William Lister's involvement was mainly financial, as although his name was sometimes added into the partnership name, and appears on contracts, it was not used very often.

January 3rd 1804 is the date of the earliest surviving letter written to Simon Goodrich by Matthew Murray on behalf of Fenton & Co (another common designation for the partnership). It was written in response to a letter from Goodrich dated the 31st December, which has not survived. The engine described was a 2 H.P. one adapted for pumping water and of a simple design:

"This Engine is completely fitted up in a Cast Iron Cistern, & needs no other fixing than bolting upon the Foundation it is to stand on & requires no Wood work whatever, the Boiler to be set in brick as in all other cases. The whole weight will be nearly two tons .... This Engine is capable of pumping with a 12-inch Pump 120 Gallons of Water 20 Ft high Pr. Minute, and more or less in proportion to the height required. The price of this Engine including the Boiler, & Door Frame & Grate Barrs for Fire Place is £175, deliv'd at Leeds .... we provide an experienced Engineer to put it up whose Wages, Board & Travelling expenses will be charged to you" [8].

No record has been found of this engine having been purchased, but the description appears to match Murray's Portable Engine as described in Nicholson's Philosophical Journal in 1805 (see below).

About this time Fenton Murray & Wood were designing a 100 H.P. pumping engine and Engine house for John Watson at Willington Colliery. A drawing survives of the Engine House endorsed *"Fenton Murray & Wood Leeds 8th February 1804"*. A later drawing shows the Engine in the Engine House, and the engine houses can be seen to be identical in the two drawings, enabling the full drawing to be identified.

John Marshall was still conducting experiments to improve his machinery, and still discussing these ways forward with Matthew Murray. In experiment 82 in March 1804, he recorded that:

"Murray's idea of the use of twisting our slivers was that the twist supplied the place of elasticity in our material, & brought those fibres which were not in action regularly forwards. It appears however that no advantage could arise from that circumstance for the fibres of flax have too many arms or branches that they are only too apt to entangle with one another, & either be brought forward before their proper time, or have their branches torn off. The more liberty it can have in the part of drawing & the more perfect the operation will be" [10].

27. Murray's 1805 100 H.P. pumping engine for Willington Colliery [9].

John Marshall then tried setting up the second drawing frame to put no twist in the sliver, but it did not work as the slivers entangled and tended to lap around the back rollers, so he reintroduced the twist at a lesser level and reduced the breadth of the slivers and achieved an improved result.

In April the Leckeys was re - mortgaged to William Hargrave (joiner), James Longley (bricklayer), John Cave (bricklayer), who presumably took over the mortgage from John Marshall. So that in the Royal Exchange Policy 210457 of April 29th the proprietors are listed as Fenton, Murray and Wood, and the mortgagees as Hargrave and Cave for:

> *"Iron Foundry with 4 rooms above (model warehouse).*
> *Steam engine house and steam engine*
> *4 storey building used as Joiners shop*
> *Building with 4 Counting Houses and a cottage*
> *Loam Foundry*
> *Brass Foundry and model chamber*
> *Building used for machine shops and model chamber (near fireproof)*
> *Another engine house, boiler makers shop and forge and engine"* [11].

Another policy 210456 existed for the circular building with the same proprietors but mortgaged with Beckett's bank. The rotunda was described as being used for workshops, to be heated by steam but warranted that there was *"no Steam Engine in, adj' to or coming with either of the above buildings"*[11]. The other buildings were 2 cottages a stable and a carthouse. It is puzzling that these two policies make no mention of William Lister.

The partnership of Marshall and Benyon was dissolved on 30th June 1804, and the Marshall Papers show some of the asset splits. They also record the amount of land which John Marshall had purchased in 1802 for David Wood, 1537 yards, and Matthew Murray, 637 yards.

In the spring of 1803 Baron Edelcrantz arrived in England. He had been commissioned by Gustav IV Adolf of Sweden to tour continental Europe and England, with a view to examining the economic situation and looking for markets for Swedish iron. He found much to interest him in England and visited James Watt who gave him an enthusiasm for the condensing steam engine. Possibly [see above] with the advice of John Watson, he ended up buying four steam engines. Three were bought from Fenton Murray & Wood and one from a London maker[12]. The engine parts arrived in Sweden in the summer of 1804. One of the erectors employed by Fenton Murray & Wood was called Samuel Owen. He was an ex-employee of Boulton and Watt, whom he had left to go to Leeds. He was thinking of moving on again, but after talking to Baron Edelcrantz agreed to go to Sweden for the firm to erect the engines. He travelled to Sweden with Edelcrantz and they arrived there in Gothenburg on May 4th 1804. Two of the engines were installed in Crown Distilleries at Kungsholmen and Ladugardsland, both in the Stockholm area. A third went to Lars Fresk's textile mill at Elfvik, which commenced operations on 24th December 1804. The fourth engine was for Baron Edelcrantz's own mill, but was underpowered for the purpose and went to the Dannemora mines as a pumping engine instead. Dannemora had some experience of an atmospheric engine ordered and put up in the late 1720s by Marten Triewald; who had worked in the Northern Coalfield in England where he had put up common engines firstly for Nicholas Ridley[13]. Triewald's engine at Dannemora had been the largest Newcomen type engine erected when put up. The later condensing engine was a 10H.P Fenton Murray & Wood engine and was erected in 1805, and a drawing made of it in 1819 survives. The engine was later adapted to draw the ore as well and survived until 1855, when it was scrapped.

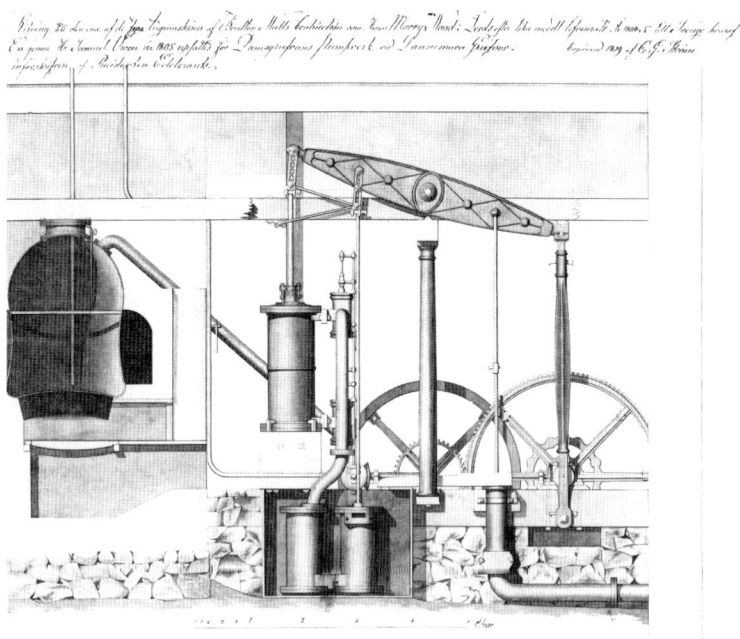

28. Fenton Murray & Co's 10HP pumping engine at Dannemore Mine [14].

In late 1804 or early 1805 Baron Edelcrantz ordered a 20 H.P. engine from Fenton Murray & Wood for his mill.

Samuel Owen made two interesting comments concerning Fenton Murray & Wood in his autobiography [12]. The first was that they were the best manufacturers of steam engines in England. The second was intended to be derogatory about the numeracy of the workforce, stating that only the clerk in the office understood the four rules of arithmetic. He added:

"Mr Murray himself made all his calculations with a carpenter's or a sliding rule".

Which actually means that Murray was very numerate, as operating the old kind of slide-rule required a firm understanding of what you were doing, and a strong grasp of mathematics.

In July 1804 Simon Goodrich once again travelled up to Mr Grimshaw's ropery in Sunderland, and he broke his journey north at Mr Heppenstall's (actually Mr Heptonstall) spinning works at Doncaster and on the way south visited Mr Porthouse's flax mill at Darlington.

On August 4th he was in Leeds and called at Messrs Fenton Murray & Wood, where he saw over the works, and:

" *Mr Murray sent his foreman with me to show me Messrs Benyon & Bages New Flax Mill which is upon the same plan as Mr Marshalls they having parted partnership. The Mill has arched floors, Iron Girders supported by Iron Pillars 9 in. in the width about 10' as under.*

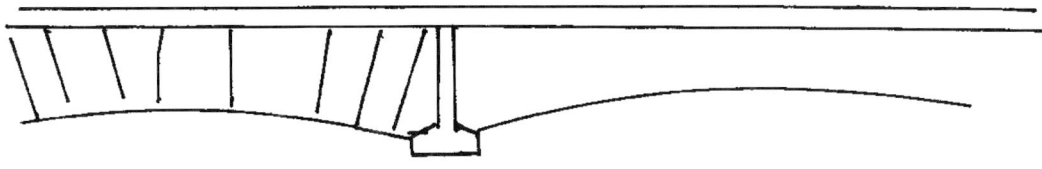

29 a. End view of a beam -the flutes project about 1 ½".

The pillars are starred and connected with one another and with the beams in this manner. The beams are about 12" deep in the middle less at the ends and straight on the underside with two tapering flanges to receive the brick floor or flagstones.

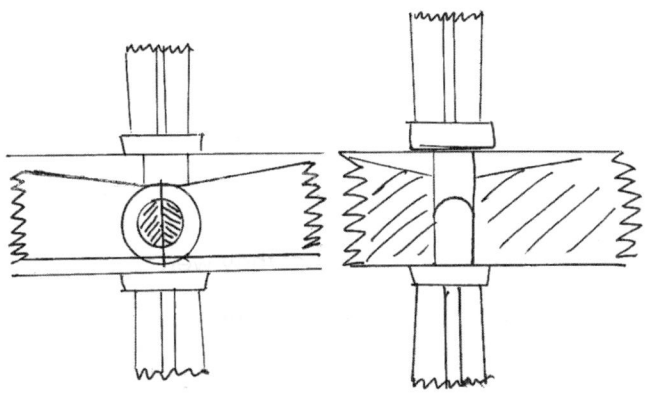

29 b. column connections.

A 50-horse engine gave motion to the mill, which was put up by Fenton & Co. 2 horizontal shafts run through the third floor making 60 revolutions per minute. The diameter of the bevel wheels which worked these shafts was three feet both iron cogs without trimming the shafts about 4 $^{1/2}$ Inches Square.

Sketch of the mode of coupling these shafts with single bearings was thus by two half boxes bolted together upon the ends of the square shafts which swell into a pyramidical form in order to prevent the boxes drawing off the base being made to fit them.

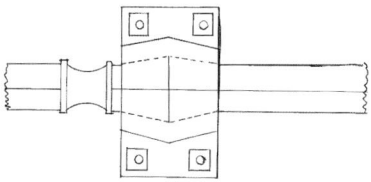

29 c. Mill shafting connector.

Six or eight inch straps were tightened upon the drums by two tightening rollers pressing them against the drums.

The flywheel was hung upon the second motion 1wheel 16-foot diameter 12" face into a 4' wheel both iron.

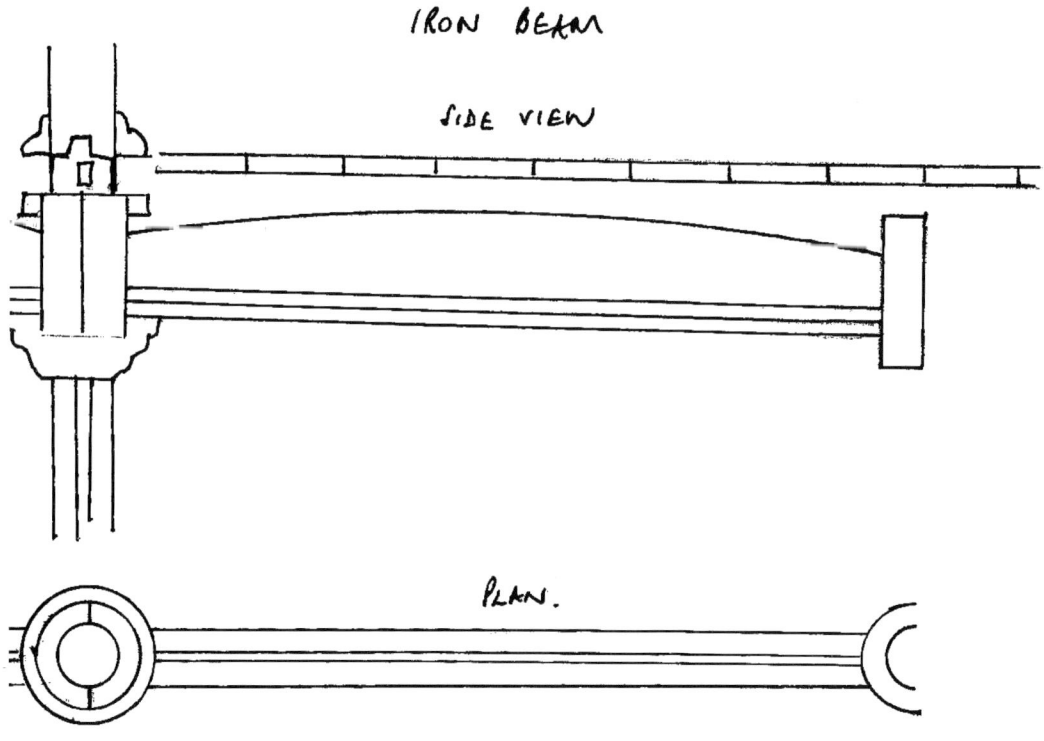

29 d.Iron Beam side view & plan

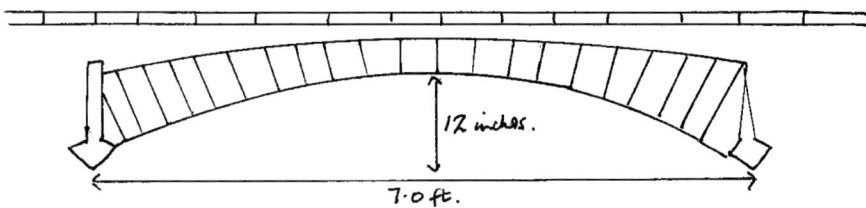

29 e. Section of an Arch & Beams

Length of the beam 9 ft.

A ¾ Bar of wrought iron goes from Pillar to Pillar just above the beam and is cottered at every other Pillar thus

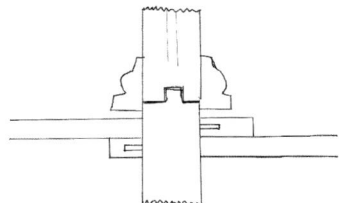

29 f. Connection of tie bars to columns.

A strap of iron is likewise connected with the ends of the beams in the sidewalls in the same manner.

Each room has a chimney to serve as a ventilator, the opening into the room being like a fireplace about 4' high from the floor.

Many large shafts run upon wood in lieu of brass. [Tough Ash].Iron Roof

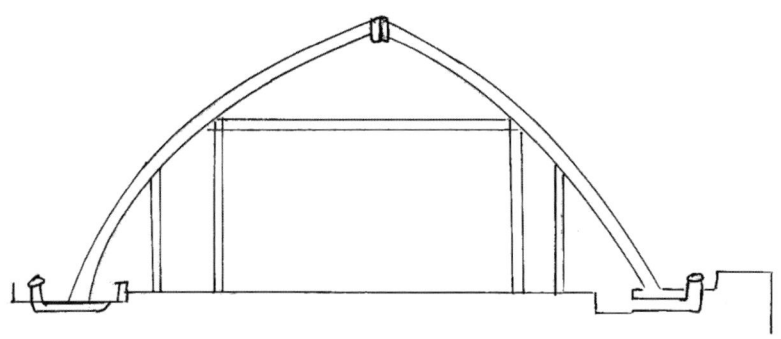

29 g. Roof.

The gutters on the outside are always full, the water runs over the top of a pipe into a gutter on the inside which is lower than the other and conveys away the water.

Water Closets –

A pipe be wash down at stated periods from a cistern above which is always full. Panopticon building 84 ft Diam divided into 8 – 3 floors- room in the centre about 18 ft Diam – 3 windows in a room – inner walls 14 inches – the outer wall 2 ft thick" [15].

The description is important in demonstrating the development of Fenton Murray & Wood under the direction of Charles Bage, who had gone with the Benyons when Marshall and Benyons partnership terminated, the firm supplied the iron work for this early fireproof mill and may well have contracted and sub-contracted to do the whole erection. Fenton Murray & Wood became one of the leading constructors of fire-proof mills for the next two decades, and in their entries in directories and surviving letterheads they listed as one of their trades "Constructers of Fire – Proof Buildings". Skempton and Johnson in their article "The First Iron Frames" [16] suggest that this was the first mill in which cast iron beams were designed in a rational manner. They also suggest that this may have been the first mill to have a cast iron frame roof. It would be fascinating to know how much understanding of beam loading and the structural strength of the columns Matthew Murray learnt from Bage in the construction of this mill, knowing that Murray was numerate and used a slide rule, and had a penchant for acquiring directly or indirectly data on new technologies.

Simon Goodrich also noted on this visit Fenton Murray & Wood's prices for various components, such as couplings at 18/8d, and that the prices of steam engines were as before. Other information he noted in his journal were that the firm would make machinery at prime cost plus 25% for overheads and profit, freight from Leeds to London was 30/- a ton, a 50 or 100 horse engine would take from order to loading on ship 3 months, and that flax spinning cost around 2 guineas per spindle. If a drawing of a machine was sent to Mr Murray he would supply terms to execute the work and suggest any improvements from his experience [15].

In an entry dated 16th August 1804 Simon Goodrich recorded comparative prices for steam engines:

"Boulton & Watt
Including all the metal materials and an iron beam and exclusive of putting up and woodwork in the framing and condenser cylinder.
30 H.P. - £1418}
24 H.P. - £1276}
20 H.P. - £1083}

Trevithick
Includes patent rights & fixing to work, but no brickwork
1 H.P. - 120 guineas
2 H.P. - 210 ditto
3 H.P. - 270 ditto
4 H.P. - 330 ditto
5 H.P. - 380 ditto
6 H.P. - 420 ditto

7 H.P. - 450 ditto
8 H.P. - 470 ditto
9 H.P. - 490 ditto
& rise 20 guineas per H.P. for all powers above. [It can be calculated that a 20 H.P. would be 710 guineas].

Fenton Murray & Wood

Not including boiler
20 H.P. - £600
25 H.P. - £715
30 H.P. - £830
35 H.P. - £938
40 H.P. - £1045.
(The prices for Fenton Murray & Wood were the same as recorded on Tyneside in 1802) [15].

In his journals Simon Goodrich described a visit to Mr Cliff's Iron and Steel Works at West Bromwich on January 23rd 1805 [17]. There he saw a double Engine lately erected by Murray of Leeds in a superior style with a 24 inch cylinder, a 6 foot stroke, which worked at 18 strokes per minute. There was one cast iron pillar to support the Engine beam with cast iron cap to carry the spring beams. It had an 18 foot diameter fly wheel. The first shaft was hollow cast iron and about 2 feet 6 inches in diameter. It had 3 cams about 5 foot diameter, one for an Iron Helve Lift Hammer with 6 Nogs, the other two with 20 Nogs for the two tilt hammers. There is also a description of the shears, but no indication of their maker.

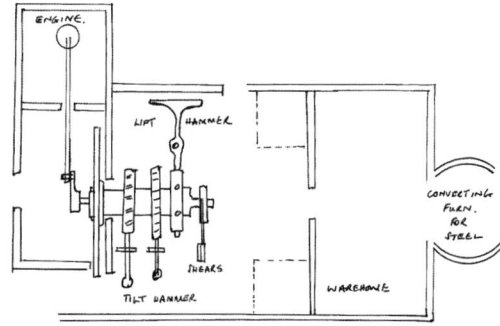

30. Goodrich's plan of Mr Cliff's Iron and Steel Works [15].

In February 1805 John Buddle received a quote from Fenton Murray and Wood for a steam machine to be erected at Elswick colliery.

		£	
"For a patent Steam Engine 20 Horse Power		£	650 – 00-00
Boiler, 12 feet diameter			100 – 00-00
Metal House supposed to weigh 18 to 20 tons	}		250 – 00-00
including the Iron Beam, supporting	}		
Pillar Cross Plate. Iron Framing for the	}		
Rotative Wall, the Drum and axle, Balance	}		
Plates Bolts &co &co	}		
Deals for the Roof Flooring & co			21 – 00-00
Joists for the Floors			12 –10-00
Planks for the Drum, with Brake & Linings			21 – 00-00

Pit Head Frame with Stays & Rollers		24 – 00-00
Pulley Wheels with Brasses &co		22 – 00-00
Condensing Cistern		17 – 10-00
The whole of mason work with Stone	}	
Bricks Lime & Co	}	70 – 00-00
Workmanship in fixing Materials	}	
Mr Murray will undertake for	}	60 – 00-00
Freight suppose 31/6d per ton	}	
		£ 1248 – 00-00

NB Murray will undertake to have the whole of the Articles they are to find shipped at Leeds in June next; and also to remove the Machine as often as may be required to any other Pit for the sum of Fifty pounds (except Leading or shifting the materials)at Elswick new Colliery.

For Fenton Self & Company
Matthew Murray" [18].

The offer to move the machine was in complete contrast to Boulton and Watt's restriction, which had so concerned John Marshall in his dealings. Fenton Murray and Wood also built the engine house and installed the engine in it, as the firm also had, on the evidence, at Willington. Buddle recorded three pages later in detail the costs of a:

"20 Horse rotative engine erected by Messrs Fenton Murray Wood & Co including a cast iron house, which totalled £1503/5/4d" [19].

The cost of the engine materials and cast iron house were exactly as estimated.

John Farey recorded that it was in 1805 that Arthur Woolf fixed, set up and ran his first experiments with his twin cylinder high pressure engine at Meux's brewery in London (Clerkenwell Road), where Woolf was employed as engineer. Farey also informs us that the engine Woolf adapted was 6 H.P. one made by Fenton Murray and Wood and that the new parts to try his patent were also made at Leeds [20]. They were no doubt made by Fenton Murray & Wood, as in the Letter by Matthew Loam referred to above, it is stated

"Woolf and him (Murray) was well acquainted, I have heard Mr Woolf give anecdotes of him, but when I was at Leeds I had all I have said confirmed, and more, nearly all the adaptations which Mr Woolf used in gear work etc, when I was there I saw them as Murrays, and had reason to believe it" [21].

Loam served as an apprentice to Woolf and later became his main assistant [22], he was also familiar with Leeds.

In May 1805 Matthew Murray wrote to Mr Nicholson with a description of his portable steam engine which was published along with a plate in the Philosophical Journal:

31. Mr Murray's Portable engine 1805. [23].

The two most important aspects of this design were firstly its compactness, which did however make maintenance difficult, and the fact that the beam was underneath the engine. The latter fact had a major effect on the centre of gravity of the engine and became a fairly standard feature in early marine engines.

In July Samuel Bentham the Inspector General of the Navy was despatched to Russia, along with a number of artisans, to represent the Government in a plan involving Russia building some ships for the Royal Navy, although in the end the politics around the Napoleonic alliances stopped the plan coming to fruition. In his absence he proposed to the Admiralty, and they approved, that Simon Goodrich should act for him. Bentham and Goodrich had agreed some proposals for Portsmouth Dock Yard which were laid out in a memorandum and signed by Bentham. One of the proposals was to replace a 12 H.P. Sadler engine in the Wood Mills with another 30 H.P. engine. A Boulton and Watt 30 H.P. had been added to the Mills in 1800, and the proposal would provide two interchangeable engines so that in the event of breakdown or maintenance requirements neither the mills nor the dry dock would be affected [24].

September 1805 saw Henry Maudslay preparing for his first steam engine for his workshops in Margaret Street in London. The engine was to be made by Fenton Murray & Wood, and Maudslay wrote to Matthew Murray after he arrived back in London from a meeting, which had probably taken place in Leeds, as Maudslay had met David Wood. The letter reflected on the discussions about his engine and enclosed a plan of the area for installation. He needed the strokes per minute of the engine to be reduced to 30 from around double that value to match the rest of his set up [25]. Among other possibilities it may be that Fenton Murray & Wood had been recommended to Maudslay by Simon Goodrich, as Maudslay was at the time working on the machinery for the Block Mills at Portsmouth. Maudslay in turn recommended Fenton Murray & Wood to others.

Simon Goodrich continued to work through the procedure to gain approval for a second 30 Horse engine for the Block Mills at Portsmouth, and sent a detailed rationale to William Marsden, the secretary to the Admiralty on the 24th September which resulted in a letter from the Admiralty to the Navy Board on the 28th instructing them to erect the engine, provided they had no objection [24].

In August 1805 John Buddle had advertised for an engine for drawing (raising) coals at Lambton Colliery in County Durham, listing 10 terms or conditions. In his surviving papers there are four responses, two local; Phineas Crowther and Reay and Bailey, and two from Yorkshire; John Curr and Fenton Murray & Wood. Phineas Crowther's bids were between £1350 and £1700 with an 8 month period from order to erection. Reay and Bailey quoted between £1518 and £1568 with 7 months for the making. Mr Curr wanted between £1250 and £1700 and from 8 to 11 months to complete. Fenton Murray & Wood's price was £1250 with completion in 4 months and contains more detail than the others. So as was often the case Fenton Murray and Wood's quote was the lowest. However their letter was marked "came to hand 2nd Oct" and all the other letters seem to have arrived in September. It is not known which firm obtained the business [26]. Crowther had the contract for the pumping engine.

During October 1805 Simon Goodrich was busy undertaking further drawings for the proposed new 30H.P. engine and compiling the list of metal parts required to be supplied. The matter was complicated as an engine could not be purchased off the shelf as it was to fit the space occupied by the existing 12 H.P. Sadler engine. The Navy Board then asked for his estimate, and Simon Goodrich supplied an amount of £1650 including two boilers. In his letter sent to the Navy Board with the estimate, Goodrich recommended Whitmore & Son of Birmingham (who had made the engine for the metal mills at Portsmouth Dock Yard) and Messrs Fenton Murray & Co of Leeds as proper persons to be employed. On the 28th October Simon Goodrich spoke to Sir John Henslow, a surveyor of the Navy and member of the Navy Board, on the matter and Henslow agreed on receipt of two copies of the drawings a set would be sent to each manufacturer. Whitmore replied by the 8th of November and fearful of a mistake Simon Goodrich wrote to Fenton Murray & Wood to find out if they had received the drawings. The Admiralty received a letter from Fenton Murray & Co on 23rd November, and three days later the firm informed Simon Goodrich that they had had the drawings and responded. Fenton Murray & Co quoted £1845 and 6 months, against £1656 from Whitmore, however Whitmore did not provide a time for delivery [27].

In November 1805 John Buddle noted the prices for Murray and Wood's engines, delivered at Hull:

"10 Horse	£400
15 Horse	525
20 Horse	650
25 Horse	750
30 Horse	850
35 Horse	950
40 Horse	1050
45 Horse	1150
50 Horse	1250
55 Horse	1350
60 Horse	1450
87 Horse	1870
100 Horse	2012 " [28].

The increases from the 1804 prices show a much higher level for the lower powered engines with 8% for the 20 Horse and a mere 0.5% for a 40 Horse.

Simon Goodrich's journal of 3rd December recorded his decision over whom he would recommend to the Navy Board for the new 30 H.P. engine for the Block Mills:

"Write to the Navy Board and recommend Fenton Murray & Co in preference to Mr Whitmore notwithstanding Mr Whitmore 's Estimate was 189£ lower than the other as I was really afraid that Whitmore would do no credit to the Engine and that he would disappoint us in time – He had specified no time. Murray had specified 6 months for the delivery of the materials and also is I hope from what has been experienced of the other much more to be depended upon.
Besides the various blunders in the Copper Mills machinery and constant disappointments in time promised, another circumstance contributed materially to set me against Mr W, which was lately discovered namely that he had made the steam cylinder of the Metal Mills Engine 6" longer than the stroke of the piston so that about 1/30th of the whole steam or about 2 horses power was uselessly wasted – This was great want of attention to principle – The principal reason I urged to the Navy Board was time and carriage in favour of Murray. The present engine being much in want of repair as soon as it can be spared by the erection of another" [29].

His letter to the Navy Board was dated the same day. The Navy Board wrote to Messrs Whitmores for a delivery period, without advising Simon Goodrich, but only obtained one of 7 months deliverable at Deptford. The Navy Board wrote to Fenton Murray and Wood on the 17th December 1805 to accept their offer of the 23rd November and required them to deliver the engine direct to Portsmouth Dock Yard [30]. The rejection letter to Whitmore was sent on December 14th.

Goodrich recorded on December 20th that he had received a letter, dated the previous day advising him that the Navy Board had ordered the 30 HP steam engine from Fenton Murray and Wood [29].

At the end of December 1805 Fenton Murray & Co wrote to the Navy Board advising that when the materials for the engine were ready they would like them examined prior to shipping in case any alterations were necessary.

At some point in 1805 a Mr James Ford of Finhaven had erected a flax mill at Montrose, the mill was extended in 1813/4 [31] and advertised for sale in 1817 as a fire proof mill with two steam engines (25 H.P. and 12 H.P.) by Fenton Murray & Wood of Leeds [32]. In studying the sale of fire proof flax mills in Eastern Scotland in this period it can be shown that mills advertised with the firm's engines also were frequently built by them. Although there is no direct evidence, Ford's mill is candidate for the first fire proof mill erected by Fenton Murray & Wood in Scotland, as they certainly carried out extensive improvements at a later date. Whether the firm were involved in this first spinning mill at Montrose, or only the enlargement, it is known that the first engineman at the mill was George Stephenson, who had temporarily left the Northern Coalfield after the death of his first wife and new born daughter. Unfortunately this mill has not survived having eventually been purchased by Richards and Co. along with Broadford Mill in Aberdeen (where the original mill had also had Fenton Murray and Wood ironwork). Richards and Co. consolidated their business in Aberdeen in 1881 and the Montrose Mill was demolished in 1883 [33]

In February and March 1806 Simon Goodrich's Journal recorded correspondence with Fenton Murray & Wood over some changes to the 30 HP engine on order, and he requested the firm to hasten the delivery [34].

In April of that year John Marshall noted in his third experiment book that they had tried a drawing engine of David Wood for tow with a "sheet doffer" and had found it a real advantage [35].

In May 1806 Simon Goodrich recorded the details of an export order to Russia. The order was instigated by Charles Gascoigne for His Imperial Majesty of Russia and managed on his behalf in England by Baron Henry de Bode, who had obtained on order from the Lords of the Committee of Council for Trade and Foreign Plantations allowing the goods to be exported, for these goods an Act of Parliament was normally required [36]. Simon Goodrich called on Baron de Bode on 11th May and described the machinery in his Journals, some of which had been purchased from Fenton Murray & Wood and some from Fox of Derby. The Fenton Murray & Wood machinery consisted of:

1. A Forge Mill, which itself comprised:

15 H.P. steam engine complete with cast iron beam, pillar and cross piece for £525, a beat iron boiler for the engine for £ 100, a solid square shaft for working the forge, forge helve of cast iron and other parts for £ 541. The fly wheel was 18 foot diameter and the shafts were to work the tilts and a pair of shears. The shears were supplied.
 A blowing cylinder, regulating cylinder pipes etc for £150

2. A boring machine from the same patterns as Fenton Murray & Woods, for £300 (*a most excellent machine*)

3. A screw cutting machine as Fenton Murray & Woods with dies

4. A 6 H.P portable steam engine for working the boring and screw machines with the necessary shafts and wheels for £403.

All described as very good and cheap.

Fox supplied a 10 H.P. steam engine, boiler, boring turning screw machinery, and a slide lathe [34].

Baron de Bode also had to request in a letter dated 24 April 1806 permission for two people to go out to St Petersburgh to erect the machinery, one from Fenton Murray & Co and one from Fox. Artisans going abroad especially in the middle of the Napoleonic conflicts was considered a serious matter, of national import. The Lords of the Treasury granted permission and it was endorsed by the Privy Council [38].

The vision of all this output from Leeds is presumably what prompted Simon Goodrich to write to Fenton Murray & Wood on the 13th May asking when the engine for the Block Mills would be ready. The Navy Board were advised by the firm on 19th that the machinery was ready to be inspected. It was proposed to send Mr James Linaker, the master millwright at Portsmouth up to Leeds to do the inspection. Simon Goodrich supplied James Linaker with introductions to Mr Strutt (Belper) and a Mr Hughes of Manchester, and James Linaker left for Leeds on the 8th June and was back on the 1st of July [34].

The Swedish Ambassador wrote to the Government in May asking for permission for Fenton Murray and Co to send three workmen out to Sweden to erect the steam engine purchased by the Swedish Government in 1805. The matter was handled by Mr Fox and as with the matter for Russia the Privy Council and the Treasury gave the necessary permissions [37].

In a letter dated 3rd of July 1806, Samuel Bentham wrote from St Petersburg to Simon Goodrich, requesting him to order a steam engine of about 20 H.P. to a given design made so that it occupied as little space as possible. He stated:

"*You would I suppose order the Engine of Murray & Wood if so they would be the most proper persons to apply for the leave (permission to export)... I would also wish to have at the same time or before if possible a small Engine of the power of one or two horses simply to raise water for the use of the house & and garden, and if Murray*

& Wood choose to send out another or two such small engines I dare say I could find a sale for them. They should be complete in every respect with boiler etc as we saw them when we were together at Leeds" [38].

James Linaker's satisfactory report having been duly lodged from his visit to Leeds, on the 7th of July the warrants had been issued to Portsmouth to support receipt of the engine materials and to make the necessary labour available for removing the old 12 H.P. and to erect the new engine [30].

On the 9th of July 1806 Simon Goodrich called on Mr Maudslay and looked over his shop. Mr Maudslay had received his 4 H.P steam engine from Fenton Murray & Wood, to work his lathes [34].

During the rest of July and early August letters were exchanged over the steam engine's delivery (the ship was delayed at Hull). Fenton Murray & Wood's erector arrived at Portsmouth, the terms for whom had yet to be agreed. During this correspondence the business of Samuel Bentham for St Petersburgh was also discussed. By 19th of August the materials for the 30 HP for the Block Mills had arrived at Portsmouth apart from one of the boilers which had been shipped separately due to lack of space on board. By the end of August, having received a satisfactory estimate from Fenton Murray & Wood, Simon Goodrich ordered the 20 H.P. engine for Samuel Bentham. The other items could not be manufactured in time to send to St Petersburgh before ice closed the Baltic, and would need to follow on. After some debate over payment, settled when Samuel Bentham sent Simon Goodrich a letter of credit, Fenton Murray & Wood agreed on 11th October to dispatch the 20 H.P. against Simon Goodrich promising to accept their draft (bill of exchange) for £1347-19-6d [39].

Reflecting the patterns of those times, The Leeds Intelligencer carried the following paragraph;

"William Fawcett, an articled servant to Messrs Fenton, Murray and Wood was committed to Wakefield House of Correction for one month, for leaving his work in an unfinished state" [40].

It is possible that during this period working up to the delivery of the engine for the Block Mills that Matthew Murray had met another influential engineer of the period either through Simon Goodrich or Henry Maudslay, or at the very least Murray had been recommended by one of them (possibly as indicated in Henry Maudslay's letter, above) and that was Marc Brunel. Marc Brunel was designing all the machinery in the Block Mills, dealing with Simon Goodrich; and the machinery was made by Henry Maudslay. Marc Brunel, in partnership with a Mr Farthing, was in the process of setting up a veneer saw mill in Battersea (although the site was not yet agreed or procured). A steam engine was required for the saw mill and this had been ordered from Fenton Murray & Wood. Marc Brunel wrote to them on 29th October 1806, asking them to let him know when the steam engine was ready, as he intended to direct it to Maudslays, and advising that they would be paid through Mr Farthing. He also asked for a sketch to show the location of the boiler, flue and smoke burner as he had not erected an engine like theirs [41].

On 5th November Marc Brunel wrote to Mr Farthing, wanting the purchase of the Battersea site finalised, and encouraging him over some disappointing trials. At the end of the letter he remarked

"Mr Murray is still in Sweden".

On the 9th November M Brunel advised Farthing that he had heard from Leeds, and the engine had been tested and was being dismantled for shipping to London [41].

At some point Murray had also been active in Bateman and Sherrat's territory over the Pennines. An Iron Works was advertised for sale in November 1806. It was known as the Mersey Wire Works and was situated at Bank Quay Warrington. There were two engines mentioned in the advert, one to power the forge of 25 HP for which no maker was named. The Wire Mill was however:

"worked by an excellent new Steam Engine, constructed by Messrs Fenton, Murray and Wood, fully equal to Twenty Horses power, with one large boiler" [42].

Notes

1. Boulton and Watt Papers MS3147/3/79/4.
2. Local Records or Historical Register of Remarkable Events by John Sykes vol. 2. Newcastle 1833, reprinted 1866 page 5.
3. The Morning Post 28th September 1803, 29th October 1803.
4. The London Gazette March 1806
5. Land Registers Vol. EO page 470, WYAS Wakefield.
6. Watson Papers Wat 2/8/340. The North of England Institute of Mining and Mechanical Engineers [NEIMME].
7. The Leeds Intelligencer 5 December 1803.
8. Goodrich Papers A114. Science Museum Library.
9. Tyne and Wear Archives & Museums, Discovery Collection and Art and the Industrial Revolution by F D Klingender London 1947.
10. Marshall Papers MS200 folio 56 Experiment Book. Brotherton Library Leeds.
11. Royal Exchange Policies No's 210456/7, MS7253 vol. 35. London Metropolitan Archives.
12. For the Swedish Engines see: The Small Giant: Sweden enters the industrial era by Carl Carl G Gustavson, also Fenton Murray & Wood, unpublished thesis by Trevor Turner especially appendix 1 an Extract from Samuel Owen's Autobiography.
13. Description of the Atmospheric Engine 1734 by Triewald. Translation Extra Publication No 1of the Newcomen Society 1928.
14. One of the four steam engines of Boulton and Watt construction which Messrs Murray & Wood in Leeds built and delivered in 1804 to Sweden, this one was erected by Mr Samuel Owen in 1805 for pumping at the Dannemora mines. The Technical Museum Stockholm.
15. Goodrich Journals B6. Science Museum Library.
16. The Architectural Review March 1962.

17. Goodrich Journals B6. The journal is somewhat confusing as although clearly labelled 1805 the item precede others which are dated and confirmed elsewhere as 1804. E. A. Forward in his Newcomen paper 'Simon Goodrich and his work as an Engineer Part 1' (26 October 1922) favoured 1805.
18. Buddle Papers, Bud/24/32. NEIMME.
19. Buddle Papers, Bud24/35. NEIMME.
20. A Treatise on the Steam Engine Vol 2 John Farey. David and Charles 1971 page 57 also the original in the Science Museum Library.
21. Letter from Matthew Loam (1794-1875) to his nephew Matthew Loam (1819-1902) helping him prepare for a lecture on the steam engine, Cornish Studies Library, Redruth, Cornwall. Ref t7994044 and Mooo1948CC.
22. Arthur Woolf: the Cornish Engineer by T. R. Harris, Truro 1966.
23. Nicholson's Philosophical Journal vol XI.
24. Admiralty Papers ADM/Q/3323 Correspondence between Admiralty and Inspector General's Office, W Marsden 1 Aug. 1805, S Bentham to W Marsden 30 July 1805, and Admiralty to Navy Board 25 Sept. 1805. National Maritime Museum (N.M.M.), Coad Library.
25. Letter Henry Maudslay to Matthew Murray 14 Sept 1805. MS1261. Science Museum Library.
26. Buddle Papers, Bud/13/ pages 110-5. NEIMME. Also 'The Steam Engine on Tyne side in the Industrial Revolution' by Jennifer Tann. Newcomen Society 13 January 1993.
27. Goodrich Journals B10 and 11. Science Museum Library.
28. Buddle Papers 24 page 24. NEIMME.
29. Goodrich Journals B12. Science Museum Library.
30. Abstract of Navy Board Out Letters ADM49/166. National Archives Kew.
31. Factories Inquiry Commission Part II printed March 1834. Page 125 Richards & Co Montrose.
32. Edinburgh Advertiser 9 September 1817.
33. The Port of Montrose Ed. G Jackson and S Lythe. The Georgic Press and Hutton Press 1993.
34. Goodrich Journals B13. Science Museum Library.
35. Marshall Papers MS200 folio 56 Experiment Book. Brotherton Library Leeds.
36. Privy Council Papers PC 1/3684, PC 1/3693. National Archives Kew.
37. Privy Council Papers PC 1/3711, PC 1/3714. National Archives Kew.
38. Goodrich Papers A193. Science Museum Library.
39. Goodrich Journals B16. Science Museum Library.
40. The Leeds Intelligencer 15 September 1806.
41. Marc Brunel's Letter Book LBK 54. N.M.M. Coad Library.
42. The Leeds Intelligencer 24 November 1806.

Chapter 8.
A Busy Time Sweden, Mills and Water Works

The engine for Baron Edelcrantz's four storey combined flour and textile mill at the east end of Kungsholmen Island, Stockholm was delivered sometime early in 1806. Samuel Owen had returned to Leeds in December 1805, debriefed Fenton Murray & Wood, resigned, and gone to work with Arthur Woolf in London. In May 1806 Owen received a letter from the Baron stating that he had had no response to letters to Fenton Murray & Wood asking them to send someone to erect the engine, and proposing that Samuel Owen should return to Sweden and erect the engine. As shown previously the lack of response was no doubt due to the necessity of gaining Government approval to officially send engine erectors outside Great Britain. The list of jobs would then have to have been prioritised once permission was granted. Samuel Owen duly went to Sweden, only to find Matthew Murray and his erectors already there doing the job. Two contemporary references to the engine survive. The first described the action:

".....worked by a steam-engine, erected by a Mr Murray, of Leeds. Mr Brown visited this mill several times. ..... On one of these visits..... a conversation took place relative to the engine, whose power was great, but whose action was so even and steady that it made scarcely any noise" [1].

The second description was recorded by the American geologist William Maclure in 1810 and described two ancillary features. These were a mercury tube for showing the pressure in the steam and the other was a regulator fitted to the boiler. The mercury tube was fitted alongside the barometer used to measure the vacuum, and consisted of a u-tube with mercury that registered plus or minus in relation to atmospheric pressure. Maclure states that this device was fitted by Edelcrantz, but possibly after discussion with Murray. The second device was operated by a float in a cylinder attached to the boiler, as the float rose or descended according to steam pressure it adjusted the air supply to the furnace. The principle used is similar to that described in Murray's 1799 patent but uses a float rather than purely mechanical application [2].

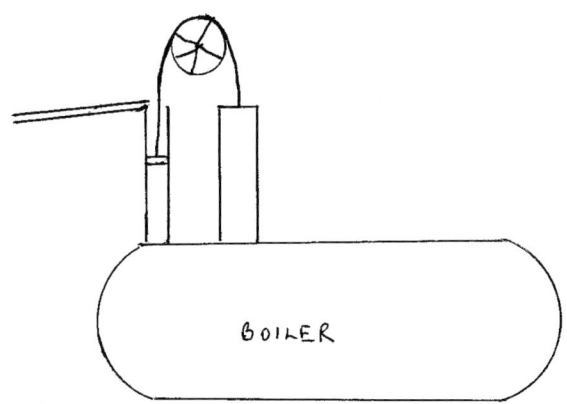

32. Author's redrawing of Maclure's sketch of furnace regulator.

Samuel Owen found employment in Sweden and went on to become a major influence in introducing steam power to Sweden, among his achievements he built the first Swedish steamboat.

The main purpose of Matthew Murray's visit to Sweden is unknown, but it can hardly have been to erect engines during an extremely busy period for the firm. There are some clues. Whilst in Sweden Murray visited Thomas Stawford, an Englishman, in charge of the technical aspects at Hoganas Stenkelsgruvor Coal & Iron Co in the Skane area of Sweden (south). Stawford is another candidate for being the recipient of John Watson's letter of November 1803. Stawford recorded the meeting with Murray in his day book for the 6th December 1806.

"Returned from Helsingborg (near Hoganas in Skane) where I had gone with Mr M Murray partner and manager with Fenton Murray & Co Engine makers at Leeds who had at my request as well as some of the owners come to see our engines as well as the situation of our colliery. Mr Watson of Willington had desired him to come this way also for that purpose. He told me he did not expect either to see so much work done or such good engines in Sweden that except theirs, or Boulton and Watts there was none so good in Newcastle" [3]

Fenton Murray & Wood continued to supply engines to Sweden, and so an element of the visit may well have been to ascertain the market potential. Mr Tyas in his paper, 'Matthew Murray, A Centenary Appreciation', mentions that Murray received a gold snuff box from the King of Sweden [4]. This item is also referred to in Matthew Murray's life in Leeds Worthies [5], however no mention was made of it in Matthew Murray's will.

The Society for the Encouragement of Arts Manufacture and Commerce received a letter dated 25th November 1806 from Baron Edelcrantz which advised that he was sending them a model of his telegraph by Mr Murray of Leeds. Edelcrantz won a silver medal for it in 1807.

By the 30th of December 1806 Messrs Brunel & Farthing had acquired the Battersea site for their veneer works, and Marc Brunel asked if the steam engine could be fixed in the existing buildings as the whole works could then be ready more quickly [6]. Simon Goodrich wrote in his journal (14 December 1806) that Messrs Brunel and Farthing were buying an engine from Fenton Murray & Wood:

"Spend the evening at Brunel 's who is now engaged in planning arrangements for his manufactory for cutting veneers by a new mode for which he has lately taken out a patent and ordered a Steam Engine of Murray & Co of six horses power" [7].

1806 saw the first flax mill erected in Scotland (Dundee) with which the firm can be definitely linked, with the provision of the structure, the flax machinery and the steam engine. This was the erection of Bell Mill called West Ward Mill which was in Dundee on the corner of Guthrie Street and Blinshall Street and was put up for James Brown of Canonsyth. William Brown one of the sons of James described the mill in 1862. The plans and machinery were from Leeds. The building was 97 by 40 feet, 4 storeys with an attic and mostly fireproof. The machinery included a 25 H.P. steam engine, 40 spinning frames for flax or tow, with some twisting frames and ample preparing machines. The structure was handsome and substantial, above all others at the time. Attention had been given to the external appearance. It attracted a lot of attention including a retired admiral who fell into

"arguments with the head engineer, an irritable man from Leeds".

The building cost £7,000 and the machinery £10,000, and the whole was unmatched in Dundee for 20 years. Spinning commenced in May 1807, and on the flax side was soon up and running. However there was an issue with the tow working, which would not produce better value goods than hand spun items:

"The maker of the machines from Leeds, David Wood, considered first rate in his trade, attended himself for a time to put things right, but not withstanding the new and costly mill and machines he failed to produce yarns equal in value by 2d or 3d per spyndle equating to the poor local works.
Some feelings and reflections, not of the pleasant kind, were expressed on the occasion – and the Leeds man was blamed for not understanding and being able to work rightly machines of his own construction" [8].

The mill was demolished in1959, but a surviving description of the lower floors gives them as brick arched on cast-iron beams and cruciform columns [9]. The cruciform columns match Simon Goodrich's description of Benyons Mill in Leeds, and also the columns in Bage's Ditherington Mill construction. However the allusion to the plans and head engineer both coming from Leeds, with David Wood being called in over the tow machinery point to Fenton Murray & Wood. The engine was stated to be by Fenton Murray and Company when the mill was put up for sale during the slump years in flax spinning in 1811. Other surviving mills built later will support the hypothesis. The construction of the factory can be considered as one of the first instances of in modern parlance 'a turn-key programme' with a prime contractor, Fenton Murray and Wood, managing the provision of all the ingredients to produce a flax spinning mill, building, machines and power, all in one. Accompanied with the usual vagaries of having to make the tow machinery perform satisfactorily.

The West Middlesex Water Works had agreed to advertise for two 20 HP steam engines for their new water works in December 1806. Tenders were received from Boulton and Watt, Bramah, Binns and Fenton Murray and Co in early 1807. The first bid from Fenton Murray and Co was delivered early in January by their agent Mr Stainburn and this was followed up by a visit from Mr Murray on the 16th January 1807, when a regular tender was submitted along with a sketch [10].

On the 30th January 1807 Simon Goodrich visited Maudslay's factory which he went round with "*young Field*" (Joshua Field a respected engineer who was to become a partner in Maudslay Sons and Field) and commented in his journals that the Fenton Murray & Wood engine was at work in the cellar in "*fine order*". He also noted that the cogged wheels previously used had been removed, due to a degree of rattling and shaking, and leather straps and iron wheels had been substituted and produced no noise or shaking. Henry Maudslay was out, but Simon Goodrich caught up with him when he revisited the factory in the afternoon, when Maudslay had returned from work at setting up Marc Brunel's veneer works at Battersea, including the Fenton Murray & Wood steam engine.

Communications were going awry in the matter of the engines for the West Middlesex Water Works as in early February Fenton Murray & Wood wrote, obviously in Matthew Murray's absence that they had submitted all the data through their agent Mr Stainburn, and were not aware of any changes that may have resulted from discussions between their Mr Nicholson and Mr Murray. They had written to Mr Stainburn to attend upon Mr Sloper of the Water Works to find out what information was required [11]. However the issues settled down and by the 3rd March Mr Nicholson was authorised to give orders to Fenton Murray & Wood to manufacture the engines when he was satisfied with the details.

On the 9th March Simon Goodrich mentions that Fenton Murray & Co wished to know if they could send the other boiler and blowing cylinder to Hull to go by the first ship. It is unclear whether these are supplies for the Royal Navy or General Bentham in Russia [12].

In a letter dated 16th March Fenton Murray & Wood supplied the West Middlesex Water Works with an estimate for pumps, it having been agreed that the engines would be two 20 H.P. ones with 24 inch cylinders.

"We have duly received yours of the 12ᵗʰ Inst. Annexed we hand you Estimate of the Pumps, Air Vessels & Mains as described in the Plan, as I suppose Mr Nicholson's alterations will not make much difference, but for your information I have drawn out the parts separately with the numbers, prices & weights to each, & we are agreeable either to take them for the whole amounts, or separately at what they may weigh & come to at these prices. With respect to the sum for fixing the whole of this pump work and pipes together at Hammersmith, as a great deal will depend upon the Situation & nature of the ground, although we suppose it at £100, it probably may considerably exceed that sum, we therefore wish Mr Nicholson would take this part into his own Hands, and we will furnish him (if required) with proper Engineers for that purpose.

The responsibility for accidents in the Pump work we cannot undertake (further than delivering the goods sound & perfect) as numbers of unforeseen accidents may happen to this part of the Works, viz such as sinking or giving way of some parts of the ground, the freezing up of pipes and their bursting by sudden expansion of the ice, the incautiously shutting off stop valves, & in short any mismanagement of the Person who has the care of the Pumps.
I am etc, M Murray
PS Freight of materials to London 30/- per ton.

List of Materials	Cwt	Qtr	Lbs		£	s	d
4 Setts of Pumps 18 in Diam 1 1/2 inthick, bor'd seated }							
turn'd in Flanges }	14	15		30/-	442	10	
6 Buckets & Clacks, turn'd		16	2	Do	24	15	
Cotters, Hoops, Cross bars & Lids for Do		4	1	1/?	23	16	
4 Wrought Iron Pump Rods, turn'd, to Buckets		6		18	50	8	
4 Do Do Connecting Rods with steel o Joints}							
Sockets & Cotters}	1	4		10	112		
400 Bolts, Nuts & Loops	1	9		9d	121	16	
13 Nine Feet lengths of Main Pipes 12 1/2 in Diam & 1 1/4 in thick	10	8		17/-	176	16	
13 Do Do Do 9 in Diam & 11/4 in thick	7	16	2	Do	133	-	6
8 Do Do Do 8 1/2 in Diam & 11/4 in thick	4	5		Do	72	5	
8 Square Connecting Pipes	3	2		Do	52	14	
1 Air Vessel & Receiver with Bottom}							
Partition & Nozzles}	6	7		Do	107	19	
10 Hanging Valves for Sqr Pipes & Loops & Nuts		3	2	1/?	19	12	
4 Spindles, Tumblers & levers for shuting of the}							
Reservoir Water from the four Pumps}		4		1/2	22	8	
3 Stop Valves for the 3 Mains					45		
Fitting up the whole at Leeds					120		
				£	1524	19	6

Deliver'd at Leeds payable 6 mo's from Delivery" [11].

In April an interesting advertisement appeared in the Leeds Mercury:

"Building Ground
To be sold at 3s per yard, eligibly situated near to Messrs Benyon and Co's Manufactory, Meadow Lane.
Enquire at Messrs Fenton, Murray and Co's Iron Foundry.
N.B. The above premises contain a valuable bed of clay, and if required the purchase money may remain two or three years upon interest.

Leeds April 3 1807."

Presumably this was either an additional service to their clients Benyon & Co, or most probably the sale of the clay beds used to furnish bricks for building the mills which Fenton Murray and Co no longer required, the mills having been finished [13].

The contract negotiations continued with the West Middlesex Water Works over bonds and guarantees for performance. Fenton Murray & Wood were uncertain whether they should correspond with Mr Nicholson or Mr Sloper. Matters were not helped when one of the Leeds letters was dated April when it should have been May. Mr Nicholson reported in early May that one of the steam engines would be shipped from Leeds on or before July 1st and the other within the following month. On the 27th May a draft of the contract (given at appendix 3, as the most complete surviving contract in which the firm was involved) was delivered from Leeds. It contained a 4 year guarantee for the engines in relation to materials, proportion (suitability for the purpose?), and erection. The remedy was given that the supplier would return the engines to good working order at their expense, but would not be liable for the operational losses incurred by the Water Works due to the engine being out of action [11]. A warranty similar to some found in the contracts of today. The pump work was also ordered on the 27th May. Fenton Murray and Co acknowledged the order on the 10th June with thanks, and advised that the work would be put in hands immediately (the origin of the modern 'to put in hand'). They also stated that the first steam engine had been shipped on the 1st June and that the other would be finished by the 23rd June, and enquired about the state of the engine house as they had an engine erector ready to begin when the house was up. The correspondence shows that there had been some confusion over matters between the Water Works, Mr Nicholson and Fenton Murray and Wood, but the extent was not revealed until Mr Nicholson's report dated the 12th June, the day after he was dismissed as Engineer, the relevant parts of which are quoted below:

"Tenders for Steam Engines were had from Messrs Boulton & Watt, Bramah, Binns and Fenton Murray & Co, the earliest offer of these last Gentlemen, was made by their Agent Mr Stainburn in that month; but it was not till the 16th of January that Mr Murray himself came to town and made a regular Tender of two Steam Engines accompanied by an outline Sketch, which was laid before the Board at Freemasons Tavern and referred to the Engineer. Mr Murray engaged to stay in town and complete his Drawings, and Contract; and on the 21st January the Well was ordered to be forthwith commenced.

Mr Murray not having appeared, and no satisfactory explanation having been given to repeated Applications made to Mr Stainburn, I wrote to the House at Leeds (Feb 9th) say three weeks and more after the Tender, - remonstrating against this conduct explaining the Terms of bargain and requesting Plans for our government. In this answer (Feby 11th) the Partners express their surprise that Mr Murray was not as they supposed, busied on the Company's work, respecting which they possessed *no information, and on the 10th Mr Stainburn having called and acquainted me that Mr Murray was then setting of for Leeds I gave him a Plan of the Works, with some Instructions.*

 On the 14th of February Mr Murray wrote objections to a Plan respecting the Pumps, which for urgent reasons had been offered for his consideration in my letter of Feby 9th, and in my answer of Feb16th. I admit his objections but direct a System of Pumps to be followed, which has been adopted instead of a plan suggested by him.

Feb 23rd no Drawings having yet been sent the ground plan &c was promised, and on the 24th it was announced by Mr Murray that they were sent and that <u>the Well will require to be altered</u>, because the Pumps must be supported on the bottom of the Suction pipes. .... enquiries were made whether the suction pipes of the Pumps, could not by the Patterns of their House be led by two quarter bonds into the Pit; and whether they could not give a direct stability to the whole Engine, by framing (similar to that adopted in place of the Lever–wall support) instead of suffering the working Cylinders, the Pumps, the air Vessel and the ponderous balance Wheels to depend upon separate masses of brick-work, any subsidence or change of which would alter the positions of the working parts, and were to be considered is the chief reasons, why such great difficulties and even failure of performance had been experienced in Steam Engines. He(Murray) *replied in the affirmation as to the Suction pipes, but said that no provision had been made for giving firmness to large Engines by counter-framing.*

 March 10th working Drawings were sent ...... the Well was then nearly finished. In my order I .... directed the manner of placing the Pumps....

 When Mr Murray came to town he required the alteration of the Well to be made and said that their House could not be answerable for the effects of large Pumps, unless placed as in his Drawing. it only remained therefore...... to fix the responsibility on the Engine Makers, by allowing the small Expense of alteration and placing the Pumps in their usual manner.

 This expense consists of removing about 4 Rod of Brick-work (£20) and of a few cubic yards of excavation ......" [10].

Only one side of the argument has come down to us but Fenton Murray & Wood were not always the easiest firm to deal with, and Matthew Murray had very fixed ideas over the arrangement of pumps, which probably came from his experience fixing pumping engines in the Great Northern Coalfield.

The Leeds Intelligencer in the last two weeks of June held an advert for the sale of Flax and Tow machinery which had been lately erected with a capital 6 H.P. engine by Fenton Murray & Co, and no doubt they had made some of the flax machinery as well. The mill was identified in subsequent adverts as being known as the Brewery at Hunslet-Moor Side. The sale was through a Mr Hurst, a watch maker at Market Place Leeds.

Simon Goodrich wrote to the firm on 30th June:

"Write to Mr Murray to forward the 3 steam engines as fast as possible and to prepare the cast iron pipe large stop cock and forcing pump send him a sketch for ditto" [15].

Goodrich did not clarify whose engines were to be forwarded, but there is no record of three engines for the Royal Navy in 1807. The number of engines would indicate that it was the follow on requirement expressed by General Bentham. Bentham by this time knew that his mission in Russia would not succeed, the Russians having concluded a treaty with Napoleon in February 1807 [16]. But Bentham may have intended to sell those engines already agreed for as a piece of private enterprise.

On the 5th August Simon Goodrich recorded receipt of a letter from Fenton Murray and Co answering a query on boilers, and at the end of August the Navy Board wrote to Goodrich:

"We have rec'd yr letter of 22inst and acquaint you we have desired Messrs Fenton Murray & Co to provide two wrought iron boilers to be in readiness to replace those of the 30 H. St Eng. at Pmth. Yds. agreeably to your proposal upon the terms therein mentioned which they are to send by water to Pmth Yd.
And as you have also recommended that a Copper Boiler shld. be provid'd immediately to replace one of the 2 wrought Iron Boilers of the said engine, in order to ascertain whether it may not be more economical and convenient in future after the Iron Boilers in Store are worn out to use copper ones in their stead we have agreeably to yr request desired Messrs Thoyts & Co to provide a copper at the price therein stated and to deliver it at Deptford Yd referring them to you for particulars as to its dimensions" [17].

Among the other letters in August was one from the Navy Board querying a charge from Fenton Murray & Co for £157/14/6d for the engineer to erect the 30 H.P. engine [17].

On the 1st September Fenton Murray & Co wrote to the West Middlesex Water Works to advise that they would be drawing a bill on them for £1000 at two months for the two engines, which was agreed [11].

On October 31st Fenton Murray &Wood wrote to Simon Goodrich advising that the first boiler had been shipped:

"We have the pleasure to inform you that we shipped the first boiler for the engine at Portsmouth according to the annexed account on Board the Hopewell Capt Palmer to Deptford on the 21st Inst. We were contrary to our expectations obliged to send it in two parts but have sent you the Rivetts necessary for the joints and any common Smith will be able to put it together with ease. It is an exceeding good one & hope it will give you satisfaction. The other is in hands, and will be finished soon.

The Honble Commisn[s] of the Navy
*   B O of Fenton, Murray & Co*
1807
Oct 28th   1 Long Boiler with square Ends, and internal flue rising out of bottom
*   & Stays for D[o].*

Cwt	Qtrs	lbs	s	£	s	d
87	3	25 @ 44		193	9	3" [18].

The West Middlesex Water Works Minutes of Directors from the 3rd November shows that they were pleased with their steam engines:

"The two twenty-horse steam engines made by Messrs Fenton, Murray & Co, of Leeds, have long since arrived, and are in a great state of forwardness; the beams, fly wheels and many other parts of both engines being already fixed in their places; and the whole of them will be fixed up and completely finished in about a month or six weeks. I cannot pass over this subject without observing that the contrivance, materials and workmanship of these engines reflects the highest credit on the house which has furnished them. The pump work is coming from the same place, and being nearly finished is expected to arrive by the time the engines are completed" [10].

The 1807 Poll Book for the County of York survives, and the successful candidate was the Right Honourable Charles William Wentworth Fitzwilliam, Viscount Milton standing as a Whig (liberal). He was opposed by William Wilberforce Independent and The Hon. Henry Lascelles Tory. The four partners in Fenton Murray & Co were listed as follows:

"	freehold	WW	Lascelles	Milton
Mirror Matthew - engineer (sic)	*Holbeck*			*1*
Wood David - iron founder	*Holbeck*			*1*
Fenton James – iron founder	*Leeds*			*1*
Lister Wm	*Bramley*	*1*	*1"*	

In the middle of November Viscount Milton paid a visit to Leeds after the recent election and the report in the papers included:

"After his Lordship had visited the White Cloth Hall, a large proportion of the rejoicing populace dispersed..; His Lordship afterwards proceeded to the manufactory of Messrs Benyon, Benyon, and Bage, which in point of ingenuity in the erection of the building, and extent of the works and machinery, is exceeded by few similar establishments in the empire. Through these works he was attended by Mr Benyon, and Mr Matthew Murray, the mechanic, who explained to his Lordship, as well as so short a visit would admit, the whole process of the manufacture" [19].

Simon Goodrich recorded the itinerary of a journey north at the end of 1807, in an un-inked part of his journal, and it included the cryptic *"Leeds - Fenton & Co"* [20].

The oldest pottery factory in Sweden, Rorstrand, received a 10 H.P. engine from Fenton Murray & Wood in 1807, which was erected and working in the factory by late in the year, drawings of the engine house survive [21].

January 1808 found Fenton Murray and Co. writing to the West Middlesex Water Works for agreement to draw a bill for £800 at three months for the pump work. The letter continued, in a manner that would be often repeated, to state the firm's need for cash:

"Having been disappointed of receiving a large sum from Scotland, which we were promised this Month, we are the more earnest that they (committee at the Water

Works) *would be so good as to grant our Request, for from the above circumstance we are really much pressed for money to discharge our Christmas Notes"* [11].

In February a quotation was made to John Christian Curwen for a large pumping engine for his Isabella pit in Workington, Cumberland. J. C Curwen was a well-known agriculturist and was M.P. at different times for both Carlisle and Cumberland. The saga of this engine has been recorded before [22], and the correspondence and references, which lasted until 1814 are preserved in the Curwen Papers in the Cumbrian Record Office at Whitehaven. The initial quotation, while long, is given here as it demonstrates the parts of the steam engine.

"Messrs Murray & Co Estimate Cost of the large Engine for Isabella Pit

4th Feb 1808.

Answer 9th Nov 1808.

Curwen Esq MP

Estimate of a Steam Engine 160 Horse Power, 66 Inch Cylinder with 9 Feet Stroke and 7 feet in Pumps capable of raising 14 inch column of Water 150 fathoms at 12 strokes per minute consisting of the under mentioned materials to be furnished by Fenton, Murray, Wood & Co Leeds.

Beam of Cast Iron in 2 plates	*Brass valves &co &co and 4 spare valves*
Gudgeon for D°.	*Metal Plate and Plungers for D°*
2 Pedestals & large plate for centre wall	*Globular condenser in 2 parts*
2 Plummer Blocks, & Caps for D°	*Bolts and Nuts for D°*
2 Brasses for	*Brass barrel for Air Pump*
8 Bolts for [lost] & 4 D° Wall Plate	*Metal Top & Bottom Parts D°*
2 Parallel Motion	*Lid with Stuffing Box Gland & co*
Brass work for D°	*Bolts Loops & Nuts for D°*
2 radius Plumr blocks Brasses & bolts	*Air Pump Rod*
2 Universal Ends for Beam Ends with }	*Bucket for D°*
Beat Iron Centre Pin & co }	*Brass valves for D° Foot & delivery de*
Piston Rod, socket for D°	*and rings for Stuffing box*
Piston and Cover	*Connecting Pipe between Air Pump }*
Iron Work for D°	*and Condenser }*
Brass & Bush for D°	*Bolts for D°*
Steam Cylinder Casings for D°	*8 bolts for holding down Air Pump & Condenser*
Cylinder Top Bottom & Ring	
Brass Work for D°	*4 Glands for D°*
Bolt & co for Cyr Top Bottom & casings	*Iron Rod & works for connecting*
Mettle plate for D°	*Air Pump to the Beam*
Bolts for [lost] Cylinder & D°	*Education Pipes*
Side Pipe & valve	*Bolts for D°*
Bolts for [lost]	*Inspection cock & valve*
[lost]	*[lost]*
Index Plate	*Metal pipe for D°*
Blow Pipe	*Hot Water Pump & Apparatus D° Working }*
Blow valve and seal	*Barrel, Bucket & Clack of Brass work }*
Bolts for D°	*Iron Work complete }*
8 links for Top & Bottom of Spear	*Lever for working Plug of Metal*
26 bolts & nuts for D° and plates	*Gudgeon for D°*

16 Brasses for Dº
V Bob of cast iron
Gudgeon for Dº
2 Plummer blocks & Caps for Dº
4 Brasses for Dº
8 Bolts & nuts for Dº
2 lengths of Steam Pipe
20 Bolts for Steam Pipes
2 Feed cisterns for Dº
Iron Work for Dº
Brass Work for Dº

2 Plummer blocks & Caps for Dº
4 Brasses Dº
8 Bolts & nuts Dº
Iron Work for connecting Plug to Beam
Steam Gauges and Barometer
Copper Pipes Gauge Cocks & co
Wire Pipe and Wire
Safety Pipe Iron and Brass work Dº
Manholes [lost] Bolts & co
Packing Cases
*NB The Boilers [lost] for Fire Place
and Bolts Nuts & Plates for condensing
Cistern are not included.*

*The above materials delivered at Leeds for the sum of £2700
One third to be paid on their arrival, one third when the Engine
Is set to work and the remainder at 3 months after*

As agreed I Have made the necessary calculation for your engine and find that it will require one of 160 Horse Power, which will be then loaded 14lb upon an inch of the area of the cylinder, or 7lb an inch on each side of the piston, this we think the best load an engine can have though it is capable of doing more. Should the Terms meet your approbation we shall be happy to receive your early order or give you any additional information you may require. We think it unnecessary to say any [more] about these engines until you see your servant Mr Penrice who I hope will be able to give you satisfaction when you see him, about those large engines at Newcastle. Waiting your reply I remain etc Mattᵂ Murray.
PS We provide an experienced engineer to assist and supervise the erection of the engine whose wages and expenses form a separate charge, we also furnish without further charge the necessary plans for the direction of the workmen in the erction of the Engine House &co" [23].

During February and March 1808 the Leeds papers contained adverts for good Greensand Moulders required for Fenton Murray Wood & Co [24].

In March Fenton Murray and Co were again in correspondence with Simon Goodrich concerning the boilers for Portsmouth Dock Yard.

In April the West Middlesex Water Works approved further payment and the Director's minutes for 3ʳᵈ May recorded:

"I have now the pleasure to inform you that your engine house is finished, and that the two steam engines with their boilers and apparatus are also completed, and were both worked, in the presence of your Directors, on the 30ᵗʰ January last.
  The cast iron air vessel, together with the pump work and apparatus for the completion of the engine work and reservoirs, together with the iron pipes for drawing the water out of the reservoirs, and for filling them are already fixed; so that nothing now remains to finish the engines but the completion of the large pumps for supplying the water, which would have been now done had they not been detained by contrary winds upwards of five weeks" [10].

A reminder that dependency on wind could have a severe impact, when they were contrary, on the ability of vessels to enter or leave harbour.

On May 27th, as part of an ongoing exercise, which was approved by government, to see whether saw mills should be established at military establishments other than Portsmouth, Marc Brunel wrote to Colonel Cuppage at the Woolwich Arsenal to set a meeting for the following Tuesday. Brunel had received a letter from the Board of Ordnance asking him to quote for a saw mill at the Royal Arsenal at Woolwich [6].

Fenton Murray & Wood were advertising for a Loam and dry sand moulder in June [25].

In the middle of June, after clarifying the requirements for the Royal Carriage Department at the Royal Arsenal, Marc Brunel addressed a letter to Fenton Murray & Wood (to Mr Wood):

"I will thank you to inform me on what terms you would undertake to furnish me with a Steam Engine of 18 Horse Power to be erected in the neighbourhood of London, and to have it fixed in place ready to work.
The masonry will be prepared according to your directions.
I am particularly interested in having the exact expense ascertained. Let me know also your price for Shafts from 2 ½ to 4 ½ inches in diameter. Spurs and Bevel Wheels, wrought iron shafts of about 2 in diameter with Carriage and xxxx.
I shall also want some heavy castings such as Iron Columns 12" in diameter 16 feet long hollow in the middle screw Bolts with Nuts.
You will oblige me by sending me a table of the dimensions of patterns of wheels you have in readiness" [6].

On the 28th June 1808 the minutes of the West Middlesex Water works record the concerns of the new engineer over the fact that the engines were incapable of acting independently because they only had one air-vessel. They also show that Matthew Murray had recommended a Mr Allen Hodgson be taken on to superintend the laying of the iron distribution pipes. They also authorise the acceptance of a bill for £1000 drawn on Fenton Murray & Wood, despite a previous ruling against it [10].

In July 1808 Hughes, Bough and Mills had a new steam dredger available in London. Using the hull of a dredger put together by Trevithick which they had purchased in 1807, they had installed new equipment in it including an engine by Fenton, Murray and Wood. The vessel was described and illustrated in the Edinburgh Encyclopaedia (1814) by John Farey, who gives the engine power as 16 H.P (see below). Simon Goodrich who described the dredger at Woolwich in 1809 gave the engine power as 20 H.P. and this is also the power quoted in a paper published in 1844 [26]. It had two chains of buckets to raise mud from 28 feet and Goodrich reported that he had been informed it could fill 15 barges with mud in a day.

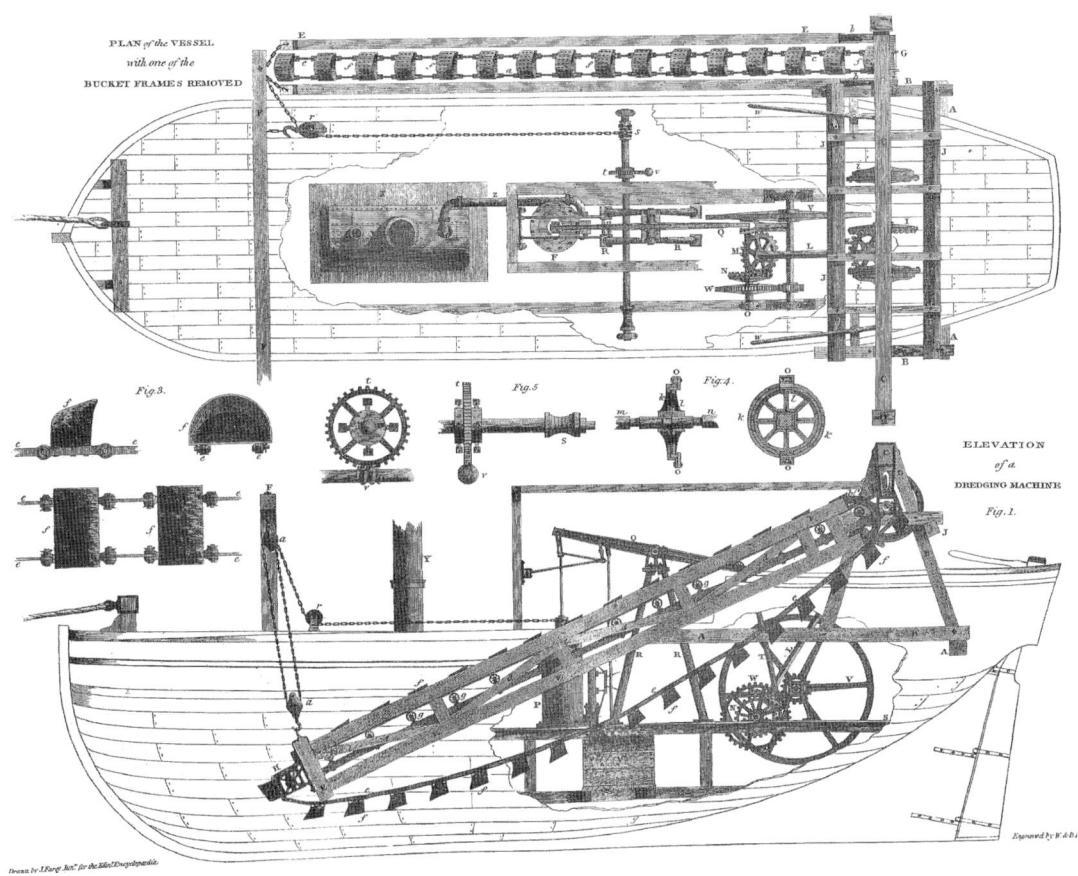

33. Hughes, Bough and Mills Dredger with equipment by Fenton Murray & Co.

On July 1st Marc Brunel wrote to the Board of Ordnance to give them the plans he had drawn up after consultation with Colonel Cuppage and another officer, who were in charge of the Royal Carriage Department at the Arsenal. He also advised that their requested 12 H.P. engine would have insufficient power and that he had substituted one of 18 H.P. He was also working on some of the erection expenses, and would provide them if the plans attached were approved. On July 20th he was again writing to Fenton Murray and Wood with details of his plans, dimensions of the wheels and shafts to be driven by the engine, etc. He requested that they send as soon as possible their costs for making, delivering, and erecting the engine; he was very short of time. He wrote again on July 30th to chase their answer, without at any point informing them of the projected use. Marc Brunel submitted his proposal and terms to the Board of Ordnance on the 3rd August including a 20 H.P. engine, which allowed 4 H.P. for operating some machinery other than the saw mill. The quote included the wheels and shafting, the sawing apparatus and columns and drums. The mill was capable of hauling the logs 600 feet to the mill. The price was £6000, or £5900 if the Board elected to proceed with an 18 H.P. engine. It would appear from the correspondence that the steam engine was to be supplied by Fenton

Murray & Wood, it is safe to assume that the sawing machinery was to be supplied by Maudslay & Co. The columns and mill work probably to be supplied from Leeds as they had been asked to quote for structural columns, but it could well have been split between the two firms [6]. The building was to be erected to the plan by the officers at the Arsenal. However the plans were postponed, probably due to a desire to properly plan the extension of the Arsenal to encompass the proposed works.

There was a brewery in Margate owned by the Cobb family, and they considered installing a steam engine over a number of years, and received many quotations. One of the letters was from Fenton Murray and Wood dated September 14th 1808:

"At the request of Mr Adams, Plumber who has seen some of our Engines here, I take the liberty to solicit your order for an 8 Horse Engine the cost of which including 1 Beat Iron Boiler, with Cast Iron Cistern, Iron Beam, & four diagonal Pillars to support d° & the whole of the Engine except the Boiler fix'd upon one large Cast Metal foundation Plate, the whole of the Materials of the newest & most improved Construction for the sum of £506 del.[d] at Margate payable in a Bill at 3 Months from such delivery. I beg further to observe that these Engines are set to work & tried before they are sent off which greatly lessens the expense of refixing as no Millwright Work is necessary in the Erection. Should you favour us with an Order could send one of these Engines off from Leeds in One Month. Should you favour us with your Communications please direct
> *Fenton Murray & Co*
> *Steam Engine Manufacturers*
> *Leeds*
or should you wish for further Information or to see a Plan of these Engines by your writing a Line to me at the Saracens Head Snow Hill (where I shall be for a few days) I will come down & see your situation.
I am etc. M Murray" [27].

An early form of cold contact, but of interest because it states that the firm tested their engines prior to shipment. The solicitation did not generate an order, and Fenton Murray & Co answered a direct query from Cobb's in 1818, when they were still considering whether to go steam powered.

In September John Hick, the brother of Benjamin Hick, left Fenton Murray & Wood, after four years with them, to take up the position of traveller and book-keeper with Sturges & Co at the Bowling Iron Works [28].

In this period John Marshall was looking for ways to improve and mechanise the flax hackling process. He described an early machine in Experiment Book 3, where he dismissed a proposal of Matthew Murray where the flax would hang down from clamps and all the cutting motion would come from the moving heckles (spikes). The machine, incorporating Marshall's ideas which he was working on with his then partner Hives, he recorded did not answer [29].

The new engineer at the West Middlesex Water Works, Ralph Walker, included in his report for improving the water works proposed alterations to the steam engines which included adding another Air Vessel and placing two smaller working pump barrels on the suction pipes and cleck boxes, etc, to avoid the expense of a new engine [10].

In October John Marshall described a hackling machine under Experiment No 98 Patent Hackling Machine. The description shows that Marshall and Hives were coming around to Matthew Murray's idea of hanging the flax and cutting it with rotating hackles.

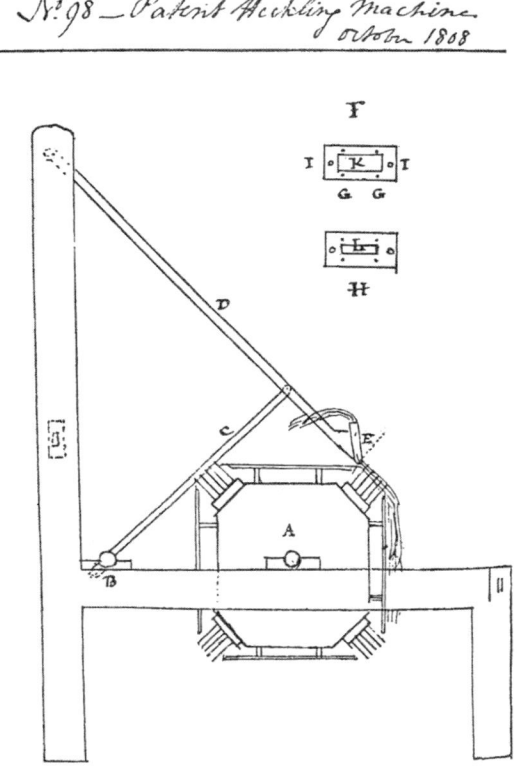

34. Marshall & Hives flax hackling machine.

A was a cylinder with four sets of heckles on it. The flax was suspended from a holder rail *E* which moved 100 times to each 25 revolutions of the cylinder. The machine was put to work with the holder filled with 8 oz of flax, the holder moving to allow each row of heckles to break the flax. When one side was broken the holder was turned. Then a finer set of hackles was introduced and the process repeated. There were 4 sets of heckles [29].

On 14th November 1808 an acknowledgement for the order of a 160 H.P. engine was sent to Mr W Swinburn who was Mr Curwen's agent or manager. Not all the letter remains legible, but it is clear that the plans were to be drawn up and sent.

There was also a discussion over the manner of controlling the pumps. The usual practice at this time was to use a diagonal spear and V bob, which had drawbacks but does not appear to have been replaced by a system of counter balances for a couple of decades [30]. The letter stated that a diagonal spear and V-bob (a secondary beam) would be used, as Fenton Murray and Co had experimented without a diagonal spear on an engine erected at South Shields, but it had not answered as well. This pumping engine was erected for Mr King in 1805, who was also the proprietor at Jarrow [31].

During 1808 Fenton Murray and Wood had supplied the iron framing, machinery and steam engine for a flax spinning mill at Broadford in Aberdeen for the firm Scott Brown & Co, and it is probable again that this had been a turnkey operation, with the Leeds firm managing the whole process. The building still stands as the oldest iron framed fireproof mill surviving in Scotland, albeit with later additions all around.

35. Broadford Mill Aberdeen Second Floor of the 1808 mill.

Robertson Buchanan published his work entitled 'An Essay on the Teeth of Wheels' in which Murray and the firm had three mentions. The first was to state that Mr Murray of Leeds had found a way of boring the hardest cast-iron. The second was an exposition of the theory of hypercycloidal motion as:

Corollaries:

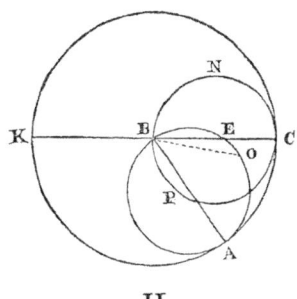

II.

36. Diagram of hypercycloidal motion.

When the generating circle CNP revolves within the circle of its base, and has for its diameter the radius BC of its base, the point C, the place of the style during the revolution of the generating circle, will always continue in the diameter CBK. Hence the epicycloidal described by the style C is a straight line, and a diameter of the circle of its base.
Upon this principal a parallel motion has been constructed. It is used by Messrs Fenton, Murray, and Wood in some of their smaller steam engines.

The third relates to the length of teeth of wheels:

"It has been mentioned to me, that the following rule, in order to determine the length of teeth of wheels, is employed by the ingenious Mr Murray of Leeds. I am informed it was communicated to him by the late Mr Rupp of Manchester, a native of Germany. Perpendicular to the line of centres CD, draw the line AB, a tangent to the pitch lines. Take half the pitch, that is, half the distance between the centres of two adjoining teeth, within a pair of compasses, setting their points upon the pitch lines, E and F, parallel with the line of centres CD, draw the line ab, and where that is cut by the line AB, at c, gives the points of the teeth of wheel and pinion.

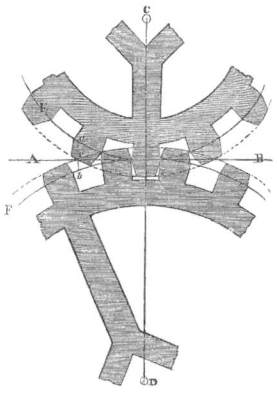

37. Teeth of wheels.

On this rule, I beg leave to remark, that it does not seem to me to be founded on any satisfactory principle; were the pinion, at all times the conductor, I should not perhaps differ from Mr Murray, because the action of the teeth would be, in that case, generally after their arrival at the line of the centres.
But in the case the wheel were the conductor, the action of the teeth would generally be almost entirely 'in approaching' the line of centres" [32].

The data used for the book will have been compiled some time before 1808, as has been seen from Goodrich, Murray was aware of the shortcomings of the hypercycloidal motion for any but the smallest engines as early as 1803. The reference to Mr Rupp of Manchester, supports the contention that the data was compiled some time earlier as Rupp died in June 1806. Theophilus Lewis Rupp was a cotton-spinner operating in the Manchester area, who made some improvements in Chemical Bleaching [33] and his work was known to John Marshall who had read his article when he was investigating bleaching in 1798/9 [34]. Rupp, who was then part of the firm of Scarth, Marshall, Rupp and Co, was involved in the building of a new cotton spinning mill around 1799/ 1800 [35]. This mill appears to have been powered by a Boulton and Watt engine. However the partnership was dissolved on the 8th August 1801. In the Boulton and Watt Collection there are some drawings of early cast iron engine beams including examples from Bateman and Sherrat and Fenton Murray and Wood 30 HP. The Fenton Murray and Wood drawing is annotated 'Mr Rupp', which if it were for a subsequent mill for Mr Rupp, would make it a very early cast iron beam, possibly dating to 1802/3 which would equate to the period when Boulton and Watt were introducing such items.

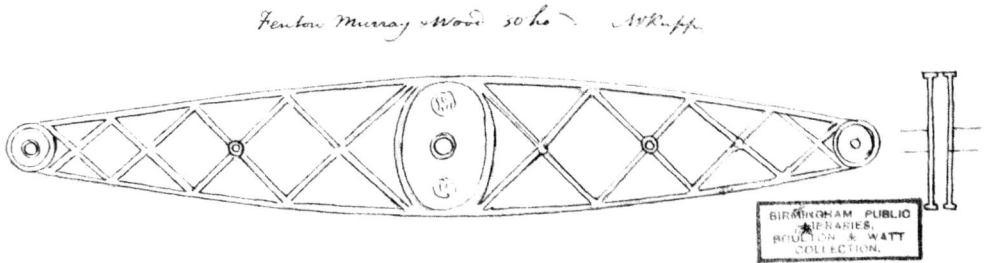

38. Fenton Murray and Wood iron beam in the Boulton and Watt Collection [36].

In January 1809 nearly 13000 square yards, with defined exceptions including 1941 square yards to Matthew Murray, of land were mortgaged by Fenton Murray Wood and Lister to Edward Sage and Richard Jennings, trading as Sage and Jennings and Sage Rawdon and Jennings, as mentioned earlier in Chapter 7 [37]. This arrangement, although it presumably altered over the years seems to have continued one way and another as R Jennings appears in the firm's accounts regularly until 1825, the year of the last surviving Customer Account Ledgers [38].

In the same month a progress letter was sent to Mr Swinburn over the 160 H.P. engine for the Isabella Pit at Workington, and, presumably in response to a point from Mr Swinburn, stated that the use of the diagonal spear was the:

"first and only plan to be observed for so large an engine" [23].

It is worth noting that at the time of erection the Isabella Pit engine was the largest in the world, but it was far from the first steam engine to work in the Curwen collieries at Workington, they had 4 or 5 Boulton and Watt engines and 3 engines by Heslop (a local engineer). The first engine purchased in 1788 had been from Boulton and Watt.

It is not known why J C Curwen turned to Fenton Murray and Wood for the engine for the Isabella Pit, but it can be surmised that their reputation was high in the Northumberland and Durham coalfields, so it was probably through a recommendation. By 1814 Curwen was consulting with John Buddle over the management of his coal concerns.

In January Matthew Murray submitted a model of his hackling machine as his invention to the Royal Society for Arts, and was awarded a gold medal for it the following year, under cover of the following letter:

"Having invented and put in practice a machine for the purpose of hackling flax and hemp in an expeditious and more perfect manner than hath hitherto been done either by my hand or machinery, I have taken the liberty of sending a model thereof to the Society. From its national importance as a powerful auxiliary to the state of our linen manufactory, which is at present cramped by the refractory conduct and inability of hand hecklers, it has been represented to me that this machine is extremely desirable. I believe that I have completely effected the end, and hope that this machine will deserve the Society's attention.
I have etc. Matthew Murray" [40].

39. Original model (somewhat modified) and gold medal reunited briefly in 2009.

Matthew Murray was aware that Marshall and Hives were developing a hackling machine as his opinion had been sought on its operation by Marshall, who had then originally stated that Murray's method would not work (see above). Murray was no doubt also fully aware that they had adopted his idea and intended to take out a

177

patent, which was indeed granted in Hives' name in August 1809. As Trevor Turner pointed out, this was a dilemma for a manufacturer of textile machinery, to have had the idea that gave success to the machine and not be able to make and sell the machines without patent infringement and upsetting his old patron and a major customer into the bargain! It therefore may well be that the submission to the Royal Society was Matthew Murray's get out, as it demonstrated clearly in the public domain that he had the idea before the patent [41]. If this was his motivation, he may have felt some satisfaction when in 1818 Hives took the Benyons to court for infringing his patent. The Benyons produced witnesses to attest that the firm of James Tennant had been operating such hackling machines made by Fenton Murray and Wood before the grant of the letter patent and that the machines (as made by Fenton Murray and Wood and Hives) were not distinguishable in principle. James Tennant had been one of those who certified the benefits of the machine to the Royal Society in 1808, prior to Matthew Murray being awarded a Gold Medal by the Society in the Mechanics section (one of four) in June 1809.

Following complaints from a Vice-Admiral Wells to the Commissioners of the Admiralty [42] concerning the slowness and unreliability over watering ships lying at the Nore (a well-used anchorage for Naval ships in the Thames Estuary, infamous for its mutiny), the matter was put under investigation and in February 1809 Simon Goodrich submitted his report to the Navy Board which described the three wells involved, the Victualling Yard well at Queensborough, the subscription well at Sheerness and the Navy Well at Sheerness . The Navy Well used horses to raise the water. Goodrich recommended the installation of a small steam engine, as the horses could not raise all the water in the well [43]. In the minutes for the Navy Board of 4th April 1809 Commissioner Bentham proposed writing to Vice Admiral Wells over the siting of the additional wells he had proposed in his letter to the Admiralty and that meanwhile the Board should proceed with the repairs proposed to the existing Navy Well and that a ready-made steam engine of 4 to 6 H.P should be installed with a pump. A minute was agreed for Mr Goodrich to "look out for a steam engine" [44]. At some unknown point the work for the steam engine and pump was given to Fenton Murray and Wood, as on the 6th July 1809 Goodrich wrote of a day in his London office that:

"Mr Murray of Leeds calls busy about the Sheerness Water Works" [45].

A drawing of a 6 H.P. engine made by Benjamin Hick while working for Fenton Murray and Wood dated April 18th 1809 in the Goodrich Collection was provided at this time. The engine was virtually identical to that which the firm proposed in 1810 to John Watson as a ballast engine (illustrated below).

The Navy Board index of correspondence noted letters from Fenton & Co for an iron boiler for Portsmouth Yard and the agreement for the same on 28th July [47].

Throughout most of 1809 Fenton Murray Wood & Co were advertising for Loam Moulders in the Leeds papers [48].

On September 22 Fenton Murray & Co wrote to Mr Goodrich with the estimate for the pump work for the Navy Well at Sheerness, which was detailed by part and had a total of £425 –0-7 1/2 d [49]. The engine (6H.P.) estimate followed on September 29th at £554-14-9 and was annotated as approved on 6th April 1810 [50]. In response to their query as to whether to insure the engine in transit, Fenton Murray and Co were instructed not to.

In early October 'The Times' and the Morning Chronicle carried advertisements that the partnership of Messrs Bradley and Coxen at the Eagle Foundry in Southwark was to be dissolved and therefore there was a capital 4 horse power engine by Fenton Murray and Wood for sale complete with all the apparatus including brickwork and timbers [51].

In December the Leeds papers record the Jubilee Collections (George III became monarch in 1760) with Fenton Murray and Wood donating 4 guineas.

Also in December, Wm Swinburn, Mr Curwen's agent, wrote to the firm to inform them that the diagonal shaft was completed and the engine house was "considerably above the ground". They therefore needed immediately the plans for the boiler and boiler seats. He would advise the depths of the different lifts for the pumps and believed the working barrels of the pumps would have to be brass as the water would be corrosive (the pit went under the sea and therefore the pumps would be working sea water). He also advised that they needed the engine to be working by early summer 1810 [39].

In Whitehaven a few miles down the Cumbrian coast from Workington where the above engine was destined, the year 1809 saw the construction of a flax mill on the Castle Meadows by Joseph Bell and John Bragg. The Mill survives today, known as Catherine Mill or Barracks Mill, and has been restored to provide accommodation. The Iron structure of this fireproof mill is of the pattern that has become known as the hallmark of Fenton Murray and Wood [53], and as was usual in these circumstances the 30 H.P. steam engine was also made by that firm and the correspondence with the Royal Society on the flax hackling machine notes that the hackling machine had been ordered by Bell and Bragg of Whitehaven. Another example of the provision of a turnkey mill, (mill, engine and machinery) by Fenton Murray and Wood?

40. Catherine (Bell and Braggs) Mill Whitehaven in 2010.

The surviving records for the beginning of 1810 start with a complaint from Mr Swinburn dated 13th January 1810 that he had not had any response to his letter of the 13th December 1809 and that if there were losses Mr Curwen would have to consider who should bear them. The levels of the pumps are described and specifications for a cast iron cistern are requested. On the 15th January Fenton Murray Wood & Co replied that they needed to view the layout situation and expected to visit within the month and the engine had been put into hands etc. etc. with every exertion to be used to get it to work in the summer and the letter then went on to say that they did not understand the language about where the losses might lie, and did not deserve such anyway. There was then a clarification that Workington were to make the pumps or to place an order on Leeds. The plans for the cistern would be put in hands. Mr Swinburn replied promptly on the 17th January, stating that if his letter had been answered quickly he would have sent off the plans of the layout, obviating the need for a visit. He confirmed that the pumps would be made locally but thought they needed to know the details to confirm the engine arrangements, and could Fenton Murray and Wood advise the best arrangements for fixing the pumps [39].

Simon Goodrich when writing his journal for the 1 February 1809 included the following piece, which demonstrated that Murray was still keeping abreast of what was new, even in London:

"*Obtained at an interview with Mr Matw Murray the following information about Gas Lights:*

100lbs coals produce 400 cubic feet of Gas in 4 hours each cubic foot is equal to 4 candles of 8 in the pound – Half the cokes from each ……. Distillation is sufficient to perform each subsequent distillation. Thus if the half of the cokes that will remain for other purposes be equal in value to half the coals used it may be said that 400 cubic feet of Gas is obtained at the expense of 50lb of coals = 1 cubic foot of Gas ¼ lb. of coals = half a pound of candles or taking the actual quantity of coals distilled it seems curious that half a pound of coals distilled should produce as much lights as half a pound of candles.
Mr Murray advises me to go and see Mr Geo Alderson's Works Thames Bank Kings Arms or No16 great Marlborough Street" [54.]

On the 6th February Matthew Murray, on his return from London, wrote to Mr Swinburn smoothing things along saying that the plan for the boilers and seating would be sent as soon as possible now that they had Mr Swinburn's plan of the situation, the plan for the cistern would also be done soon. He also advised that he was not sure of the best plan for fixing the pumps but he was due in Newcastle and would enquire into their best manner of fixing pumps and inform Mr Swinburn. Murray stated that they had cast for the engine one of the best cylinders "*I ever saw*" [23]. Reading the correspondence between Workington and Leeds emphasises the difficulties of exchanging data at that time when letters and meetings were the only available methods, and these were complicated when one of the principals was fairly often travelling as Matthew Murray was.

Activity in February and into March, when adverts appeared in London, with Murray himself visiting Aberdeen, demonstrated that Matthew Murray was a salesman who travelled. They advertised premises lately occupied by Mr C S Millward at Bromley 3 miles from London, on the River Lea, which included recently erected buildings and a steam engine of 36 H.P, by Fenton Murray & Co. The engine being capable of working any description of machinery and having unusually strong foundations [55]. Unfortunately nothing else is known of this building, but the area was developing as a major industrial centre on the edge of what was then classed as London.

On March 8th 1810 Messrs H and H advertised for roomy premises for a manufactory within seven miles of Blackfriars Bridge, with plenty of water to supply a steam engine. They also sought:

"Any person having a steam engine of eight or ten horse power to dispose of, on Fenton Murray and Co's principle, may apply as above" [56].

The end of March found Matthew Murray back in Aberdeen at Dempster's Hotel from where he replied to a letter from Mr Curwen stating that the boiler plans had been sent to Mr Swinburn, offering assurances for completion early in the summer, and promising a visit by "*our Mr Wood*" in April [23].

In October 1809 the Victualling Board had approached the Admiralty with a plan to erect a steam engine at Deptford to replace the use of horses for pumping water for the brewery and also to provide clear water for ships. In addition it was proposed to use the engine to mash the malt and obtain a better extract [57]. The proposal was accepted and between April and August 1810 Fenton Murray and Wood erected a 10 H.P. engine in the Victualling Yard at Deptford for the Brewery. Simon Goodrich and Joshua Fields visited the yard in 1813 and Fields sketched the engine in Goodrich's journal, showing it to be a development of Murray's portable engine of 1805.

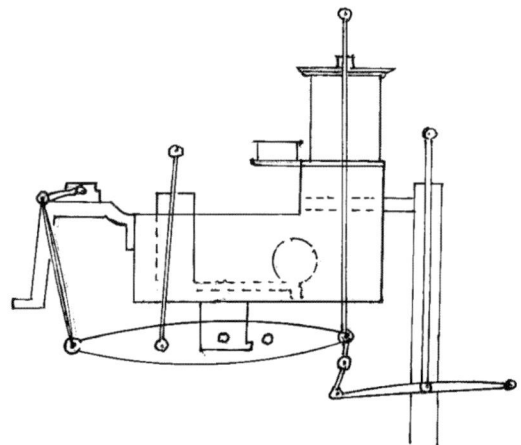

41. Fenton Murray & Co 10 H.P. engine at Deptford after J Fields.

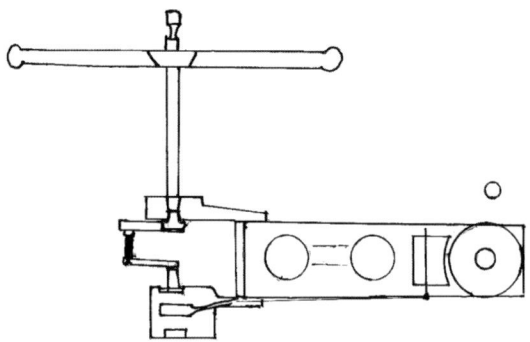

42. Plan of Deptford engine after J Fields

It was described, alongside the sketches, as having the steam cylinder 3 feet 10 7/8 inches long in the bored part with a diameter of 17 inches. The steam entrance hole into the cylinder was 2 inches wide, 3 ½ inches long and ½ inch from the top cylinder flange. The cylinder top had 12 bolts 7/8 inch and about ½ inch from the edge of the flange. The steam cylinder and case was cast as one piece and the steam came in by a round hole. The piston was made with a pinion and wheel with a screw on the piston rod to tighten the packing and was about 7 inches deep [58].

The invoice, including charges for erection was for £1018-9-5 [59], and was referred by the Victualling Board to the Commissioners of The Navy for the attention of the Civil Architect and Engineer on the 5th November 1810 [60]. The copy invoice is marked:

"*The rates of charges in this Bill appear reasonable*".

Sometime in April 1810 it appears from letters written, that Marc Brunel travelled north and visited both Fenton, Murray and Wood in Holbeck and Fox of Derby. A letter to the Leeds firm dated 25th April reads as follows

"I will thank you to put in hand Two apparatus agreeable to the first drawing I have left with you differing only in the size of the Chuck or wheel which may be reduced to 9 feet instead of 10 feet and the thickness of ¾ of an inch, from the centre to the extremity. It will therefore be parallel in every part. Should you request any further direction be so good as to let me know. I hope the two small ones are in hand and very forward. I will thank you to inform me of your terms for a Steam Engine 10 or 12 Horse Power on your most improved plan and of the time at which it might be ready" [6].

During 1810 Marc Brunel's preferred supplier Maudslay was fitting out his new factory in Lambeth, and other manufacturing sources were needed. Brunel was expanding his own facilities at Battersea and supplying an uprated sawmill for Borthwick at Leith. As is clear from later letters, both these saws (or as Brunel called them apparatus), and the steam engine were for Battersea. Borthwick's mill was to be powered by a waterwheel (eventually made by Bryan Donkin) [6].

Fenton Murray & Wood were still doing work for the Northern Coalfields and on May 3rd they wrote about an order for an 8 H.P. ballast raising engine to John Watson. Mr Overton who was a senior engineer at Fenton Murray & Wood had been up to discuss some changes [61].

Shortly thereafter a letter (written on May 7th) was received from Wm Swinburn with detailed calculations for the pumps and working barrels for the Isabella Pit at Workington, giving the volumes of water to be raised in each set of pumps and the necessary diameter of the pumps etc [39].

Marc Brunel wrote on the 9th May chasing a steam engine and his saws. This was followed by a further letter on 17th May requesting the firm to put in hand without delay a 16 H.P. engine which he needed in four months. He also wanted progress with his circular saws. And again on the 25th he wrote:

"In my last letter of the 17th instant I requested you to let me know your Terms for a 16 Horse Steam Engine which I desired you to put in hand for me. I will thank you to favour me with an answer, stating also the expence of freight for London and fixing the Engine. It is necessary for the disposition of my works here that I should know how soon one or both of the small apparatus you have in hand for me will be ready to be forwarded. You will much oblige me by informing me of the state in which they are, and when I am to expect the large one" [6].

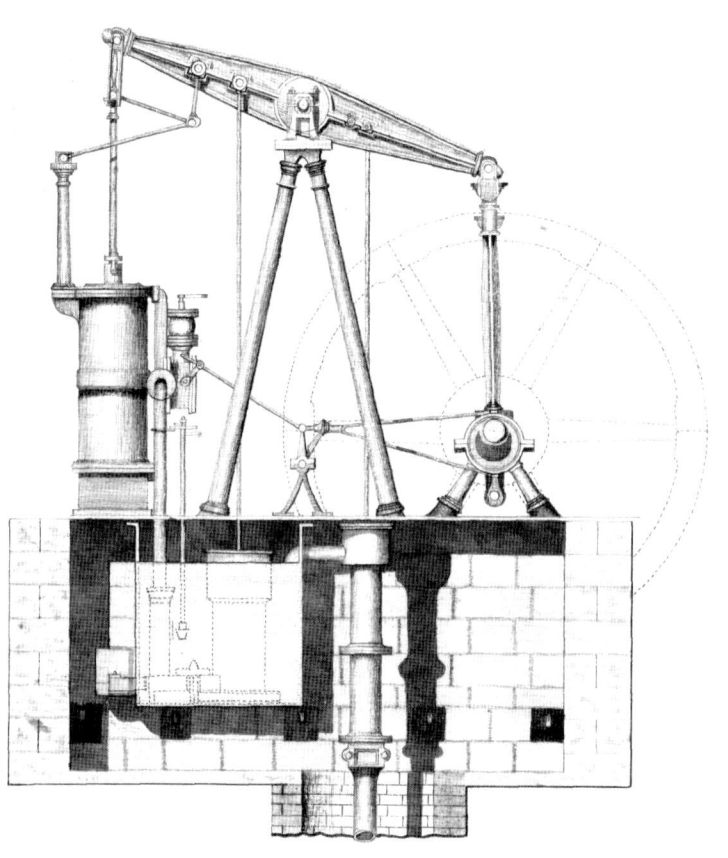

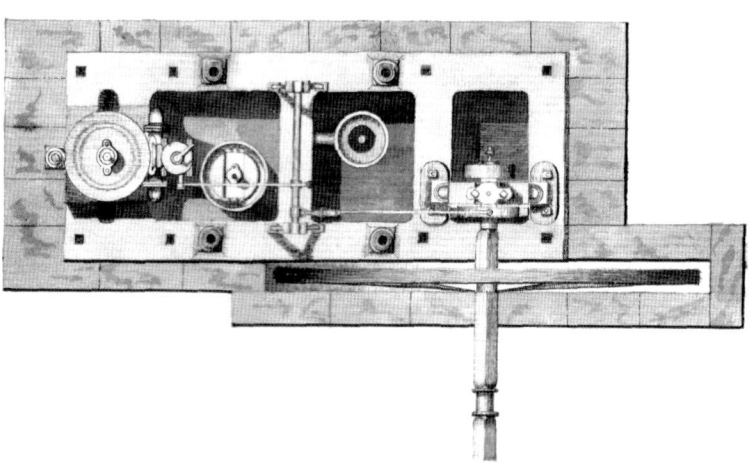

43. Drawing of 8 HP ballast engine as sent to John Watson 3rd May 1810 [61].

By the 26th May Matthew Murray was writing his reply to Wm Swinburn with a very different proposal to that made on May 7th as he had no doubt visited the collieries at Newcastle and learned the best practice. He proposed sending Mr Overton to resolve the matter, but Mr Overton was delayed with a bad leg. After describing the proposed layout for the pumps, the price for brass working barrels was given at 2/4d per lb payable in three months. The short credit being explained due to fact that Fenton Murray & Wood had to pay cash for all their brass and copper. The letter then stated that the Brass air pump had just been bored and was the best of the kind ever seen. A shipment of materials should be sent in two or three weeks. A delay was then admitted to:

"by our being obliged to make a new boring mill for boring the cylinder as we thought the one we had too slender, to make it a perfect".

Detailed calculations for the pumps were attached. The letter was acknowledged by Wm Swinburn on the 31st May. Swinburn pointed out some errors on both sides and made some clarifications, and hoped to see Mr Overton shortly [39].

Fenton Murray & Wood had at last responded to Marc Brunel albeit the letter does not survive. Marc Brunel wrote to the firm on June 4th advising that the engine area of the mill was not yet fixed and he would be guided by their plans for fixing up the engine. He went on to express surprise that the two small items ordered were not ready, and if he had known how long it would take he could have got them made in Town and:

"As to the two larger Apparatus I shall want the plancher or chuck only from the casting having here the patterns for the shafts, frames and plummer block which can be put in hand immediately".

On the 14th he again wrote asking whether they could make a waterwheel with shafting etc. (before he had contacted Donkin) and how soon. Also he wanted an answer to his letter of the 4th. The waterwheel and shafting was for Borthwick's in Leith. On June the 20th he acknowledged a reply which had informed him the work was not ready and went on to voice his dissatisfaction:

"I have hitherto complained of the slow progress of the work we obtain in London, but I find you are even worse".

And further he added:

"I am sorry that your prior engagements interfere so much with my work. I have more I wish to have done".

In between voicing his complaints Brunel agreed to a change to the movable collar on the back of the chucks. His patience finally expired after a further bout of silence and Brunel wrote to the firm on July 12th that his partner Mr Farthing was coming to Leeds to personally progress the work, and added:

"Lest the water wheel, shafts etc I have written to you about lately should retard the work you have for us I must not at present add to the stock of your orders".

A week later Marc Brunel answered a letter from Mr Farthing at Leeds, stating that he had the plans (steam engine?) and went on to say he imagined the six foot saw had been turned and that the others could be turned in London. He then says that he would forward the dimensions for two spur wheels and the other shafts for the saws [6].

In his journal for 1st August 1810 Simon Goodrich noted that the 10 horse engine at Deptford needed 10 gallons of water a minute and it was proposed to take it from the canal by pipe to a tank by the brew house [62].

Marc Brunel contacted Mr Farthing on 7th August with more dimensions, including changing those for the two larger apparatus.

Fenton Murray Wood & Co were still working on the engine for Mr Curwen's Isabella Pit and on 18th August they wrote to Wm Swinburn:

"We wrote to you on the 6th saying that we were much obliged to Mr Curwen for his kind offer of advancing us the first installment & that we should be glad to receive it - Money is uncommonly scarce with us & would now be exceedingly acceptable - Not having received any Answer to our letter we are apprehensive that it has somehow or other miscarried" [23].

The bills of exchange duly arrived and were acknowledged with sincere thanks by Matthew Murray on August 23rd. They were followed up in late September by a letter of concern from Wm Swinburn that nothing had been heard from Leeds and the engine house was completed. On the 1st October Matthew Murray advised Wm Swinburn that the goods should be put on board that week, but by the end of the month wrote again to admit that 40 tons had gone to Selby as that was the most the river boats could ship. The additional 12 tons was waiting to be shipped. Murray stated that he had been ill and confined (to bed). He stated they could make Swinburn a Fly Punch and would do so if he ordered one.

On November 8th Fenton Murray Wood & Co sent a detailed shipping list of the goods despatched amounting to 52 tons 19 hundredweight 3 quarters and 13 pounds. They had been shipped on the Providence with Captain Joseph Foster. Fenton Murray and Co advised Wm Swinburn that the outstanding materials would be shipped via Liverpool, and that they were in the process of making his Fly Punch. This missive was acknowledged by Wm Swinburn on the 12th November, when he asked whether they had as requested taken out insurance, and for how much, and at what premium.

He was informed, dated 19[th,] that the sum of £2400 had been insured in London for £3.3 per cent, and that the Fly Punch would be ready in 4 or 5 days [39].

Further letters passed regarding the insurance which had been arranged by a Mr Grace and Fenton Murray and Co were asked to look after the policies and settle any commission with Mr Grace. Wm Swinburn was advised that there were no fees from Mr Grace and that the London brokers were called Grigg, Green & Co and a list of thirteen underwriters was given. Captain Foster had only sailed from Selby to Hull and there the ship had been taken over by Captain Wright.

In early December the details of the shipping of the Fly Punch with detailed list totalling to 30 hundredweight one quarter 1 ½ pounds was sent to Wm Swinburn [39].

Joseph Glynn, an eminent engineer of the day in the field of steam engines, described Matthew Murray's crane, which was credited elsewhere to be in operation in 1810, as follows:

"The application of a column of water to lift weights was made, many years ago, by the late Mr Matthew Murray of Leeds, who employed it to raise the heavy boilers he manufactured for the spinning mills in that district.
His mode of using it was very simple and effective. From a cistern placed upon a lofty building, a water pipe communicated with a cylinder set upright upon the top of "a triangle" formed of three stout trees, and fitted with a piston and rod, which passed through a collar of leather or stuffing box in the bottom of the cylinder; on the end of the piston rod was a loop or shackle, to which strong chains were attached, for suspending large boilers, engine beams, &c.
By admitting the water between the cylinder bottom and the piston, the load was lifted sufficiently high to allow a waggon to pass under it, and, by allowing the water to escape, the weight descended upon the carriage. … Thus in Mr Murray's lifting apparatus, if the cistern were 60 feet high, and the cylinder 40 inches in diameter, the pressure upon the piston would be sufficient to lift a weight of 24 tons" [63].

1810 had proven to be a busy year, just on the surviving records of three main pieces of business, in Cumbria and London and had taken Matthew Murray to Aberdeen. If the full records of such a year had survived it would have been fascinating. Marc Brunel in writing to Borthwick stated that all the shops were very busy, and it was difficult to get work done. Marc Brunel's surviving letter book unfortunately ends in April 1811 and so far no further letters to Leeds on how this business turned out have been found.

Early in 1811, on January 15th Wm Swinburn wrote to inform Fenton Murray and Co that the Providence with its 52 tons of engine parts had been driven on shore on the east coast of Ireland near Donaghadee, which is just south of Belfast Lough. They were to advise the underwriters. This was duly done and the message passed back that the insured must do the best in their power, using local agencies as necessary as the underwriters could take no action at that stage.

From February up until September 1811 and for a couple of years thereafter the Scottish papers ran an advertisement for the sale of West Ward Mill Dundee. As seems to have been the norm the engine was attributed to Fenton Murray & Co, but not the machinery or fire proof mill. One exception was the sale advert for Broadford Mill Aberdeen which appeared about the same time which did state that the steam engine of 20 HP and the greater part of the machinery were by Fenton Murray Wood and Co, but still did not credit them for the fire proof construction [64]. The flax industry was going through a slump at the time, and many spinning mills were put on the market.

On the 15th Wm Swinburn was again in flow expressing concern at the state of the engine if salt water had penetrated, and as Mr Curwen still wanted the engine up by early summer would Fenton Murray Wood & Co send an engineer to Carlisle to meet their Mr Penrose (their engineer), and travel with him to the accident spot to see what remedial work was needed. Mr Overton was duly despatched [39].

Mr Penrice reporting on the state of the engine to Wm Swinburn and to Mr Curwen stated much damage to the cylinder, the ornamental parts, the piston rod, cylinder pedestal and side pipe and that Mr Overton was clear that unless any replacements parts were made by Fenton Murray & Co, they would not be liable for the performance of the engine. The upshot of his reports was a letter from Wm Swinburn dated March 18th to Fenton Murray and Co stating that Mr Curwen wished to surrender the whole engine up to the underwriters and get settlement. Matthew Murray responded, dated 21st March, that he had understood the request and they would fulfil it if asked again, however they wished to point out that the only part requiring to be remade was the cylinder due to the salt water damage, and a broken flange and pedestal. All the rest could be cleaned at the underwriter's expense. A new cylinder would be about £200 after deducting the metal of the old one. The ship was being repaired to take the parts on to Workington. However if they still wished it all to be abandoned so be it. Fenton Murray and Co wished to inform Wm Swinburn that they could not make another engine on the same terms partly from an increase in the costs of materials, but mainly due to the experience of making the first one which far exceeded their estimates [39].

Mr Lloyd of Uley in Gloucestershire advertised in March for a 6 or 8 HP steam engine made by either Messrs Boulton Watt and Co or Messrs Fenton Murray Wood and Co within the last three years [66]. The firm's reputation had reached the point when two adverts in two years asking for second hand engines had asked for them by name.

On 31st March 1811 Marc Brunel submitted, at their request, a revised quotation to the Ordnance Office for a saw mill and machinery for the service of H.M. Royal Carriage Department in the Woolwich Arsenal. This was the basis for an order to proceed with the mill in June 1811. While his original estimates in 1808 had been based on a steam engine with shafting and other work by Fenton Murray Wood & Co, it is clear that his new quote was based on a steam engine and machinery by Maudslay and it is unlikely that there was any input from the Leeds firm. It was probably still too soon after the difficulties over getting the steam engine and sawing equipment for his own Battersea plant produced at Leeds the previous year. The engine at the Royal carriage saw mill was long thought to be by the Leeds firm, but when writing to the Royal Ordnance in January 1828 over some comments about the performance of the steam engine supplied with his saw mill for Berbice in the Caribbean, Marc Brunel wrote:

"It is to be remarked that at the Woolwich Saw Mill, the Steam Engine, which is likewise made by Mr Maudslay, is 20 horse power which is sufficient to work six frames etc. etc." [65].

On March 23rd Wm Swinburn wrote back to Fenton Murray & Co, and while not fully legible today the gist can still be made out. Swinburn agreed with the Leed's logic [*to get the engine up and running*], and asked that Fenton Murray and Co should not communicate with the insurers until they heard from him again. Swinburn had written to Mr Curwen. This was followed by two further missives in March stating that Mr Curwen had misunderstood the state of the engine and finally that he wished Fenton Murray & Co to proceed with the necessary replacement parts as quickly as possible [39].

In May the Leeds papers showed that Fenton Murray Wood & Co had paid 10 guineas subscription for British Prisoners in France, apparently while the Government could not provide assistance it was possible for private individuals to do so [67].

In June the Leeds papers announced an item, which would lead to possibly the most well-known product made by Fenton Murray and Wood:

"Mr John Blenkinsop of Middleton near this town had obtained a Patent for certain mechanical means by which the conveyance of coals, minerals, and other articles is facilitated and the expence attending the same rendered less than heretofore" [68].

Another 265 hundredweight of components for the engine for the Isabella Pit were shipped on 27th June aboard the Volunteer, Captain John Hulmes, consigned through Liverpool.120 cwt of the weight was accounted for by the new cylinder. The shipping list was dated July 3ʳᵈ and the covering note requested news of the whereabouts of the components from Donaghadee [39].

The death of William Lister at the age of 48, after a long and lingering illness, on the 26ᵗʰ June was announced in the Leeds papers on July 6ᵗʰ [69]. The most common name for the firm then changed from Fenton Murray Wood & Co to Fenton Murray and Wood. Lister's will was witnessed by R K Bingley book-keeper (who also witnessed Matthew Murray's will) James Overton, Overlooker, who has been seen visiting Mr Watson near Newcastle and working on the engine for the Isabella Pit, and Henry Aveson Iron founder. An obituary for Henry Aveson appeared in the Leeds Intelligencer on 26 February 1829 which said he had been for 20 years book-keeper and manager for the firm.

Wm Swinburn advised by a letter dated July 19ᵗʰ that he had been informed that the engine parts had been loaded at Donaghadee, he then proceeded to ask Fenton Murray and Wood to inform the insurers that proceeds from the ship's (Providence) insurance should be paid to Mr Curwen as he had paid for the ship's repair and the owners had surrendered the policies to him as security. A further note on the 24ᵗʰ announced the arrival of the parts from Donaghadee and asked when they might expect the engineer to erect the engine and get it going. Matthew Murray replied that it would be 2 or 3 days before the necessary people returned home and then they would come over and would orders please be sent for the collection of the cylinder from Liverpool. Wm Swinburn 's next letter dealt mainly with the insurance on both the engine and ship, but he then stated that the delay in shipping the cylinder from Liverpool was due to not being able to find a ship with a large enough hatch to take it. Mr Overton responded for Fenton Murray and Wood to say it was not worth them coming over until the cylinder arrived from Liverpool. At the beginning of September Wm Swinburn at last advised that the goods were expected the middle of the week [39].

In the preceding months Fenton Murray and Wood had been erecting another engine, this time in the south west. This was for the tobacco firm of Wills then at Redcliffe, Bristol and was to be used to drive machines to cut the tobacco [70]. Payments were made in September and November 1811 to Fenton Murray & Co amounting to £639-11-9d with £42-18-9d being paid to the carrier plus £6-6-0d for the erecting engineer's board in October and £17-15-0d for his board in November and his present [71]. Details of the engine do not survive in the Wills' archive, but it is possible it is the same engine that was recorded for sale in in March 1861:

"For Sale Price £120, a 12 horse condensing beam engine (with boiler) by Fenton Murray & co of Leeds, in excellent working order. For particulars and view of the engine at work apply W Short Engineer St Johns Bridge Bristol" [72].

The next surviving communication from Wm Swinburn` was on 12 October and requested costs for the cylinder and other new parts for inclusion in the insurance claims. On November 15th Fenton Murray and Wood sent the consignment note /shipping list for the balance of the engine, a further 196 hundredweight. This brings the total shipped to around 75 tons. The original document survives, and used the old headed notepaper [39]: As can be seen this carries the classic view of the Round Foundry and includes among the firm's activities the item 'Constructors of Fire Proof Buildings'.

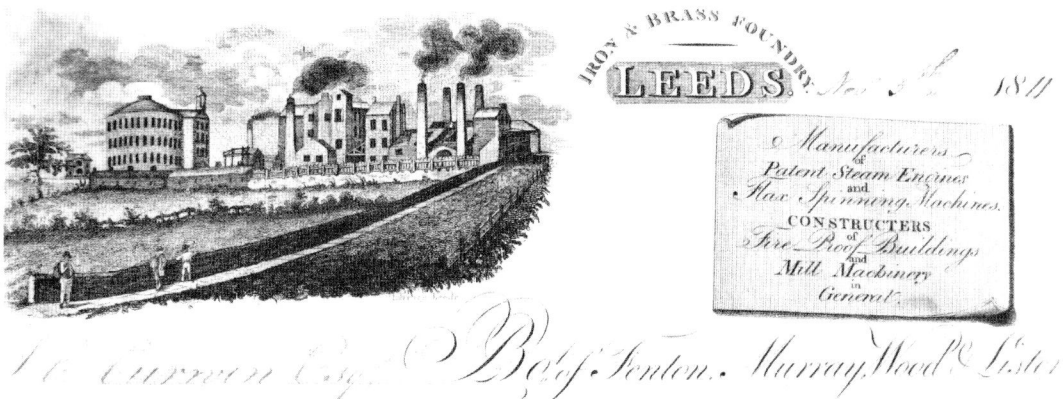

44. Surviving letter head for Fenton Murray Wood and Lister in Cumbria Record Office.

The Admiralty abstract of letters to the Navy Board for 5th December 1811 stated:

"Steam Engine for sawing timber to be erected in Chatham yard employing Mr Brunel to erect it" [73].

Wm Swinburn wrote on the 8th and 9th of December to inform Leeds that the goods despatched on the 5[Th] November had been received. He then switched subjects back on to insurance and in a somewhat complex arrangement said that Mr Curwen would be obliged if Fenton Murray and Wood would take the ship's policies as security and reimburse Mr Curwen for the amounts he paid out on the ship and then proceed to resolve all the ship matters with the ship's owners and the underwriters [39].

The Navy Board Index for 17th December 1811 stated:

"Saw Mills - Ground to be pointed out to Mr Brunel for the erection of one in Chatham Yard" [74].

Matthew Murray wrote to Wm Swinburn on the 18th December 1811 stating that they had done what was necessary for the engine valuation for insurance but in regard to the proposals of them sorting out the ship, he could not advise as Mr

Fenton was in London and they were not in a position to advance any money at that date. He added that he had met Mr Clark, the ship's owner, and after that discussion:

"the whole Business is such a Mistery to us, that I am convinced we cannot serve Mr Curwen in it, without injury to ourselves, a thing Mr Curwen I am sure will not urge. Our situation at present is affected by the decease of Mr Lister one of our Partners & whose capital we are urged by his Executors to payout, together with some very heavy jobs which has been long in Hands by unforeseen causes which renders our situation at present extremely critical" [23].

Notes:

1. The Northern Courts; containing original memoirs of the Sovereigns of Sweden and Denmark. Vol 2 by John Brown London 1818 page 341.
2. The European Journals of William Maclure by Maclure and J Doskey, The American Philosophical Society Philadelphia 1988.
3. The Day Book of Thomas Stawford, available on the internet in Swedish or see Fenton Murray and Wood by Trevor Turner.
4. Matthew Murray A centenary Appraisal by G F Tyas, Newcomen Society 24 February 1926.
5. The Biographia Leodiensis by Rev R V Taylor, John Hamer 1865.
6. Marc Brunel's Letter Book LBK 54. N.M.M. Coad Library.
7. Goodrich Journals B16. Science Museum Library.
8. Reminiscences of flax-spinning by a flax-spinner long in the trade at Dundee, William Brown, Dundee 1862.
9. Jute and Flax Mills in Dundee by Mark Watson, Hutton Press 1990 page 32.
10. West Middlesex Water Works Company ACC 2558/WM/A/1/01, London Metropolitan Archives, minutes of 11 December 1806 and Wm Nicolson's report dated 12 June 1807.
11. West Middlesex Water Works Company ACC 2558/WM/A/03 W9.1, London Metropolitan Archives.
12. Goodrich Journals B17. Science Museum Library.
13. The Leeds Mercury 11 April 1807.
14. The Leeds Intelligencer 22 June 1807, The Leeds Mercury 12 September 1807.
15. Goodrich Journals B19. Science Museum Library.
16. The Life of Brigadier-General Sir Samuel Bentham Ksg by M Bentham (wife) originally published Longman Green London 1862.
17. Navy Board Correspondence with Inspector General of Naval Works ADM 106/2539, National Archives Kew.
18. Goodrich Papers A222. Science Museum Library.
19. Various see The Leeds Mercury 14 November 1807.
20. Goodrich Journals B18. Science Museum Library.
21. Information by courtesy of Carl-Henrik Ankarberg, joint author of Rorstrand i Stockholm [the story of Sweden's oldest ceramics factory] Stockholm 2007. The engine house drawings are in the Technical Museum in Stockholm.
22. Matthew Murray's Pumping Engine for Isabella Pit, Workington 1803-1813. Dr Blake Tyson, The Cumbrian Industrialist Volume 1 1998.
23. Curwen Archives D/Cu 2, Letters 1760-1849, Cumbria Record Office Whitehaven.
24. The Leeds Intelligencer 29 February 1808.
25. The Leeds Intelligencer 6 June 1808.
26. Goodrich to the Navy Board 2 February 1809 ADM 106/3198 National Archives Kew and Quarterly Papers on Engineering Volume 1 by Weale London 1844.
27. Cobb of Margate Papers EK-U1453/B2/40/224, East Kent Archives.
28. Histories of Bolton and Bowling, William Cudsworth, Bradford 1891.
29. Marshall Papers MS 200 folio 56. Brotherton Library.

30. Treatise of the winning and working of Collieries, Matthias Dunn, Newcastle upon Tyne 1848, covers the point and many general principles.
31. An Historical, Geological and Descriptive View of the Coal Trade of the North of England, Matthias Dunn Newcastle upon Tyne 1844 page 46.
32. An Essay on the Teeth of Wheels, Robertson Buchanan revised by Peter Nicholson London 1808.
33. Science and Technology in the Industrial Revolution, Musson and Robinson, Manchester University Press, 1969 page 82.
34. As note 33 but page 331.
35. As note 33 but pages 82/3.
36. Boulton and Watt Papers MS3147/5/1378.
37. Land Registers Vol. FI pages 127/8, WYAS Wakefield
38. The Royal Bank of Scotland Group Archives, BEL 3/3 Customer Account Ledgers 1813-1825.
39. Curwen Archives D/Cu 2, Letters 1760-1849,for letters from Fenton Murray and Wood and D/Cu/3/43 for William Swinburn 's Letter Books.
40. Journal of the Royal Society for the Arts volume xxvll pages 148-153.
41. Fenton Murray & Wood, unpublished thesis by Trevor Turner
42. Victualling Board Out Correspondence ADM 110/59, to Lords Commissioners of the Admiralty 15 January 1809, National Archives Kew.
43. Goodrich Papers A267. Science Museum Library.
44. Navy Board Minutes ADM 106/2672, National Archives Kew.
45. Goodrich Journals B19. Science Museum Library.
46. Goodrich Papers C132. Science Museum Library.
47. Navy Board Index of Correspondence 1808-1811, ADM 106/2748, National Archives Kew.
48. For example The Leeds Intelligencer of 6 June 1809 and 28 August 1809.
49. Goodrich Papers A283. Science Museum Library.
50. Goodrich Papers A284. Science Museum Library.
51. The Morning Chronicle 10 October 1809.
52. The Leeds Intelligencer 11 December 1809.
53. Industrial Archaeology and the RCHME, K Falconer and R Thornes, Industrial Archaeology Review ix volume 1 Autumn 1986 and Fireproof Mills - the Widening Perspectives, K Falconer, Industrial Archaeology Review xvi volume 1 Autumn 1993. Generally informative on Fenton Murray and Co Mills.
54. Goodrich Journals B21. Science Museum Library.
55. The Morning Post 22 February 1810.
56. The Morning Post 8 March 1810.
57. Victualling Board Correspondence ADM 110/60 to Admiralty 31 October 1809, National Archives Kew.
58. Goodrich Journals B29. Science Museum Library.
59. Goodrich Papers A293. Science Museum Library.
60. Victualling Board Correspondence ADM 110/62, to Navy Board, National Archives Kew.
61. Watson Papers 2/2 page 308. NEIMME.
62. Goodrich Journals B23. Science Museum Library.
63. Rudimentary treatise on the construction of cranes and machinery, Joseph Glynn, London 1849.
64. See Aberdeen Journal 6 March 1811 which contains adverts for both mills.
65. Royal Ordnance Correspondence WO 44/457 National Archives Kew.
66. The Morning Post 20 March 1811.
67. The Leeds Mercury 8 June 1811.
68. The Leeds Intelligencer 10 June 1811.
69. The Leeds Mercury 6 July 1811.
70. W.D. & H.O. Wills and the development of the UK tobacco industry 1786-1965, B Alford, London 1973.
71. Wills Archive 38169/F/6/3 Bristol Record Office.
72. The Western Daily Press 23 March 1861.

73. Admiralty Abstract of Correspondence ADM 106/2094, National Archives Kew.
74. Admiralty Index of Correspondence 1808-1811 ADM 106/2748, National Archives Kew.

Chapter 9

Locomotive Engines and Steamboats

On the 31st January 1812 the Chatham Officers wrote to the Navy Board about a wall to surround some new land being incorporated into the yard and mentioned the Navy Board Warrant dated 29th January 1812 advising them that Mr Brunel had been instructed to proceed with the Saw Mill at Chatham [1].

In February the Navy Board letters recorded an instruction that the saw mill at Chatham was to be completed without delay [2].

Meantime the correspondence over the insurance claim for the Mr Curwen's engine was proceeding, the insurers corresponded with Mr Edward Grace (Fenton, Murray and Wood's insurance agent in Leeds), who passed the letters to Fenton Murray and Wood for onward dispatch and resolution with Mr Swinburn. Interspersed with this business, Mr Swinburn imparted the news that the pumps had been erected, and the spier rods would shortly be added and the engine would soon be ready to work. He also requested a reliable man to work the engine and pumps [3].

At some point the cross mills at Marshall & Co.'s mill at Ditherington, outside Shrewsbury, had been destroyed by fire and in 1812 John Marshall recorded the expense of rebuilding the Hackling Shops as £2816-06-00d [4]. The iron work in the rebuilding bears the distinct hallmark of Fenton Murray and Wood and is distinguishable from the iron work in the main mill (cast at Hazeldine's foundry in Shrewsbury). The Flax warehouse which Marshall recorded as costing £2360 in 1811, has the Fenton, Murray and Wood pattern of ironwork as well. It is clear that the firm were as involved in the Shropshire Mill as they were in Marshall's mills at Leeds.

Concurrent with the time of the rebuilding of the 'hackling shops', the Leeds Mercury recorded at the end of February 1812, the erection of a patent chain foot bridge at Marshalls & Co in Shrewsbury but has no detail as to the maker [5].

During the years 1809 to 1812 the highest prices for raw flax (much of which was imported from Riga and the Baltic area) were recorded for the whole period between 1805 and 1840. This was due to the European wars and Napoleon's ban on trade with Britain and resulted in great pressures within the industry and a slump in the trade.

45. The Cross Mill at Ditherington.

As a result during 1812 three flax mills in the Dundee area were on the market and advertised in the papers, with engines from Fenton Murray and Co of Leeds; in March, June, July and November, Tay Street Mill with a 12 H.P. engine; in May, June and October Lochty Mill in Dundee with an 8 H.P. engine; and in July and October a mill at the Townhead of Arbroath with a 16 H.P. engine. The latter was of fireproof construction, they were all on offer with their machinery, no doubt much of these provided by the engine maker. A fourth share in another mill at Brechin advertised (October 1812) machines of the newest construction, made at Leeds, further possible work from Fenton Murray and Wood. The Brechin mill was water driven [6]. The mills were not selling and the prices were coming down.

The Morning Chronicle carried an advert at the end of April concerning the sale of a Cloth Factory near Spa Road Bermondsey, which had the machinery run by an 8 HP Fenton Murray and Wood steam engine.

In May Fenton Murray and Wood received a second part payment of £764-3-3d for Mr Curwen's Workington engine, Mr Curwen had endorsed a bill of exchange received from the insurers over to the Leeds firm [7]. Fenton Murray and Wood acknowledged the payment on the 3rd of June. Their letter added that they were glad to hear that the engine was nearly completed, and asked that their erectors should be told to finish off as soon as possible as they were needed in Leeds [3].

On the 20th of June the Officers of Chatham Yard acknowledged the receipt from the Navy Board of five plans for the buildings to be erected for the saw mills [8].

At the end of June 1812 the Leeds press carried the articles, well known in railway circles, concerning the public trials of the Murray Blenkinsop locomotive, using Trevithick's high pressure steam. John Farey described in Rees's Encyclopaedia the previous trials which had been carried out using a small condensing engine, but the water became so hot little was gained by condensation. This probably supports the existing evidence that Matthew Murray was not a proponent of high pressure steam and he only used it when there was no option. The description in the Leeds Mercury is the fullest for the public trial:

"On Wednesday last (24th June) a highly interesting experiment was made with a machine, constructed by Messrs Fenton Murray and Wood, of this place, under the direction of Mr John Blenkinsop, the patentee, for the purpose of substituting the agency of steam for the use of horses in the conveyance of coals on the iron railway from the mines of J C Brandling Esq at Middleton, to Leeds. This machine is, in fact, a steam engine of four horses' power, which, with the assistance of cranks turning a cog wheel, and iron cogs placed at one side of the rail way, is capable of moving, when lightly loaded, at the speed of ten miles an hour. At four o'clock in the afternoon, the machine ran from the Coal staith to the top of Hunslet Moor, where six, and afterwards eight waggons of coals, each weighing 3 ¼ tons, were hooked to the back part. With this immense weight, to which, as it approached the town, was super added about 50 of the spectators mounted upon the waggons, it set off on its return to the coal staith, and performed the journey, a distance of about a mile and a half, principally on a dead level, in 23 minutes, without the slightest accident. The experiment, which was witnessed by thousands of spectators, was crowned with complete success; and when it is considered that this invention is applicable to all railroads, and that upon the works of Mr Brandling alone, the use of 50 horses will be dispensed with, and the corn necessary for the consumption of, at least, 200 men saved, we cannot forbear to hail the invention as of vast public utility, and to rank the inventor amongst the benefactors of his country" [9].

On the 10th July Fenton Murray and Wood submitted a further account to Mr Curwen for the steam engine, fly punch, replacement cylinder and elbow steam pipe, and the expenses in travelling to Donaghadee (James Overton) and the erectors Allen Hodgson and John Sewell, amounting to £3256-6-3d deducting payments made of £1680-19-3d leaving a balance owed of £1575-7-0d. The accompanying letter requested £500 on account as the firm "*are very much distressed for money just now*". They also advised that Mr Clark had left Leeds to visit Mr Swinburn to settle the insurance claims. Mr Swinburn replied promptly remitting the sum requested and advising that he had seen Mr Clark but still thought that the value of the claim was understated. He therefore requested that Fenton Murray and Wood should send the policies they held to his solicitor Mr Falcon in London for the attention of Mr Thompson. Fenton Murray and Wood sent warmest thanks for the bill for £500 and advised that they had issued necessary instructions through Mr Grace for the London policies and had written to Hull concerning the insurance held there [3].

In July both the Leeds Mercury and the Leeds Intelligencer published a drawing of the Murray Blenkinsop locomotive with the patent specification granted to John Blenkinsop.

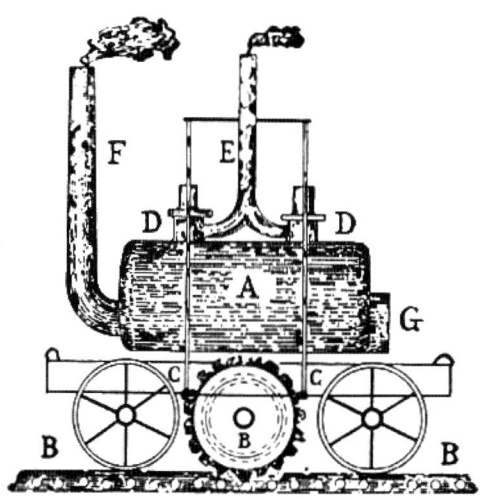

46. Earlist depiction of Murray/Blenkinsop locomotive

A Boiler

B, B, B Mr Blenkinsop's Patent Road Rack and Wheel

C, C Crank Rods

D, D Steam Cylinder

E Discharging Pipe

F Smoke chimney

G Fire door.

It is clear from reading Blenkinsop's patent that it is confined to the method of conveyance using the rack rail and does not cover the design of the locomotive, although many books loosely describe his patent as for a locomotive rather than for the track and rack.. The innovations in the locomotive, the main one being the use of two cylinders, will in the main have come from Matthew Murray. The use of two cylinders had been proposed in Murray's patent 2632 of 1802. The other patent involved in the locomotive was the use of high pressure steam. Trevithick and Vivian had been granted a patent, number 2599 in 1802 for:

"Methods for improving the construction of steam engines, and the application thereof for driving carriages, and for other purposes".

It is unclear who held the licence for high pressure steam from the consortium or partnership who had shares in the patent. A letter from William West, a relation of Mrs Trevithick, who held one of these shares has survived. He was answering a letter of Trevithick's over a dispute over the patent accounts and he wrote:

"Now sir, in the first instance, what right have you to make me debtor to you for £40 received of Wood and Murray? I hold a copy of your answer to them, saying you held no shares in the patent at that time when they wrote to you respecting the engine, but recommended them to W West whom you sold a share to, saying Wm West would license to erect engines on the patent… The part of patent money you allude to, received from Mr Rabey, I suppose is settled in our patent accounts, as I never received a single sixpence from the patent before that from Messrs Wood and Murray; then I made a present of £1 to your children, because you refused making a charge for the drawing sent to Leeds.."[10].

A study of the correspondence between Blenkinsop and Watson indicates that Fenton Murray and Wood did indeed hold the licence in regard to locomotives they built which Blenkinsop was licensing to use his rack wheel patent (Blenkinsop to Watson of 18 November 1814). There are many references to the locomotive being built to Trevithick's patent, true it used high pressure steam and so to some extent the references are correct but it ignores the more important aspects of the machine. Fenton Murray and Wood did not build all the locomotives to the plan, it is known that two or three operated on the Orrell Colliery in Lancashire and were built at the Haigh Foundry under the direction of Robert Daglish by arrangement with John Blenkinsop. It is unknown whether any license existed for the Orrell locomotives with Trevithick/West/Vivian etc. Two engines were also built in Germany to the general plan but different dimensions and detail, but were not a success. As there was so much of Fenton Murray and Wood's technology in the locomotive, it had to be likely that John Blenkinsop paid the firm for all the development work on the understanding that this gave him the right to make arrangements elsewhere, or that they had an agreement whereby John Blenkinsop shared any royalties he received. With Matthew Murray's later statement to the Stockton and Darlington Railway that building locomotives was a disruption to his shop, and his statements to Watson at the time on the subject, presumably he was also not overly worried about them being built elsewhere.

Following the announcement in the Leeds Mercury on the 1st August that "Mr Blenkinsop's machine" was in service, Blenkinsop wrote to John Watson on Tyneside to inform him that his "patent steam carriage" was working daily and was capable of moving 20 wagons weighing 3 ½ tons each at 3 ½ mph, being 74 tons including the locomotive, (making the locomotive weigh in at 4 tons). A round trip of 3 miles with 20 wagons took 50 minutes. The value of coals consumed in a day he reckoned at 2/9d and the water consumption was 55 gallons an hour. To minimise water and coal consumption Blenkinsop stated that he was going to replenish the boiler with hot water, thereby avoiding having to let off all the steam prior to filling the boiler. He promised a further update after another week of working [11].

On the 20th of October 1812 the Navy Board sent Marc Brunel's invoice for the metallic elements of the steam engine, for the saw mill to the Officers at Chatham. They advised the officers:

"to take every precaution in landing the packages and cases [and] *to shelter them from the wet"* [12].

The 30 H.P. engine for Chatham was made by Fenton Murray and Wood as advised by the Surveyor of Buildings to the Navy Board [13] in 1827 when the engine needed repair and they were seeking an increase in power. The engine in the saw mill at Chatham has long been assumed to have been made by Maudslay along with all the original sawing machinery, however as he was also fully involved with the machinery and engine for Woolwich Arsenal, Marc Brunel may well have considered it advisable to pick up the old quotes for Woolwich made by Fenton Murray and Wood and have them updated to undertake the similar work at Chatham. It may also be that he had more confidence in the track record of the Leeds firm when it came to making structural iron for buildings, because it is almost certain that the cast iron girders, columns and cast iron floors of the Chatham Saw Mill were made by Fenton Murray and Wood along with the engine.

The Navy Board sent additional drawings to Chatham for the saw mill buildings in November, and on 18th of that month the Chatham officers reported that Mr Brunel had delivered a large quantity of the machinery for the saw mill and in their opinion this would justify an advance payment to Brunel of £3000 [14].

On the 12th November Blenkinsop wrote again to John Watson concerning the use of the rack system at Coxlodge and Fawdon collieries (John Watson was the viewer [manager] at Coxlodge as well as a shareholder), with a full financial justification and offering an incentive to him if the business went ahead to the extent of one eighth of Blenkinsop's royalty receipts. The financial justification is given below and reflected the very high cost of feed for horses resulting from the Napoleonic wars:

"Estimate of Leading the Coals from Coxlodge & Fawdon Collieries a distance of 5 ½ miles, on a vend of 30,000 chaldrons each:

Food for 81 Horses at £50 per annum ea	*£4050.00.00*
Trapping for 81 Horses at 3 Gns[guineas] each	*255.03.00*
Farriers Wages, Drugs & shoeing for 81 }	

Horses at 1s each Horse per week or annum	}	210.12.00
Decay of Horses, say 8 at £45 each		360.00.00
Greasing & Ballasting the road 5 ½ Miles at £150 per mile		825.00.00
Waggonmens wages for driving, on 60,000 Cha^n at 1s	} }	3000.00.00
8 Horsekeepers Wages @18/- per week ea		374.08.00
House Rent & Fire Coal for 81 Waggoners & 8 Horsekeepers at 2/6 per week each		578.10.00
		9653.13.00
Deduct for Manure Brid by 81 Horses		200.00.00
Total Expense of leading the Coals		£9453.13.00

on 60,000 Cha^n is 3s 1 $^{3/4}$d per chaldron

Estimate of Conveying 60,000 Cha of Coals 5 ½ Miles by Patent Steam Carriages calculating on Engine to take 16 Chaldrons & perform three Journeys per day will require:

5 Engines & one man to each Engine at 25/- per week each man or per ann	325.00.00
1 Engineer per annum	100.00.00
6 men House & Fire Coal @ £6.10 per	39.00.00
Coals, Tallow, & oil 5 Engines @ £50	250.00.00
Wear & tear of Engines supposed at	300.00.00
4 Horses to be kept to remove the Empty Waggons from the Engines at £ 50 }	200.00.00
Grass for D^o 3 Gns. each	12.12.00
4 men wages at 18/- per week each	187.04.00
House Rent & Fire Coal for D^o	26.00.00
Farriers Wages, Drugs & shoeing the 4 Horses at 1s each per week	10.08.00
Decay of Horses	18.00.00
	£1468.04.00
Deduct Manure had by 4 Horses	10.00.00
Expense of leading by Patent Steam	£1458.04.00

Carriages exclusive of the Patent Right or 5 $^{3/4}$d per Chaldron.

Expense in altering the Road for Steam Carriages:

1 Yard of Cogged Rail with Pedestal 64lbs per yard at 14/- per Cwt is per mile £704, on 5 1/4 Miles	3872.00.00
Labour in altering one Side of the Road £ 50 per Mile	275.00.00
7 Steam Carriages @ £300	2100.00.00
	6247.00.00

By the sale of 77 horses @ £45 each	4465.00.00
Supposed Stock of old hay 500 tons at £5 per	2500.00.00
One side of the present road to be sold as old Metal 31 tons per mile =170 1/2 Tons at £7 per ton	1193.00.00
	8158.00.00

Difference in the value of the Colliery Stock in favour
of Steam Carriages £1911.00.00

Expense in Conveying Coals with   Horses as per estimate 9453.13.00
Expense in Conveying Coals by the patent Steam Carriage <u>1458.04.00</u>

Savings to the owners £7995.09.00

*From my estimate it appears the Coals will be led at £300 each Engine which
includes Horses etc - but I think the expense will not exceed the following*

	£	
One man	*60*	*per annum*
Coals	*40*	*D°*
Oil & Tallow	*10*	
[Repair] of Wheels	*20*	
Grate Bars	<u>*10*</u>	
	140	
deduct for old metal	<u>*10*</u>	
	130	*per annum. I am not aware of*

*any other expense provided no accident happens the Engine and should the Coals
be led at night as well as during the day 3 Engines will be capable of conveying
60,000 Chaldrons.*

*The Patent Right at £10 per cent is £800 per annum - Boulton & Watts charge
was 33 $^{1/3}$ per cent per annum, or one third of the savings.*

J Blenkinsop" [15].

Sometime in October or November Fenton Murray and Wood had written to Wm Swinburn asking for a further payment on the Isabella Pit engine, which received a reply dated 20th November reporting that Mr Curwen had suffered much from the failure of Wood & Co bank. The first payment for the engine had been made prior to the contractual time and they should prepare to be patient in regard to the next (the terms had been that one third of the price was payable on delivery of the materials, one third when the engine was set to work and the balance three months later). Wm Swinburn reported that he hoped to have the engine working in two months' time. He had obviously not made the progress expected when he wrote in February that the engine would soon be ready [3].

On the 7th of November the Leeds Intelligencer reported that a shocking accident had occurred on Tuesday 1st December on Hunslett Moor when George Butler who was employed on the Middleton Colliery locomotives fell from the platform while supplying the fire with coal and his right hand was severed from his body.

Simon Goodrich made an interesting entry in his journal at the end of December, but its context and detail have not survived:

"On Mr Bramah proposals for a Murray Engine & Hydrostatic press for bending planks" [16].

The ledgers of the Middleton colliery for December 1812 record for the 31st December two entries for Fenton Murray and Wood, first:

"By cast Iron goods £1724-5-3d"

and secondly:

"By two patent steam carriages £700-0-0d" [17].

The cast iron goods were most probably cogged rails and in the same year there are recorded purchases from Messrs Gotthards, a known supplier of the rails, valued at £1359-11-10d.

During the year an order was received related to the provision of a steam dredger for the Port of Dublin. George Halpin, the Inspector of Works for the Dublin Ports Board, made a proposal to purchase a steam dredger for the port's use, similar to the one used by the Port of London. The engine and machinery was supplied by Fenton Murray and Wood of Leeds, together with plans for a suitable boat to accommodate the machinery. The machinery was an endless chain of buckets. The vessel was built in Dublin and known as the 'Steam Dredging Vessel Patrick', but she was not self-propelling, she had four anchor points enabling her to change position once she was at the site to be dredged. The operation had four lighters of 35 tons each. The first job was to dredge the Ringsend Gut near the Grand Canal Docks in 1815 [18].

On the 6th of January 1813 Blenkinsop sent a letter to Watson wherein he proposed a royalty of £50 per annum to use the 'Patent Steam Carriage' from either Coxlodge or Fawdon to Stoddon Bridge at Hadericks Mill. If this was acceptable to the Colliery, he wanted it all up and running on a good rail by June that year [19].

Wm Swinburn wrote on January 11th with an offer from Mr Curwen to pay before the three months were up as he expected a large sum shortly. However, as the coal trade was dull it was not convenient to accept their draft at the moment, but he would keep it by as a convenient remittance when the time came [3].

Towards the end of January John Blenkinsop replied to a new enquiry from a Mr Bevan jnr. of Messrs Lockwood & Co. of Morriston near Swansea, he gave some costings (along the lines of the ones shown above for the Coxlodge Colliery) for a 36 mile railway, assuming that during a 12 hour day 100 tons would be transported, which showed an annual saving of £3270. He asked that the first engine should be made in Leeds and that a good rail road be laid as all depended on the strength and firmness of the rail [20].

Wm Swinburn sent off Mr Curwen's acceptance of Fenton Murray and Wood's draft for £600 on the 27th of January and a corresponding entry was made in the payments book for Chapelbank Colliery. Receipt was acknowledged from Leeds on the 29th [3 &7].

On that same day Simon Goodrich visited Deptford Victualling Yard with Joshua Field when the drawings and description of the steam engine made for the yard were made (see above 1810).

In March and April Fenton Murray and Wood were once again looking for a skilled man and unusually were more specific as to their requirements:

"To Foundry Men, Wanted, One Green Sand Moulder, a person that has been accustomed to heavy castings in the steam engine and millwright line. A good workmen will meet with constant employment and good wages may have the work by piece, with boys to assist him" [21].

On March 4th Matthew Murray replied to a request for information from Simon Goodrich on the application of steam engines to corn grinding. At the time Goodrich was designing the Victualling Mill for Portsmouth. Murray confirmed that a pair of stones 4 feet diameter would grind 6 bushels of corn an hour and a six horse engine would work a pair of such stones. A 20 horse engine would grind and dress 20 bushels an hour, however a 30 horse engine with 5 pairs of stones, 4 working and one standing was the set-up for the best mills. He suggested that Goodrich's working hours per week at 142 did not leave enough time for cleaning and repairs. Murray then gave their prices for engines noting that engines above 12 horse were fitted with D valves. He proposed a flywheel rather than the use of multiplying wheels. Sketches of ground plan and elevation of 20/30 horse were given. He proposed that if salt water were used then the engine house should be 3 or 4 yards from the water so it would enter the condenser under atmospheric pressure, without using cisterns or cold water pumps. He agreed with the proposal for copper boilers but suggested that the latest thinking was to have three small ones, two working together to equal the engines requirements and the other as spare. The advantages were that the layout used less coal, made more regular steam, could more easily changed, and when only 1 or 2 sets of stones were to be used only one boiler was necessary [22].

Horse Power	Price of Engine Materials	Price of Single Iron Boilers	Price of Single Copper Boilers	Length of Stroke	Strokes per Minute	Add't'nl Cost eg Brass Air Pump Bucket Copper Rod &c	Injection Water Requ'd per minute
	£	£	£	Ft. In		£	
6	400	40		3' 0"	36		
8	468	60		3' 6"	32		
10	557	70	315	4' 0"	28		
12	644	82	"	"	"		
14	716	100	"	4' 6"	25		
16	770	100	"	"	"		
18	827	130	630	5' 0"	22		
20	894	140	"	"	"	140	60galls
22	963	150	"	"	"		
24	1028	160	"	"	"		
26	1083	175	900	6' 0"	19		
28	1134	190	"	"	"		
30	1186	200	"	"	"	200	90galls
32	1228	215	"	"	"		
34	1266	230	"	"	"		
36	1304	240	"	"	"		

Goodrich Letter No 438 Table of Prices.

Four days after writing to Simon Goodrich, Matthew Murray wrote to John Watson, with whom he had now been dealing for a decade, on the subject of the colliery railway, and it is clear that this was not the first correspondence on the matter they had had. Murray was replying to a letter from Watson and stated that his objections were the same as made the previous July and since then he had improved the method of joining the rails for Mr Blenkinsop which "*made them both simple and perfect*". The rails of those days were much shorter and required more joints. Murray goes on to say the engines exceeded his expectations and claimed the system as the most valuable improvement in the coal business for 50 years. He went on to state his preference for having the rack in the centre of the track. He attached his own estimate of the costs and profits, which were only based on saving 40 horses, and offered to visit Watson in Newcastle to discuss the execution of the work if he decided to proceed [23].

John Blenkinsop was continuing his correspondence with Mr Bevan in Swansea, and discussed the implications of varying the width of the track on the wagons and more importantly the engine. The expenses and profits were again enumerated this time for 1 ¾ miles. He proposed a visit to Swansea and mentioned a letter from a Mr Alex Ruby from near Swansea concerning the locomotive (whether for acquiring or building is unclear) [20].

On the 23rd John Blenkinsop replied to a letter from John Watson and offered a license at a royalty of £200 per annum for taking the coals from Coxlodge (north of Newcastle) to the Tyne. He claimed to have one person contracted at £50 per mile and to be negotiating with three others for higher royalties. He advised

Watson that the 3rd engine for Middleton was being made and he would be happy to divert it to Coxlodge if it suited. The width of the road and breadth of the rail surface would need to be set to suit the engine. He referred to Chapman's scheme (his first; in which the locomotive hauled itself along on chains) as larceny.

Presumably Watson demanded some better terms as the next surviving document from Blenkinsop which was dated the 5th April, showed a royalty of £50 for the first year and £100 per year thereafter for the term of the patent. Watson wrote some notes on the letter:

"Weight of Common Rail now used per yard 46 lb. including pedestal
  "   of Cogged Rail   "   64 lb. including D°
Weight of Engine including water   --------- 5 ton
The Engine will go 5 mile without filling the Boiler
Mr Blenkinsop to have the Width of our W Way sent him also breadth of the top of our present Rail.
55lb pressure upon every cubic inch.
Price of Engine £350" [24].

Further letters followed in April concerning the proposed course of the railway over Mr. Brandling's Northumberland estates, which John Blenkinsop attempted to gain Brandling's permission for while he was visiting Yorkshire (Brandling owned the Middleton Collieries which John Blenkinsop managed for him). Blenkinsop closed one letter, dated 12th April with a hope that Watson had written to Matthew Murray about the engine for Coxlodge, as Murray would rather make a new engine and he would need progressing to get it done quickly [25].

The buildings at the Chatham sawmill were progressing and on 12th April the Navy Board dispatched further drawings of the railway and the tunnel from the mast pond to the reservoir to Chatham [26].

On 17th April Matthew Murray replied to a letter from John Watson saying that he would have been glad to see him in Leeds if time had allowed as a face to face meeting could have resolved many things. Murray explained he was committed to going to Scotland via Carlisle and did not know how long he would be away. He went on to describe some of the things he wanted to discuss. Firstly that John Blenkinsop had offered Watson the spare engine, which was a necessity when there were two engines working all the time, due to a broken engine having to stand while spares were made. He had not the least objection (to sending the spare Middleton engine), but he recommended changes to the engines for Watson as the boiler was sized for the Middleton line and would require extra filling at Coxlodge. He then returns to the subject of putting the rack wheel and line in the middle of the track. He was quite candid, that the current configuration caused the wheels of the engine and the rails to wear away very fast, the oblique action or side pull was very detrimental going up moderate rises and for going around bends:

"The side rack did very well as a cheap method for trying the Scheme but certainly is not calculated for practice".

He informed Watson that the engine would be ready in 6 to 8 weeks and the price was £350 plus delivery and was to be paid in 2 months by a bill at 2 months. He went on to say that they were tested before leaving the factory. He was making some side rack rails for Blenkinsop and if Watson wanted would send him two with their chairs (14 pairs per cwt), and this was the weight he would recommend. Credit was tight as they were still paying out Mr Lister's considerable capital. He went on to disclose that he was considering making the boilers of wrought iron rather than cast which would reduce the weight and increase the volume of water carried. Then Murray surprisingly added that for level ground where only 10 or 12 wagons were to be drawn the engines could be made with one cylinder. A very interesting letter, which demonstrated Murray's dislike for something that worked, but could be made so much better. The firm's financial position at the time presumably precluded any in-house investment [27].

The continuing shortage of money had prompted another letter to Wm Swinburn seeking the balance on the Isabella Pit engine to which they received a reply on May 13th from Wm Swinburn. Mr. Curwen had desired him to tell them that he had not expected to be asked for the balance so soon, as he had done more than fulfill his contract. He did not wish to be pressed. He apparently had 20,000 unsold wagons of coal leaving him short of ready money [3].

On the 15th May the Chatham Officers acknowledged receipt of the drawings of the plan and section of the engine and boiler house for the saw mill and that for the section of the South East Wing of the building which showed the cast iron girders, columns, joists and stone floor. This was followed by another letter on the 24th to state that the erection of the steam engine had progressed to the point where the cast iron floors were required, and could they be sent without delay [28].

Blenkinsop was in touch with Watson on the 1st June mainly to get answers to some questions which his lawyer, who was drawing up the license agreement, had requested. He told Watson that Murray was back in Leeds, but that Murray was now saying he could not execute an order in 2 months due to a large order for Russia and an engine for Mr. Buddle (to be fixed underground at Percy Main). Murray had told him that Watson could have the engine in 3 to 4 months and had promised to write to Watson. Blenkinsop then commented that Murray frequently devoted Sundays to letter writing and therefore not to expect the letter before Monday. Blenkinsop then offered components of his third engine provided they were replaced by October and said that Watson must have a wrought iron boiler with a double iron tube. So at this point presumably Watson was thinking of having an engine bespoke for Coxlodge [29].

By the middle of June Fenton Murray and Wood had replied to John Watson's letter of the 4th which had included the dimensions of the rail and railway. It was pointed out that using parts of Blenkinsop's engine would not save any time and merely increase the cost of both engines. They apologized for the delay but had to dispatch a large engine to St Petersburg which had to be put on board a ship to sail in August and which would incur stiff penalties if delivered late [30].

At Chatham saw mills the Officers had been advised by the Navy Board that Brunel would provide all the necessary pumps. On the 18th June the Officers wrote to the Navy Board that the erection of the steam engine was at a standstill due to the non-delivery of the cast iron floors [31].

In June the Leeds Press announced the advent of a steam boat on the River Aire [32]. Fenton Murray and Wood were fitting it up in the basin of the Leeds and Liverpool Canal for Mr. Richard Wright of Stockwell and Mr. John Wright of Rochester to provide a service between Yarmouth and Norwich. The Wrights had purchased a captured French privateer named L'Actif at Yarmouth, and it was sailed to Leeds to have a high pressure engine of 8 H.P. installed to drive paddle wheels. The Wrights' original intention had been to work an engine using the explosive force of hydrogen, but the apparatus was unsuccessful. The Fenton Murray and Wood engine had an 8 inch diameter cylinder with 2 foot 6 inch stroke and a cast iron boiler 6 to 8 feet long [33]. The boat returned to Yarmouth and started a service from there to Norwich on the River Yare, which was probably the earliest regular passenger service in England operated under steam power. Fenton Murray and Co had entered a new line of business, which they would remain in until they closed down.

A report dated the 28th June 1813 survives in Simon Goodrich's papers summarizing the total outlay made on the Naval Well at Sheerness. Goodrich had been questioned on the expenditure and the report was submitted to Mr. Nelson the Secretary to the Commissioners of the Navy. It recorded a first payment to Fenton Murray & Co for the steam engine and spare parts of £554-14-9 on 29 September 1809, a second for the pump and machinery on 3rd May 1810 of £461-8-10d and a third on 15th January 1811, for pump work, machinery and the erectors time fixing them all up, of £269-14-10d. Lloyd and Ostell (major Navy contractors) were also paid to repair the horse wheel apparatus in case of stoppage of the steam engine and some spare parts for the pump in the sum of £216-14-10 and a local plumber Mr.

Greenwood £17-14-5 ¾ d. The total amount was £1519-18-5 ¾ d. Copies of the detailed billings for the first and third payment to Fenton Murray & Co also survive in the Goodrich papers [34].

In July 1813 a large delivery of cast and wrought iron for the saw mill was brought to the attention of the Chatham Officers, which had been delivered in May. The Navy Office requested the Chatham Officers to confirm delivery and to report separately the weights of the cast iron and the wrought iron. The only parts identifying their use were for the elevated railway, and so the delivery was probably by Maudslay on behalf of Marc Brunel [35].

On the 22nd July John Blenkinsop wrote to a John Clark at Birchfield Liverpool. John Clark had written to him after Blenkinsop had met with Mr. Holding and Mr. Dagleish. Apparently Clark had offered Blenkinsop £500 for the rights to use his patent throughout Lancashire, but having refused this offer in no uncertain terms Blenkinsop offered to travel to Wigan or Liverpool to discuss usage at Clarks works, the Orrell Colliery. This piece of business, referred to previously, came to fruition. This letter was also accompanied by a detailed costs and benefit statement [20].

The next day Blenkinsop was writing to John Watson mainly about the draft agreement, and there appeared to be a difficult clause, as the Coxlodge side wished to be able to lay aside Blenkinsop's patent at 3 months' notice and adopt any other system they chose. Blenkinsop naturally was not in agreement with this. He also reported that Murray had told him the engine would be ready in three months. He acknowledged that although he had sent a cogged wheel 5 or 6 weeks ago by carrier to adjust the track, it did not seem to have arrived. Probably it can be assumed from this that the track at Coxlodge was being laid [36].

Murray wrote to Watson on the 2nd August and was not quite so upbeat about the delivery, stating that they were "*almost working day and night*" to get the engines ordered by Russia completed for the end of the month, and that after it was finished Watson's engine would be "*one of the first jobs we begin*". They hoped to complete it in 2 months but the engines were not straightforward to make yet and were a difficult job [37]!

In August the Norfolk newspapers reported that the L'Actif, (now renamed Experiment) had, on August 9th undertaken successful trials on Braydon Water with Sir Edmund and Lady Lacon and party on board, and had travelled from Yarmouth to Norwich the following day. An advertisement advised the public that the steam packet:

"will regularly leave Turner's Bowling Green Yarmouth at seven o'clock every morning, and leave Norwich every afternoon at three o'clock" [38].

On the 26th August the Chatham Officers wrote to the Navy Board to inform them that the roof of the saw mill was progressing well but in a few days would come to a stop as they were still awaiting the cast iron columns, girders and joists for the south east wing and that the whole building could suffer damage if they were not received soon [39]. On the 1st September the Navy Board replied that they had been told by Mr. Brunel on the 18th August that he had been advised that the items had been shipped from Selby [40]. Selby at this time was one of the ports for Leeds, as items could be shipped on the canal from Leeds to Selby, and there transshipped probably in this instance via Hull. As the engine came from Fenton Murray and Wood, with their history of supplying Marc Brunel, it is fairly certain that the other structural items for the steam engine house originating in Yorkshire came from the same source.

Fenton Murray and Wood initially delivered the Middleton spare locomotive to the Kenton and Coxlodge Colliery as on the 2nd September the first official running took place and 70 tons were hauled. A convivial dinner was enjoyed by the guests afterwards to celebrate the event [41]. The occasion was reported far and wide including the Morning Chronicle in London [42].

September was a busy month for correspondence concerning the Chatham saw mill, by this time the steam engine had been erected, and the boiler installed with its fireplace, however the Navy Board and Marc Brunel wished to know why the chimney had not been started. The chimney was started and was the subject of much debate and angst, for some time [43]. The Officers wrote to the Navy Board on the 24th September repeating their message that the saw mill roof would soon be stopped as they still had not received the cast iron columns, girders and joists for the south east wing [44].

John Blenkinsop reported to John Watson on the 4th October that Fenton Murray & Co were working on a steam engine for Percy Main and the replacement for Willington (the name given to the Middleton engine that had apparently in the end been sent to Coxlodge as their first engine). Blenkinsop added that the second engine for Coxlodge, which would be made rapidly, did not appear to have been ordered and Watson had better write to Murray and confirm his orders. He then asked Watson for £30 for the Trevithick licence for high pressure, which had been asked for. Watson was also asked to settle with him for Willington as Murray had charged it to Mr. Brandling's account, and the accounts were to be closed at Christmas. He then described the silencer he had had made to lessen the noise of the steam discharging. It basically consisted of a wooden box, where the discharging steam blew against the top of the box and then vented through a wooden pipe which was offset in order to allow the steam access to the top. A small drain pipe was fixed at the bottom of the box to drain any condensed water away from the cocks [45].

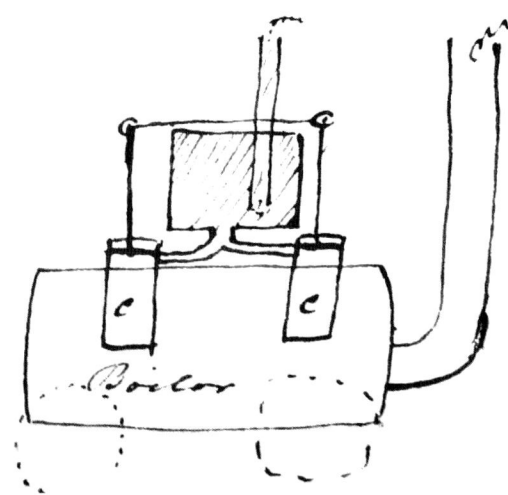

47. Blenkinsop's silencer as described above

Possibly due to the amount of business Fenton Murray & Co were undertaking, with maybe an additional number of workmen, on the first of October the firm increased its subscription to the Leeds General Infirmary from 6 guineas a year to 9 guineas [46].

Interest in the locomotive engine was spreading, and John Watson received an enquiry, dated the 6th of October 1813, over the success of the experiments, both in Leeds and at Kenton. The questions covered; the number of wagons that were drawn, the cost of the engine, the cost of the track, whether the system coped with bends or curves or sharp turns, how many men were needed to tend the engine and wagons, and at what speed could the engine and wagons travel. The enquiry came from John Brailey who was the agent for the Duke of Portland, who had a large colliery in Ayrshire. Watson sent a detailed response on the 11th of October which is summarized here. The engine was of 4 H.P. and cost £380 of which £30 was the licence fee for Trevithick. Lightly loaded the machine achieved 10 mph, but when loaded on a level way the speed was between 3 ½ and 4 mph. Watson reckoned to save 5/6 ths of the costs incurred using horse transport. He calculated these costs at 2/- a chaldron for his 5 ½ mile journey, and reckoned when they had their new cogged way laid this would reduce to 4d per chaldron (the 5/6ths reduction). They transported 36000 chaldrons a year which would have made a saving of £3000, a very large amount at the time. At Leeds the weight of the full wagons was 84 tons, with their level way, but due to there being an ascent in the Kenton line they were planning a load of 64 tons. For a new line Watson calculated that having one side cogged increased the cost of the track by 4/- a yard. He also advised that curves should not exceed 1 in 10. He considered the introduction of the system in Ayrshire would give considerable savings. He added in a postscript that he felt he should add that Blenkinsop had not finally fixed the level of his licence fees, and he had heard that there was an agreement with a gentleman in Wales to pay £50 per annum per mile of track.

48. Sole representation of Murray-Blenkinsop locomotive on Tyneside. [47].

Matthew Murray wrote to John Watson on the 14th October, and acknowledged the receipt of his order for two locomotives. He explained that they were busy making the replacement locomotive for the Middleton colliery as Middleton needed it for their busy winter period. The castings for Watson's order would be made before castings for any other order. He then re-iterated that the manufacture of these engines was by no means straightforward, and that no firm delivery date could be promised. The letter then went on to state·

"We have had an Engine of larger dimensions but of nearly the same description in hands these six months for Mr Buddle, it is now nearly finished, we believe it is for drawing coal upon an inclined plane" [48].

Murray then dealt with Watson's requests for some spare parts including pistons, and stated that these should not be made until the cylinder diameters were known exactly and that they would try and ensure that all four cylinders were one exact size. He also stated that the engines would have a bigger fireplace than those for Blenkinsop, so Murray was resizing the locomotives to suit the Kenton operation.

The context of the letter indicates that Murray is referring to a locomotive engine, as he mentions the work in order to reinforce his statement that he cannot give a reliable delivery date as he has been making this one for Mr Buddle for six months. The whole letter is solely about locomotive engines. It is not unlikely that Fenton Murray and Wood would receive such an order from Buddle, they had been supplying him with engines since 1805 (Elswick). Buddle was closely involved with William Chapman in trying to achieve a viable locomotive at the time. At this distance in time, it may never be certain, but the basic layout of the frame, boiler and cylinders, and the drive to the wheels is the same between the Murray/Blenkinsop locomotive and the Steam Elephant. The rack principle was a major driver in the Murray/Blenkinsop system, but Murray's correspondence shows that he was not necessarily a fan. He may have been quite happy to try for a version of the locomotive with direct traction, provided it was paid for, to see how it turned out. Such a locomotive may then have been replicated.

It has been mooted that Murray was in fact referring to a high pressure or Trevithick engine, not a locomotive. Mathias Dunn refers to such an engine at Percy Main in his view book in January 1816, but it is uncertain, through lack of detail, whether this engine can be the Murray one. Trevithick engines were available in the north-east, and a stationery one is more likely to have been made by someone other than Murray [48a].

Fenton Murray and Wood were also pursuing another payment for the engine at Workington and were advised by Wm Swinburn that none was yet due but that he would lay the matter before Mr Curwen on his return from Ireland. He also advised that the engine was working very well though the local water was not good to use for injection and they would need to divert some river water [3].

John Blenkinsop acknowledged the receipt of £30 from John Watson on 20th October and, advised that Murray was proceeding with the engines for Kenton. He also said that payment for the transferred engine could be delayed until February, and mentions that Mr Brandling has told him the Kenton engine was not operating as there was an issue over crossing the turnpike. He asked for the address of the Duke of Portland's agent, obviously concerned that the enquiries had not been addressed to him. He opined that neither Chapman nor Hedley would succeed with their locomotives. He asked for the return of the draft agreement [49].

The positions of Inspector General of The Royal Navy and Chief Mechanist had both been abolished in December 1812. Samuel Bentham remained a member of the Navy Board for a few years and Simon Goodrich was retained on a somewhat ad hoc basis. So it was that in November 1813 Goodrich went to Bridlington to look at a job rebuilding Bridlington Harbour, and was also considering the post of Engineer to the East London Water Works. While in Bridlington he noted that he was preparing a letter to Murray about rolling copper [50].

The Chatham Officers advised the Navy Board in response to their request of the 20 November that the Saw Mill was progressing to the point where four saw frames would be operational by the following February. They stated that the wings of the saw mill were up to the parapet, but the iron roof tank was still to be completed. The centre of the building was roofed. The underground water tanks were about half finished. The tunnel and reservoir were substantially complete, but the railway was not begun. The chimney continued to go up. Frost was identified as a risk to progress [51].

On the 9th December Simon Goodrich left Bridlington for York where he boarded the Mail to Leeds and arrived there at 4 a.m. and went to bed. On the 10th he called on Mr Murray and found him at the factory. They had breakfast and Goodrich moved his stuff into Murray's house, and was then given a tour of the factory. The following day Murray introduced Goodrich to Mr Wilson a stonemason, with whom he arranged to view two quarries the following day, presumably for stone for Bridlington Harbour. He then went to see the:

"bang up Steam Engine, a flax spinning mill and another steam engine, the gas light apparatus etc"

Murray told him about the Workington engine and Goodrich recorded:

"Workington new coal pit belonging to Christian Curwen Esq. MP in the sea, is 300 yards deep 13 foot diameter, has 160 horses engine work double by a diagonal spear. 66 inch cylinder, 10 foot stroke in the cylinder, 8 foot stroke in the pumps, works 6 lifts of 17 inch diameter, 50 yards each lift, diameter of piston rods 6 ½ inches".

On the 12th Goodrich went to Bramley Fall Quarry and Scotts Hall Mills Quarry where the products, carriage and prices were set down in great detail. He recorded information on Boulton and Watt's prices for Gas Lights including retorts (from Murray) and Murray's prices for the lamps and details of the pipes to be used. Other data was also set down relating to the use of wrought iron to suspend weight. On Monday the 13th after writing some letters and having another look at the gas lighting he departed for Sheffield on a typical Goodrich trip home, visiting Sheffield Park Iron Works, Mr George Strutt in Belper, Mr Fox in Derby and Mr Whitmore and the Eagle foundry in Birmingham [52].

The Middleton Colliery Ledgers updated for the 31st December 1813 record purchases from the firm of 2 Patent Steam Carriages for £700 and cast iron goods of £952-6-10d, so presumably a good quantity of rail as well as the engines. On the other side Fenton Murray & Co purchased £429-6-3d worth of scrap metal [17].

In this busy year the textile side continued to generate business and the Marshall papers record the following work for Fenton & Co:

From January to June 1813:

10 double drawing frames for Leeds

1 line roving machine for Leeds

2 double drawing frames for Shropshire

1 line roving machine for Shropshire

2 short tow spinning frames for Shropshire

From July 1813 to January 1814:

4 first drawing frames for Leeds

11 spinning frames for Leeds

5 first drawing frames for Shropshire

2 second drawing frames for Shropshire [53].

Notes

1. Chatham Officers to Navy Board ADM 106/1822, 31 January 1812.
2. Admiralty Abstract of Correspondence ADM 106/2094, National Archives Kew.
3. Curwen Archives D/Cu 2, Letters 1760-1849,for letters from Fenton Murray and Wood and D/Cu/3/43 for William Swinburn 's Letter Books. Cumbria Record Office.
4. Marshall Papers MS 200 folio 1. Brotherton Library.
5. The Leeds Mercury 29 February 1812.
6. The Dundee Advertiser e.g. Tay Street Mill March 20, June 5, and July 3 1812.
7. Curwen Archives D/Cu Payment Account Book. Cumbria Record Office.
8. Chatham Officers Out Letters CHA/B/18 April 1812 – February 1813. National Maritime Museum (NMM).
9. The Leeds Mercury 27 June 2012.
10. William West to R Trevithick 7 September 1815 as quoted in Life of Richard Trevithick by Francis Trevithick London 1872, reprinted Adam Gordon 2006 page 237.
11. Watson Papers Wat 3/13 Blenkinsop to Watson 2 August 1812. NEIMME.
12. Navy Board to Chatham Officers 20 October 1812. CHA/E/105. NMM.
13. Navy Board Index and Digest ADM 106/2164. National Archives Kew.
14. Navy Board to Chatham Officers. CHA/E/105. Chatham Officers to Navy Board. CHA/B/18. NMM.
15. Watson Papers Wat 3/13 Blenkinsop to Watson 12 November 1812. Covering letter Watson Papers Wat 3/112/6. NEIMME.
16. Goodrich Journals B27. Science Museum Library.
17. Middleton Colliery Ledgers 1809-1822 MC34 WYL 899. WYAS Leeds.
18. A History of the Port of Dublin by H A Gilligan. Dublin 1988.
19. Watson Papers Wat 3/112/7. NEIMME.
20. Copy of Blenkinsop/Embleton Letter Book. Northumberland Record Office or WYAS Leeds.
21. E.g. The Leeds Intelligencer 1 March 1813.
22. Goodrich Papers A 438. Science Museum Library.
23. Watson Papers Wat 3/13/114. NEIMME.
24. Watson Papers Wat 3/13/115. NEIMME.
25. Watson Papers Wat 3/112/9a. NEIMME.
26. Navy Board to Chatham Officers. 12 April 1813. CHA/E/108. NMM.
27. Watson Papers Wat 3/13/116. NEIMME.

28. Chatham Officers to Navy Board. 24 June 1813. CHA/B/20. NMM.
29. Watson Papers Wat 3/112/10a. NEIMME.
30. Watson Papers Wat 3/112/11. NEIMME.
31. Navy Board to Chatham Officers.16 June 1813. CHA/E/109. Chatham Officers to Navy Board. 18 June 1813. CHA/B/20. NMM.
32. The Leeds Intelligencer 21 June 1813, The Leeds Mercury 26 June 1813.
33. Journal of the Society of Arts. 30 March 1877. Pages 445/6.
34. Report – Goodrich Papers A 464, detailed invoices A 284 and A 311. Science Museum Library.
35. Navy Board to Chatham Officers.21/28 July 1813.CHA/E/109. NMM.
36. Watson Papers Wat 3/112/14. NEIMME.
37. Watson Papers Wat 3/112/15. NEIMME.
38. The Norfolk Chronicle and Norwich Gazette 14 August 1813.
39. Chatham Officers to Navy Board. 26 August 1813. CHA/B/20. NMM.
40. Navy Board to Chatham Officers.1 September 1813.CHA/E/110. NMM.
41. Local Records or Historical Register of Remarkable Events by John Sykes vol. 2. Newcastle 1833, reprinted 1866 page 74 for 2 September 1813.
42. The Morning Chronicle 7 September 1813.
43. Memoir of Sir M I Brunel by R Beamish London 1862. Pages 111 and 112.
44. Chatham Officers to Navy Board. 24 September 1813. CHA/B/20. NMM.
45. Watson Papers Wat 3/112/16. NEIMME.
46. The Leeds Intelligencer 4 October 1813.
47. Coal Certificate c. 1815 for Riddel's Walls End Coal. Newcastle City Library. Also see Steam and Speed by Andy Guy Newcastle 2003 page 33.
48. Watson Papers Wat 3/112/17. NEIMME. 48a Mathias Dunn view book, copy held at Newcastle Central Library. See also Richard Trevithick & Pioneer Locomotives by Jim Rees and Andy Guy in Early Railways 3 page 213/4. Would a Locomotive be used to haul waggons 450 yards, or would it be a stationery engine?
49. Watson Papers Wat 3/112/18. NEIMME.
50. Goodrich Journals B31. Science Museum Library.
51. Chatham Officers to Navy Board. 23 November 1813. CHA/B/20. NMM.
52. Goodrich Journals B32. Science Museum Library.
53. Marshall Papers MS200 folio 28 Brotherton Library.

Chapter 10

1814 to 1817.

The correspondence between the Navy Board and the Chatham Officers explains that as the saw mill was being prepared to function the task was proceeding to employ an Engine Keeper and fit persons to operate the saw frames [1].

The Marshall papers included an update on the price of steam engines at January 1814 with the following prices given: Boulton and Watt for a 32 H.P. with boiler £1830 as against £1443 for Fenton & Co, other prices were given without boilers with a Fenton & Co supplied 26 H.P. at £1083, 36 H.P. at £1304 and 50 H.P. at £1487 [2]. So it would appear that the Leeds firm was still more affordable than Soho.

The Leeds Intelligencer carried the sad story in January of Mr Jacob Wright a painter and stainer of glass who having lost one eye early in life had recently got a splinter in his good eye "plunging him into total darkness. Among the subscribers was Matthew Murray at £1 [3].

On February 15th 1814 the payments book for the Chapelbank Colliery in Workington recorded a remittance to Fenton Murray & Co of £484 being the balance of account for the Engine at the Isabella pit. This was promptly acknowledged and a full receipt issued to J.C. Curwen for £3256-6-3d for the metallic materials for a 160 Horse engine and other materials per accounts, along with warm letter to Wm Swinburn. Six years after the original quotation dated 4th February 1808, the business was finally concluded, after all the complexities surrounding the shipwreck and subsequent insurance claims [4].

On the same date, 15th February, Simon Goodrich noted in his journal that Mr Haliday had called with a small fly wheel from Mr Murray, which Mr Haliday was presumably delivering to Goodrich [5].

Fenton Murray and Co were still busy, as through February, March and April the Leeds papers carried their adverts for Green Sand Moulders accustomed to the casting of Engine and Millwright work [6].

On the 12th of March Matthew Murray received his final patent which was number 3792 'New methods and improvements in the construction of Hydraulic Presses for pressing cloth and paper, and for other purposes'. An illustration from the patent is shown below on the top.

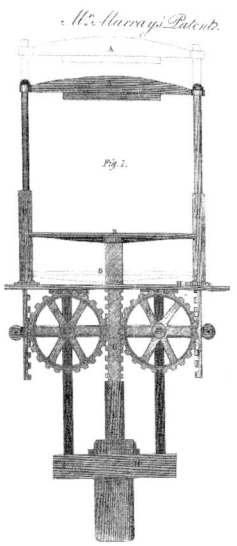

49a. Murray's patent no. 3792 of 1814.

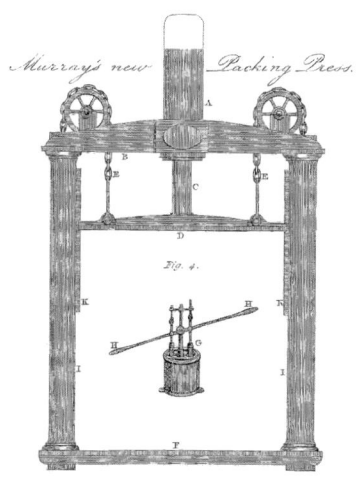

49b. from The Mechanic or Compendium of Practical Invention 1816.

The most important advance in this design was the fact that both the top and bottom part of the press moved towards each other so that when the pressing was achieved the object was at the middle height and still relatively easily handled by the machine operator, as opposed to a press where only the bottom or top part moved. The method of fixing the frame together was stronger than previous designs. The third improvement was the provision of a pressure indicator showing the pressure at any time exerted by the press.

It would appear from an 1816/8 publication "The Mechanic" that Murray continued to improve the packing press and the lower illustration above shows his further improved version [7]. A major innovation in the second design was the introduction of a body of air above the water in the hydraulic cylinder. The compound action of the two mediums provided a much more uniform action, the presses driven solely by water had a jerky action by comparison. The action has also moved all to the top, with a descending press which again provided a finished article at an ergonomic height for removal. From correspondence and other scraps of data it is clear that the subject of hydraulics was of great interest to Matthew Murray and seems to have been the subject of his study from around 1810 (cranes) to the end of his life.

John Blenkinsop was continuing to publicise the locomotive and had an article and illustration published in 'The Monthly Magazine or British Register' for 1814. He advised that any gentlemen wishing to see the carriage in operation could do so at the Middleton Colliery Leeds, Orrell Colliery Wigan, or the Kenton and Coxlodge near Newcastle upon Tyne which all had daily working.

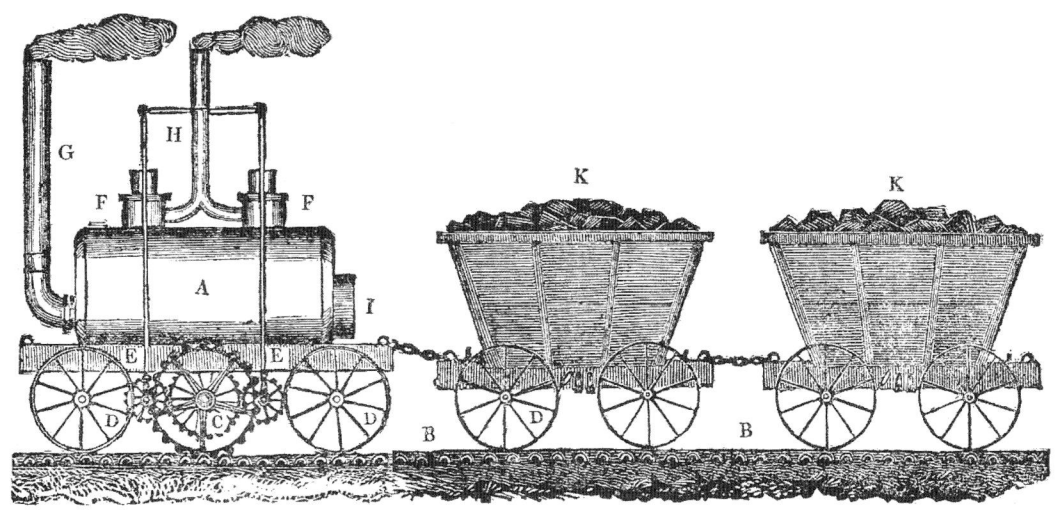

50. Murray-Blenkinsop locomotive from Monthly Magazine or British Register 1814.

In April the Leeds papers recorded subscriptions raised for the relief of distress in Germany and contrary to their usual practice of donating as a firm each of the partners gave separately; Mr Wood £2-2-0, Mr Fenton £2-2-0, Mr Matthew Murray £1-1-0 (assuming that the Messrs Fenton and Wood are the right ones) [8].

The largest record of work done by the firm survives in the form of newspaper sale notices, when the mill or object changed hands. Unfortunately these do not always record the date that the work was carried out, but for completeness many of the initial adverts are included in this text. Many repeat adverts appeared for a lot of the mills, and in April 1814 following the bankruptcy of a Mr George Cobb there was a sale of a Flax and Tow Mill at the Steander near Leeds which included a steam engine of 10 H.P., made by Messrs Fenton Murray and Wood together with all the fixtures, machinery, movements and going gear [9].

Mr Goodrich recorded in his journal on the 30th April that he had his warrant appointing him as Mechanist and Engineer to the Navy Board at £600 per year [10], so he was back in a substantiated post. He was also to have a draughtsman and a clerk to support him.

In May 1814 the flax spinning mill, washing mill and bleaching works at Trottick on the Dighty Burn just to the north of Dundee, but now on the northern edge of the city, were advertised for sale with the usual list of machinery, some of which may well have been made by Fenton Murray and Wood, but with a 20 H.P. engine stated to be by them [11]. It is worth noting that at one time James Brown, who had employed Fenton Murray and Co to build him the Bell Mill in Dundee had had an interest in Trottick Mill [12]. It is known that the Dighty Burn, which had many other mills and bleach works upon it, was a fairly small stream with a tendency to dry up in the summer, which accounts for the use of both water and steam power.

Throughout June and July Fenton Murray and Wood were advertising for smiths to forge steam engine work, with the usual offer of liberal wages and constant employment [13].

The books of the Middleton Colliery recorded the receipt of £354-2-4d on June 24th 1814, for 'a steam carriage had of the Middleton Colliery', and the cash book advises us that the payees were Williams and Watson (the owners of Coxlodge Colliery), and the machine redirected to Kenton and Coxlodge had thus at last been paid for [14].

By the 22nd of August Marc Brunel considered that the metallic part of the machinery at the Chatham saw mill was finished and in a working state as he wrote to the Navy Board requesting the balance of his price (£3000), which notification they passed to the Chatham Officers for comment [15].

By the end of that month Brunel was no doubt feeling in want of the funds, as a report dated Tuesday August the 30th in the Gentlemen's magazine reported the destruction of his veneer sawing equipment at Battersea by fire. The really unfortunate part was that due to another fire at Bankside only two fire engines responded, and they were hampered by a lack of water from a low tide. The stocks of wood, the right wing of the building including the steam engine were all that was saved [16]. A description of the veneer works was published in 1817 :

"At a few yards from the toll-gate of the bridge, on the Western side of the road, stand the work-shops of that eminent, modest, and persevering mechanic, Mr Brunel; a gentleman of the rarest genius, who has effected as much for the Mechanic Arts as any man of his time. The wonderful apparatus in the dock-yard at Portsmouth, by which he cuts blocks for the navy, with a precision and expedition that astonish every beholder, secures him a monument of fame, and eclipses all rivalry. In a small building on the left, I was attracted by the solemn action of a steam-engine of a sixteen-horse or eighty-men power, and was ushered into a room, where it turned, by means of bands, four wheels fringed with fine saws, two of eighteen feet in diameter, and two of them nine feet. These circular saws were used for the purpose of

separating veneers, and a more perfect operation was never performed. I beheld planks of mahogany and rose-wood sawed into veneers the sixteenth of an inch thick, with a precision and grandeur of action which really was sublime! The same power at once turned these tremendous saws, and drew their work upon them. A large sheet of veneer, nine or ten feet long by two feet broad, was thus separated in about ten minutes, so even, and so uniform, that it appeared more like a perfect work of Nature than one of human art! The force of these saws may be conceived when it is known that the large ones revolve sixty-five times in a minute; hence, 18 x 3, 14= 56, 5 x 65 gives 3672 feet, or two-thirds of a mile a minute; whereas, if a sawyer's tool give thirty strokes of three feet in a minute, it is but ninety feet, or only the fortieth part of the steady force of Mr Brunel's saws!" [17].

The steam engine is described as of 16 H.P. which was the power ordered from Fenton Murray and Wood in 1810, and therefore the probability is that some of the sawing equipment that was destroyed in the fire came from the Leeds firm and that the steam engine that survived was from the same place (remembering that at the time the mill was built Maudslay was moving into larger premises at Lambeth).

Brunel was at Chatham at the time of the fire, writing a report justifying the changes he had made in the construction of the Saw Mills compared with the drawings. His final paragraph read:

"You will be pleased to observe, in regard to the execution of the whole of the mill, and of the steam engine, that nothing has been spared that would in any way contribute to the general perfection of the workmanship" [18].

On the 12th September the Chatham Officers informed the Navy Board that the saw mills could cut more scantling and deals than the yard required and should they consider doing such work for other yards [19].

Although not published until 1818 [Repertory of Arts vol 33] a series of questions on the conveyance of coals on railways were posed by Sir John Sinclair, an eminent politician, ex MP and member of the House of Lords, and were answered by John Blenkinsop on the rack rail and by a J.B. from Newcastle (dated 5 October 1814) for the other forms of locomotion being used [20]. Dendy Marshall proposed John Buddle as a candidate for the undeclared J.B, due to his support for Chapman's chain pulled locomotive [21]. Much of the data is repetitious of that given above concerning costs, speeds and tonnage to be carried; but the new aspects were the questions relating to the possibility of long distance use from London to Edinburgh. Blenkinsop advised that the lines could be put in alongside the roads but there would need to be a system for passing through towns.

In the middle of November John Blenkinsop told John Watson that Mr West had written to Murray to say he only wanted £10 each for 4H.P. engine and Blenkinsop suggested that Watson paid it. He also said he had written to West saying he was not paying any more fees as he did not believe using his patent made any fees due on Trevithick's patent (provided high pressure steam was not used). He then asks for an update on locomotive news, were the Coxlodge Managers or the Fawdon managers going to use the Killingworth scheme (Stephenson), and did Watson consider that Stephenson's engine would go uphill in the wet or when it was frosty [22]?

On the 30th November Simon Goodrich noted that he was going to write to Mr Murray about Hydraulic Presses, and he obviously took the opportunity to tell Murray that he was formally back in office[23]. He received a reply dated December 3rd wherein Matthew Murray started by saying that he was glad Goodrich was comfortable in his position and perhaps the Navy had recognised the need for him to be given the position due to his absence. He then said that he was happy to share his knowledge about hydraulics, it was a subject for which he had a patent, on which he had expended a great deal of thought on improvements. He claimed that he was making improvements very often (daily). He expressed no doubt that it was the best way of providing the large amount of power needed to test the strength of rope. He suggested Goodrich worked on a machine to provide 1000 tons of power, which could be done with a ram of 16 (inches) diameter operating on a forcer of ½ an inch. Or he could have different sized forcers which would operate at different powers, and could all be fixed in one cistern. A gauge to show the power exerted at any one time would be needed, such a gauge was included in his patent and Goodrich was welcome to use it. In order to stretch a rope to test it, he attached a sketch with an upright cylinder, he had rejected using a horizontal one, which would require a chain and wheel to convert the pull into a horizontal one, and it could all be accomplished with a little stone work and included in a small building. He sent by the night mail the specification of his patent, asking that it be returned, and a section of a forcing pump, as shown below.

The cost was dependant on size, but Murray estimated that the forcers fitted on cisterns with levers would be between £25 and £30, the ram and cylinder would be sold by weight and would cost about 42/- per cwt. Murray then admitted that he had not yet made one of his pressure gauges, but was sure it would do the job, probably better than anything else. A cylinder and ram of the dimensions given, with a 4 foot stroke, he estimated to weigh 5 tons, using a 12 inch (rather than 16 inch) ram would give half power and be cheaper [24]. Goodrich noted the arrival of the letter and plans on the 6th December, and that he wrote to Murray on the 7th, but this letter does not seem to have survived [23].

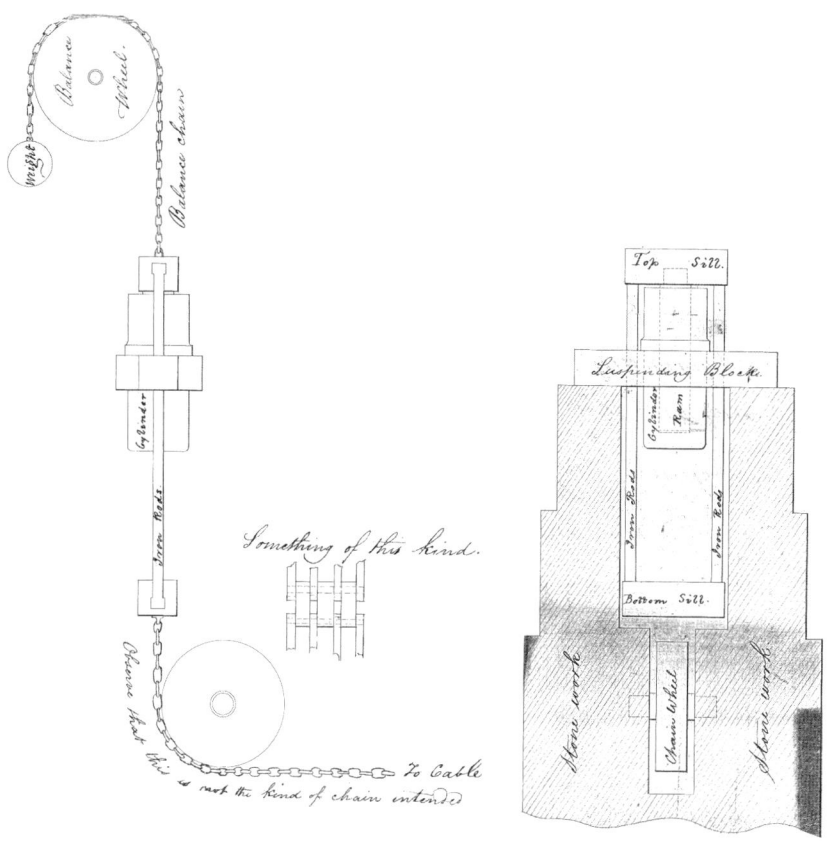

51. Murray's drawings of hydraulic rope testing apparatus as sent toGoodrich.

In 1814 J G May the Factory Commissioner of Prussia toured England, and he recorded that he visited, in regard to machine building works, Boulton and Watt, Fenton Murray and Wood (Morro and Wood), Bramah and Maudslay. He stated that steam engines were also manufactured in the workshops of Fenton Murray and Wood who employed over 100 workers. This firm also made steam locomotives which drew loaded waggons, gas lighting appliances, spinning machines for flax and wool and a variety of other machines. Later under the heading of Gas Lighting he included a full description of the plant at the Round Foundry:

"The engineering works of Fenton, Murray and Wood in Leeds are lit by gas and here I had an opportunity of examining the apparatus which makes the gas as well as the method used to supply various parts of the plant with this form of lighting. I was given detailed information concerning the success of these gasworks. The gas is made from bituminous coal. The best and most compact coal is used for this purpose since this type of coal contains most gas. To extract the gas the coal is put into airtight furnaces. A fire is lit under the furnaces to heat the coal. The gas is extracted from the coal and passes through pipes into a large iron cylinder which stands in a vertical position. Here the tar is separated from the gas. The tar flows into an underground sump from which it is later pumped out and sold. The gas now passes through a cooling chamber into a cylinder filled with water. In this cylinder there is an airtight condenser made of sheet iron. The acids in the gas combine with the water and the gas is now quite pure. The gas collects under the condenser and – aided by a counterweight – it presses the condenser upwards. Several pipes above the level of the water carry the gas to all parts of the building which need to be lit. The pipes are provided with taps and if a tap is closed the gas cannot flow beyond it. It is possible

for some taps to open and others to be closed. The last taps are near the ends of the various pipes. The ends of the pipes are closed but some have three little holes in them. The gas flows through these holes when the tap is opened. This is done in the evening when enough gas has accumulated to raise the condenser so that only one or two inches of the condenser are still in the water. The gas is lit and it burns with a pleasant bright flame without any smoke. Nothing more has to be done. As the gas burns so the condenser sinks and in this way the gas is pressed into the pipes. Eventually there is no more gas and the condenser has returned to its original position in the water. I was told that 600 cubic feet of gas can be made from 1 hundredweight of the best coals. This is sufficient to maintain 126 burners in action for three hours"

In the same year, 1814, a Mr Campion erected a new flax spinning manufactory in Whitby which contained 12 spinning frames, each having 30 to 36 spindles, carding frames, preparing frames and was all driven by a 12 H.P. Fenton Murray and Wood steam engine [26]. The mill was of fireproof construction and the pillars and cast iron work bear the hallmark of the Leed's firm. It is therefore likely that this mill was another turnkey mill with building and machinery by Fenton Murray and Wood.

52. Campion's Mill Whitby.

The History of the Earthenware Works at Leeds, tells that the Leeds Pottery installed a Fenton Murray and Wood steam engine in 1814 to grind the flints used as a raw material. The work was previously done by horses [27].

The Leeds and other papers carried a very interesting advertisement in 1837 which offered for sale a 12 horse power steam engine made by Messrs Fenton and Murray in 1814, and used ever since for drainage; with a nearly new boiler rated at 24 horse power. The engine and boiler were at Sutton Saint Edmunds in Lincolnshire in a good state and would be sold a bargain [28]. Sutton Saint Edmunds was the site of the first permanent steam powered pumping facility in the Fens, which had by this date been made redundant by the cutting of the North Level main. Mr Hinde's supposition, in 'Fenland Pumping Engines' that the engine was made by the firm is therefore confirmed, with possibly an even earlier date for starting than he assumed [29]. The yearly costs of the engine had been £200 [30].

In January 1815 one James Gray, a Greensand moulder and articled servant of Fenton Murray and Wood went absent. In the practice of the day the firm advertised this in the press with a description (26 year old Scot, 5 foot 3 inches with a dark complexion) and advised other firms not to employ him as his contract would be enforced. The advertisement notice was signed Leeds Foundry [31].

Simon Goodrich had been offered the job at Bridlington Harbour, but due to his new contract of employment with the Royal Navy was seeking someone else to undertake it and offered it to Matthew Murray. On February 9th 1815 Murray wrote stating that he should have been happy to take it had

> "his present engagements permitted, but as I am situated at present, it is impossible for me to undertake any thing, unless mechanical, as I am totally confined to our manufactory here" [32].

He also commented on a couple of other candidates mentioned by Goodrich. He went on to say that he had worked out how to use the machine for cable testing horizontally and enclosed a sketch. He also offered to look over any drawings that Goodrich prepared [32]. Goodrich recorded receipt of the letter on the 13th.

On 11th February the Leeds Papers carried an advert regarding another Fenton Murray and Wood engine for sale. This was an 8 H.P. working for Robinson Dearlove & Co in Knaresborough and had been made in 1806 [33]. It is believed this is the same Dearlove that had previously been in partnership with John Marshall.

William Losh, the Newcastle iron-founder who knew and worked with George Stephenson and funded some of his early ideas, took out a patent for fireplaces and furnaces dated April 8th 1815. From this it is learnt that one of the engines he tried his equipment on was an engine for drawing coals at Killingworth Colliery made by Fenton Murray & Co, which he described as well constructed with a 30 inch diameter cylinder. The boiler was having problems raising enough steam to haul up forty score corves of 20 peck capacity in fourteen hours. With the new furnace etc. the engine performed very well [34].

On the 3rd of May the recent death of Master Wood son of David Wood, was announced after a long and tedious illness [35].

On the 26th June the eldest daughter of Matthew and Mary Murray, Margaret married Richard Jackson in the presence of Ann Murray (sister), Charles Gascoigne Maclea, and the bride's parents. The Leeds Intelligencer contained the information that the bridegroom was late of Bradley Iron Works Staffordshire [36]. This was the introduction of a future partner in the Round Foundry into the inner circle of the partnership.

Matthew Murray took up his pen to Simon Goodrich, on a new subject between them, namely the use of steam boats by the Royal Navy. Marine engineering appeared to have entered the firm's horizons with the success of the engine in the L'Actif/ Experiment. Murray basically proposed a twin cylindered, 40 to 60 H.P. engine to power a tug with four paddle wheels, and no sails, to manoeuvre ships of war in and out of port in all winds, and to move stores around naval facilities and to ships. He also proposed adopting them for towing fire ships. It is worth remembering that in the age of sail, if the winds were adverse they could keep a ship waiting to enter port for a considerable time. Goodrich received the letter on the 21st; however any response he made has not survived [37].

In August there was an appeal for the relief of the families of the men killed at and for the wounded of the British Army at the battle of Waterloo. All three partners contributed individually [38].

In the Marshall papers there is a note that Fenton Murray and Wood charged between 21/4d and 30/- for iron ends for spinning frames, no doubt dependant on the size, and that wooden ends cost 25/4d. There is also the list of machines ordered for the year:

January to July, all for Mills A and B:

4 long tow spinning frames with 30 spindles and a draw of 8 £35/15/1d

1 line roving frame with 12 spindles and a draw of 5 £50/3/7d

All the machines listed for the second half of the year are for the new Mill C for which Fenton Murray and Wood would also have been busy in 1815 making all the cast iron frame and possibly prime contracting the whole construction:

12 line spinning frames with 30 spindles, brass bosses, 5 in. pitch, 20 in. ratch. Draw of 18 with carrying Ro[ller] and cyl[indrical] pinion £37/0/7 ½ d

11 first drawing frames £16/0/0d

There were also noted prices for many individual pieces for machines which may have been spares or parts for a new model of machine. The prices for the machines appear to be the unit price [39].

So 1815 appears to have been a year spent mainly on the home front building Mill C and starting to make its machinery.

On the 6th January 1816 Ann Murray, Matthew Murray's second daughter married an employee of the firm Charles Gascoigne Maclea, in the presence of Richard and Margaret Jackson and Mr and Mrs Murray. Charles Maclea was born in Edinburgh on the 24th November 1792 to Duncan Maclea and Anne Rennie [40].

Duncan Maclea became a Colonel in the Russian Imperial Army, and with his connections and emigration to Russia he probably knew Charles Gascoigne who had been the managing partner of the Carron Ironworks in Scotland. Gascoigne was persuaded to go to Russia in May 1786, which may have felt like a good prospect as he had been made bankrupt in 1783. In Russia Charles Gascoigne managed the Olonets Factory, the Aleksandrovsk textile factory and established the Koncherzersky iron foundry [41]. Friendship with or patronage of Duncan Maclea may explain the given names of Charles Gascoigne Maclea, and explain how an engineer born in Edinburgh ended up working for Fenton Murray and Wood, with their own trade connections with Russia. The exports to Russia in 1806 through Baron de Bode had been on behalf of Charles Gascoigne.

According to William Hirst, who did much to bring the Leed's Woollen Trade up to the standards of the West Country, he started using Fenton Murray and Wood's hydraulic presses in 1816, with the result that they went into general use:

"When I began my factory in School Close, in 1816, Messrs Fenton Murray & Co told me that if I would use the hydraulic presses, they would let me have them at almost any price, as they wanted to get them into the hands of some party who would bring them into operation, so as to give them a fair chance. The result was as they expected; as soon as it became known how well they worked, they were soon brought into general use. Mr Gott had his brought into operation and very soon added others, for they were found to be the best and most powerful presses that had been constructed. Mr Murray himself said that they would probably have failed altogether had it not been for my perseverance in overcoming the difficulties which were at first experienced in working them. The workmen at that time had a very foolish objection to the introduction of any kind of new machine, and they thought even the hydraulic press might injure labour; and consequently they would render as little assistance as possible in working them" [42].

The reference to Gott having a hydraulic press, was to an earlier period when Hirst had been working for him, the press had been made by Bramah, but put aside as the workmen would not or could not work it.

In May 1816 John Marshall recorded an experiment using wet flax rovings, the idea had been developed by David Wood, and with adjustments it showed some improvements over the first attempts but remained uneven [43].

The Leeds Mercury carried an advert on the 20th July

"Wanted. A sober steady man, to take the executive charge of the Castings and Models of an Iron Foundry. He must write a legible and quick hand. No person need apply but such as have been employed in an Iron Foundry, in the above situation, and can bring a good character.Apply to Fenton Murray & Wood, Leeds; if by Letter, Post paid".

The Hereford Journal carried a description of a journey to England by a J C Fischer which had taken place in 1814 which included a visit to the Middleton Colliery in Leeds where he was invited to ride on the locomotive and described the engine man increasing the pressure until the pistons were operating at 80 strokes a minute which caused Fischer to be wary of an explosion. After the train was slowed down:

"I rejoiced to enter Leeds seated in this triumphal car of human ingenuity (for so I would call it), where the elements confined within so small a compass themselves impel 23 wagons laden with 80 cwt of coal each" [44].

Fischer, who ran the Schafhausen Iron Works in Germany, had an introduction to Murray from George A Lee of Manchester, and called on Mr Gott trying to find Murray. He failed to meet Murray on this trip but forged a lifelong friendship with Gott [45].

In December Bryan Donkin, having heard that he was likely to shortly receive an order for a steam dredging machine, wrote to his wife from Durham asking that Mr Wilks should immediately send the drawing of the engine and machine to him at the works of Fenton Murray and Wood [46]. He explained that he proposed to try the Leeds firm as he expected that they would obtain an engine as good as one from Boulton and Watt and at a much lower price. In his journal he recorded on December 10th:

"Called upon Messrs Fenton Murray and Wood and ordered a 10 horse engine for the dredging machine for Prussia. Mr Wood showed me an improved method of making a pair of screw stocks. Mr Banks, pencil maker, this week recommended by Mr Wood of Leeds".

And on the December the 11th he added:

"Mr Murray having promised to make a sketch of the steam engine by this evening, I remained and dined with him. Found he could not complete it in time, he engaged to send it off tomorrow evening by the coach along with his own drawing of a dredging machine which he had made for Ireland" [47].

The dredger ordered from Donkin was for the Prussian Government for dredging the harbour at Stettin

The Grand Duke Nicholas of Russia visited England and Scotland and in December 1816, was travelling through the country. On Tuesday the 10th of December he arrived in Leeds and in the morning with his attendants visited the Cloth Hall, and Wormald Gott & Co (Woollen manufacturers) where they were given a cold collation. Afterwards they went to a Shearing Works belonging to Mr Manks, and on to the Linen Works of Marshall and Hives. They then visited the foundry (or Great Iron Works as some papers put it) of Fenton Murray and Wood. They finished their visit by seeing the locomotive at the Middleton colliery [48].

At some time or times in his life Matthew Murray was given a diamond ring by the Empress of Russia and a gold medal by the Emperor of Russia. They are mentioned in his will. The Emperor's gold medal was passed by Matthew Murray

Jackson (Murray's grandson) to his son William Jackson in his will (dated 17th April 1891) .No details of the occasions or reasons for the presentation of the ring has come down, however the firm exported machinery to Russia over most of its existence. The medal was presented in August 1818, see Chapter 11.

Textile machinery for Marshall's new Mill C continued to be made in 1816 and comprised:

> *"4 line spinning frames*
> *2 long tow – Do   30 spindles*   *£35-13-7 each*
> *   with upright traverse rod.*
> *12 carriages – 2 frames Tow Roving*   *£58 each frame.*
> *1 Heckling machine*
> *6 Long Tow Spinning frames*   *£ 37. 0. 7 ½*
> *   extra cylinder wheel*   *18. 6 ½*
> *2 Line roving frames*
> *6 Line spinning frames with extra cyl'r wheels*
>
> *4 short tow spinning frames*
> *   with upright traverse rod &c*   *£ 33.13. 4*
> *   extra cyl'r wheels*   *18. 6 ½*
> *   metal beam & ends*   *5. 1. 0*
> *and a list of component / spare parts"* [49].

A book published in 1816 by Mr A Nesbit a master at the Commercial and Mathematical Academy in Bradford set its readers, among many exercises, the following task. The diameter of a circular building at the iron foundry of Messrs Fenton Murray and Wood in Leeds, measures 73 feet 3 inches; how many square yards of paving are contained in the ground floor? The answer was also given at 468.23475 square yards [50].

Dr S H Spiker, the librarian to the King of Prussia, travelled through England in 1816, visited the Middleton Colliery railway and left a description of the visit, he also visited Mr Hood's cloth manufactory in Leeds which he described as follows:

"The shearing, preparing, and weaving the cloth, is performed by mechanism, put in motion by a steam engine of eight horse power, made in a peculiar manner by Messrs Fenton Murray and Woods, of Leeds; for which the makers obtained a patent, and which cost 1500 pounds sterling" [51].

The Middleton Colliery ledgers show items purchased from Fenton Murray and Wood in 1816 consisting of castings etc. (another ledger describes it as 'a boiler and castings') £1547/8/6d and a locomotive engine at £310. On the other hand the colliery sold Fenton Murray and Wood £892/14/6d of coal and £53/13/10d of scrap metal. In the same year the Middleton Colliery purchased nearly £900 of castings from Gothards, so were presumably still laying the railway line. The Middleton Colliery had now been charged for 5 locomotives, and repaid by John Watson for the one diverted to the Kenton and Coxlodge Colliery [14].

In Baines Directory of Leeds dated 1817, the foundry of Messrs Fenton Murray and Wood at Holbeck was described as being "*on a very extensive scale*" and it is mentioned that there were several smaller foundries in the town. Under the alphabetical listing of firms Fenton Murray and Wood were described as manufacturers of steam engines, flax spinning and mill machinery, constructers of fire-proof buildings, water presses and gas light apparatus. Clustered around the Round Foundry were to be found many of the existing and future management of the firm; David Wood Iron Founder, Richard Jackson, and Charles Maclea all living in Water Lane, and Matthew Murray Iron Founder with a house near Water Lane.

In February 1817 the firm advertised for sale a 16HP engine with boilers nearly new and working, and in the same month were recorded as donating £21 for the relief of the Leeds poor [52].

An advert for the sale of Moore Tennat & Co's mill at Bishop Monkton near Ripon began appearing in April, subsequent to the bankruptcy of the partners. The mill was set up for Flax preparation and spinning and included a newly erected 12 HP engine and boilers by Fenton Murray and Wood.

On Good Friday 1817 the sister boat to the Courier (which had replaced the Experiment/ L'Actif but re-used its Fenton Murray and Wood engine and boiler), the Telegraph (not Fenton Murray and Wood engine or boiler) was leaving the Foundry Bridge in Norwich when the boiler exploded. Nine people were killed including William Nickerson the steersman. The engineer John Diggens survived. He was trying to out sail a new boat that had started in competition that day. The Telegraph had previous to this date been sent for sea trials around the Medway, operating between Sheerness and Chatham, on her return to the Yare it was found that one end of her boiler was damaged probably by salt corrosion, and a repair to the wrought iron boiler had been carried out with cast iron (one end had been replaced). The repair and the likelihood (never proven) that the engineer had held down the safety valve to outbid the competition were deemed the likely causes of the explosion. This catastrophe resulted in a number of investigations and culminated in the Report of the Select Committee appointed to consider the means of preventing the mischief of explosion from happening on board steam boats…. This was published on June 24th 1817 and was the first of a series of legislative safety measures for steam engines and boilers, it introduced inspectors, double safety valves (one to be inaccessible to the engine man), and stated that boilers for passenger boats should be made from wrought iron or copper [53].

A copy fragment of a letter/notes remains of a communication from Bryan Donkin about and to Fenton Murray and Wood dated 17th April 1817, *"the engine has arrived from them,"* which refers to the engine for the Stettin dredger. And part of a letter referring to a further order for a dredging machine

"I am at present making a design and estimate for another dredging machine to be worked by an engine of 4 horses power. Will you have the goodness to send me an Eye sketch with the dimensions figured upon it of an engine of that size and similarly constructed to the one you have sent us, together with the boiler etc....." [54].

In April the Caledonian Mercury had an advert announcing the commencement of the seasonal programme of sailing for the Morning Star of Alloa, a steam boat plying from Alloa near Sterling to Newhaven next door to Leith, the port of Edinburgh. The news of the Norwich explosion having been widely reported, the ship's owners saw fit to include the following paragraphs in the advert:

"The proprietors take this opportunity of assuring the public, that the engine on board of the boat is the common condensing engine, constructed upon Messrs Watt and Bolton's principles, and not the high pressure engine, which is known to be very dangerous, and liable to accidents, owing to its requiring the steam to be raised to an uncommon degree of strength; the pressure of steam in the one case being equal to 80lb to each square inch, and in the other only 4lb to the square inch.
They also beg leave to assure the public that the boiler used for their engine is of the best maleable scrap iron, (not cast iron) made in pieces, constructed, and put together with rivets, by Messrs Fenton Murray and Co of Leeds and is entirely new" [55].

The Leeds firm only made the boiler (at 44 shillings per cwt), the original no doubt needed replacement after two years sailing in the saline Firth of Forth. This was quite common for early boilers exposed to saline conditions. It is uncertain who made the original engine in 1815, it probably came from Glasgow and it was replaced by David Napier in 1818/9 [56].

In May an announcement in the Leeds Press showed that there was still rivalry with Boulton and Watt and may well contain a quote from Matthew Murray:

"POWER - TO BE LET CHEAP
For a term of years
Two spacious Rooms with shafts drums and Going Gear complete – The Ground and Second Floors in the Black Dog Mill, Far Bank, Leeds with ten or twelve horses' power from a Patent Steam Engine, by Bolton and Watts, and at present undergoing a thorough Repair by Messrs Fenton, Murray and Wood Engineers –
When completed will be equal to any of their own Manufacture." [57].

Later in the same month a 12 HP engine built by Fenton, Murray and Co. was up for sale along with St Peters Well Mill, a Gig-Mill in Leeds [58].

On the 28th June the offer for sale was announced of Mr James Ford's flax mill at Montrose, the mill has been discussed earlier in Chapter 7 being the mill George Stephenson worked in during his absence from Northumberland. The mill was advertised as being on five floors including the attic, as being fireproof and driven by steam. The flax machines are listed and stated to be of the most approved construction (but most machinery was described as such in 19th century adverts!) and there were two steam engines one of 25 and one of 12 HP made by Murray Wood and Fenton [sic] [59]. The mill had recently been extended by Fenton Murray and Wood, but James Ford had gone spectacularly bankrupt and gone abroad leaving the firm unpaid. The mill was acquired by John Maberley, an English M.P. and banker, who also by then owned Broadford Works in Aberdeen [60].

In July an interesting letter passed from a Wm Warre[n] of the Elsecar Iron Works in South Yorkshire to Mr Chambers of Newton Chambers Iron and Steel Company in Sheffield, it is dated 4th July.

"I am sure you will excuse me when I tell you the reason I did not see you according as I proposed in my last respects to you – I was taken extremely ill the same day and was obliged to keep my bed 3 days and as soon as I was well Mr Darwin would send me to Leeds and for your Government Fenton Murray & Wood are completely out of Metal and have been buying some from Smiths of Sheffield No1 at 6/15 I saw the Invoice – it is foolish letting them come into the market – they will give 7£ per ton 12 [wks] Cr. Your Friend Mr Butterfield desires his respects to you and says they are going to have a meeting soon to raise the price, which you would like perhaps to attend – Mr Hallday of Wakefield would give 5/5 mone[y] for no 3 – but the price is too low.
The reason I cannot see you now is I am subpoenaed to London on a trial there between Crawshays & Butler and shall be at the old House, the Angel Inn – if I can do anything for you shall be very happy to hear from you" [61].

The information was responded to with alacrity as the accounts of Newton Chambers show sales to Fenton Murray and Wood dated 10th July of 7 tons 14 cwt. of No1 pig iron at £6-10-0 per ton and 15 tons 8 cwt. of No 3 pig iron at £5-10-0 per ton [61].

The Leeds General Infirmary was trying to expand and acquire its freehold, and was conducting the necessary fundraising. Fenton Murray and Wood donated 12 guineas, against the rough average of 10 guineas [62].

A further flax mill was up for sale locally at the beginning of September, namely Hill House Bank in Leeds owned by James Tennant & Co. The sale included a 30HP Fenton Murray and Wood engine with two boilers [63]. James Tennant also purchased textile machinery from Fenton Murray and Wood and he was one of those who endorsed Murray's hackling Machine as part of the submission to the Royal Society of Arts, and as shall be seen he also used these machines [64].

The Leeds Intelligencer for November 10th announced the plan to light *"this now gloomy town"* with gas. It contained the following advert:

"We the undersigned considering the Advantages likely to result to the Public from the Estabishment of a GAS LIGHT COMPANY for the Purpose of Lighting the principal parts of the Town and such factories, Shops, Houses, Inns & c as the Proprietors may be willing to contract for do engage to advance the Sums subscribed

opposite to our Names for carrying into Effect such Measures as may be determined upon by a Committee to be chosen from the Subscribers

John Hill	500	David Nell	100
George Banks	300	F Chorley	200
J B Charlesworth	300	John Cawood	100
James Fenton	100	Charles Coupland	100
Matthew Murray	100	Joseph Fawcett	100
Jonathon Wilks	100	James Holdforth	200
John Atkinson	300	John Wilkinson	100"

As well as James Fenton and Matthew Murray the list includes another well known Leeds iron founder who had the capability to install gas systems, John Cawood.

On the 15th December sale notices appeared for Holbeck Moor Mill with cottages, house, machinery and a 16HP engine by Fenton Murray & Co [65].

In December the Leeds Gas Company succeeded in raising its target of £20,000 and announced that application would be made in the next parliamentary session for leave to bring a Bill for Lighting with Gas the Town and Neighbourhood of Leeds [66].

The list of machinery ordered by Marshall and Co during 1817 survives:

<u>Jan – Jul 1817:</u>
Mill A 2 Hackling machines }
Mill C 4 Hackling machines }   £ 22.17.10
[each]
       4 [Second] card Engines   68.17.10
<u>Jul – Dec 1817:</u>
Mill C 3 Hackling machines   no8 inclusion   £ 22. 0 . 0
[each]
       9 First drawing [engines]   20 ditto   15.13. 0
       1 Line Roving   [engines]   3 ditto   49. 9. 7
       2 Second Cards [engines]   6 ditto   69.10.10
       4 Long Tow Spinn'g f[rames] 10 ditto   { 34.18. 9
                                        { 23. 4. 5
       3 Shorts Tow Spinn'g [frames] 3 Do   { 33.13. 7 ½
                                          { 24. 9. 5
       13 Line spinn'g [frames]   35 Do   { 36. 5. 2
                                          { 24.12 . 8 ½
Mill B 2 Short Tow spinn'g [frames]   { 32.10. 7 ½
                                          { 22.19.11
Salop 3 Double hackling machines
       2 second Cards [67].

A William Marshall, the Inspector of Linen and Flaxseed for the Port of Londonderry, made a report in 1817 to the Right Hon. the Trustees of the Linen and Hempen Manufactures of Ireland, and was called to give evidence from this report to other Government Committees. His main terms of reference were to look at improving the methods of scutching (cleaning off the plant parts from the flax) carried out in Ireland by examining the methods employed in England and Scotland. He visited most of the principal flax manufactories in England and Scotland, and asked the owners what improvements in the cultivation of flax in

Ireland might make it more suitable than other sources. While in Leeds he visited Fenton Murray and Wood and discussed the various improvements made in flax spinning. He had the opportunity to visit the shops and see the machines being made. He added of their machinery:

"which appears to me much superior to that in use in some of our Spinning Establishments in Ireland; and I have been informed, by several of the Scotch Manufacturers, that their work was much superior to any other in the trade."

They promised to make him a breaking machine with rollers, and to describe the improvements in their machinery since their invention. Whether these last two items happened is not known [68].

Notes:

1   Various letters CHA/E/111 and CHA/B/19. NMM.
2   Marshall Papers MS 200 folio 28. Brotherton Library.
3   The Leeds Intelligencer 31 January and 7 March 1814.
4   Curwen Papers D/Cu Correspondence and D/Cu Payment Account Books. Cumbria Archives.
5   Goodrich Journals B32. Science Museum Library.
6   E.g. The Leeds Intelligencer 21 February 1814.
7   The Mechanic or Compendium of Practical Inventions by James Smith Vol 2 Liverpool 1816. Pages 391/2 and plate C1.
8   The Leeds Mercury 2 April 1814.
9   E.g, The Leeds Mercury 23 April 1814.
10   Goodrich Journals B32. Science Museum Library.
11   The Dundee, Perth and Coupar Advertiser 13 May 1814.
12   Reminiscences of flax-spinning by a flax-spinner long in the trade at Dundee, William Brown, Dundee 1862.
13   E. G. The Leeds Intelligencer 20 June 1814.

14   Middleton Colliery Ledger 1809-1822 MC34, WYL 899. WYAS Leeds.
15   Navy Board to Chatham Officers 24 August 1814. CHA/E/112. NMM.
16   The Gentlemen's Magazine September 1814. Page 286.
17   A Mornings Walk from London to Kew by Richard Phillips, London 1817.
18   M I Brunel to G Parkin (Master Shipwright, Chatham) 31 August 1814. CHA/B/20. NMM.
19   Chatham Officers to Navy Board 12 September 1814. CHA/B/21. NMM.
20   Repertory of Arts Volume 33, 1818.
21   A History of Railway Locomotives down to the year 1831 by C F Dendy Marshall. London 1953. Pages 43 and 71.
22   Watson Papers Wat/3/112/19. NEIMME.
23   Goodrich Journals B32. Science Museum Library.
24   Goodrich Papers A545. Science Museum Library.
25   Industrial Britain under the Regency 1814-1818 by W O Henderson. Frank Cass 1968.
26   History of Whitby and Streonshealh Abbey by Rev. George Young. Whitby 1817.
27   Leeds Pottery a History of the earthenware works at Leeds by L Jewitt as reviewed in the Art Journal vol. 4 1865.
28   E.g. The Leeds Mercury 23 September 1837.
29   Fenland Pimping Stations by K.S.G Hinde Landmark 2006. Page 162.
30   The Drainage of the Fens by Rev. Dr R L Hills Landmark 2008. Page 96.
31   The Leeds Mercury 14 January 1815.
32   Goodrich Papers A568. Science Museum Library.
33   The Leeds Mercury 11 February 1815.
34   The Repertory of Arts vol. 28 1816. Page 88.
35   The Leeds Mercury 13 May 1815.

36   The Leeds Intelligencer 3 July 1815.

37   Goodrich Papers A602 17th July 1815. Science Museum Library.

38   The Leeds Intelligencer 28 August 1815.

39   Marshall Papers MS200 folio 28. Brotherton Library.

40   The Clan Maclea/Livingstone Forum Website.

41   Dictionary of National Biography.

42   History of the Woollen Trade for the last sixty years by William Hirst. Leeds 1844.

43   Marshall Papers MS200 folio 38. Brotherton Library.

44   Hereford Journal 11 September 1816.

45   J C Fischer and his Diary of Industrial England 1814-1851 by W O Henderson. Frank Cass 1966.

46   Bryan Donkin Letter Book. Donkin to Mrs Donkin 1 December 1816. Donkin Archive, Derby Record Office Matlock. I am grateful to Mr T Woodhouse for advising me of the existence of these records.

47   Extracts from Donkin's Journal 1816-1817. Donkin Archive, Derby RO.

48   The Leeds Mercury 14 December 1816, The Morning Post 16 December 1816 and many more.

49   Marshall Papers MS 200 folio 28. Brotherton Library.

50   Practical Mensuration by Nesbit. 1816 Edition. Longmans London.

51   Travels through England Wales and Scotland in the year 1816 by Dr S H Spiker. London 1820.

52   The Leeds Mercury 15 February 1817, The Leeds Intelligencer 17 February 1817.

53   Journal of the Society of Arts. 30 March 1877. Pages 445-7. The Norfolk Chronicle and Norwich Gazette for 5 April and 12 April 1817.

54   From notes of the Bryan Donkin Letter Books kindly provided Mr T Woodhouse.

55   The Caledonian Mercury 19 April 1817.

56   Contrary to the author's assumption in the article 'The Marine engine output of Fenton Murray & Co Leeds 1795-1844' in the Mariners Mirror vol.98 no1, 2012. The Falkirk Herald and Linlithgow Journal 29 October 1874, History of the Steam Navigation of the River Forth.

57   The Leeds Mercury 10 May 1817.

58   The Leeds Mercury 31 May 1817.

59   The Caledonian Mercury 28 June 1817.

60   The Port of Montrose Ed. G Jackson and S Lythe. The Georgic Press and Hutton Press 1993.

61   Newton Chambers Papers TR. Sheffield Archives

62   The Leeds Mercury 30 August 1817.

63   The Leeds Mercury 6 September 1817.

64   Transactions of the Society for the Encouragement of Arts Science and Commerce vol. XXVII 1809. Page 150.

65   The Leeds Intelligencer 15 December 1817.

66   The Leeds Intelligencer 22 December and 29 December 1817.

67   Marshall Papers MS 200 folio 28. Brotherton Library.

68   Report from the Select Committee on the Laws which regulate the Linen Trade of Ireland in The House of Lords, The Seasonal Papers 1801-1833 vol 165 and elsewhere.

Chapter 11.
1818 to 1821

In 1818 John Hives (John Marshall's partner) took Thomas and Benjamin Benyon to court for the alleged infringement of his patent for a Hackling Machine (Patent number 3257 of 12th August 1809). The prospect of such an action had maybe been behind Murray submitting his hackling machine to the Royal Society. Some of the affidavits taken out by the defence have survived. The Benyon's had started using a new model of hackling machine on the 20th August 1817. Their defence relied upon the fact that machines nearly similar in construction and principle were in use prior to the date of Hives' patent (a fact that would normally prevent any patent being granted or render it invalid). The issue has been briefly described in chapter 8 above, because the earlier machines were made by Fenton Murray and Wood based on Matthew Murray's submission to the Royal Society (Gold Medal). Hives' patent had used some of Murray's ideas which both Marshall and Hives had initially rejected.

The first affidavit was from John Sharp who had examined the Benyon's machine. He stated that in 1809 when working for James Tennant & Co he had been sent to Messrs Matthew Murray & Company to learn the management of a machine for dressing flax, and then returned to his employers to superintend the machines which had by then been set up there. The machines at James Tennant were similar to those described in Hives' patent except that Hives' machine used a crank instead of Murray's eccentric.

The second affidavit is from the machine maker Joshua Wordsworth (an eminent Leeds machine maker) stating that he had examined the specifications of both Murray's machine and Hives' patent and found Hives' patent machine to be an alteration of Murray's but working mainly on the same principle and containing some of the most essential parts including the cylinder with hackles attached.

The third surviving affidavit was from a former colleague of John Sharp, one Bernard Brown. He claimed to have written to Hives stating that the machine he was about to patent was already in use. He added that by the end of 1809 James Tennant & Co were operating 8 of these hackling machines. It is understood that Hives did not pursue the case [1].

Meetings of the Leeds Gas Light Company had commenced at the beginning of 1818, and appointments were made to the necessary committees, including John Marshall as chairman of the Committee for Mechanics. At a meeting on the 2nd February the first three resolutions that were passed are of interest:

1.   Marshall was to request Murray to send a competent person to Exeter at the company's expense to obtain data on a Mr Phillip's plan and information from other sources on the way.

2.   Marshall was to ask Murray to put forward his plan for executing the gas works.

3 Mr Garsed to ask Mr Cawood to put forward his plan for executing the gas works [2].

The competition for the works was to be between two well-known Leeds firms of iron founders.It is unclear how much experience either party had in the erection of gas works. Fenton Murray and Wood had installed a system in their own works and in Marshall's Mills and no doubt in other mills as well. That Murray had understood the principles was seen in the explanation he had given to Simon Goodrich in 1810. The firm was manufacturing parts by 1812 as demonstrated by the prices given to Goodrich. The Round Foundry was lit by gas by 1814 as has been described by Mr May (see above).They had no doubt installed a number of systems in mills as they had for some time advertised as makers of gas apparatus. Cawood's put an advert in the Leeds papers in February and March 1819 which stated that Cawood had installed gas works in 6 mills in the last year, letting everyone know that they were still in the business [3].

On the 2nd March 1818 Richard Jackson, Murray's nominee, reported back on his visit to Exeter and also provided information from London, Bristol, and Birmingham [2]. Mr Phillips at Exeter had a patent for purifying coal gas using nearly dry lime, and it seems to have been one of the better ones around in 1818 [4]. On the same day Murray made and explained his proposal for the gas works to the committee [2].

On Monday the 2nd March the papers announced the MOST DREADFUL ACCIDENT. On the last day of February one of the locomotives at the Middleton colliery, named Salamanca, exploded killing the engine man Hutchinson.

"On Saturday Afternoon last, George Hutchinson, who was one of the men employed in conducting the steam engines, used in conveying coals from Middleton to this place, stopped at the usual place on the road with a number of loaded waggons brought down, and was waiting until a sufficient number of empty ones had arrived there, to be conveyed back again. During the time he was waiting he put a quantity of coals upon the fire, under the boiler, and then left it to burn up, without having taken off the spring or fastener by which the safety valve is kept steady, during the time the engine is in motion, and whereby the boiler is relieved by the steam blowing off as soon as the pressure arrives at a certain point, without the least possible danger. – On Hutchinson's return to the engine, after an absence of about an hour and a half, it was discovered and pointed out to him by another man that the steam in his absence had got so high as to be issuing through the cocks and apertures in the boiler, and though he was requested to move on and thereby give relief to the boiler, he had the

imprudence not only to reject this salutary advice, but to make the only cock by which any of the steam could escape, still tighter; and then go to break up the fire he had previously made. The fatal consequence was, that the boiler burst, as it inevitably must do from the length of time the engine had been standing, and the increased pressure of steam, occasioned entirely by the safety valve being close down; and Hutchinson being then unfortunately standing at the end next to the fire, he was carried by the violence of the explosion to a very considerable distance, where his body was afterwards found in a most mangled state, quite dead. These are the facts of this melancholy case, as they have been carefully collected. They may be relied on. It is therefore quite evident there is no blame imputable to anyone but to the unfortunate individual who has fallen a victim to his own negligence, in leaving the boiler standing with a large fire under it with the safety valve made fast down for a considerable time and then further overcharging in defiance of the prudent caution given to him by his Companion. The explosion was heard at the distance of a mile and a half. The Iron wheels of the carriage on which the engine was placed, were broken. The damage is estimated at some hundred pounds. It is some consolation under the painful impression produced by this shocking accident that Hutchinson had no family" [5].

At the inquest it was learnt that one engine was named Salamanca and another Lord Wellington. Richard Jackson appeared for the firm, stating that he had examined the remains and the castings appeared sound and the metal good, he also advised that the engines were tested to 60lb per square inch [6].

At a committee meeting of the Leeds Gas Light Company on March 6th Mr Cawood made his presentation for the gas works. The following day the committee decided to ask both contenders to submit their prices with regard to the brickwork, masonry and all ironwork by Friday 13th March. On the 9th of March the company were asking for tenders for pipes in the newspapers. And on the 13th March both contenders for the works duly submitted their estimates; however Murray's did not include the costs for the walls and gates. The original site for the gas works was to be in Meadow lane, which was objected to by Mr Benyon and an alternate was examined in Marsh Lane but Mr Murray stated that it was not large enough, and recommended Busk's field in Meadow lane. On the 16th March the committee accepted Fenton Murray & Co's estimate for the ironwork and asked them for a full quote for the brick and stone work. On the 23rd it was reported that the Land Committee had on Mr Murray's advice agreed on a piece of land in York Street. At this meeting it was also agreed that Fenton Murray and Wood should be contracted for the erection of the whole of the gas works for their estimated sum of £8000. A tender from Messrs Darwin & Co (of Elsecar Ironworks) was also agreed upon for the pipes and Murray was to agree the weight and thickness of the pipes with them before they were formally accepted. On the 24th March the Company wrote to

Chambers and Newton, who had also tendered for the pipes, advising them that their tender had not been selected [2].

The Caledonian Mercury of 28th March announced the start of the sailing season for the Morning Star of Alloa for 1818, and reiterated the previous year's information about the wrought iron boiler with double safety valves by Fenton Murray & Co.

Fenton Murray & Co wrote to Newton Chambers on April 7th about pipes:

"Having engaged to fit up the Gas Light works at Leeds, and thinking we could procure some of the Pipes & other castings from you that we have to furnish, if you have time to do them - But as this could not be properly explained without an interview, could your Mr Chambers the active manager come over here, and we will agree with him if possible and give him an order for the quantity of Pipes & other castings that are in our Contract. Your answer by return will much oblige.

P.S. Your Neighbour Mr Darwin has got an order for some of the main Pipes from the Committee" [7].

The Committee of the Leeds Gas Light Co. met on 4th May and among other business agreed that the contract to be made with Fenton Murray and Wood was to include a forfeiture of £100 per week, to be paid by them for every week the works were not completed after 1st November 1818 [2].

An Englishman working in New York theatre production management called Edmund Simpson was over in England in 1818 trying to recruit actors to go to New York. One of his interests was the use of gas lighting in theatres. He visited Leeds and made the following entry in his journal for the 30th May 1818:

"Went to Murray & Wood –Iron Founders Water Lane to enquire about an estimate for a Gas Apparatus- [they] could give none now without a plan of the building & without knowing how many retorts would be wanted. The gasometer containing 1800 Cubic feet would cost £300- they are now erecting a Gas Works to supply the town of Leeds. Their Gasometers will contain 48,000 Cubic Feet & 40 retorts. Could not give them any account of the retorts or their sizes & so came away as full of information as I went" [8].

On the 1st June the Leeds Gas Light Co authorised payments to be made to Mr Jackson of £30/8/6d for the expenses incurred in his visit to Exeter, and to Fenton Murray & Co of £1000, as the first payment for the gas works. By the 15th June Newton Chambers had started raising invoices for the pipes delivered to Fenton Murray & Wood; the first consignments consisted of:

8 Iron Pipes 9 inches by 9 foot long

12 Iron Pipes 14 inches by 9 foot long

1 Iron Pipe 14 inches with ribs cast in loam 8 foot 8 inches long

1 cylinder [cast] in loam

The value of the invoice was £106/9/3d [9].

A 4 HP engine by the firm was advertised for sale on June 15th at Bent Mills, Wilsden, near Bingley, Yorkshire [10].

On the 29th of the month the Leeds Gas Light Co sent the following letter to Fenton Murray and Co:

"As Chairman of the Committee of the Gas Light Co, meeting this day, I have the pleasure to inform you, that confiding entirely in your honourable intentions and under the conviction that you will complete the Works now commenced in a manner satisfactory to their and your own credit, they have come to the unanimous resolution of accepting the Agreement as drawn up by Mr Murray with a slight alteration only of the last clause which they wish to have worded as follows;
'And further to allow James Fenton, Matt[w] Murray, and David Wood, and their workmen, full possession of the premises above mentioned until the Works are finished without being compelled to admit any person or persons, <u>excepting the Managing Committee, their immediate workmen, or any one empowered by them as a body'</u>
The underlined sentence you will perceive is the addition required. The Committee will meet again on Business next Monday at 12 O'clock at which hour, I have to request your attendance to give your joint assent to the Agreement, and put the same into a proper form for immediate execution.
On Mr Murray's recommendation, the Committee have resolved to send Mr Boulby (the manager of the gas works) to London and Birmingham, to inspect the laying and jointing of the main pipes, and they request you to give him such instructions and letters, as will further his enquiries on these points.   J Wilks Chairman" [2].

Newton Chambers ledgers show further deliveries of pipes, cylinders and round plates of 139 tons 3 cwt. 7 quarters on July 1th and 68 tons 3 cwt. on July 11th [9].

On the 13th July the treasurer of the Leeds Gas Light Co was authorised to make a second payment of £1000 to Fenton Murray & Co, and on the 10th August the third payment, also £1000 was authorised [2].

It is known that His Imperial Highness the Grand Duke Michael of Russia visited Leeds towards the end of July, and visited the Cloth Halls, the different manufactories and the coal engines of Mr Blenkinsop. While it is not stated which manufactories were visited, the outline programme is so similar to that undertaken by the Grand Duke Nicholas 18 months previously, that there must be a strong probability that the Round Foundry featured in the visit [11].

Fenton Murray and Wood had approached a brewer named Cobb in Margate in 1808 (see chapter 8), having heard that they had a requirement for a steam engine. The matter was obviously under further consideration as Messrs Cobb and Sons wrote to them on 21st August 1818, and the reply has survived:

"In answer to your favour of the 21st Inst, our price of a 6 Horse Steam Engine and Boiler for the latest and most approved description is £ 475 - -, 8 Horse Steam Engine & Boiler £ 570. Beg to remark that these Engines are constructed upon an Iron Cistern which stands above the floor and consequently saves considerable

expense in the Stone Work this arrangement also renders the Air Pump and every other part of the Engine easy to get to, which is very convenient.

These Engines are fitted up with the D slide valve, Governor &c &c and are tried here for about a week with their full load upon them which ensures their acting with certainty.

The above prices are for the Engines delivered here & to be paid for in 3 months by Bill at 2 months from such delivery" [12].

So far as can be made out, Messrs Cobb continued to obtain quotes, including from Boulton and Watt, without making a purchase, for many years.

The Durham County Advertiser for August 29th 1818 reported that Matthew Murray had been awarded a superb gold medal weighing 5 ounces by the Emperor of Russia for his great and skilful improvement in various articles of machinery. The medal was despatched via Count Lieven, the Ambassador to the British Court.

On September 7th the Leeds Gas Light Co made a fourth payment of £1000 to Fenton Murray & Co. On the 16th it was noted that Mr Charlesworth was to consult Mr Murray on the subject of the glasses for the burners [2].

The 19th of September edition of the Leeds Mercury carried an advert addressed to Iron Founders and Model-Makers:

"Wanted. A steady sober active man, to overlook and manage an Iron Foundry. Also one good Green Sand Moulder, who has been accustomed to make clean castings for Steam Engines. And one good Model Maker who has been in the Steam Engine business.
Constant Employment and liberal wages will be given to good workmen of the above description.
Personal Application would be preferred; if by letter, post-paid" [13].

The advert implies either the loss of a manager or a reorganisation.

On the 14th of October 1818 the firm lost one of its clerks Thomas Mallinson to typhus fever. He was a young man of considerable mathematical and literary attainments, and his amiable private character secured him the esteem of all with whom he was connected [14].

Matthew Murray junior had been on a visit to St Petersburg, where he had met with General Alexander Wilson, the manager of the Aleksandrovsk Textile Factory, a customer of Fenton Murray & Cos. On his return he had written to Thomas Telford and had enclosed a letter from Wilson concerning some technical drawings of projects in Russia, including a bridge over the Neva. General Wilson had entrusted them to Murray junior to carry home [15].

Newton Chambers were still delivering pipes as recorded by their books on 14th November, albeit this time only 22cwt. and 3 quarters [9].

An American Mr Francis B Ogden [16] had obtained a United States patent for expansive steam in a double cylinder marine engine which was dated 31 December

1813. After some early attempts to build engines in the U.S.A, Mr F.B. Ogden came to England in 1816, and among the eminent engineers he consulted was James Watt. According to Ogden, Watt had agreed that the use of expansion with double cylinders would produce the required results. Boulton and Watt quoted for producing the engine, but were so expensive that Ogden applied to Messrs Murray & Co of Leeds. The firm did indeed make an engine for Ogden and it was probably delivered in 1818. It was a double acting engine, consisting of two cylinders, 30 inches diameter, 4 foot stroke, with a double condenser, air-pumps etc. The whole weight of the engines and boilers when complete was 68 tons. The narrative continued

"Mr Murray was an ingenious, practical man; he took great pride in the job; nothing to be compared with it in magnitude for marine purposes having yet been undertaken in England. But he never could be made to understand the principle of expansion, insisting to the last that wire-drawing the steam would produce the same effect. It was only through Mr Ogden's positive orders that the throttle valves and cut-offs were introduced" [17].

On 26th of November 1818 Ogden wrote to Fenton Murray and Wood[18] from Norfolk Virginia, primarily on behalf of his friend Henry Heth [19] who he had recommended to apply to them for an engine of about 8 HP to work in his collieries. Henry Heth's application is given below. At the end of his letter Ogden advises that he has nearly completed erecting the engine Fenton Murray and Wood had built for him and it was to his entire satisfaction. He hoped to raise steam about a fortnight after writing, when he would write again (if he did, the letter has not come to light). He continued by stating that everyone admired the execution of the work, and he then enquired about Mr Gordon's application and an order from New Orleans. The latter enquiries are interesting as there were suggestions made that Ogden's designs had been pirated (perhaps by Fenton Murray and Wood) and operated elsewhere in the U.S.A. This letter appears to be evidence that Fenton Murray and Wood may well have fulfilled other orders that were made with Ogden's agreement and knowledge. In 1824 Ogden purchased second hand an engine made by Fenton Murray and Wood to the same pattern as his 1817/8 engine in New Orleans. He had this engine placed in a tug on the Mississippi, which he described as going upriver against a 3 ½ mile current, towing two vessels of 300/400 tons along with a couple of smaller ones [17]. Ogden's original boat in which he installed the original engine per his letter may well have been the Roanoke, which is mentioned in a survey of American steam boats by a Frenchman Marestier [20]. Marestier described the steam engine and this matches Ogden's patent and the description of the Fenton Murray and Wood engine made for Ogden. It was an unusual engine for the U.S.A where most steam boat engines were high pressure.

The Committee of the Leeds Gas Light Co. met on the 7th December with Mr Marshall in the chair and resolved that the following letter should be sent to Fenton Murray and Wood:

"The Committee having made every preparation for the commencement of lighting the town with gas on the 1st January next and being pledged to many of the inhabitants in Briggate and other principal streets to supply them with gas at that time it will be a great disappointment to them as well as a very considerable loss to the Company if Messrs Fenton Murray & Co should not have the Gas Establishment completed so as to commence working at that time. The object of this communication to Messrs Fenton Murray & Co is to impress upon them the necessity of them using every exertion to forward the Works so that they may not subject themselves to a still more heavy loss and responsibility than what they have already incurred from their undertaking not being completed at the time specified in their contract" [2].

On 8th December Henry Heth wrote to Leeds about his need for an 8HP engine in Black Heath Virginia. The engine was both for drawing up the coal and pumping out water. The water in his mine was very corrosive on iron and therefore he needed copper boilers and all the parts that would be in contact with the water would need to be of brass or copper. He wanted the engine shipped in one month either direct or via New York. His two current steam engines were often out of order. Funds would be placed with Hayes and Story of Liverpool. He advised that he had been made aware of Fenton Murray & Wood and handed their price list by Edmond Taylor & Co. Heth said that he had seen the very superior engine furnished to Ogden [18].

Once again the Marshall papers give the list of machinery ordered from Fenton Murray and Wood for 1818:

Mill C

2 Heckling machines including No. 10		£ 22. 0. 0
1 First Card	No. 1	66. 7. 9
5 Second Card	No. 11	69.14. 8 ½
2 Tow Roving	No. 4	46.11. 5 ½
9 Long Tow Spinning	No. 19	{ 35. 5.11
		{ 23. 6. 1 ½
4 Short Tow 24 spindles	8	{ 32.14. 5

		{ 23. 9. 6
7 Line Spinning	42	36.12.10 ½
		24.14. 8 ½

July Dec 1818

Mill C

2 Heckling Machines & 6 sets tools		27. 2.10
3 first cards	including No 4	66. 7. 9
1 tow roving	5	46.11. 8 ½
5 Long Tow spinning	24	{35. 5.11
		{24.11. 3 ½
1 Gill frame first drawing		55.14. 3

Salop

2 Second Cards		69.14. 8 ½

An attempt to total these on a conservative basis gives a value of around £1600, a great deal of money at the time. Where there are two values against a machine this valuation uses the lower, as it is not clear why there are two values [21].

In January 1819 the Leeds Intelligencer announced that it understood that the Gas Works would be completed in the middle of February or early March at the latest [22]. Newton Chambers supplied a further two pipes to Fenton Murray & Wood in January [9].

It would appear that the firm had completed some work in Aberdeen as they received a payment from Mr Maberley in February 1819 of £59/1/3d [23]. Unfortunately in the surviving bank records most of the receipts are aggregated together for a day, and it is only when a single payment arrived that it bears a name. This is probably a part payment, with the remaining payments being aggregated with others.

On the 2nd of February Fenton Murray & Wood replied to Henry Heth. Fenton Murray & Wood acknowledged Heth's letter but advised that with their current orders it would take at least 4 months to fulfil his order, and that would be after agreeing the requirement. They would need to know the length of the shaft for the drum, whether it would be fixed on the fly-wheel shaft or a separate one on wheels to change the revolutions for the 8 HP engines from its normal 40 revolutions a minute. How did the pump work, did it connect to the crankshaft or the beam, what length of chimney was required etc.? They advised a probable cost delivered at Leeds of around £1300 to include all the brass and copper parts and 3 copper boilers. They added that any order would need to be placed through some respectable house in England, to whom they could deliver, send their invoice to and receive payment from [18]. They enclosed a plan and elevation of one of their 8 HP engines:

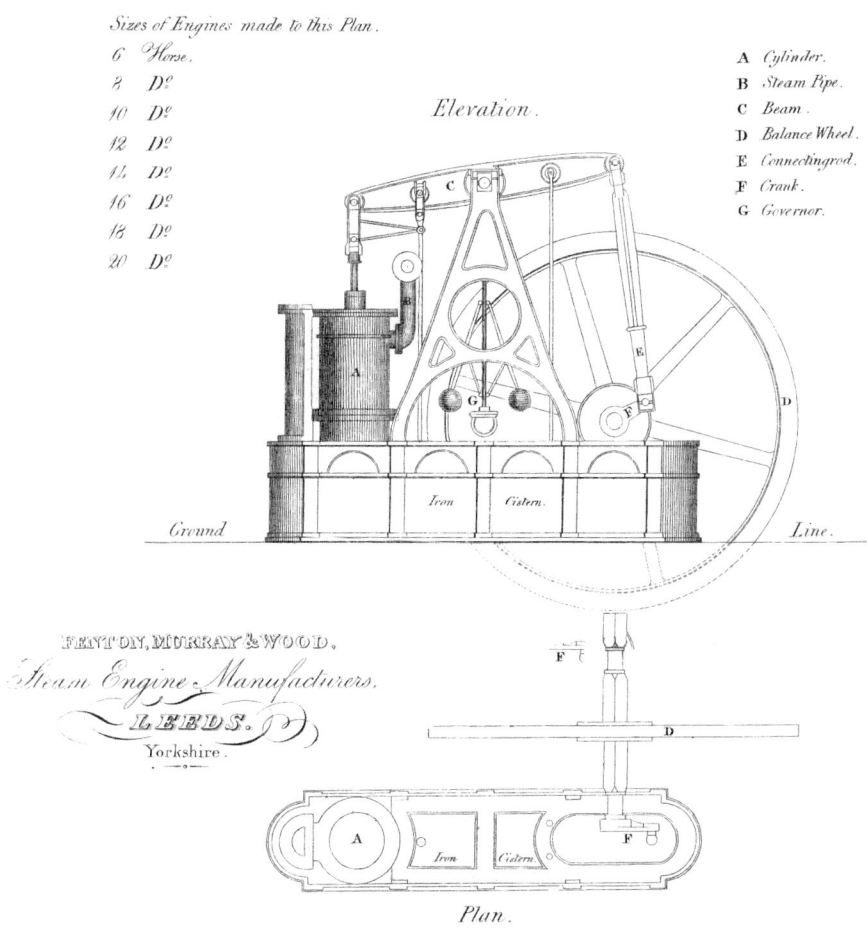

Sizes of Engines made to this Plan.
6 Horse.
8 Dº
10 Dº
12 Dº
14 Dº
16 Dº
18 Dº
20 Dº

Elevation.

A Cylinder.
B Steam Pipe.
C Beam.
D Balance Wheel.
E Connectingrod.
F Crank.
G Governor.

Ground   Line.

FENTON, MURRAY & WOOD,
Steam Engine Manufacturers,
LEEDS.
Yorkshire.

Plan.

53. Plan of 8HP engine sent to Henry Heth [18].

The Committee of the Leeds Gas Light Co resolved on the 3 February 1819 that as they had been unable to supply gas to the town that morning a sub-committee would meet at the Works every morning to confer with Mr Murray as to when the gas could be sent into the pipes [2].

On the 8th of February the Leeds Intelligencer announced that the gas lights in the town had been lighted up for the first time on the evening of the previous Thursday, which was the 4th. Initially while the air within the distribution pipes was cleared, the light was disappointingly dull, however by 9 o'clock the degree of brightness was better and on the evening of the 5th, they were sufficiently bright to be described as:

"Leeds Gas Lights will equal those of any other town in the kingdom."

On the 1st March the treasurer of the Gas Light Co. was instructed that Fenton Murray and Wood were to be paid a further £1000 [2].

Steam Engine work was continuing as attested by a request dated March 15th to Newton Chambers, asking if they were interested in making 12 large fly wheel felks (segments). They were to be 12 inches broad, 6 or 7 inches thick and 10 feet long, in a clean casting and of metal that could be easily chipped and filed. The felks should weigh about 1 ¼ tons each. If they wanted the business, the models would be sent from Leeds [24].

The Gas Light Co decided to write to Messrs Fenton & Co on the 5th April, to request the immediate completion of the gasometer [2].

In April 1819 Simon Goodrich went to visit Hawke's in Gateshead and having to visit a Mr Proctor in Sheffield on the return journey, broke his travels at Leeds, probably on the 18/19th. Goodrich noted in his journal that Murray's slide lathes were made with a square frame for the bed and the poppets slid upon a cross bar. He recorded that he saw a simple screw cutting machine.

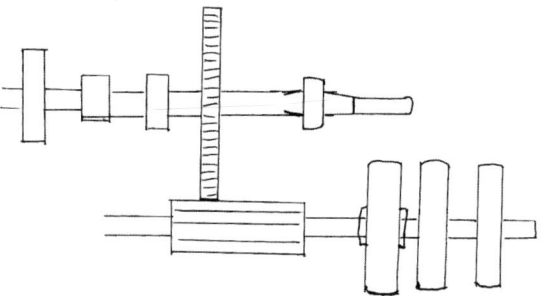

54. Fenton Murray & Co lathe seen by Goodrich 1819.

He added that Murray recommended water being let into a steam boiler at around 12 inches below the surface to prevent boiling over or too much disturbance in making the steam, and to let it in by a large pipe stopped at the bottom with 4 taps on its sides of about 1 inch diameter to direct the water to the four corners. Murray turned his crank pins "*globular*" which had the pleasant effect of reducing noise.

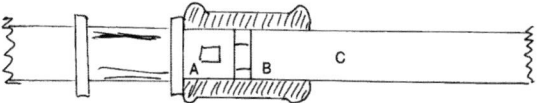

55. Murray's coupling, as seen by Goodrich 1819.

Goodrich drew Murray's coupling where the square coupling was keyed by "**A**" on the end of the square shaft made pyramidical [.... ...] there was a brass "**B**" which fitted close by the square inside of the coupling. There was a turned pin to the shaft "*C*" which fitted a round hole in the brass "*B*" and the other part of the coupling fitted on the square of the shaft "*C*" slackly [25].

The Caledonian Mercury carried a sale notice for the Flax Spinning Mills, Washing Mills and Bleaching Works of Trottick [26].

Demand for gas in Leeds was good to the extent that a committee of the Gas Light Co. meeting on the 18th of May desired some of its members to consult with Mr Murray on the best method of increasing the number of retorts [2].

Fenton Murray and Wood's outstanding debt of £2122 /1 shilling in regard to Ford's flax mill in Montrose had remained unpaid. The debt had been payable by 8 bills of exchange all dated the 9th October 1815, with the last of them payable in December 1816, however none of them had been met. William Ford the father had died and James Ford the son, who had been a joint acceptor of the bills, had become bankrupt. Fenton Murray and Wood had pursued payment in the Scottish Courts and judgement had been given in their favour, apart from a demand to justify their expenses, by Lord Gillies on the 8th June 1819 [27]. It is unlikely they ever saw their money due to the completeness of James Ford's financial failures, which resulted in further legal cases involving other parties in the Scottish Courts for some years.

Fenton Murray & Wood wrote in exasperated tones to Newton Chambers on the 9th June asking where the fly wheel felks were. Their Mr Jackson had expected them 3 or 4 weeks ago. They needed either some felks or their patterns back immediately as they would now have to work night and day to meet their orders. A reply by return of post would oblige [24].

Presumably the additional retorts for the Gas Light Co. had been ordered from Cawoods instead of Fenton Murray and Co. or some of the original orders had been, because on the 16th June the Gas Light Co decided that the retorts set up to Mr Cawood's plan should be reset to Mr Murray's plan [2].

The issue with the fly wheel felks appeared to have been resolved quickly as on the 30th of June Fenton Murray and Wood wrote to Newton Chambers for 80 pipes for re-laying the gas works at the Round Foundry. 80 pipes were duly entered in Newton Chambers ledger to their account on the 15th July at just over 159 cwt. and costing £100/14/2d [24].

In August the firm donated £10 to the fund for the distribution of soup to the unemployed poor in Leeds [28]. It is interesting that the name above them in the listing is Robert Owen Esq. of New Lanark. It is possible that Owen had been visiting the partnership, as he later wrote an article on some of David Wood's work (see below), and the cast iron pillars supporting the balconies in the school room at New Lanark have some resemblance in pattern to those designed at the Round Foundry.

In September the Leeds intelligencer stated that the Leeds Gas Light Company were building another gas depository of nearly the same size as the original due to the demand. On the same date the Gas Light Co. wrote to Fenton Murray & Co. regarding the repairs to the gasometer pits. They expressed their concern that the walls of the gasometer tanks had been pulled down and remained in that state for 2 weeks. They pointed out that they needed to be put back in use due to the large amount of gas wanted in the town. They also wanted the new retorts built on the new construction to be ready as soon as possible, in case of any accident to those currently in use. These matters could not be delayed any longer without causing "material detriment" to the company. By the 15th of October a letter was to be sent to Fenton Murray & Co asking for an estimate for 20 complete retorts, a condensing vessel (on a plan to be explained) and two purifiers. However on the 20th October it was decided to accept Mr Cawood's plan and estimate for twenty retorts and the necessary other apparatus, the contract to be based as far as applicable on that for Fenton Murray & Co. Messrs Fenton Murray & Co were asked to make four apertures in the retort house roof to allow the smoke to escape [2].

In November Simon Goodrich again travelled to Sunderland, and although the date is unclear, he again visited the Round Foundry, probably on his return journey [25].

In the month of November 1819 the South London Water Works Company, having decided that their existing engine was badly in need of repair and inadequate, were looking to buy a 40 H.P. steam engine and wrote to Mr McMurdo (13th November), Boulton and Watt (17th November) and Woolf and Edmunds (13th November) for estimates. Again there is evidence that Fenton Murray and Co had excellent sources of information, for the records of the South London Water Works record that they received a letter from Messrs Fenton Murray and Wood dated the 16th of November asking to be permitted to give in an estimate for either a 30, 35 or 40 H.P. engine. Fenton Murray and Wood duly submitted a quote dated 25th November [29].

Newton Chambers were looking to buy a steam engine and received a letter from Fenton Murray and Wood dated 11th December. They were advised that Mr Jackson would come to see them about an order the following Tuesday or Wednesday, and that they could only take payment in metal for half of the value plus a bill at 4 months for the balance, as their normal terms for such engines was payment when it was set to work. They quoted for an 8 H.P. engine excluding boiler, grate bars, bearers and fire plates at £364 and a 10 H.P. at £420. The price for the 6 pillared, cistern framed engine seen by Mr Chambers was £600. All prices were delivered at Leeds but included wages for erecting the engine but not the board and lodging for the engineer. The lead time for an 8 or 10 horse engine was from 6 weeks to 2 months as they would shortly have their annual stocktake which took 2 to 3 weeks [30].

The Leeds Intelligencer reported on December 13th that the gas retort at Messrs Marshall Hives and Co had exploded on the previous Friday night, fortunately with no casualties.

The year 1819 saw the publication of a book which described some of the highlights of modern systems adopted at the Pauper Lunatic Asylum at Wakefield where:

"Much manual labour is saved…by a steam engine of two horse power, erected by Messrs Fenton, Murray and Wood of Leeds. It is employed for pumping the water, and performs the labour of washing, and mangling linen. The steam boiler is of much larger size than is required for the engine, in order to supply the kitchen, and baths, and to heat water in different parts of the building" [31].

In a paper on the Horse Power of Steam Engines included in the minutes of the Institution of Civil Engineers for 1850/1, the early types of indicators used on steam engines were described. Mr Watt's simple indicator was defective and became effective when improved by James Lawson their one time agent and engineering representative. The improvement involved the addition of a 'moving tablet' that recorded the fluctuations measured by the indicator. The paper states that the details of the apparatus were kept secret by all who were entrusted with its use. Fenton Murray and Wood however knew of it, as the paper went on to state that John Farey obtained his knowledge of it when he visited Russia in 1819 from an indicator sent out there with an engine by Fenton Murray and Wood [32]. The mechanism is described in Farey's Treatise on the Steam Engine [33].

Marshall Hives and Co. were still buying textile machinery from Fenton Murray and Wood in 1819 for Mills A, B and C and Ditherington or 'Salop'. This was the period when they started introducing the so called Gill frames into their mills, which were stated in the Marshall Papers to effect a great improvement in drawing fibres of unequal length and enabling the spinning of finer linens. Marshall obtained the detail of these frames from a Daniel Dakeyne who had a small mill in Darley Dale. Dakeyne had followed the scheme of Busk or Hall's specifications and a trial machine had been ordered jointly by Marshall Hives & Co and Titley Tatham and Walker. The latter company performed the initial trials and it was then installed at Marshall Hives and showed such promise that John Marshall junior and David Wood went to Derbyshire (probably in 1818) to see them at work in a small mill. They returned home and started to introduce them, and of the 34 or so machines made by Fenton Murray and Wood for Marshall Hives & Co in 1819 year all bar one were Gill machines. The introduction of the Gill machines was complicated as it affected the processes after hackling and required a reorganisation of the production line [34].

Youngman's History of Norfolk published in 1819 describes the silk mills of Messrs Grout Bayliss & Co at Yarmouth as having a 24 H.P. steam engine by Fenton Murray and Wood, but with additional space being prepared; the whole factory to be then powered by an 80H.P. engine by Boulton and Watt.

By the beginning of 1820 the Gas Works were working reasonably. At a committee meeting of the Leeds Gas Light Co. on February 11th 1820 it was decided to pay Fenton Murray & Co the balance on their account of £1492/12/9d, which sum was duly recorded in February in their account with Beckett's bank, one of the few named deposits. The Gas Light Co. also noted that if they wished to retain the steam engine in the gas yard it would cost them a further 100 guineas [2].

After a severe winter in the area subscriptions were collected for the relief of the poor, and the following contributions may be noted; Fenton Murray & Co. 12 guineas, Mr M Murray 2 guineas, Mr J Fenton 3 guineas and Mr Wood £2 [35].

At the end of February the Royal Lancastrian Free School in Leeds published some accounts in the newspapers which recorded a payment from Mr Matthew Murray of 1 guinea. Perhaps Matthew Murray junior was educated at this school [35].

In March 1820 Fenton Murray & Wood purchased 25 tons of number 3 pig iron from Newton Chambers at £8 per ton [9].

In March Fenton Murray and Wood advertised for sale, along with a Mr Ingledew of Gainsborough, a second hand 8 H.P. Boulton and Watt engine, which could be seen at work in Gainsborough [36]. Presumably it was replaced with an engine from the Round Foundry.

A further delivery of number 3 pig iron from Newton Chambers followed in April but this time it was only 2 cwt 14 quarters. This was followed in July by a couple of pipes some lead and a block etc with a value of £3/17/0d [9].

In June the Gas Light Company advised that the steam engine on site was no longer needed and requested that it be removed by Fenton Murray & Co [2].

In October the Leeds and Newcastle papers announced the death on October 1st of David Wood, the original partner of Matthew Murray. The brief obituary summed up what little was known of the man:

"As a mechanic, he was a man of genius – and in his private character he was upright, humble, humane and charitable" [37].

This man, who was of such character, that when Boulton and Watt had been desperate to wreck the business in the early days and instructed James Lawson to offer David Wood a job at Soho, Lawson had decided it was not worth even making such an offer. One suspects that Wood was greatly missed as the steady man who looked after much of the day to day technical running of affairs, who gave Marshall's the attention they needed as number one customer, and played the foil for his more flamboyant and rumbustious partner.

In the Leeds papers it was recorded that Matthew Murray had a new personal subscription to Leeds General Infirmary of 2 guineas [38].

During 1820 Fenton Murray and Wood continued to supply the new Gill machines to Marshalls, in number about 45, with one hackling machine and 6 line spinning machines [21].

1820 also saw Fenton Murray and Wood supply the steam engine and dredging equipment for a dredging machine for the Aire and Calder Navigation. The engine is variously reported as 6 H.P. or 10 H.P [39].

By February 1821, the firm, now styled Fenton and Murray, was seeking to sell cheap a second hand 28 H.P. by Boulton and Watt [40].

In a letter dated 15th February 1821, addressed to the editor of the London Journal of Arts and Sciences, Matthew Murray described a system for burning smoke:

"Sir,
Being an admirer of the plan adopted in your Journal, of giving to the public an early account of new inventions, and as the burning of smoke at present engages the attention of many, I request you to insert in your next, if consistent with convenience, a new invention for that purpose. The contrivance will be best understood by inspecting the drawing which accompanies this. (see below). The most effectual method yet known for consuming smoke, is by the admission of a large quantity of air to the hottest part of the fire, at the time the smoke is bursting from the recent charging of coal. The necessary quantity of air to be admitted ought not to be less than may pass through an aperture of four square inches for each horse power, that the boiler or fire is equal to. This will consume the smoke in from three to five minutes, according to the quantity and quality of coal put on each time. The times of charging being not more than five times in an hour, nor less than three. The air's rushing into the flue is the moving power, for giving motion to the new regulating machine, which continues in motion during the consumption of smoke, but no longer. By this method there is no unnecessary loss of heat, as when the aperture is left open, or shutting it off is intrusted to the uncertainty of neglect, which is the case if regulated by hand, and from whence a great loss of fuel is the consequence. The opening of the fire door to admit the fuel, puts the machine in a state for measuring off the quantity of air to be admitted after each charging. In giving this a place in your useful Journal, you will much oblige, Sir
Your's truly, MATTHEW MURRAY.

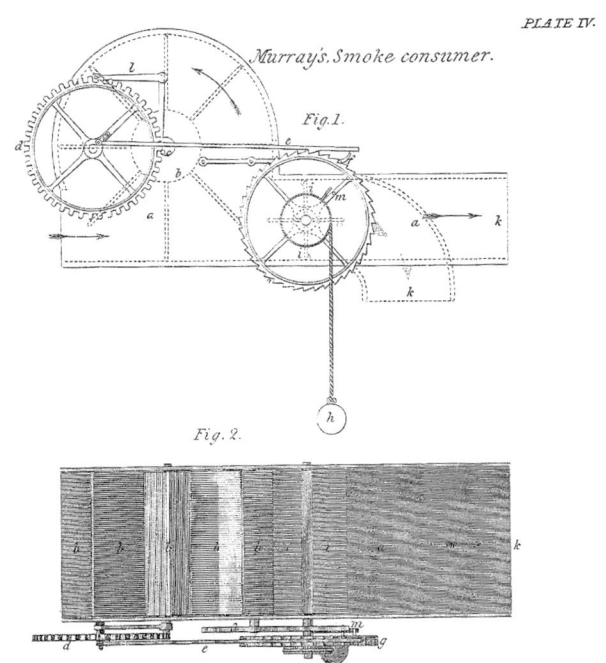

56. Illustration of smoke burning apparatus from London Journal.

Explanation of the Air Regulating Machine.
The same letters refer to the same parts in each of the figures, Nos 1 and 2, plate IV.
—A,a,a, is a sheet iron box, the end k, set in the brick-work, which communicates with the fire-place; b,b, a light fan wheel (shewn by dots in figure 1,) freely moving in this box, and put in motion by the air, in the direction of the arrows; c, a one-toothed pinion upon the axis of the fan-wheel b; d, an iron-toothed wheel, which is turned round by the one-toothed pinion c.; e, a catch rod, worked by an excentric pin near the centre of the wheel d; f,f, a pall catch, for discharging the rod e, from the wheel g; g, a ratchet wheel, upon the axis of the turn valve i; h, a weight and pulley for turning the wheel g, and valve i; k,k ends of the box, to conduct the air to the fire under the boiler, either of which forms may be used as circumstances require; i, a crank for locking the wheel d, during each half revolution of the one-tooth pinion c; m, a pinion on the wheel g, to strike against a finger on the axis of the turn valve i.
The action of this machine will be as follows;-
When the fire door is opened to take in fresh coal, it discharges the pall f,f, by means of a wire and slip catch connected to the door, but not shewn; the weight h, being then at liberty, turns the wheel g, one revolution and a quarter, which places the valve i, in the direction of the arrows, or horizontally; and thus leaves a free communication between the atmosphere and upper side of the fire. In this state of rest the machine remains until the fire door is shut. The rushing in of the air to consume the smoke, turns the fan-wheel b, rapidly round, and, by the revolution of the pinion c,c, drives the cog-wheel d, by means of the catch rod e, carries round the wheel g, by one revolution. It then brings in contact the finger and pin m, which gradually shuts the turn valve i, by bringing it in a perpendicular position. The smoke being consumed, the fire continues burning until a fresh supply of fuel is necessary: when the fire door is opened, the whole operation is repeated. By increasing or lessening the weight h, the time allowed to consume the smoke may be increased or diminished.
It is obvious that this principle may be applied in various ways, to obtain the purpose of a measuring regulator"[41].

252

In 1819 the firm had submitted a bid to the South London Water Works, where the selection process was proceeding and at a special meeting of the directors the company's engineer, Thomas Gregory was questioned:

"Questions from the Chair;
What is your opinion of the Company's Engine?
Answer; I think the Engine is too small for the supply required.
Question; What is your opinion of the state of the Engine?
Answer; I think the Engine is in danger, particularly when working upon the high services, and at all times on account of raising the Steam beyond the calculation or power of the Engine, the joints are then strained as well as the valves and Levers which are constantly giving way, the Engine beam which was only cased after the fire in 1807 springs too much in consequence of which the Engine is very much impoverished and in great danger. The Cylinder stands upon a foundation of Brick and Wood, the Wood is very much decayed which occasions the Cylinder to move at every stroke of the Engine, the upright columns are Wood and rest upon Wood beams which are very much decayed, the foundation of the pumps is Wood which is also very much decayed. The Brick and Stone work which supports the main pipe from the pump to the air vessel has given way.
Question; What Engines have you been accustomed to work?
Answer; Both Bolton and Watts, & Murray & Woods Engines the former for Messrs Mitchells & Holt at Holt Town near Manchester, One Engine 45 Horse Power & One 16 Horse Power two years, the latter for Messrs Copeland and Wilkinson a 36 Horse Power Engine at Hunslett near Leeds for two years.
Question; Did you think the Engines of both the Companies did their work well?
Answer; They did their work well and effectual whilst I was there.
Question; What is your opinion of the difference of the goodness of the work between Bolton and Watt's Engines and those of Murray & Wood?
Answer; I do not think there is any difference.
Question; If the prices of the two Company's Engines were equal, to which would you give the preference?
Answer; To Bolton & Watts
Question; And for what reason?
Answer; More from general character than anything else.
Question; If Murray & Wood would furnish an Engine for £400 less than Bolton & Watts to which would you give the preference?
Answer; To Murray & Wood
Question; If £200 less?
Answer; To Murray & Wood.
Question; If £100 less?
Answer; I cannot tell which.
Question; In what state is the Company's small Engine?
Answer; The small Engine is now totally useless for serving the tenants" [42].

The Board then considered its three estimates which were from, Boulton & Watt, Fenton Murray & Wood and Mr Jonathan Dixon and unanimously selected the 40 H.P. by Fenton and Murray, which included two wrought iron boilers, etc and included the erection costs for £2200 [42]. The decision, with a requirement that the engine had to be ready for use by 1st September next, was communicated to Fenton and Murray on February 26th along with a request to undertake the necessary brick and stone work for the foundations for a fixed sum. Fenton and Murray were also asked to provide their best extended payment terms and to visit the site for an examination of the engine house and instructions for a back pump to pump the water from the canal into the reservoirs [43].

The Economist of March 17th published an explanation of 'The New System of Society projected by Robert Owen' (New Lanark Mills), within which is found evidence of the work of David Wood:

"Among the economical arrangements which might be introduced, and as an instance of the truly advantageous manner in which the introduction of scientific power would always operate beneficially for the inhabitants of these villages, I will here notice a valuable application of steam, for the purpose of cleansing linen, invented by Mr Wood, of the house of Fenton, Murray and Wood, of Leeds; and which Mr Wood informed me had been for a considerable period used in his own family with complete success. He has erected an apparatus, by which the action of the steam alone, without any friction, and with half the usual quantity of fuel and of soap, effectually cleanses linen in a much better manner than can be done by the present laborious and expensive method; and though the cost of such an apparatus will for ever exclude it from the dwellings of the poor, under the present arrangements of society, it might be introduced most beneficially into co-operative societies. Independent of the saving in soap, and in the prevention of great waste of linen by friction, in all the present modes of washing, this simple but valuable improvement would save to the establishment the constant daily labour of from fifteen to twenty women; which number I am informed, would be required to wash for a population of from 1,000 to 1,500 persons" [44].

Unfortunately there is no remaining evidence that the system was employed at the New Lanark Mills, despite the revolutionary improvements Owen introduced in operating his mill and workers village.

In March and April letters from the South London Water Company were sent asking when Fenton and Murray would visit the site, especially as Fenton and Murray had advised that the brick and stone work would need to be finished by May. A letter from Fenton and Murray had advised that the visit had been delayed by Mr Murray's unexpected absence, but a further three weeks had elapsed and there was still no attendance at the site [43]. This prompted a response and on the 26th of April Richard Jackson attended a South London Water Company board meeting. Jackson advised that Fenton and Murray would supply a double 40 H.P. with two boilers etc. as requested for the amount quoted, although Mr Murray had assumed that the supply would be for a single acting engine. The engine was to be fixed and ready by October 1st, which date the board accepted. Fenton and Murray had already offered payment terms of £1200 three months after the engine started work with the balance by a bill at two months date. The back water pump was to be extra, and Jackson promised the plans for the bricklayers work in a fortnight [42]. It is interesting to note that the requirement for a double acting engine came from the company, it was not usual to have such an engine for pumping water, and when the Vauxhall Water

Works (as the South London Works became known) amalgamated with the Southwark works the whole engine house appears to have been disposed of in 1846 to the Phoenix Gas Company, with new Cornish engines being ordered to operate the network from another location. One of these engines was probably sold on and it was probably the 40 H.P. Fenton and Murray engine advertised for sale in The Times on 23 August 1850.

The South London Water Works acknowledged receipt of the plans by letter dated May 26th. They asked Fenton and Murray to review the boiler size which appeared to them rather large if they were going to use Messrs Parkes' patent smoke consumer, which they noted was endorsed by Fenton and Murray. Fenton and Murray's view on the use of Messrs Parkes' system was requested. The letter also stated that Jackson during his visit the previous week had found the Water Work's mason at rather a loss over where to procure the necessary stone, and Jackson had proposed that the mason and stone should come from Leeds, and that the directors had approved this measure. In the engine house stone was specified for the top of the cylinder, the top of the column wall, the top of the rotation wall, the top of the pump wall, around the fly wheel shaft and the entablature plate [43].

It would be very interesting to know the full story of the letter written by Matthew Murray to Simon Goodrich dated May 31st 1821 as he asked whether Goodrich could recommend a man to work in a copper mill in Holland, owned by a M. de Heus of Amsterdam [45]. While Goodrich recorded that he wrote to Murray in his journal on for 7th June, the content is unknown. His entry read;
"wrote to Mr Murray, to Mr Morgan about references on Tuesday".

By June 21st the South London Water Works were chasing the stones and stone mason from Leeds, and saying they needed the new engine as soon as possible as they only had one engine operational. The chimney was half up and the engine foundation dug [43].

Goodrich's journal for July 27th 1821 is headed *"On board the Hero steam boat – Built by Mr Murray"*, and he was referring to the engines. The firm had at last produced some marine engines to operate in a boat on one of the prime routes in England. Hero was produced as a package boat to ply from London to Margate. She was a vessel with a tonnage of 427 with two 50 H.P engines by Fenton and Murray. Her draught was 6 feet 4 inches. The Paddle Wheels had a diameter of 14 feet and a breadth of 8 feet and their speed at the extremity was 15 mph, they extended 18 inches deep into the water. The Hero's speed was 11 ½ mph making her the fastest boat on the Thames at the time. She consumed 2240 lb of coal an hour. The engines had cylinders of 38 ½ inches diameter, a stroke of 3 foot and could do up to 33 strokes per minute. The Paddle Wheels could turn 22 revolutions per minute [47].

In August the South London Water Works (Mr Marshall Chief Clerk) wrote to Fenton and Murray regarding the positioning of the air vessel as discussed with Mr Jackson, and proposing to move it. In a postscript they advised that Mr Murray had called at the works that afternoon and approved much of the alteration. A further letter at the end of the month acknowledged the plans for the Pump House and said they had been given to Mr Whitehead the stone mason. The letter then goes onto explain the sequencing of the water from the river and the pumping requirements. The interesting part today is that the cleaning of the Thames Water consisted solely of allowing it to settle in the basins. On the 31st August Mr Marshall wrote again with another proposal that two of the three main water circuits operated by the company could possibly be operated together using the large pump, and would the engine be able to provide the necessary power, as the idea would save considerable time and fuel. He acknowledged their letter of the 28th with the list of materials shipped on board the ship 'John and Joseph' Captain Arnold. On September 4th having received Fenton and Murray's comments, further arrangements needed to be put to the directors before they could be finalised.

The next letter at the end of September was on a different note and from a Mr Wm Heath who had taken over from Mr Marshall. It states that time is running out to make any more major alterations and what they need is a branch nozzle separate from the working barrel and a common leather bucket for the main pump. He then went on to indicate that Mr Whitehead needed more stone but that this requirement needed to be justified. The next item covered the advisability of having common sets of (fire) bars and asked for some dimensions and generally for a letter to explain again what the South London Water Works were to receive in terms of Engine, Pumps and Fire bars. The whole of this letter is quite confusing, but probably reflected that Heath had just taken over from Marshall [43].

On the 7th October William Peppercorne, one of the directors, wrote in answer to a letter from Fenton and Murray dated the 4th. He confirmed that Mr Marshall had quit. He noted that the remainder of the parts of the engine could be expected at Vauxhall in a few days, and therefore it was necessary that Fenton and Murray should send an engineer down. Whitehead had stated that if the engineer came the stonework would soon be finished. He then went on to query the amount of stone used by Mr Whitehead:

"quantity that has surprised most persons who have seen the foundation".

He then half apologised for Heath's confusing letter, but did manage to convey the message that there would need to be an adjustment to Mr Whitehead's accounts and Fenton and Murray's help would be needed to effect this. An addendum from Mr Heath went on to expound the various places where they considered too much stone had been used; the cylinder pillar was an immense depth and there was waste in flagging the whole area of the boiler seating [43].

By October 11th the South London Water Works had been advised that the engine parts had been despatched from Leeds on October 4th. On the 20th of that month Mr Heath wrote to state that they needed a visit from Mr Jackson to discuss the arrangements agreed with Mr Marshall, in order to understand the foundations and the water routings. He also advised that Mr Parkes had called about his smoke consuming apparatus, but no order had been placed, and he assumed that the boiler would proceed without. The bargeman had arrived with the engine components [43].

On the 20th October the Leeds Mercury carried a very unusual item, which was an advert by Fenton and Murray. It has to be assumed that their order book was not as good as they wished. Under the Royal Arms the wording read:

"*Steam Engine Manufactory, Leeds.*
Fenton & Murray beg leave to inform their friends and the public that, in consequence of the great improvements they have made in Steam-Engines, and in the application of their extensive machinery for manufacturing them with accuracy and despatch, they are enabled to sell their superior engines on as low terms as any house in the trade.
Fire-proof buildings and mill-work executed in the best manner by estimate if required, of which accurate plans and specifications can be made out.
Iron and brass castings, of every description, on reasonable terms.
F & M have erected an extensive and convenient building, for the purpose of keeping in perfect order a large and select assortment of wheel models for all sorts of mill-gearing. 20th Oct. 1821".

On the 8th November the South London Water Works decided that they would write to Mr Parkes declining his smoke consuming apparatus as they would proceed with the proposal made by Fenton and Murray [42].

Zebulon Stirk, a small iron founder and engine maker of Leeds, announced at the beginning of December 1821 that he was going into partnership with a William Horsfield who had spent the last 3 years as foreman to Joseph Shaw (another Leeds iron-founder) and prior to that he had been a foreman at Fenton Murray and Woods for 17 years [48].

The South London Water Works wrote to Fenton and Murray on 7th December and asked for the price of stone flags of various thicknesses at Leeds and an urgent answer. They noted that J Millar was getting on well. This is Millar's first mention but he stayed with the firm until it closed, he mainly worked erecting engines at customer's sites. It was recorded that Millar and his assistant had been paid up to that date £17/4/3d [42].

On the 20th December a Mr Wade appeared at the South London Water Works to discuss his account for stone delivered by order of Fenton and Murray. He was told that they had come to an adjustment with Mr Whitehead and that he was the only person they were dealing with in regards to stone as he had been recommended by Fenton and Murray [42].

The textile machine side remained busy supplying Marshall & Co. with 24 gill machines, 24 spinning frames, 2 hackling machines and 2 carding engines in the year [21].

Notes.

1.   Marshall Papers MS 200 folio 15. Brotherton Library.
2.   Leeds Gas Light Co. records. WYL 359. WYAS Leeds.
3.   The Leeds Intelligencer 1 March 1819.
4.   Fenton Murray & Wood, unpublished thesis by Trevor Turner.
5.   The Leeds Intelligencer 2 March 1818.
6.   The Leeds Mercury 7 March 1818.
7.   Newton Chambers Papers TR 278. Sheffield Archives.
8.   Theatre Notebook Vol. 13 Oct 1958-July1959, Edmund Simpson's Talent Raid on England in 1818 (3) by P H Highfill.
9.   Newton Chambers Papers TR 10. Sheffield Archives.
10.   The Leeds Mercury 13 June 1818.
11.   The Lancaster Gazette 8 August 1818.
12.   Cobb of Margate Papers EK-U1453/B2/40/224. East Kent Archives.
13.   The Leeds Mercury 19 September 1818.
14.   The Hull Packet 27 October 1818.
15.   Telford Archive item 157. Institute of Civil Engineers.
16.   Francis B Ogden was a nephew of Aaron Ogden (who was an early operator of steam boats around New York, circa 1811). Francis B Ogden had been an aide de camp to General Andrew Jackson at the Battle of New Orleans in the American War of Independence. He took a very active interest in steam engines and steam boats. F B Ogden was later appointed the U.S. Consul at Liverpool and subsequently Bristol. He married an Englishwoman and is buried in Bristol. Ogden supported John Ericcson, one of the inventors of screw propulsion, and the boat used by Ericsson to demonstrate his principles to the British Admiralty was the S.S. Francis B Ogden. See Appleton's Encyclopaedia of American Biography, 1888.
17.   Mechanics Magazine No 994 Vol. 37 27 August 1842. Who first introduced the use of expansive steam for marine engines? (From The American Repertory of Arts, Science and manufactures).
18.   Papers of Henry Heth MS38-114. University of Virginia Library USA.
19.   Col. Henry Heth known as Harry (emigrated from England in 1759), not to be confused with his grandson of the same name and nickname, General Henry Heth. The younger Heth is credited with prematurely starting the Battle of Gettysburg, when his reconnaissance force made contact with Union Cavalry, against the orders of General Lee, which were to avoid unnecessary contact with the enemy. American Civil War Trust website and Wikipaedia.
20.   Memoire sur les Bateaux a Vapeur des Etats-Unis d'Amerique, M. Marestier, Paris   1824. Page 159.
21.   Marshall Papers MS200 folio28. Brotherton Library.

22. The Leeds Intelligencer 18 January 1819.
23. The Royal Bank of Scotland Group Archives, BEL 3/3 Customer Account Ledgers 1813-1825.
24. Newton Chambers Papers TR 279. Sheffield Archives.
25. Goodrich Journals B40. Science Museum Library.
26. Caledonian Mercury 25 April 1819.
27. Decreet, Messrs Fenton Murray and Wood, and John Forman, WS, their mandatory v William Ford. Ref. CS40/32/22. National Records of Scotland.
28. The Leeds Intelligencer 16 August 1819.
29. South London Water Works Letter Book SV/1/163/1 and Board Minutes SV/1/3. London Metropolitan Archives.
30. Newton Chambers Papers Letter Books TR. Sheffield Archives.
31. The Philosophy of Domestic Economy adopted in the Derbyshire General Infirmary, Charles Sylvester, Engineer, Nottingham 1819. Appendix, Page 60. Also A year in Europe comprising a Journal of Observations, John Gibson 2nd edition New York. Vol.2 page 186.
32. Minutes of the Proceedings of the Institute of Civil Engineers Vol X Session 1850-51. London 1851.
33. A Treatise on the Steam Engine by John Farey, Longman Greene & Co 1827, David & Charles 1971. Chapter 6 page 481.
34. Marshall Papers MS200 folio6. Brotherton Library. Also Fenton Murray & Wood, unpublished thesis by Trevor Turner.
35. The Leeds Intelligencer 28 February 1820.
36. The Lincoln Rutland and Stamford Mercury 24 March 1820.
37. The Leeds Intelligencer 9 October 1820.
38. The Leeds Mercury 23 December 1820.
39. The Aire and Calder Navigation, part 4 by R G Wilson, in, The Bradford Antiquary, New Series Part XLV June 1971 page 337 (6 HP), and A History of the Steam Dredger 1797 -1830, by A W Skempton, read at Institution of Mechanical Engineers, 7 May 1975, item 19, table 2 (10HP).
40. The Leeds Mercury 17 February 1821.
41. Newton's London Journal of Arts and Sciences 1821, pages 122-124 and Plate 4.
42. South London Water Works Board Minutes SV/1/3. London Metropolitan Archives.
43. South London Water Works Letter Book SV/1/163/1. London Metropolitan Archives.
44. The Economist No 8 17 March 1821, pages 115-116.
45. Goodrich Papers A951. Science Museum Library.
46. Goodrich Journals B 46. Science Museum Library.
47. Goodrich Journals B 47. Science Museum Library.
48. The Leeds Mercury 1 December 1821.

Chapter 12.

1822 to 1826

On January 9th 1822 the South London Water Works' chief clerk William Heath wrote to Fenton and Murray in his usual cryptic style, requesting them to make a valve:

"We would be obliged by your putting in hand a valve of description, Piston & Box, Viz A Common square box with a stout door &c, a branch at the top bored, with a copper or brass short cylinder placed therein about 6 inches long, a common seating & stout fixed valve brass 8 in., a 6in. branch at the side short. Place the piston of about 2 ½ in dup & 7 in dia. truly upon the same rod & cause an iron pipe of 1 ¼ in dia to be fixed to the side as dotted to carry the water upon the top of piston to set it at a nearer medium.
The new piston for Main Pump, hope is forward. We are Gentlemen ..."[1].

On the 11th Heath sent Fenton and Murray another letter stating that the directors wanted them to send their account in, which was to include the back pump. He also asked for their opinion of the equalising valve from his previous letter and an estimate [1].

Either Fenton and Murray's workload was increasing or they had lost a number of skilled men as during January and into early February 1822 they advertised in the Leeds papers for 2 or 3 greensand moulders and a dry sand moulder who were to be experienced in engine castings and mill work. They were also seeking one or two workmen to erect steam engines and undertake millwright work [2].

Dated the end of January some notes on textile machinery costs are to be found in the Marshall papers. It is a price agreement with The Foundry, which usually means Fenton and Murray, and gives the price for a 40 spindle spinning frame at £88/15/2d. It also includes the price for cutting brass spur wheels at 2d for twenty teeth for each 1/0 inch material thickness, or 3d if they were bevelled. Wrought iron spur pinions, again per 1/8 inch, were 4d plain or 5d bevelled. All odd wheels and pinions were to have a minimum charge equal to that for twenty teeth even if they had fewer teeth. Wrought iron boilers were 27/- per c.w.t without tubes, and 28/- with tubes, although if the tubes rose out of the bottom the charge was 29/-. Nautical boilers cost between 40/- and 60/- per c.w.t [3].

In early February it was reported that on Saturday the 4th, Leeds had been subjected to violent winds and heavy rain and that cellars and ground floors in Water Lane were flooded. Both Water Lane and Hunslett Lane were impassable on the Sunday morning [4].

During late February and early March the South London Water Works prepared their accounts in regard to the steam engine, including all monies paid out to Fenton and Murray's servants (J. Millar and company), goods ordered on their behalf and any charges in connection with the repairs to the cold water pump rod, which had been carried out by a local smith. The Board had instructed that these amounts were to be charged to Fenton and Murray [5].

In April a letter arrived for Mary Murray, Matthew Murray's youngest daughter, from Matthew Murray junior who was in the process of establishing himself in Moscow. He had either procured an order for Fenton and Murray or erected an engine for them as he thanked his sister for the shirts and handkerchiefs sent out with the engine [6].

In his journal for 1823 Simon Goodrich made further mention of the Hero, from information which he had obtained from the 1822 publication of the report from the Select Committee on Holyhead Roads. As Holyhead was the port from which the mail was sent to Ireland, the committee had extended its enquiry into the best practices for steam packets etc. They had called up, among many others, Captain John Percy of the Hero. He provided a general description of the boat and explained that they had been experimenting with between 12 and 16 paddles on a wheel, and that twelve worked very well. The usual time the Hero took between London and Margate was seven and half hours, as she travelled at between 11 and 12 miles per hour. Captain Percy was asked if he was familiar with Boulton and Watt engines, to which he replied that they had none in his company. He added that he knew vessels with Boulton and Watt engines, which he thought were the best as they did not have as many breakdowns as they did. When asked to expand he stated that the cross heads of the Hero had broken twice in the last season, several bolts had "gone" and the steam pipe went and was replaced with a copper one. Captain William Rogers, another witness, who commanded one of the steam packets that plied between Holyhead and Dublin with the mail, which both had Boulton and Watt engines, also described their cross heads breaking so the condition was not peculiar to any particular make of engine. As part of the Select Committee's enquiry, letters had been written to leading engineering firms soliciting answers to a standard set of questions. Fenton and Murray were one of the firms addressed and some of their response was included in the report:

"*Query 5 - Supposing the length of voyage to be twenty leagues from port to port, across a sea exposed to strong tides and heavy gales, how many vessels, in your opinion, would be necessary to be maintained, so that two should sail every morning, that is, one from each port throughout the whole year; except on days when a strong cutter of one hundred tons burthen would be obliged to put three reefs in her mainsail?*
Answer - In our opinion it would require six vessels, that is, a spare one at each port undergoing such repairs as the repairing or putting in a new boiler, or undergoing a

general repair. The other four vessels to change alternately, and would have a day for cleaning and slight repairs. Fewer than these would not be safe to trust to.

Query 6 – What description of vessel as to tonnage, form, strength of building, masts and sails, in your opinion, is the best for sea navigation?

This question must refer to the particular purpose of the vessel, and will be best obtained by a reference to vessels built particularly for that sea it is intended for. We believe the deep and sharp formed bottom is best for a rough or boisterous sea, as the Channel, and the flat or round formed bottom for the North Sea; but ship-builders are not agreed on this point.

Query 7 – What description of engine as to power, materials, and general form and arrangement, and what description of boilers and paddles in your opinion are the best?

The double engine with two cylinders is the best at present known for a steam vessel; the power most easily managed, is from twenty to sixty horse power, according to the purpose and tonnage of the vessel. We cannot describe the general form and arrangements without plans, better than by saying that these engines have their beams working below, with side rods, and cross heads, all of which ought to be made of the toughest description of wrought iron. The boiler ought to be what we call a combined boiler, viz. three distinct boilers put together to form one boiler, with the fire passing three times through each, and so constructed as to be taken up and down a hatchway without pulling up or destroying the deck. The paddle wheels, shafts and arms of the toughest wrought iron, and we think wood paddle boards for the best, for sea.

Query 8 – In what parts do engines most frequently get out of order or broke?

Cross heads, beams and paddle wheel shafts have generally broken from being made of cast iron, and ought to be made, as stated above, of the best wrought iron. In general, things that are difficult to repair should be made so strong as to ensure them against breaking, and nothing ought to be suffered to be near the breaking point but what can easily repaired.

Yours etc., Fenton and Murray" [7].

The boiler described in the answer to query 7 above may well be the one illustrated by Goodrich in his description of the Hero, which shows the 3 parts mentioned in the letter above:

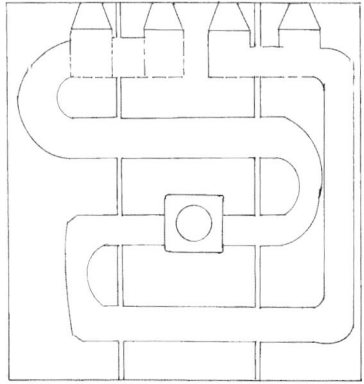

57. Boiler for the Hero from Goodrich Journals [8].

In May Fenton and Murray were advertising for a boiler maker [9].

Fenton and Murray received another brief missive from the South London Water Works which was dated 22nd May:

"I rather imagine you will find an order for a piston bucket gear'd with cupp'd leather on examining my letter which I am very desirous of receiving at your earliest convenience. Waiting your compliance – for the South London Water Works Co" [1].

On June 3rd Fenton and Murray wrote to the South London Water Works requesting payment for the steam engine and ancillary items, which they considered were then due. They had a reply dated the 8th, which included a list of queries and complaints, but also stated that they had been paid £1200 through Glyn & Co the bankers. The South London Water Works stated that they were looking into the charges paid to Fenton and Murray's men, carriage paid from Vauxhall wharf to the works, and the labour costs incurred for the back pump. They expressed their surprise that no responsible person, from Fenton and Murray, had been to inspect the steam engine and confirm the fuel consumption as proper, etc. Further there was concern in regard to fuel consumption as the engine was only operating at two-thirds power. There were issues with the valve seatings on the main pump, the grate bars allowed coals to fall through and the poor functioning of the condensing pump at low speeds had required a further pipe to be fitted. The schedule of materials omitted the iron base plate for the air vessel etc.

Fenton and Murray do not appear to have responded to these issues, at least nothing survives, and the next news was contained in the Minutes of the South London Water Works which recorded a visit from Richard Jackson at the beginning of July asking for the balance of the engine and offering 2 months discount for cash. Having been told by the Board that until the matters addressed in the letter of July 8th were satisfactorily settled there would be no further payment, Jackson had gone to Mr Peppercorne's office in London and asked him for the payment and had not told him of the position of the Board. A letter was written to Leeds, dated July 13th, in which it was explained that while Mr Peppercorne had allowed Mr Jackson to draw up a bill, because of the previous refusal of the directors to pay, Mr Peppercorne had been unable to obtain the necessary three signatures to the bill. Mr Peppercorne was unhappy that Mr Jackson had not given him all the facts, and in consequence of "*Mr Jackson's error*" the bill had been cancelled. The bucket did not work and a full report on the correctness of the engine and other matters connected with it was to be received by Thursday week, and then the directors might be able to give a more satisfactory response. On the 18th of July Mr Peppercorne handed a letter to the South London Water Works which he had received from Mr Jackson asking for the bill promised at the beginning of July [1]!

In 1822 the Royal Navy were actively pursuing the introduction of steam powered ships into the fleet, however to date the full details of the complete process remain uncertain. However some valuable snippets survive in the Digests of Navy correspondence. In August there was a letter from Fenton and Murray requesting to know whether the Navy requirement was for engines of 25 or 50 horse power. Whether a letter was sent to manufacturers, which would have been expected to stipulate something of this importance, or Fenton and Murray's intelligence network was responsible for them writing, is not known. However a response was duly sent informing Fenton and Murray that the requirement was for two engines of 50 H.P. Following this exchange the Navy Board did issue letters to Boulton and Watt, the Butterley Company, Mr Murray of Leeds and Mr Maudslay requesting their terms for supplying two steam engines of 50 horse power. The Navy Board observed to Boulton and Watt that they had had an offer which was 25% less than the price so far submitted by them. The Navy Board asked Mr Murray and Mr Maudslay what steam engines they had fitted in boats. Very cryptically in the Navy Board Digest, on August 8th it was noted that Fenton and Murray had given in their price [10].

At the York Summer Assizes Fenton and Murray's name cropped up, although not for any wrongdoing of their own:

"Hall v Coe, a case of a missing Bill
...and as to the bill for £170 (which plaintiff attempted to charge the defendent with) it was drawn by Day & Coe, in their life time, upon, and accepted by Messrs Fenton Murray & Co for a partnership debt, the amount of which was received by the witness" [11].

They were also mentioned during the same month in a case that was heard by the Leeds magistrates:

"Whereas one George Speight, of Meadow Lane, Leeds, a labourer at Messrs Murray & Co's Works of Leeds, has falsely sworn against me and my wife at the Leeds Borough Sessions on Tuesday the 6th instant, that he had seen in my house women of ill fame dancing in impropor undress, by looking through the Key hole or lock hole of my Door for ten minutes together whereby the Jury have found me and my wife guilty.....The Door has been viewed by a magistrate & other worthy persons – no key hole. George Speight to be indicted for his wilful and corrupt perjury at the next Assizes at York"[12].

The Leeds Intelligencer for September 23rd reported a meeting of the inhabitants of Leeds on the subject of smoke nuisance, which noted that great inconvenience and considerable injury was being met by the people of Leeds from smoke from steam engine furnaces. The same copy included an advert by Mr Josiah Parkes for his smoke consuming system which was operating to great effect at Fenton and Murray, Hirst and Bramley, Benyon & Co, Titley Tatham and Walker, Lord and Robinson and Mr Wilks etc. Enquiries were to be addressed to Fenton and Murray or Mr Parkes (Albion Street Leeds).

By October all the tenders for the steam engines for the navy's new boats had been received at the Navy Board and the Surveyors of Buildings laid before the Navy Board a comparative statement of the tenders for two fifty horse steam engines which said that the lowest offer was from Messrs Fenton and Murray. Shortly after Fenton and Murray reported which other marine engines they had fitted and the names of the vessels and the owners. The owners were contacted to state the consumption of coal [10].

At the end of October Simon Goodrich noted that he was preparing a letter to Murray on the subject of boilers, but no detail has survived [13].

On the 21st of November the directors of the South London Water Works considered another letter from Fenton and Murray that requested the settlement of their account. The directors instructed Mr Heath to write to Leeds and request an amended account to reflect an abatement to offset the costs of the extra coals consumed and the replacement fire bars. The letter put a sum of £100 on the abatement [5].

The West Yorkshire press carried a notice of the sale, which was to take place on the 25 to 27th of November, of the Worsted Mill of Messrs Turney, Turney and Bates of Salterhebble near Halifax due to bankruptcy of the partners. In the equipment included in the sale was a 20 H.P. steam engine by Fenton Murray & Co [14].

By the beginning of December the Navy Board was receiving the owners coal consumption figures for the nominated vessels, Pearson Darley & Co for the Yorkshireman with engines made by Fenton and Murray [15], Mr Duke for the Sovereign whose engines were made by Maudslay (they already had data on Maudslay's engines for the Engineer) and Gibson and Co for the Hero (Fenton and Murray) [10].

On the 12th of December the South London Water Works authorised a cheque for £1000 for Fenton and Murray being the balance of their account in full of all demands [5].

The Navy Board digest of correspondence contains a particularly tantalising entry concerning steam engines on 27th December 1822:

"*Steam Engines Competition in the supply*
The Board considers it expedient to admit a competition in the supply of Steam Engines for Vessels but that it should be confined to persons of established reputation as Steam Engine makers. They determine to employ Messrs Bolton & Watt to make the Engines for one of the vessels building at Deptford and Mr Maudslay for the other" [10].

It sounds as if they had had advice from above, either formally from the Board of the Admiralty or informally from some of its members.

The Marshall Papers contain a different topic for the year 1822, which was a notice of a school being set up by the company:

"The School lately erected by Messrs Marshall Hives & Co being now nearly completed, the Inhabitants in the Vicinity are hereby inform'd that it is intended to be ready for the reception of Children, Boys & Girls, immediately after the Christmas vacation.
With a view of rendering the Institution as useful as possible, to the surrounding population it is proposed that a day School be opened for the instruction of Children in reading, writing & accounts on the following terms.
Boys   { Reading & Writing per Week 2d.
*   { Writing & Accounts ditto at the Rule of Three 3d.*
Girls ....Reading, Writing, Accounts & Sewing per week 2d
It is also intended that there shall be an evening school at the same place which may be very convenient and useful to those young persons who are engaged in the Manufactories and other Employments during the course of the day, from 7 to 9 O'clock, at 3d per week finding candles at 2d ditto   without light.
And also that a Sunday School will be kept where the Children will be taught gratuitously, the Rules to be observed in the management thereof, as in similar institutions.
The following persons are requested to constitute a Committee for the superintendence and management of the forementioned School.
John Marshall Esq
John Hives Esq
James Fenton Esq
Matthew Murray
Richard Jackson
C G McLea
Joseph Taylor
Thos Wordsworth
Geo Eddison
Thos Roberts
Subscriptions Annual
Marshall Hives & Co - £10.0.0
Fenton & Murray - 6.0.0" [16].

The Marshall and Hives purchase of textile machinery from Fenton and Murray was recorded for 1822, and amounted to 55 machines at between £20 and £104 each [17].

Early in 1823 the relationship between Fenton and Murray and Josiah Parkes seems to have ended, as a new advert in the local press announced that applications should be addressed to Samuel Green who was a bricklayer. The explanation, which was that Fenton and Murray seem to have found a better product, may be read into the details of the visit described below.

In February 1823 the Round Foundry had another visit from an old customer with whom relations had at times been strained. Marc Brunel noted in his private diaries for the 27th February:

"Fenton & Murray Leeds (Their Works Admirable)
Visited their works in the town. Mr Murray took me round and showed me everything.
I was highly pleased with all his implements some of which were new since I was
here. His lathes are very fine and upon the whole exceed anything I have seen.
Saw two Steam Engines of their making, they are remarkable for their beauty and
superior workmanship.
Their Cog wheels are upon a new pitch, that is, the height of the tooth is 1 1/3 of the
pitch, by which they work so smooth"[19].

Murray informed Mr Marshall of the visitor and Mr Marshall showed Brunel around his flax mills, where the most remarkable items found were:

The 70 H.P. engine (by Fenton and Murray) and the long cogs on the bevel.

The arrangement of the first motion in the mill.

The machinery for all the operations, and that everything was clean and the building fireproof.

Brunel disclosed his plans to Murray for his marine steam engine, and had Murray's first opinion of it (unfortunately not recorded). After a visit to the Bowling Iron Works where he had business, Brunel returned to Leeds on March 5th and revisited Fenton and Murray, where he met Mr Brunton. Mr Brunton showed Brunel his fireplace which was considered truly ingenious and likely to become a standard.

Brunel travelled on to Sheffield to visit the Milton ironworks who he had contracted to build the so called Bourbon Bridges as they were themselves contracted by the French Government for the Ile de Bourbon (Reunion). By now Isambard and Marc Brunel were partners, and it is probable that Isambard was with Marc for at least some of the trip and was introduced, and possibly stayed in Yorkshire to manage affairs after Marc returned to London. There was a further visit to the Round Foundry on passing through Leeds for York on the 22nd April.

In May Brunel visited Chatham and went to see the saw mill and found everything in good order. His private diary recorded on June 10th that he returned to Mr Murray of Leeds five pounds, borrowed by Isambard from Mr Murray, by hand of Mr Alderson (possibly the author of Essay on Steam?). So it is learnt that Isambard Brunel had visited the Round Foundry on his own [19].

In October discussions took place regarding a problem with the engine in the Navy Well at Sheerness. Sheerness had by then started its major rebuilding to the design of John Rennie, which was being undertaken under his son's supervision. After sending an engineer to measure the articles requested Fenton and Murray suggested by a letter dated the 12th November that a new cylinder could be introduced into the engine, and on the 22nd of that month they stated their terms for increasing the power of the engine, and the officers of the yard were asked to comment by the Navy Board [20].

The newspapers of December 11th carried reports of major storm damage in Leeds and Hull [21].

On December 16th the Sheerness Officers reported that the engine in the Navy Well was strong enough to take the cylinder proposed by Fenton and Murray, and that the works would be strengthened by the repairs and improvements. The procurement of spare parts was also recommended. The Navy Board approved Fenton and Murray's proposal of November 12th and ordered accordingly along with the spare parts, advising Fenton and Murray and the Sheerness Officers of the position [20].

The tally of textile machines made by Fenton and Murray for Marshall & Co for 1823 was 36 machines for Leeds and 9 for Ditherington near Shrewsbury [17].

1823 saw the installation of a 4 to 8 HP Fenton and Murray steam engine at the Westminster Gas Works, which worked two pumps and two purifiers. The load was reckoned as 4 HP [22].

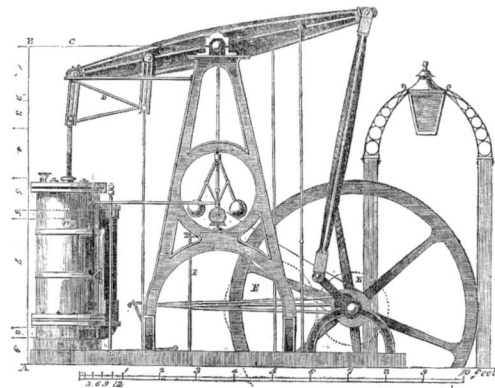

58. Steam Engine at the Westminster Gas Works [22].

The Sheerness work seems to have been managed efficiently for the Navy Board records show receipt of the bill from Fenton and Murray for the cylinder on January 26th 1824 [23].

In January 1824 a meeting took place in Leeds to set up the Mechanics Institute:

"At a meeting of Gentlemen friendly to the Establishment Of a Society for the more extensive Diffusion of Knowledge amongst the Working Classes of the Community in this Town & Neighbourhood held at the Court-House, on the 28th day of January 1824 the following Resolutions were Proposed and adopted
1.That an Institution to be established, to be called the MECHANIC'S SCHOOL OF ARTS somewhat on the Plan of that now existing at Edinburgh, and which has been attended with as much Success as its most sanguine Supporters anticipated.
2.That this Plan shall include a Library, a School, with a Master for teaching Mathematics (in which common Arithmetic, Geometry and Algebra are included) and at least One Course of Lectures Yearly on Mechanics, and Chemistry.
3.That a Prospectus be drawn up and circulated, under the direction of a Committee, previous to a Subscription being entered into for carrying these Measures into Effect.

4. That the following Gentlemen be appointed and compose such Committee, and that five be empowered to act

Mr Gott, Mr Marshall, Mr J Marshall, Mr Baines, Mr E S George, Dr Hunter, Mr Cawood, Dr Williamson, Mr Murray, Mr J Atkinson, Mr Joshua Dixon, Mr Rawson, Mr West, Mr T Benyon – Jun, Rev Mr Walker, Mr Todd, Mr Waites Jun, Mr Stansfeld, Mr Holdforth.

5. That the Resolutions be printed and published once in the Three Leeds Papers" [24].

Fenton and Murray were doing a considerable amount of work for Hives and Atkinson, a partnership formed to spin flax by two of John Marshall's former partners: Mr Hives had been the man who took out the patent for a hackling machine while he had been in partnership with John Marshall and Mr Atkinson had managed Ditherington Mill near Shrewsbury. They were building and equipping their new mill with up to date (gill) machinery and Fenton and Murray received £2000 from them in January 1824 and a further £2000 in February [25].

Fenton and Murray had submitted their further bill, for fitting the new components to the engine at Sheerness, to the Navy Board by March 3rd [22].

Matthew Murray took up his pen and wrote to Simon Goodrich on the 16th of April 1824. He started by stating that he should have written earlier but had been away for a month. Goodrich had asked him for an opinion on Jessop's metallic elastic spiral packing for steam engine cylinders, for which Jessop had received a patent in March 1823. It had received a lot of attention in the technical journals, as an alternative to the usual packing which was hemp. Murray explained that they had tried them on an engine for a week but had used a lot more coal than with the hemp packing, as they could not make the ends fine enough to prevent steam escaping. They were waiting for one of Jessop's men to come and fix the packing his own way to see if it improved things. He explained that he had made his own experiments with metal packing but always gave it up. He observed that using vegetable matter as well as being tighter caused a lot less wear than metal on metal. He used the D valve as an example:

"the flat surface of which is brass, rubbing upon iron, but the back or circular part is iron, rubbing against a hemp packing – now although the pressure upon each face (back and front) is equal, the brass and iron faces wear out very fast, and the other seems not to wear at all."

Murray mentioned that he was making a large hydraulic cable tester for General Alex. Wilson in St Petersburg with a capacity of 400 tons. He promised Goodrich some drawings and returned those that Goodrich had sent him. Goodrich had written to him in June 1823 with drawings and a description of a cable testing machine in Sunderland as well as details of some of the other machines in existence.

The only improvement Murray had made for steam engines that he described to Goodrich was to build a 60H.P. with two condensers, one for the steam on top of the piston and one for that under the piston, which produced a better vacuum. They had just completed such an engine for Leeds which was very satisfactory [26].

When Matthew Murray had visited Sweden in 1806, there was an indication that he had been in discussion with John Watson, the Northumbrian Viewer, concerning some of the details, and while in Sweden he had met one of the managers over there named Thomas Stawford. The Watson Papers in Newcastle contains a series of questions and answers dated 3 May 1824 with the questions put by Stawford and answers from Murray. Stawford asked 8 questions, but Murray only gave detailed answers to the first 4. The answers are quite interesting as showing Murray's prejudice against high pressure steam.

"Questions submitted to Matthew Murray respecting the late Improvements in Steam Engines with regard to Fuel &c and answers thereto.
Questions submitted to Matthew Murray Esq.
1st Question. Do you find that since the year 1810, or still later from 1816, any particular saving of Fuel has taken place, in large (or as they are generally termed heavy Pumping Steam) Engines built upon the principle of Messrs Bolton and Watts; and if so, to what amount?
Answer. There has been no material improvement in Bolton & Watts Engines except in the Article of mechanically feeding the fire with Coals by Mr Brunton's fire regulator. The saving in Coals is about 25 per Cent and a further considerable saving in the Boilers – as they are rendered more durable and the Worst Coal will answer the purpose.
2nd Question. If such improvements exist, to what extent and (about) at what expense can such alteration be made as near as can be calculated in proportion to that saving?
Answer. The best of these fire regulators may be stated at about £300 each for a 70 or 80 Horse Engine and £150 for 10 or 15 horse.
3rd Question. Admitting there is a saving; pray is that accomplished by the altitude of the Chimney, by consuming of the smoke or oxygenising the air by some means previous to the entering of it into the grate?
Answer. The Altitude of Chimney is of great advantage and we think for a large Engine it ought not to be less than 46 yards high, and 5 or 6 feet square within at the base, diminishing to about 2 feet at the top with a round hole, An iron Pipe 5 or 6 feet long then placed at the top is of much service.
4th Question. Please give me your opinion on high pressure Engines, I mean with respect to large or heavy pumping Engines, if you think them applicable, and if so with what saving of fuel or any other advantage over the condensing Engines on Bolton and Watts Plan.
Answer. We do not recommend High Pressure Engines of any kind on account of the Danger which inseparable from them besides there is no advantage in using them, the consumption of Coal is equal to if not more than that required for condensing Engines and the expense of Hemp Packings &c is greater owing to the excessive heat of the Steam."

The remaining questions refer to specific engines in various locations in Sweden, and Murray merely stated:

"To the following questions we can give no other reply than merely to state the consumption of coal for engines in this country which is from 12 to 16 lbs weight of good coal per hour for each horse power of the engine" [27].

In May 1824 two Fenton and Murray engines were to be found on the second hand market, one at Mr Atkinson's Worsted Mill in Bradford of 14H.P. [28] and the other was to be enquired about by a letter to the Leeds Mercury Office and was of 8 H.P.[29].

In July the firm advertised for two or three greensand moulders promising liberal wages [30].

In July or thereabouts Simon Goodrich received a letter from General Wilson in which he described the chain testing machine he had ordered from Fenton and Murray:

"I long ago expected to have been able to address you & to have sent you a drawing of the machine for proving cables which I ordered from Mr Murray of Leeds but I have been disappointed in getting this machine finished & am apprehensive that I shall lose another season in waiting for it - When it is done I expect that it will be a very useful tool - The Hydraulic part is fixed at one end, & there is to be a combination of levers, to serve as a counter check, which is to be moveable, so that even short pieces of chain, bars etc. can be proved by it & at the same time even a whole cable may be stretched in case the [sides] are extended to a sufficient length or a mass is [built] to fasten on to" [31].

The Chatham saw mills had now been operating for 10 years and the boilers were in bad repair, even the replacement one just purchased. On the 27th of July Cunningham the Commissioner of Chatham Dockyard advised the Commissioners of the Navy that the bottom and sides of the new copper boiler (supplied by Mr Lloyd) at the Saw Mill had given out and that the bottom of the iron boiler would not last much longer. He requested that Mr Lloyd was to send a copper smith immediately to repair the new boiler and that Messrs Murrays should be asked to send a proper person to put a new bottom in the spare iron boiler, originally made by them. The Commissioners agreed to write to both contractors, but with an additional question as to whether such work could not be carried out in-house [32].

In August Fenton and Murray were advertising for a loam moulder, as usual the advert stipulated that only good steady workmen should apply, and as usual promised liberal wages [32].

On the 1st September Cunningham wrote to the Navy Board to advise them that Mr Richard Jackson from Messrs Fenton and Murray had, following the Navy Boards letters, visited Chatham to examine the spare iron boiler. Mr Jackson and the Superintendent of Machinery had agreed that the boiler also needed new sides and a new tube. This would cost three quarters of a new boiler and the joint recommendation was that a new boiler should be procured, to be made larger than the present one and of either copper or iron as the Navy Board decided [32]. On September 20th the Navy Board noted that the Commissioner at Chatham was recommending a new boiler instead of repairing the old one. Fenton and Murray had been consulted and recommended two new boilers and provided the dimensions and prices (36/- per c.w.t for iron boilers). The Board decided to seek competitive quotes from Mr Lloyd and Mr Maudslay, and instructed a holding letter to be sent to Fenton and Murray [34].

Two of Fenton and Murray's staff were considering setting up on their own, the first was Charles G Maclea, the spouse of Murray's second daughter Ann, and the other was Joseph Ogdin March who was courting Murray's third daughter Mary. They had presumably decided that as Richard Jackson, married to Murray's eldest daughter was being groomed for management that they stood better chances in their own partnership. In September 1824 the Union Mills in Holbeck came up for auction with a 20 H.P. engine by Fenton and Murray [35]. The sale date was October 13th, and as there does not appear to have been any subsequent advertising for sale, it is reasonable to assume that this was when they purchased the mill, that was to become the home of Maclea and March. They opened for business the following year, and the firm were highly successful manufacturers of textile machinery and machine tools and remained in business until the 1890s.

It is interesting to note that by this time many of the estimates for new colliery engines on Tyneside found in the records, instead of going into detail on the engine quote Fenton and Murray as a price source for the power of engine required, and then added estimates for the specific pumps, buckets and clacks etc. [36].

On the 15th October Cunningham wrote to the Commissioners with a new proposal from the Master of the Saw Mill, that the power of the engine at the mill should be increased from 30 HP to 36 HP, and that this would be power enough and more economical than going up to 50 HP. If the proposal was agreed it would be necessary to change the steam cylinder, piston, piston rod, nozzles and cold water pump. To change these over would take at least 3 weeks, during which time it would be necessary to suspend operation of the mill [32].

In October Fenton and Murray were advertising for people again with one of the positions on offer possibly a replacement for Maclea or March;

"*Leeds Steam Engine Manufactory*
WANTED-TWO GREEN SAND MOULDERS accustomed to Engine and Mill Work.
As constant Employment and good Piece-Work Prices will be given, none need apply but good Workmen.
Also, a Person in the Capacity of YARD-KEEPER; - he must be a sober, steady Man from 30 to 40 Years of Age, who can write and have a good Character from his last Employer.

There is also an Opening for a Young Man in the Office as DRAUGHTSMAN, and occasional Book-keeper; he must be a Person who has a Taste for Drawing and Mechanical Pursuits, and a good Scholar. No Premium is required.
FENTON & MURRAY Sept 30th 1824" [37].

Cunningham's letter to the Commissioners was recorded by them on October 18th, and they instructed that the worn items were to be replaced, that a letter was written to Mr Lloyd and that the Commissioner at Chatham should be informed of their decision [34].

In November Fenton and Murray received a further £800 from Hives and Atkinson [25].

Simon Goodrich was due to lecture to his local Philosophical Society on the steam engine and had written to Matthew Murray asking for the loan of a model and some information. Murray replied on November 21st. The letter despite being mainly answers to questions, that have not survived, is quite interesting and informative:

"I received your favour of the 16th, & observe that you are going to deliver a lecture to a Philosophical Society on the Steam Engine. I am very sorry to inform you that I have not any Model that would be of any use to you or you should have had it with the greatest Pleasure.
The locomotive Engines that I have made, or heard of, are all moved upon Iron rail ways that are nearly upon a level. Where there is any considerable rise, it requires one of the rails to be cogged, as I believe you have seen at Leeds. I don't see any objection why they might not be made to go upon a macadamized road, if a motion was given to all the 4 wheels at onetime, & made with broad Surfaces. The objection to locomotive Engines is, that they must be worked with high pressure steam, say from 40 to 60 pounds upon an inch, above the atmosphere, & of course require strong boilers, & other Parts, which makes them a very objectionable weight - those we have already made, including the water they carry, are not less than 6 Tons for the Power of six Horses.
I am afraid nothing can be done by Mr Perkin's scheme (Jacob Perkins, who took out numerous patents, those referred to here are probably 1. 'Steam Engines, June 1823' and 2. 'Construction of the furnace of steam boilers and other vessels, November 1823') *without increasing the danger, however a steam carriage would have been the best to have made his experiment upon, as he pretends to get great power, with little room & no danger.*
 This I cannot understand & in the absence of direct experiment he cannot satisfactorily explain.
There is another, Mr Brown's Explosive Gas Engine (Samuel Brown patent 4874 of December 1823, Mr Brown did eventually get the engine to work) *- this is not original, as it was attempted at least 40 years ago although rather different. The principal was to drop, by drops a bituminous or inflammable substance on a red hot Plate, at the Bottom of a Cylinder, at the same instant an admission of air caused it to explode & force up the Piston, but this also has gone to the "Grave of all the Capulets".*
The rotatory, or circular Engine, is also a thing to be desired, but has not yet been bought to any perfection, when compared with the cylinder & Piston Engine.
I have lately made an Engine of 60 Horses Power in which I have 2 separate condensers, one condensing one stroke of the Engine, & the other, the other Stroke, by which means the condensers are kept much cooler, & a much better vacuum is obtained, than is possible by one condenser, where the water & steam are

continually rushing in; this you will plainly understand without any further observation.
(This engine had already been described in Murray's letter of April 1824).
I think Mr Brunton's revolving fire Grate, the best mode of feeding Engines with
Coals, & nearly destroys all the smoke, at the same time regulates the quantity of
fire, to the quantity of steam required & also preserves the boilers at least double the
time, than by firing the old way.
The quantity of Coal for each horse Power varies from 10 to 15 Pounds, or even 20
Pounds, from different Coal Mines; so that there can be no Established rules for
Grates, but our proportion for Newcastle coals is 1 Square foot for each horse Power.
Mr Brunton informed me that he has frequently to alter his Grates, on that account.
Our proportion for boilers, in which there are no tubes, is 20 Cubic feet of water for
each horse power, this is a large proportion but we find it best. In small cylinders
there is more friction than in large ones, of course we allow them a little larger
proportion, which is not necessary in larger Engines. Our cylinders are some-what
larger than the nominal Power, that they may work with lighter pressure of steam,
which is found to be the best, when they are working upon the Crank.
A higher steam is used for Pumping.
We calculate our Crank Engines at 7lbs. of steam upon every square inch of the
Piston.
We make our air Pump, one fourth the contents of the Cylinder, & for single pumping
Engines one Eighth.
This is the data of what we call a horse power

lbs	feet per minute	Ratio
150 x 220	or lifted at that rate	= 33000}
or 3300	raised 10ft high per minute	= 33000} equals 1 Horse power

Or any other weight & velocity producing the same ratio.

Dimensions

Cylinder 45 inches diameter   = 1590 inches Area of Piston
Piston making 16 strokes per minute x length of dbl. stroke 14 ft. =
224 feet per minute
Pressure of steam 7lbs on each Sq. inch of the Piston.

Calculation

'area of piston' 'pressure of steam' ' lbs on the piston'   'velocity per minute'
   1590   x   7lbs   =   11130   x   224 feet

*   ' ratio'   'data of 1 horse'   'Horse Power'*
= 2493120   /   33000   = 75.549.

We have not yet been able to get Mr Wilson's proving Machine finished -
unfortunately we have had two waster Cylinders (failed castings)*, and are now about*
to try the third.
It will be an immense machine, and it is intended to exert a pressure of 1000 Tons.
After it is made I intend to try a set of experiments with it, upon cast & wrought Iron,
to determine and compare the strength in different positions.
As soon as it is finished I will send you a drawing of it,together with your Plans, for
which I am much obliged.
I am sorry I have not any models of Engines, I had a small Locomotive Engine, but it
was sent several years ago to Russia, or you should have had it.
We are very busy in the Engine-trade at present, but are sorry that the price of Iron
has risen so very much lately, as to render Engines considerably dearer than they
were some time back, and I am afraid the exportation of machinery, and allowing
mechanics to go abroad will be a disadvantage to trade in general.
Enclosed I send you a sketch of our Portable Engines, which we make as high as 24
Horse Power - they are reduced to as simple a point, as Engines can be, to obtain

the full effect of a cylinder & piston Engine. We think the Beam the best medium between the Piston & crank, it is so very useful and convenient a slave to attach pumps to or any other motions; and exhibits the Engine in its naked principles. But Engines are made of all forms, witness some made at London, which are cocked up very similar to a Pagoda Temple, and difficult to be understood, besides, what is worse, they are very unmechanical in their operations, and liable to be much oftener out of order, than the plan inclosed, which I think cannot be altered for the better, for this kind of Engine. I remain etc." [38].

At a meeting on the 1 December 1824 the Leeds Mechanics Institute was formally established with B Gott as President, two vice presidents John Marshall and John Luccock, a treasurer, a secretary and 15 directors. Matthew Murray was not there; in his place was Joseph Ogdin March (Marsh in the papers) [39].

On the 23rd of December the papers tell us that a boy working for Fenton and Murray was stopped on his way home on Holbeck Moor by three men and robbed of his weeks wages of 8/-. It was also reported that due to much rain the River Aire had overflowed its banks and that most of Water Lane and Meadow Lane were under water [40].

On the 30th of the month the papers showed, among other donations, James Fenton giving £10 to the Mechanics Institute. It was also recorded that the river had risen again to nearly the same level causing much congestion on the roads around the bridge [41].

The Deeping Fen Trustees has been seeking to improve the drainage of the Deeping Fen which covered an area between Spalding Common and Crowland. They had had a report from W Jessop and John Rennie in 1800 that had recommended the erection of two steam engines at Pode Hole. This recommendation was renewed by a Thomas Pear and again by Mr Bevan (who was appointed engineer for Deeping Fen in July 1823) in 1823. An Act of Parliament was obtained in 1823 to undertake the work. The land for the pumping station at Pode Hole was not finally purchased until March 1824, and it was therefore in 1824 that Mr Bevan ordered two steam engines for the pumping. He ordered a 60H.P. engine from Fenton and Murray and an 80H.P. engine and both scoop wheels from Butterley. The intention was to have both engines working by Christmas 1824. In the event it was stated that the Fenton and Murray engine (named Kesteven) was expected to be ready in about 6 weeks, but it was only got working by June 1825, and the Butterley engine (named Holland) was nearly finished by August. The engines were both overhauled by Watt & Co between 1881 and 1883. Both these engines worked until 1928 and the engines were not scrapped until 1952. The drainage board published a commemorative booklet on the occasion which was very complimentary on the workmanship of both engines.

" *All the castings, notably Kesteven's beam, the fluted connecting rod, the finely moulded cast iron beam spanning the beam support columns, and the cylinders are extremely fine examples of the pattern maker's and iron moulder's craftsmanship, necessitated by the inadequacy of their machine tools. These castings emerged from the foundries so smooth and finished that no machining other than boring holes for various bearing pins was required".*

Photographs of both engines were provided, along with the dimensions of Holland. Some of the dimensions of the Fenton and Murray engine are also known; she had a 45 inch diameter cylinder, a stroke of 6 foot 6 inches, the fly wheel had a 24 foot diameter and revolved at 22 revolutions per minute. The scoop wheel (by Butterley) for the engine had a 31 foot diameter and contained 40 scoops each of which was 5 foot 6inches long and 5 foot wide.

The engine house still survives as a workshop, minus the chimneys.

59. Pode Hole Pumping Station today.

60. 3 views of Kesteven at Pode Hole at the time of being dismantled.

In Leeds 1825 did not start propitiously, on January 13th there was an explosion at the Middleton Colliery which killed 25 mineworkers.

At this time changes were being planned at the Victualling Yard at Deptford. The Fenton Murray and Wood engine erected in 1810 had been put up primarily for brewing beer for the Royal Navy, but the establishment of brewing at other sites was reducing the amount of time spent on this activity. It was therefore mooted that the steam engine had sufficient slack time to operate a saw mill in order to cut staves for making the vast number of barrels needed to provision the fleet. The following letter by Henry Vaughan, the Engineer at the yard, gives an idea of this engine's work routine:

"In obedience to your directions of yesterday for me to report the number of days on which the Steam Engine in the Brewhouse of this yard was employed during the undermentioned years I beg to state the same to be as follows viz:
Year
1813   144 days employed in brewing and crushing malt
*   144 do employed in pumping Beer from Square*
*   288 Total number of days employed this year.*
1814   124 days employed brewing and crushing malt
*   124 do employed in pumping Beer from Square*
*   248 Total number of days employed this year.*
1815   83 days employed in brewing and crushing malt
*   83 do employed in pumping Beer from Square*
*   166 Total number of days employed this year.*
1822   15 days employed in brewing and crushing malt
*   15 do employed in pumping Beer from Square*
*   1 do employed in pumping water from the River Thames*
*   5 do employed in pumping water out of utensils*
*   36 Total number of days employed this year.*
1823   19 days employed in brewing and crushing malt
*   19 do employed in pumping Beer from Square*

5 do employed in pumping water out of utensils
43 Total number of days employed this year.
1824   *21 days employed in brewing and crushing malt*
   21 do employed in pumping Beer from Square
    4 do employed in pumping water out of utensils
   46 Total number of days employed this year.

With respect to the power of the Steam Engine I have to observe that it is variously applied in the process of brewing, sometimes working the full power of Ten Horses, and at other times working at the rate of only Two, the power being changed by the connection of the different machinery necessary for the process of brewing, as for example; in crushing malt the power of Five Horses is required, in working the mashing machine the power of Three Horses, and in pumping the worts into the copper the power of Two Horses, at this threefold employment the full power of Ten Horses is required, but this connection of machinery to the Engine occupies but a short time not exceeding one hour which is from nine to ten O Clock in the morning on Brewing Days. The variation of the power of the Steam Engine when it is required to be extended, is hereunder stated commencing at 6 O Clock in the morning:
From 6 to 7 O Clock the power of 3 Horses is required
From 8 to 9 O Clock the power of 2 Horses is required
From 9 to 10 O Clock the power of 10 Horses is required
From 10 to 11 O Clock the power of 7 Horses is required
From 11 to 12 O Clock the power of 8 Horses is required
From ¾ past 12 to ½ past 1 O Clock the power of 5 Horses is required
From 2 to 3 O Clock the power of 2 Horses is required being the last process of the day's brewing.
On the following day the Steam Engine is employed in pumping the beer out of the square, which takes two hours, and requires the power of Five Horses.
In pumping the water from the River Thames which we have occasion to have recourse to, when we cannot get a supply from the Kent Water Works, it requires the power of Four Horses, and upon the average consumption of water in this yard including brewing, the Steam Engine would have to work Five Hours to make up this deficiency in the Liquor B[ooks]. In the summer the utensils are charged with water to prevent them from becoming mouldy and when the water becomes tainted or offensive the Steam Engine is then employed to pump it out, which requires the Engine to work for Five Hours with the power of Five Horses" [45].

The scheme, to make the steam engine available for the purpose of sawing staves, was approved by the Lords of the Admiralty on the 20th January 1825 [46].

On the 27th of January the Leeds papers published the names of the dead and injured at the Middleton Colliery explosion, 1 man left a wife and 10 children. It also published the donations to the fund which included James Fenton £5, Matthew Murray 3 guineas, the workmen at Fenton and Murray's £5/9/6d [47].

A proposal survives in the Marshall papers dated April 24 1825 for the firm to set up its own extensive machine making shop. Marshall's was no doubt suffering from the death of David Wood and the fact that the experienced Hives was now a member of a competing partnership. In the move towards spinning finer linens, in particular 4lbs per bundle it was seen that their profitability would be greatly increased by some further improvements in the machinery in the preparatory processes. The spinning machinery was seen to be satisfactory. The issue seen by

278

the author (perhaps John Marshall junior) was that it would be better to make the improved machinery in house, rather than employ their external machine makers who would soon introduce the improvements to their competitors. In setting up a new machine shop they expected great assistance from John Farey who was employed by Marshalls at the time. The plans were detailed as to the capital and numbers of each category of worker needed. To this was added the plan for a new mill of 10,000 spindles or so. Within the detail are two sentences of particular note to this study; *"Mr Murray might be kept making spinning frames as fast as he pleased and carding engines "* and *" Our foundry note* (unclear if this was their annual spend with the various machine makers in Leeds or just with Fenton and Murray) *has been I think £5000 a year for some time"*.

The tone is not content [48].

On the 14th June the South London Water Works ordered a bucket with cupped leathers which, as it was similar to the one sent, they trusted could be delivered immediately and wrote again on the 20th confirming the dimensions and that they did not need the rod. Presumably any upset over the settlement of payment for the engine had dissipated [49].

On the 16th of July Matthew Murray's youngest daughter Mary married Joseph Ogdin March at Leeds parish church [50].

On 24th September the Leeds Mercury carried the following recruitment advert;

"TO ENGINEERS, &c – WANTED,
a Person to take the Management of a Steam Engine manufactory and Millwright Work in general. A Person having a practical Knowledge of Steam Boat Engines would be preferred. Also, TWO or THREE good MILLWRIGHTS would meet with constant Work and liberal Wages by applying to RICHARD JACKSON, at Fenton and Murray's Steam Engine Manufactory, Leeds: if by Letter, Post-paid.
Leeds Sept 20 1825".

It is very likely that Benjamin Cubitt, who was then a foreman at Mr Penn's in Woolwich, successfully applied for the first position in the advert, although by that time he can have had little experience in steam boat engines as Penn's had only just started in that business. Cubitt went on to work for Rothwell & Co at Bolton, followed by a spell at the Brighton, Croydon and Dover Railways and ended up as the Superintendent Engineer of the Locomotive Department of the Great North Railway. Whether it was in response to this advert or by some other means Benjamin Cubitt certainly came to work at the Round Foundry about this time.

An even more well-known individual appears to have started work with Fenton and Murray about the same time, although probably not until 1826/7, and this was John Chester Craven, who having worked for Todd Kitson and Laird and the Railway Foundry ended up at the London Brighton and South Coast Railway.

1825 was the year that the Stockton and Darlington Railway commenced operations. The Railway only had one locomotive initially, 'Locomotion'. In the early days it had been difficult to raise enough steam for her to operate effectively, and there had been issues with her fire tube which had had to be replaced. As the only company other than Stephenson's known to have built a number of successful locomotives it is not surprising that the Stockton and Darlington wrote to Fenton and Murray to see if they would supply a locomotive. But they were to be disappointed. The answer from Leeds was to decline to take the business;

"We have your favour respecting a Locomotive Engine for the Darlington Railway Company, and after giving it mature consideration we find we shall not be able to make you one at present, neither can we ascertain the Cost till every particular be determined on .

The construction our Mr Murray proposed for these Engines (and described to Mr Stephenson, your Engineer) was to have the Engine upon one Carriage with four Wheels & the Boiler upon another Carriage, each Spring mounted, connected together by jointed Steam Pipes, this would reduce the great evil, viz, the weight of the Engine one half, and would be a great saving of the Rails, till this is done we would recommend you to use Horses, at any rate not to make any more Engines above four Tons, or they will frequently break themselves by their weight, and what is more they may eventually injure your Railway, and which at present has every appearance of being a good one" [50].

Mr Otley the secretary of the Stockton and Darlington Railway wrote again on November 7th, presumably asking for a reconsideration and explanation of their unwillingness to supply a locomotive.

On November 9th the South London Water Works wrote to Fenton and Murray urgently asking for an air pump bucket rod for the 40 H.P. engine. The letter added that they had decided not to order a larger engine for the time being [49].

Fenton and Murray replied to the Stockton and Darlington Railway on the 14th November:

"Mr Otley   Sir
We have your favour of the 7th and cannot give any other information than we did on the 26th ult. excepting that it does not suit with the present arrangement of our Business to take orders for High Pressure or Locomotive Engines, we have not made any this 8 years, and we do begin of it, it must be where they have become a regular article of sale, a new arrangement of tools &c would be necessary, in the mean time we shall not lose sight of the best construction & applications of Locomotive Engines. To your query which we suppose relates to the Engine & Boiler being separate there is no difficulty but what practical engineers and good workmen will easily surmount, but in the interim that the Public may be benefited by our improvements, we are drawing Plans that is intended to be printed in some of the Public Journals and which we hope may lead to some further improvements. Yours very respectfully" [51].

An interesting item was advertised for sale in late November / early December in the London paper Public Ledger. The metallic materials for a 12 H.P. cistern framed engine by Fenton Murray & Co still packed as delivered from Leeds lying at a wharf in Wapping. Plus 3 saw frames made to be driven by the engine, from the models of the Chatham Saw Mill, which were at a yard by Chatham Yard Gate.

If the surmise over when Benjamin Cubitt joined Fenton and Murray is correct, then by one of those curious coincidences, around the time he left John Penn's establishment, John Penn was being engaged by Captain Hill of the Deptford Victualling Yard to examine the Fenton and Murray steam engine at that yard. The saw mill had been constructed by a Mr Cooper and was not considered satisfactory, and it was now uncertain if the engine was up to the increased job. On the 5th December Captain Hill forwarded Mr Penn's reports on both the saw mill and the engine to the Victualling Board. In reading the highlights of the report on the engine below, it should be remembered that the engine was erected in 1810 and many improvements in steam engines had taken place since then. The maintenance had also been extremely poor.

"In obedience to your letter of 18th November requesting that I would examine and report the state of the Steam Engine at His Majesty's Victualling Yard at Deptford, I beg leave to state that I have carefully examined the Steam Engine and respectfully submit the following as my report on its defects.
….. the cylinder is out of a perpendicular; this causes a great strain, when the piston rises and falls; ….. and the beam centres and the parallel motion must be strained at every stroke, ….. much unnecessary friction exists;…..Upon inspecting the cylinder I found it in a very imperfect state ….. larger in some places than in others, it is also considerably oval,……. also found many small holes in the working part of the cylinder, these holes are very injurious to the packing of the piston, which being of hemp, the edges of those holes are continually catching it and tearing it to pieces, thereby causing the packing to wear out much sooner than it might….. I find further that the piston is too small for the cylinder, it ought to fit the cylinder as tight as it can be put in,…..the cylinder is between three and four inches too long which causes a proportionable loss of steam at every stroke of the Engine, this I find amounts to full one thirtieth part of the steam necessary for the working of the Engine…... casing of the cylinder has been broken in two places, and is mended with patches, …...The slide valve is upon a very old and imperfect construction, it works in the middle of the cylinder and the steam is conveyed to the top and bottom by two pipes,…..Upon examining the joints of the Engine, I found some of them leaky, I also discovered a leak in the condenser, which had been cracked. In examining the Air pump….serious defect the valves of the bucket have so little space for the delivery of water that a very great strain is necessary to force it through the valves, this causes a very great tremor all thro' the Engine, and was liable to have been attended with very serious consequences: it might have lifted the beam from its centres….. directions for its immediate alteration,…. the result was that the Engine became free from that shake and tremor,…..The Air Pump is out of truth; not being either round or parallel, and the bucket is too small. ….. this Engine not being provided with any lever or handle for the purpose of stopping or starting it. In starting the Engine the Men are obliged to

pull round the Fly Wheel until it opens the valve, and when they stop it, it is done by shutting off the steam, by the throttle valve, which makes the Engine a long time stopping, When examining the boiler, I found the water had been suffered to get below the fire, round the flues; this would in a short time have destroyed the copper…..a rapid oxidation of it ; I advised that the water should be raised to the proper level, and you gave immediate orders for the same: which is now done, and fortunately no serious injury has been done to the copper…... I consider the Engine, in general, to be in a very imperfect state; and that it consumes much more fuel than it ought to consume, according to the quantity of work it has to perform, it may go for some years; but I think it will be liable to frequent repairs, it will however never be perfect, without a very material and expensive alteration" [52].

On December 12th 1825 Matthew Murray made his last will and testament [53].

On the 22nd December there was a senior inspection at Deptford Victualling Yard, called a Visitation, which resulted in Mr Goodrich being requested to examine the saw mill.

At some point in 1825 Fenton and Murray erected the first English type of flour mill in Germany at Magdeburg, and another one soon after in Berlin [54].

On Friday 27th January 1826 an inquest was held in Leeds on the death of Joseph Mullin, a labourer at Fenton and Murray's, who had died that day. He was whitewashing the top of the millwrights shop when his smock frock caught in a horizontal drive shaft from the steam engine. He was dragged between the shaft and the wall, a gap of 18 inches, over 100 times before the engine could be stopped. He died before he could be freed. The jury returned a verdict of Accidental Death [55].

On the 1st February 1826 Matthew Murray replied to a letter of enquiry from Mr Pease of the Stockton and Darlington Railway:

"I rec'd your Letter and should suppose you could get your Waggon Wheels cast full as well at Stockton or Darlington as anywhere, it is my opinion they would be best cast out of an Air Furnace of good No 2 Iron and not case hardened but cast all in one piece without wrought Iron arms. I believe such wheels would be found most economical under all circumstances. With respect to the Axle trees, as we make a good deal of scrap iron, if you will send us one as a pattern we will then make you one and inform you at what Price we could make you a quantity, but still I must say that I think it would be worthwhile for your Stockton and Darlington Foundries to accommodate you with these every day articles, even if they put themselves to some little expense to do it.
The price we are charging here for Waggon Wheels is 16/4 per cwt and one of our Waggon Wheels weighs nearly 3 cwt with ten arms, they are not very liable to break but will wear a long time" [56].

On February 16th Simon Goodrich arrived at Deptford Victualling Yard to examine the saw mill, and in his notes remarked:

"Mill worked by 10 H.P. Murray and Wood engine, 4 ft. stroke and 28 strokes per minute" [57].

Matthew Murray died on the 20th of February in his 61st year. He was probably aware that his time was limited, which was why he had made his will the previous December. His obituary described him thus:

"a man whose mechanical abilities were perhaps inferior to none; his great improvements in the steam engine, flax spinning and other machinery, will be a lasting testimony of his unceasing labours" [58].

On March 9th the usual notice of the time appeared in the Leeds papers asking all who either owed or were owed money by Matthew Murray personally to contact the executors the Rev. Joseph Bushby or Richard Jackson [59].

The Victualling Board decided to pay Mr Cooper for his saw mill at Deptford, after receiving Mr Goodrich's and Mr Penn's reports for the saw mill and repairing the engine but with the instruction that he should not be employed in any millwrighting work again by the Board. And in consequence of these reports they decided to order a new engine for the Brewhouse at Deptford. They accepted an offer from a George Porter for an engine on Dr Albans principle [60]. Dr Alban's patent had an unusual way of raising the steam. Small pipes were injected with enough water the raise the steam to a required level and regulated by a force pump dependent on a stated pressure. The pipes were placed in a cast iron tank filled with a low melting point metal which was heated by a furnace, thus turning the water in the pipes into the necessary steam. It was determined to order a 16 H.P. engine [61].

On the 16th of March a mill at Holbeck was offered for lease along with a warehouse. The mill was Murray's and was described in his will as *"my newly erected mill on the south side of or near to my dwelling house"*. In the advert it was described as a three storey fire proof mill with good attics, the rooms 68 feet long and 24 feet wide, along with a fireproof warehouse, and application was to be made to Richard Jackson of Water Lane engineer [62].

Mrs Thompson, Matthew Murray's mother in Law, who had been living with the family for some time, died at Holbeck aged 92 on May 23rd, making it a double blow for Mary Murray.

The situation at Deptford Victualling Yard apropos the new steam engine was not encouraging as described by Captain Hill to the Board on September 7th:

"By the agreement entered into by the Board on the 10th May last with Mr G R Porter, for the erection of the steam engine in the Brewhouse of this yard on Dr Alban's plan, the engine with all its apparatus belonging thereto was to be in perfect readiness by the 1st August last. Nearly six weeks have elapsed since that period and the engine is not yet ready. The Iron Boilers for the metal were tried according to Dr Alban's original plan, and split. Separate boilers (circular) with plugs at the bottom were then made for each tube, which also failed and were thrown aside, a set of boilers were then recast without plugs at the bottom and tried yesterday, and they too failed to contain the metal. The engineer of this department reports to me, that it

283

would take nearly a month to properly replace our own Steam Engine. As the Board have ordered the Brewings of this Yard to commence on the 15th of next month, I beg leave to call their attention to the circumstances stated herein, as to what is to be done" [63].

On the 15th of September Richard Jackson, as executor, lodged his affidavits for probate for Matthew Murray's will. One of the witnesses was Samuel Exley of Holbeck, who was then employed by the firm as a bookkeeper. The probate valuation appears to have been £8000, which in today's money was a considerable fortune, running into millions. The significant legacies were

- Richard Jackson (son in law); £1000 plus the mechanical drawings and books.
- Margaret Jackson (daughter); his diamond ring from the Empress of Russia and his house and the one next door.
- Matthew Murray Jackson (grandson); his gold medal from the Emperor of Russia. Matthew Murray Jackson in turn bequeathed the ring to his son William in his will of 1891.
- William Murray Jackson (grandson); his gold medal from the Royal Society of Arts.
- Mary Murray (wife); an annuity of £225 per annum, paid in quarters, and his household goods, furniture, plate, linen etc.
- Matthew Murray jnr (son); £1000 payable 12 months after Murray's death, and a further £1000 payable after Mary Murray's death and to inherit all the household goods, furniture, plate , linen etc after her death.
- Ann Maclea (daughter); £1000 within 2 years.
- Mary March (daughter); £1000 3 years after his death.
- His Mill (see above), partnership capital and stock in trade and any other assets went in a trust, to be managed by his executors to pay his wife's annuity and subject to that payment to be shared among his grandchildren when the eldest was 21. (This would have been William Murray Jackson who was 21 on 17th October 1837)
- £100 to his sister Margaret Clayton [61].

The delays written into some of the legacy payments was probably due to the partnership agreement stipulating maximum withdrawals of capital per year on death. The books of the firm show payments still being made to David Wood's estate in March 1825, four years after his death.

Mr Porter had been granted an extra week by the Victualling Board to complete the new steam engine. Captain Hill reported on 23rd September that the engine was in parts, the boilers were leaking and the saw mill machinery had never been worked by the engine, despite it being stated as complete several times. He also reported that the yard's engineer and Mr Penn both agreed that *"our own steam engine"*, the Fenton Murray and Wood original would take three weeks to re-erect and with the time taken for the beer to settle it would be the end of October before any beer could be issued. The Board were reminded that they had instructed brewing to begin on October 15th, three weeks hence. Mr Porter was given a final chance, to have the engine working by the following Wednesday and if not he was required to remove it according to contract. On the 12th of October after further unsuccessful trials Mr Porter admitted defeat and Captain Hill employed Mr Penn to re-erect the original engine [64].

A 24 H.P. engine by Fenton Murray and Wood was posted for sale in the Leeds Mercury of October 14th 1826 by the Marshalls of Kirkgate Bradford and an alternate contact was given as Low Moor, the engine was probably being superceded [65].

In 1826 Newton's London Journal published the details of Matthew Murray's new locomotive, as had been promised to the Stockton and Darlington Railway in the letter dated November 14th 1825.

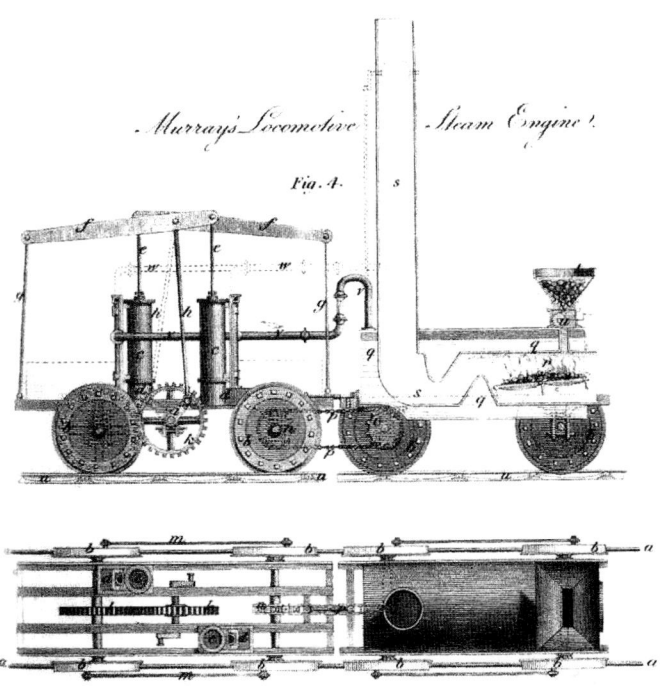

61. Murray's 1825 design for a steam locomotive.

"Mr Mathew Murray of Leeds, whose great experience and superior talent in the construction of steam engines, the scientific world are well aware of, has proposed a design for a locomotive steam engine, which he considers will be found to possess superior advantages to any that have been heretofore employed. This plan for a locomotive engine, however, is to be considered but as a design, for though Mr M. believes it to possess advantages over all others heretofore made, or at present in use, yet he is far from presenting it as a piece of machinery that is not susceptible of improvement; it is rather given to the public, as the best construction of steam carriage, according to his view, that the present stage of science will admit of.

At the close of 1826 it is time to consider the position of James Fenton, the surviving partner, who had now seen three partners die, William Lister, David Wood and Matthew Murray. Presumably he felt he had a competent set of managers around him and did not see any necessity for raising further funds at this point in time, despite having to repay Murray's capital to his estate. For whatever reason he decided to operate as a sole trader, continuing to use the style Fenton and Murray, a position he was to maintain until 1832.

Notes

1. ACC2558 South London Water Works LB SV/1/163/1. London Metropolitan Archives.
2. e.g. The Leeds Mercury 2 February 1822.
3. Marshall Papers MS200 folio 32. Brotherton Library.
4. The Leeds Mercury 9 February 1822.
5. ACC2558 South London Water Works Directors Minutes SV/1/3. London Metropolitan Archives.
6. Private Collection.
7. BPP. Reports from Committees session 5 Feb. to 6 August 1822 vol vi. 5th Report from Select Cttee on Holyhead Roads (Steam Boats etc,) 2nd April 1822 Capt John Percy page 151, Appendix 8 pages 214-5 for Fenton & Murray's letter. For additional coverage by Goodrich see Goodrich Journals B50. Science Museum Library.
8. Goodrich Journals B47. Science Museum Library.
9. The Leeds Mercury 4 May 1822.
10. ADM 106/2154 Navy Board Digest In Letters August – Dec. 1822. National Archives Kew.
11. The Leeds Mercury 10 August 1822 page 4.
12. The Leeds Mercury 10 August 1822 page 2.
13. Goodrich Journals B50. Science Museum Library.
14. The Leeds Intelligencer 23 September 1822.
15. Mr Fields Report attached to the Select Cttee Report on Holyhead Roads (see 7 above) supported by Newcastle Courant 7 March 1829 for the sale of two 40 HP engines by Fenton Murray & Co which were to be replaced with more powerful ones (by Butterley). The author apologises to the shades of Mr Field for doubting his listing in the earlier article 'The Marine Engine output of Fenton Murray & Co' in the Mariners Mirror. The article in the Newcastle Courant had not then come to light.
16. Marshall Papers MS200 folio 15. Brotherton Library.
17. Marshall Papers MS200 folio 28. Brotherton Library.
18. The Leeds Intelligencer 20 February 1823.
19. M.I. Brunel Diaries 1822-1827. University of Bristol.
20. ADM 106/2156 Navy Board Digest 1823. National Archives Kew.

21. The Leeds Intelligencer 11 December 1823.
22. The History of the Gas Light & Coke Company 1812 -1949 by Stirling Everard, A & C Black 1949. Description and Drawing – Mechanics Magazine Volume seventh 1827 No. 198 of 9 June 1827.
23. ADM 106/2157 Navy Board index for 1824. National Archives Kew.
24. The Leeds Intelligencer 29 January 1824.
25. The Royal Bank of Scotland Group Archives, BEL 3/3 Customer Account Ledgers 1813-1825.

26. Goodrich Papers A1078. Science Museum Library.
27. Watson Papers Wat 1/5/71 and1/5/72. NEIMME.
28. The Leeds Intelligencer 20 May 1824.
29. The Leeds Mercury 29 May 1824.
30. The Leeds Mercury 17 July 1824.
31. Goodrich Papers A1088. Science Museum Library.
32. ADM 106/1829 Chatham Officers, In letters to Navy Board (Commissioners) 1824. National Archives Kew.
33. The Leeds Mercury 21 August 1824.
34. ADM 106/2158. National Archives Kew.
35. The Leeds Intelligencer 23 September 1824.
36. E.g. Forster Papers FOR/1/9/132. NEIMME.
37. The Leeds Mercury 2 October 1824.
38. Goodrich Papers A1097. Science Museum Library.
39. The Leeds Intelligencer 9 December 1824.
40. The Leeds Intelligencer 23 December 1824.
41. The Leeds Intelligencer 30 December 1824.
42. Holland & Kesteven Pumping Engines at Pode Hole Spalding, Deeping Fen, Spalding & Pinchbeck Internal Drainage Board 1952.
43. The Drainage of Fens and Low Lands by Gravitation and Steam Power. W. H. Wheeler. 1888. Also as a series in 'The Engineer' 1887, see June 17 Edition.
44. The Leeds Intelligencer 13 January 1825.
45. ADM 109/80 Victualling Board 1825. National Archives Kew.
46. ADM 111/260 Victualling Board 1825. National Archives Kew.
47. The Leeds Intelligencer 27 January 1825.
48. Marshall Papers MS200 folio32. Brotherton Library.
49. ACC2558 South London Water Works LB SV/1/164. London Metropolitan Archives.
50. Rail 667/939 Stockton & Darlington Railway, Fenton and Murray to Mr Otley 26 October 1825. National Archives Kew.
51. Rail 667/940 Stockton and Darlington Railway. National Archives Kew.
52. ADM 109/80 J Hill to Victualling Board 5 December 1825. National Archives Kew.
53. Will of Matthew Murray, Yorkshire Wills Vol. 174 no.377 (microfilm no.1078). Borthwick Institute York.
54. Co-Operative Wholesale Society Annual 1887, History of Milling by R Witherington, Milling Engineer to Messrs Thomas Robinson & Son Ltd of Rochdale. Co-Operative Society Archives Manchester. Also 'Beitrage zur Kenntui des amerikanischen Muhlenwesens' 1832 Berlin.
55. The Leeds Intelligencer 2 February 1826.
56. Rail 667/983 Stockton and Darlington Railway. National Archives Kew. Presumably the local foundries could not oblige as Pease wrote to Low Moor, Kirkstall Forge and Bowling Iron Works as well as to Fenton and Murray.
57. Goodrich Journals B55. Science Museum Library. Goodrich had previously visited the site with Mr Field in 1813 when they had drawn a sketch of the engine and some of its parts, see Chapter 8.
58. The Leeds Intelligencer 23 February 1826, The Leeds Mercury 25 February 1826 and The York Herald 25 February 1826.
59. The Leeds Intelligencer 9 March 1826.
60. ADM 106/261 16 March 1826. National Archives Kew.

61. Register of Arts and Sciences 5 November 1824.
62. The Leeds Intelligencer 1 June 1826.
63. ADM 109/81 J Hill to Victualling Board. National Archives Kew.
64. ADM 109/81 J Hill to Victualling Board 23 September, 12 October, 13 October 1826. ADM 110/261 13 October 1826. National Archives Kew.
65. The Leeds Mercury 14 October 1826.

Chapter 13.

James Fenton Sole Trader 1826 to 1831.

The Kenton and Coxlodge colliery line, with locomotives by Fenton Murray and Wood, as far as can be established, had not operated the locomotives during 1815 to 1817. During this time the ownership was changing and there were disputes with the neighbouring Fawdon Colliery over the use of the locomotives on the shared railway. This railway/waggonway took the coal down to the Tyne east of the collieries, to near Walls End. It is interesting therefore to find adverts which ran from December 1826 to early 1827 which offered for sale:

"All that current-going Railroad, five miles in length, leading from Fawdon Colliery to the River Tyne, at Scotswood; together with the inclined plane, machinery & ropes, and also a condensing engine, on Boulton and Watt's principle, of 26 horses power, made and erected by Fenton Murray and Wood of Leeds; the whole constructed only eight years ago, of the best materials, and according to the most approved plans. Also about 130 chaldron waggons........
To be viewed on application to Mr Douglas Fawdon Colliery" [1].

From the wording it would appear that this railway which went down to the Tyne west of Newcastle was built around 1818. There was a previous railway from Kenton to the Tyne west of Newcastle, known as *"Kitty's Drift"* which had been described as *"a subterraneous waggonway"*, which had been unused since 1808 when the line to Wallsend was put down [2]. In the surviving records of the Watson Papers there is a paper headed Kenton and Coxlodge Colliery 22nd December 1817 which re-evaluated the economics of operating the locomotives on a limited section of the line, and the costs of using the locomotives was reckoned at £572 per annum, and for using horses at £1780 [3]. By this time the Kenton and Coxlodge had been sold to the Brandlings. As yet there is no evidence of the locomotives having been re-instated, despite this financial case and the new railway, referred to in the advert for the Fawdon coals.

At the beginning of March, the Mill and warehouse next to Matthew Murray's house were advertised for sale by auction. Information was still be had from Richard Jackson at the Foundry [4].

On the 15th March John Marshall junior wrote to his father to tell him that the coupling of the upright shaft on the 70 H.P. engine had broken and stopped the mill for three days. The time had been used to put the machinery in order. He gave the reason for the breakage as the engine being overloaded and made a proposal to change some of the machine functioning and reducing the speed to balance the steam engine. He went on to describe a visit to Hives and Atkinson's flax mill. Their wet spinning made excellent yarn and they had been told by Atkinson that the resulting cloth was good. Marshall & Co were changing some of their machinery to this process and John Marshall junior complained that the roving frame would not be delivered until Monday night and it was to be 10 days before they received the spinning frame. He added that they could not accept either the delay or the lack of exertion by Jackson (Fenton and Murray). After some examination in earlier years of their costs this is the first expression of dissatisfaction with Fenton and Murray personnel, and was probably the result of Marshall's no longer getting the personal attention they had grown used to from both Matthew Murray and David Wood. They were probably getting less attention than Hives and Atkinson who had given Fenton and Murray very extensive orders in the previous few years [5].

On April the 14th John Marshall junior was again writing to his father, and after a long explanation mainly in jargon, concluded that before they started buying the machinery for the new D Mill, they needed some 75 new machines for the existing mills in Leeds and Shropshire, which he valued at £5450. He continued by saying that this was as much machinery as could be made in the remainder of the year by Fenton & Murray and Cawood & Murfin. He proposed that Cawood & Murfin should make the 1st and 2nd Drawing Machines (40 off) and an experimental line or tow wet spinning frame. Fenton & Murray were to make the remainder. A proposed part or initial order was pencilled in after the signature which gave 9 off 1st drawing machines to Cawood & Murfin at £45 each, total £450 ?, and to Fenton & Murray 6 carding engines at £110 each, 3 tow drawing machines at £25 each, total £735 [6]. Elsewhere later in that month he noted that Fenton & Murray's price for a carding engine was £110 but he had told them he would only give £100 and wanted them to reduce the price to £95. In the same note he listed the prices for various machines from Fenton & Murray, and the 1st drawing machines were £3 more expensive than Cawood & Murfin. The list also indicated a charge of between 35/- and 38/- per spindle and on May 2nd John Marshall junior noted that Lawson & Walker would charge 28/- to 30/- per spindle "and for 2 to 3/- extra would endeavour to make as good as a job as Murray's". On May 14th he described a roving frame from Cawood & Murfin, which Marshall's purchased and another by Fenton & Murray which they also took [7]. So while Fenton and Murray were still very much in the running for Marshall's business, the younger generation at Marshall's were taking a far harder look at the competition. It is clear from the quote above that Fenton & Murray were still producing the highest quality.

In reply to a query, that has not survived, Fenton and Murray wrote to Mr Pease junior on July 27th enclosing, as requested, the working drawings of a 26 H.P.engine that they had made for Messrs Newmarch Sons & Co of Newcastle in

1818. They also advised that their engine layouts may now be totally different. This is the same engine as mentioned for sale at the beginning of this chapter as at this time Messrs Newmarch owned Fawdon Colliery, so it is possible the Pease's were considering purchasing the engine [8].

In August 1827 Joseph Wade of Crimple Mill near Knaresborough was offering for sale three tow carding engines made by Fenton Murray & Co [9].

One of many adverts for the sale of Rawcliffe Flax Mill, which was about 8 miles south east of Selby on the River Aire, appeared in some of the Yorkshire papers in September. The mill had a *"most complete steam engine of 20 horses power (by Fenton and Murray of Leeds)"* [10].

John Marshall junior wrote to his father on September 9th and reported very good results from the wet spinning frame at 60 leas (a fine linen), and the cloth makers believed it would make good cloth. Fenton and Murray were employing two teams of workmen on making Marshalls wet spinning frames, and were to put a third team on the job in a weeks' time, then a 4th and a 5th after about a month, when their other orders were to be complete. They had another team making carding engines, 2 for Shropshire and one for Leeds, and they could continue making the machines for Mill D. He adds that with these resources Richard Jackson could make their wet spinning machines fast enough, while he would not be the cheapest, the quality was better than could be had from the Manchester machine makers. Cawood and Murfin were to continue to make the preparatory machinery and were doing well. They were increasing the number of men on Marshall's work which would enable them to produce as much machinery as needed to keep pace with the spinning machinery coming from Fenton and Murray. He added that the engine house for Mill D was now at full height and Fenton and Murray should begin on their work in a week or fortnight. The cylinder was turned and ready to fix and the valve boxes were ready [11].

On the 20th of September he reported that Fenton and Murray had delivered another wet spinning frame which did pretty well but still had some gremlins [12].

In early November 1827 the Navy Board considered a request to repair the engine in the saw mill at Chatham and to increase its power from 30 to 36 H.P. There was an additional engine at Chatham engaged for pumping water while some other works were carried out and there was a proposal that this should be permanently replaced with a new engine. The Navy Board based their discussion on a letter from Commissioner Cunningham of Chatham Yard which enclosed a report from the Officers. The report stated that the engine had not yet been repaired, as it had been so busy that no time could be allowed for the necessary stoppage. The state of the engine was now such that they strongly recommended that it was given a general repair and that its power was at the same time increased to 36 HP. They observed that the engine had been made by Fenton and Murray, and in the opinion of the Officers they would be best placed to carry out the repair and upgrade. The other engine used to pump water was in such a bad state of repair that it would cost almost as much to overhaul as to buy a new engine. The Surveyor of Buildings, Mr Taylor, supported the Officers view in regard to the Saw Mill engine and for a new engine for the water works. He recommended that Fenton and Murray should be asked to quote for repairing the Saw Mill engine, upgrading the engine and the sum for providing a new engine for the water works. The Navy Board agreed and instructed that Fenton and Murray should be written to accordingly [13].

John Marshall junior's next report to his father on November 11th reported that the 70 H.P. engine for Mill D was building with the cylinder and valves in place, the beam was up as were the air pump, condenser fly wheel and shaft, and first wheel. The boilers were built and their seatings were nearly half completed. Richard Jackson, for Fenton and Murray, was anxious to get the engine running before Christmas, which he hoped to do, but Marshall was dubious that he would succeed. Marshall had ordered a new line of shafts and drums to be made jointly by Fenton and Murray and Sandfords. This he explained to his father was after Jackson had produced an elaborate and expensive design for the first wheel, which he had been desired to revise with a view to simplicity and economy. Matters were not going well, for Marshall senior was also told that the previously advised delivery of a wet spinning machine (20 September above), remained the only delivery to date. Jackson was promising one every fortnight, and if he didn't deliver, some of the work would have to go elsewhere. Cawood & Murfin had put another floor on their shop enabling them to employ about another 16 to 18 men and were looking to make another similar sized expansion in the spring. Marshall Junior added that he was going to give Taylor and Wordsworth a trial at making carding engines as they had too much tow in stock and had to get more machines to process it. Hives and Atkinson had apparently purchased a corn mill next to their mill and had ordered another 14 wet spinning frames from Fenton and Murray [14].

The Navy Board studied a response from Fenton and Murray on the works required at Chatham in the middle of November. Therein it was stated that the combined cost for the repair and the increase in horsepower would be £500 without new boilers, their standard charge for a double powered 20 H.P. pumping engine with boiler was £1000. They asked for confirmation on the state of the foundations for the 20 H.P. engine (for the water works) and whether it was to be a single or double engine with or without a fly wheel. They also requested permission to send a man to Chatham to check those parts that needed repair on the 30 H.P. engine [15].

In a book describing visits to St Petersburg the following account of the Alexandrovsk Factory was given:

"*On entering the very extensive workshops where the machinery is made, accompanied by General Wilson, with whom we conversed in English, I could have fancied myself transported into a Birmingham or Sheffield manufactory, were it not that now and then I caught the Russian sounds of the workmen's conversation. The number and high polish of the tools, and complicated machinery made here, or serving to make others; their methodical arrangements; the hundreds of operations set in motion by one mighty power; the sight of a beautiful steam-engine of sixty horse power, by Murray of Leeds, was a spectacle which I thought existed only in England, on so large a scale, and in such perfection*" [16].

The Navy Board reviewed a further letter from Fenton and Murray on the Chatham business on the 15th December. The estimate for increasing the power was now £455, and the parts requiring repair were stated. There was no estimate as yet for repairing the engine without increasing the power, and some discussion as to the framing of the 20 H.P. engine, or whether it would require columns, which would increase the cost from £1205. At a further review on the 22nd of December Fenton and Murray's best estimate for the repair only was tabled at £160, and after the Boards clarifications on the requirements for the new 20 H.P. engine that cost would now be £1050. The Surveyor of Buildings suggested that Mr Maudslay and Mr Lloyd be asked to quote for

1. Repairing the 30 H.P. engine

2. Repairing the 30 H.P. and increasing the power to 36 H.P.

3. Erecting a new 20 H.P. engine.

The latter course of action was agreed by the Board and Maudslay and Lloyd were written to [15].

A six H.P. engine by Fenton Murray & Co with an 8H.P. boiler was advertised for sale on December 27th from Milner Royd Mill near Sowerby Bridge [17].

On January 11th 1828 the Navy Board considered the responses from the other bidders for the Chatham engines. Mr Lloyd had declined to bid for any of the business. Mr Maudslay had bid £1200 for the new 20 H.P. engine and £500 to repair the 30 H.P. engine with £152 to increase its power to 36 H.P. His bid also included £192 for some extra machinery connected with the 20 H.P. engine. On February 5th 1828 the Navy Board having determined that Fenton and Murray's bid was lower, posed the following questions

1. How long would the saw mill be shut down for the repairs?

2. As the new wheels should be properly pitched and trimmed and having cogs of wood working into iron, would this result in an increase in cost?

3 Would it cost more to have the 20 H.P. fly wheel pitched and trimmed with cogs of wood working into iron?

Fenton and Murray replied that they would repair and alter the 30 H.P. engine for £455 and make the 20H.P. for £1255. The Surveyor of Buildings recommended Fenton and Murray should be contracted for the work.

On the 17th February the Navy Board agreed to authorise Fenton and Murray to repair and upgrade the saw mill engine, but that they would not authorise any purchase of a new 20H.P. engine. Fenton and Murray and Commissioner at Chatham were advised. Fenton and Murray accepted by letter received on the 23rd February.

On the 26th May1828 the Navy Board noted that Fenton and Murray had provided a list of the materials sent to Chatham for the saw mill engine. This was confirmed by the Commissioner at Chatham on June 10th and the invoice was received about the same time. On July 5th Fenton and Murray asked for a date when they could send their engineer to Chatham.

By October 18th Fenton and Murray were requesting some money for the work done and an allowance for making the cog wheels work wood to iron etc. On the 24th October 1828 the Navy Board decided to pay them £400 in part payment of their bill and asked Mr Taylor, the Surveyor of Buildings, to report on their claim for an allowance for wood working on metal [15].

This period in history was one when the emancipation of Catholics was quite a contentious topic with two bills having passed the House of Commons but been stopped in the House of Lords (1821 and 1825). The topic was often in the newspapers and the following appeared in the Leeds Intelligencer on December 18th 1828:

"THE POPISH RECRUITING SERVICE. The following facts will show the means employed by the Popish Party, to collect their friends, at the Cloth Hall, on Friday week. We make the statements on sufficient authority.

The Irish stuff-weavers, and their wives, from the Bank: a band of music being sent round the neighbourhood to collect them together.

The workmen employed in the factory of Messrs O. Willans and Son (Holbeck), *principally Irish, attended in marching order.*
Messrs Marshalls stopt their mill at twelve o'clock, and gave their men the remainder of their time as holiday, not of course that they might attend the meeting, but credit Judaeus, to celebrate the nuptials of John Marshall, jun. Esq.

Mr Dorrington (of the firmof Bruce Dorrington, and Walker) headed 300 of their workmen to the meeting, who proceeded there in military order. N.B. Mr Bruce is son in law to Mr Edward Baines.

The manufactory of Messrs Titley, Tathams, and Walker, (with which house Mr Benjamin Walker, the Radical Quaker, is connected), stopped work,and the men repaired to decide the fate of the Nation.
The manufactory of Messrs Wm. Lupton and Co. did the same.
Business was suspended at the extensive iron foundry of Messrs Fenton and Co situate at Holbeck.
The machine-manufactory of Messrs Taylor, Wordsworth and Co the same.
NO PROTESTANT PROPRIETOR ALLOWED HIS FACTORY TO STOP"

The year 1829 brought unwelcome matters to begin with, and on the 26th of February the papers reported the death of Mr Henry Aveson aged 58 who had been upwards of 20 years a book-keeper and manager at Fenton and Murray. The same day the inquest was reported on James Leach who had had his skull seriously fractured by some machinery, whilst at work at Fenton and Murray, which had resulted in his death. He left a wife and 8 children. The verdict was accidental death [18].

What appears to be a one off advert appeared in a Newcastle upon Tyne paper on the 7th March. The advert was for the sale of two excellent engines of 40 HP by Fenton Murrah and Co (sic). They were the engines of the Yorkshireman Steam Packet which plied between Hull and London since the 29th May 1822. They were replaced with more powerful engines by the Butterley Co and the boat continued on the same route. Enquiries were to be made to W Watson Engineer at Hull or Isaac Aydon Steam Engine Builder Wakefield [19].

On the evening of Thursday the 26th of March a fire broke out in the premises of Fenton and Murray, which does not appear to have been very serious and there was little injury. The usual notice of thanks for assistance received was posted by the company in the papers on the 28th March [20].

On the 1st August the Leeds Mercury carried an advert for a 14 H.P. engine made by Fenton and Murray but applications were to the Mercury Office, so no details are available. It may be connected to the item below.

John Rand and Sons of Bradford had the cylinder and valves of a 14 H.P. engine by Fenton and Murray for sale on the 3rd of October [21].

The latest improvement to the docks along the Thames in London at this time was the construction of the St Katherine Docks, and the surviving minutes of the company for 13th October show that they made a contract with Fenton Murray & Co for a steam engine for a crushing mill for £450 [22].

The Times of the 20th October 1829 carried an advert for the
Lease of premises at 57 Redcross Street, Cripplegate in London along with a 16 H.P. steam engine by Fenton and Murray.

A six horse engine with mustard machinery was offered for sale at the end of October by J. Linsley of 151 Briggate Leeds. The engine had an 8 H.P. boiler and complete foundations all by Fenton and Murray [23].

A book called 'The Fouling and Corrosion of Iron Ships' describes briefly a boat ordered by a William Gravatt stating that he had in 1829:

"designed and had constructed by the late firm of Fenton and Murray, of Leeds, an iron paddle steamboat, with a horizontal tubular locomotive boiler, the paddle-shaft being placed in front of the smoke-box, against which the plummer-blocks were bolted. The lines of this boat were taken from a fine gig built by Roberts, boat-builder of Lambeth. This is believed to have been the first steamboat in which a tubular boiler was used." [24].

The identity of this boat is unknown, but it may well have worked locally on one of the canals or rivers, although it may have been for further afield. Fenton and Murray may well have been recommended by Bryan Donkin under whom Gravatt served his apprenticeship or one of the Brunels. William Gravatt was an assistant to I K Brunel in the Thames Tunnel Work and later went on work for him at the Great Western Railway, and the Bristol and Exeter Railway. At the latter they fell out and Gravatt was dismissed, probably with due cause, as he had a bad reputation for intolerance. Gravatt went on to do important work on the 'difference engine', the forerunner of computers.

Although it is difficult to be precise on the dates and some information is conflicting, the best knowledge indicates that The Aire and Calder Navigation was operating two paddle steamers between Selby and Hull from 1826/9. They were named 'Leeds' and 'Calder', and were both built by Pearson's of Thorne. The 'Leeds' and the 'Calder' both had 30 H.P. engines made by Fenton and Murray [25]. According to Baines History and Directory of Yorkshire 1822, there was a steam packet named 'Leeds' plying the route in 1822, but whether this was the same boat or an earlier one remains unclear.

During 1829 and 1830 the Norfolk and Lowestoft Navigation was being constructed under the direction of William Cubitt. The principal contractors for the machinery for the works were Fenton and Murray, with bridge work being supplied by Seaward & Co [26].

On the 8th of January 1830 Marshall & Co advertised their 8 H.P. Fenton and Murray engine for sale as it was no longer needed [27].

At some point towards the end of 1829 or early in 1830 Hull Corporation asked for bids for a steam engine for the new Hull Waterworks. No advertisement has been found in the newspapers, so they probably contacted selected manufacturers. Fenton and Murray wrote in with their bid on the 18th February 1830.

They quoted for a double acting 24 H.P. iron foundation, framed engine with one boiler, along with three pumps and the necessary machinery to work them at £1500 erected, and delivered to the wharf at Hull. Later in the month Mr Bennet, the agent of Boulton and Watt arrived with an introduction from Mr Gott of Leeds. Boulton and Watt had provided the original engine for the old waterworks, however the prices noted on their paperwork were expensive, the charge for a 10 H.P. was £1260; a 14 H.P. was £1560; a 20 H.P. at £2030 (double?) and a note that a single engine (presumably) of 20 H.P. was £1830. The Capponfield Ironworks from Bilston quoted for an 18H.P. engine and pumps but the price does not seem to have survived. The probable winner was Isaac Aydon & Co, as Mr Aydon appears to have had an ongoing discussion with the waterworks officials. The quotes were for an 18 H.P. on the 10th April, a 20 H.P. on the 26 April, which seems to have been negotiated to £1050 on the 26th April. The file shows that a revised specification had been sent to Overton and Aydon on the 22 April. Whatever the relationship between Aydon and Overton was, it appears to have been for this piece of business only. Mr Overton was a former senior employee of Fenton Murray and Wood, who had been prominent in the operations involved with the Isabella pit engine at Workington. He had set up business in Hull some years previously with various partners and been joined by his son, and made a reputation for marine engines [28]. However that may be, it is taken that with such a low bid and the evidence of negotiation, Aydon & Co won the business [29].

In March, Simon Goodrich was travelling north to Mr Grimshaw's Ropery in Sunderland again. On the 15th of March he left Sheffield on the 2 o'clock coach for Leeds, and called on Mr Jackson. After breakfast on the 16th he went to Fenton and Murray's works and Mr Jackson showed him around and also took him over to Mr Marshall's flax mill. They dined together and then went to see the rail road and locomotive engine. Goodrich made quite extensive notes of the visit. He recorded that they were then making independent steam engines of the following sizes:

Horsepower	Diam'r of Cylinder	Stroke	Strokes/minute
2	10 inches	2 ft 6 in	44
3	12	2 ft 6 in	44
4	14	2 ft 6 in	44
6	16	2 ft 6 in	44
8	18	3 ft	36
10	19 ½	3ft 6 in	32
12	21	3ft 6 in	32
14	22	4 ft	28

16	23	4 ft	28
18	24	4 ft	28
20	25	4 ft 6 in	25
24	27	4 ft 6 in	25
26	28	5 ft	22
28	29	5 ft	22
30	30	5 ft	22.

Goodrich "*Observed their mode of turning rams for hydraulic pumps by an upright lathe. Mr Jackson says that they turn them so true and fit so well even without leather [   ] the ram is supported for [ ] by the [ ] in the cylinder.*

Went with Mr Jackson and looked at Mr Marshalls new 70 H.P. Engine made by Messrs Murray.

Cylinder 44 in. diam. & 7 ft. stroke, 21ft beam now making 18 strokes per min. proper speed 17. Fly wheel 24 ft. diam. six arms weight as follows

	Cwt	qtr.	lb.
Rim	*144*	*1*	*14*
6 arms	*49*	*2*	*22*
Centre plate	*17*	*2*	*17*

A bevelled cogged wheel on 1st motion, 12 ft. diameter and is [full] 3 ½ in. pitch. This drove a 2nd wheel 4 ft. diameter taken to pitch line outside. There was a 2nd fly wheel on the 2nd shaft 16 ft. diam.

centre plate	*8*	*2*	*11*
6 arms	*19*	*3*	*27*
6 felks or segments [rim]	*52*	*3*	*15*

They have [tried] different lengths of coggs [compared] with the pitch. Mr Jackson says they find 2/3 of the pitch for the [hold] of the cogg one over the other with about 1/8 inch clearance the best.

Mr Jackson says that Brown of Leeds makes the best Boiler Plates in Yorkshire or in the Kingdom. They paid 18£ per ton for them last year when the Staffordshire people were selling at 12£ pr. ton. These plates are made from scrap iron.

At Fenton and Murrays Works they make their own iron from Scraps bought up from London and other places, but are very careful in sorting and selecting it. They consider it the best Iron for their purposes.

Mr Jackson has undertaken to make a pair of 8 H.P. Marine Engines each not to weigh more than 3 ½ tons and the boiler and water not to exceed 8 tons, altogether = 15tons, for the Soan (Saone) in the South of France. Price 1150£ del. at Havre.([Le Havre).

In respect to D valves he observes that they have tried brass against rust and find it will not answer on account of cutting. Then with brass on the D and the face plate of cast iron, the steam wears or [rusts] the cast iron very quickly. But D of cast iron and the face plate on the cylinder lined with brass answers well.

On working an experimental pump arranged to try the power of steam engine like the water regulator for checking the descent of heavy weights in cranes when the pressure was throttled so as to produce a strong [resistance] equal 30 lb. pr. circular inch and the same water forced by the solid piston backwards and forwards for 15 hours, it was found to get hot and expand so as to exude out the iron cement parts, the vessel being quite full of water and become boiling hot. Piston dble. [leathered] that is for up and for down stroke, pump 12" bore 2 ft. stroke 11 strokes per minute [ ] forced the water thro. an opening = ¾ [Cy unit]"

Jackson also showed him a sketch of the arrangement of the engines and pumps at St Katharine Docks, which consisted of 6 pumps and two engines. The engines were two Boulton and Watt 80 H.P. ones with 7 foot strokes. Jackson considered two 56 H.P. engines would have been sufficient, and added that more than 12 strokes per minute would shake the engine house too much. After dinner they went to see the rack rail, Goodrich noted that the locomotive had 2 steam cylinders 10 inches diameter with a 2 foot stroke and that the boiler was about 3 foot 6 inches diameter and 9 foot long [30].

The Leeds Mercury carried an advert for 2 or 3 Boiler Makers for Fenton and Murray, and adverts to sell or let the Fire Proof Mill (the one in Matthew Murray's will). For the latter applicants were to apply to Richard Jackson at the Foundry of Fenton & Co [31].

A 10 H.P. engine was up for sale from April, with two boilers, all made by Fenton and Murray and it was at work at Beavers Hole near Barnsley. Applications could be made to Beavers Hole, Fenton and Murray or to Mr A Faulds of Worsborough (also near Barnsley). Andrew Faulds was a coal agent and later on his son, of the same name, was to marry one of Richard Jackson's daughters [32].

On May 8th a 13 H.P. Fenton Murray and Wood engine was advertised for sale in the Manchester Courier.

On June 3rd the papers carried the report that David Wood, the son of David Wood of Fenton Murray and Wood, had died at Dusseldorf (a city of the Prussian States) whilst returning to England. So it is learnt that the sons of both Matthew Murray and David Wood had chosen to work abroad, and not to follow their fathers into the business [33].

A fire broke out in the saw mill of Jonas Brown and Son on Tuesday the 10th of August. The saw mill was located next door to the Round Foundry across Foundry Street. Fenton and Murray promptly despatched all their fire engines to the scene and used one of their steam engines to pump water to the appliances. This action was shortly followed by the arrival of several engines from the fire companies. The fire seems to have started in the timber drying rooms about 7.30 pm and was put out around 10 p.m. the same evening [34].

In September Messrs E & J Taylor of Marsden near Huddersfield were advertising a portable 12 H.P. engine by Fenton and Murray for sale.

On Wednesday the 8th of December 1830 an auction was held above Messrs Carr's Dye House, Swinegate in Leeds, of the belongings of the late David Wood senior. It included most of his mechanical collection, lathes, drilling machines, measuring tools, saws and planes. All made by David Wood himself and of superior quality. There were also optical and mathematical instruments, two fowling pieces, a brace of pistols and a gun.

The marine engines mentioned by Simon Goodrich were indeed destined for a boat on the River Saone in the South of France. The main route from Paris to Marseille and the south west of France had for some time relied on land carriage as far as Chalons-sur-Saone. At Chalons-sur-Saone a boat was taken down the Saone to Lyon, where a larger boat was boarded to travel down the Rhone to Arles and the Mediterranean. Steam Boats had been introduced at Chalons in 1826. In 1830, a major French transport company known as Messagerie Co. employed a local builder Galline & Co to build a new improved steam boat for the journey from Chalons to Lyon. The boat was called Hirondelle (Swallow) and was the recipient of the Fenton and Murray engines which were described as twin (8 H.P. each) coupled low pressure 16 H.P. engine, the internal diameter of the cylinders was 0.457 metres, with a stroke of 0.66 metres and 30 revolutions (strokes) a minute, all running at a steam pressure of 1 1/3 atmospheres. The boiler was rectangular with two grates and 30 metres of heating surface, nine horizontal and 21 vertical surfaces. The consumption was 125 kilos of coal in an hour. The Hirondelle was a great success and plans were made shortly after she was launched for Hirondelle 2 [37].

The January 31st 1831 edition of Aris's Birmingham Gazette carried an advert for a 10 H.P. engine with two wrought iron boilers by Fenton and Co, Leeds. The enquiries were to be made with Mr Radenhurst at the Nelson Hotel.

John Watson, the colliery viewer, did not confine his activities to the coalfields of Northumberland and Durham. In 1819 he had become the viewer for the Garforth Collieries to the east of Leeds. These collieries were owned by the Gascoigne family and the estate was to be crossed by the Leeds and Selby railway. Due to then current pits being nearly exhausted Watson had decided to sink a new pit near a feature called the Hawk's Nest. The first pit in this new series was named the Isabella Pit after Mr Gascoigne's eldest daughter. The three main foundries in the area, Fenton, Murray & Co, Low Moor and the Bowling Ironworks, were asked to quote for a pumping engine and a head winding engine. The pumping engine was to raise water 140 yards using 18 inch pumps and working at 15/16 strokes a minute. The winding engine was required to bring up 45 corves weighing 6 tons each every 12 hours with an added capability of taking pumping apparatus. Fenton, Murray & Co and Low Moor put in bids for both engines, and Bowling for the winding engine. There were obviously unrecorded discussions and these led to Fenton Murray & Co providing the winding engine and Bowling providing the pumping engine. The

Bowling engine does not appear to have been finished until 1835, whilst the Fenton Murray & Co engine was started in February 1832 [38].

The Fenton Murray & Co letter which offered the winding engine was dated 1st March 1831 and described a 26 H.P. engine with one wrought iron boiler for £1000 payable by a 2 month bill drawn in London when ready to work. The price included 2 rope pulls for the pit to be hung upon the fly wheel shaft which was to be long enough to fit them and the flywheel was to have a square at the outer end to fix a crank for working the pumps, the pumps were not however included in the price. A sketch of a typical engine for that time was included.

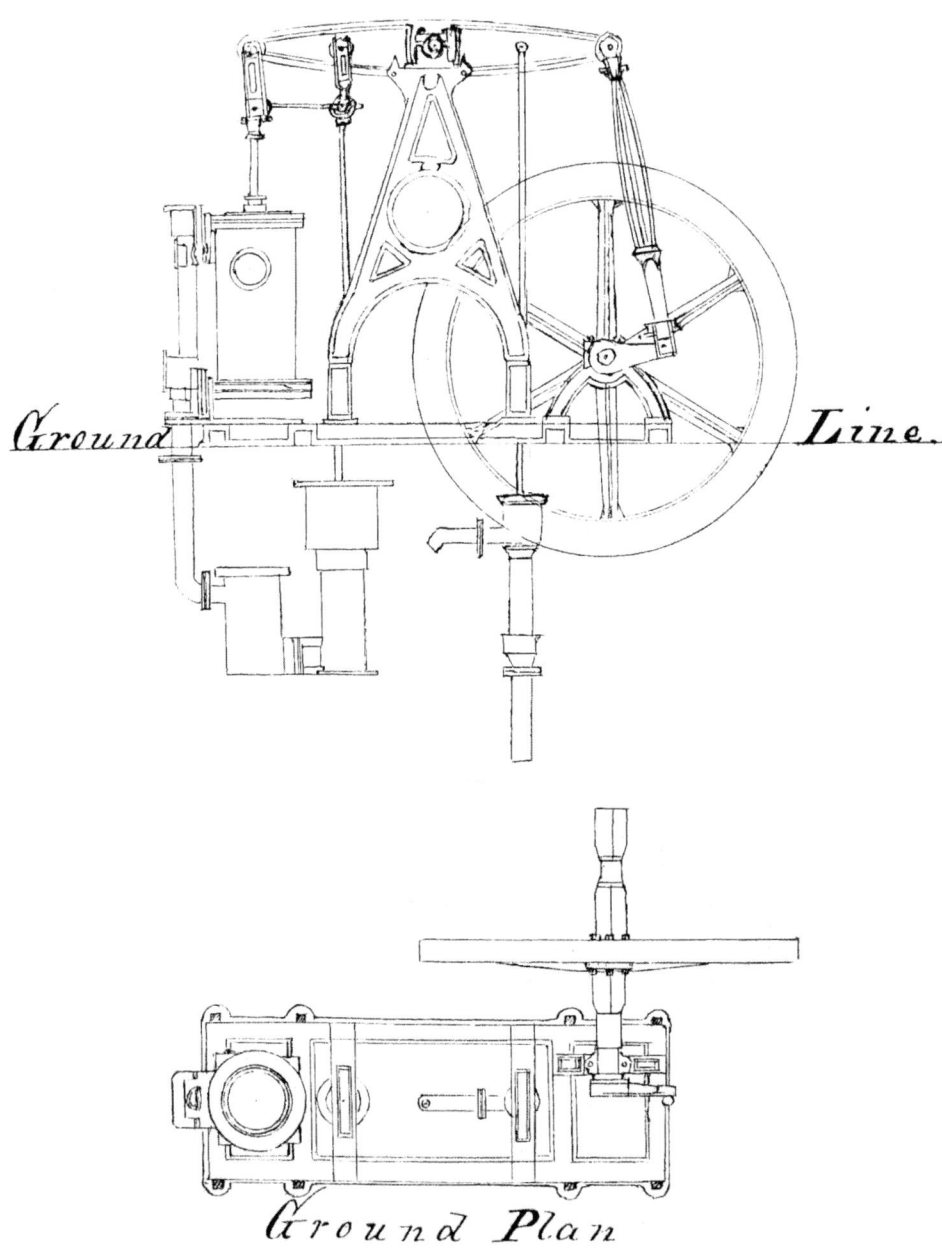

62. A frame engine as proposed by Fenton Murray & Co for Garforth Pit.

A later letter to Watson in February 1832 advising that their engineer Mr Miller would have the engine ready for a trial the following Monday but that the reservoir would not be finished by that time. Fenton Murray & Co's part of the pump work would also be ready by then or a week later [39].

The firm's quote for the pumping engine also survives, dated March 8th 1831, and followed a visit made by Richard Jackson to see John Watson. The engine they proposed was to have a cylinder with a 78 inch diameter and a stroke of 8 feet 9 inches, and all complete with 4 substantial wrought iron boilers was priced at £3000. Payment was stipulated as to one third on delivery of the materials to Garforth, one third when the engine was erected and ready to work, with the final third when the engine had been running for 6 months. A list of materials was supplied with the quotation. However this quote lost out to an offer from the Bowling Ironworks [40].

The Hull papers carried a significant announcement on March 31st for Fenton Murray & Co, with the news of the launch of the Steam Packet Transit. Transit was the second largest steam packet built in Great Britain at that date, only exceeded in size by the S.P. United Kingdom. She was 149 feet long with a 41 foot 5 inch beam, and a tonnage of 400. She had a well fitted cabin and berths for 24 passengers. The twin engines were of 80 H.P. each and the paddle wheels gave a 5 foot stroke all made by Fenton Murray & Co of Leeds. The boilers were however made at Hull. The boat was built by Messrs Pearson & Co of Thorne, for the Hamburg trade. These were the first large sea going marine engines to have come out of the Round Foundry [41].

By 1831 Robert Stephenson & Co were achieving success with their locomotive design known as the Planet type, to the extent that they felt the need to have some work carried out by other companies to their design. Fenton Murray & Co, as a previous manufacturer of locomotives and a foundry with an excellent reputation was selected as one of the companies to carry out such work. Fenton and Murray received from Robert Stephenson & Co a set of drawings for Planet type locomotives, which were dated 4th January 1831 [42]. As a result the firm re-entered the business of locomotive making and in May 1831 delivered a locomotive named Vulcan to the Liverpool and Manchester Railway or in the words of the minutes of that company for the 30 May 1831:

"The Treasurer reported that a new Locomotive Engine had arrived, made under Robt Stephenson & Co' s directions by Fenton & Murray of Leeds – which was ordered to be called the Vulcan" [43].

63. Vulcan for the Liverpool and Manchester Rlwy 1831.

On June 7th an Iron Steam Boat was launched on Humber bank, built by Mr J Livingstone a whitesmith. She was 62 feet long by 5 feet 4 inches broad with an estimated speed of 10 mph and powered by a 10 H.P. engine made by Murray and Wood of Leeds. The boat was noted as the first iron steam ship built at Hull, and was for trade between Goole and Castleford [44].

In the summer of 1831 there was a crisis in Galway and the West of Ireland of pestilence and famine, with deaths from starvation reported in Galway town. An Irish relief committee was established, and it reported the local subscriptions in early July in the Leeds papers. From this it is learnt that the following sums were contributed by the Round Foundry; Richard Jackson 10/-, Fenton and Murray £10-0-0, Fenton and Murray's workmen £4-1-0 [45].

In August Fenton Murray & Co delivered a second Planet type locomotive to the Liverpool and Manchester Railway; which was called the Fury [43].

The October 6th edition of The Leeds Intelligencer carried a piece entitled 'Canal Steam Navigation'. It described some experiments conducted over the previous days by the Engineer of the Aire and Calder Navigation. The subject was an iron steam tug just launched from the works of Fenton & Murray which was for use on the navigation between Leeds and Goole. The tug had been operated at 12 mph with the current and 6 mph against the current, she had towed 2 barges at 6 mph and 4 ½ mph on the canal. Her operation had not caused any injury to the banks of the navigation. The piece concluded with a statement that steam boats had been plying on the navigation for 2 years and that this proved that they were useful to inland canal navigation.

Notes

1. Newcastle Courant 6 January 1827.
2. Waggonways and Early Railways of Northumberland, C R Warn. Frank Graham 1976. It is also clearly shown on the map 'Tyne and Wear waggonways' pages 40/41 in The Evolution of Railways by Charles E Lee. The Railway Gazette 1943.
3. Watson Papers Wat 3/13/153. N.E.I.M.M.E.
4. The Leeds Intelligencer 8 March 1827.
5. Marshall Papers MS200 folio 17, 15 March 1827. Brotherton Library.
6. Marshall Papers MS200 folio 17, 14 April 1827. Brotherton Library.
7. Marshall Papers MS200 folio 17, 31 May 1827. Brotherton Library.
8. Rail 667/993 Stockton & Darlington Railway 27 July 1827. National Archives, Kew.
9. The Leeds Mercury 18 August 1827.
10. E.g. The Leeds Mercury 4 September 1827, Hull Packet 9 September 1827.
11. Marshall Papers MS200 folio 17, 9 September 1827. Brotherton Library.
12. Marshall Papers MS200 folio 17, 20 September 1827. Brotherton Library.
13. ADM 106/2164 Navy Board Minutes, ADM 106/1835 Chatham Officers to Navy Board. National Archives, Kew.
14. Marshall Papers MS200 folio 17, 11 November 1827. Brotherton Library.
15. ADM 106/2164 Navy Board Minutes. National Archives, Kew.
16. St Petersburg, A Journal of Travels to and from that Capital, A.B.Granville. Volume 2,2nd Edition London 1829.
17. The Leeds Mercury 27 December 1827.
18. The Leeds Intelligencer 26 February1828.
19. Newcastle Courant 7 March 1829.
20. The Leeds Mercury 28 March 1829.
21. The Leeds Mercury 3 October 1829.
22. St Katherine's Docks, Minute Book of the Court of Directors, book commencing 16 June 1825, 13 October 1829 contract, 30 March 1830 payment. London Docklands Museum.
23. The Leeds Mercury 31 October 1829.
24. The Fouling and Corrosion of Iron Ships, Charles F T Young. London 1867.
25. On the rise, progress and present position of Steam Navigation in Hull, by James Oldham. A paper presented to the 23rd meeting of the British Association for the Advancement of Science, 1853.
26. British Almanac 1830.
27. The Leeds Mercury 23 January 1830.
28. The first Hull/London steamer, the Kingston, in service 1821 with Brownlow & Co, had engines by Overton and Smith. See note 25 above.
29. Hull City Papers, CBRS/33/1-6. Hull History Centre.
30. Goodrich Journals B66. Science Museum Library.
31. The Leeds mercury, 17 April 1830 and 24 April 1830. In the letter advert note that the firm is referred to as Fenton and Co, not Fenton Murray and Jackson.
32. The Leeds Intelligencer 29 April 1830, and others.
33. The Leeds Intelligencer 3 June 1830.
34. The Leeds Intelligencer 12 August 1830.
35. The Leeds Mercury 25 September 1830.
36. The Leeds Intelligencer 2 December 1830.
37. Sur la navigation a la vapeur de la Soane et du Rhone, M W Manes. Annales des Ponts et Chaussees, 2nd serie (1843), 1st Semestre. Paris 1843.

38. The Aberford Railway and the History of the Garforth Collieries by Graham S Hudson. David and Charles 1971. Also see note 39.
39. Watson Papers Wat 2/3. N.E.I.M.M.E.
40. Watson Papers Wat 3/19. N.E.I.M.M.E. The information is in a letter offering locomotives for the Clarence Railway.
41. The Hull Packet 29 March 1831.
42. A Century of Locomotive Building, J G H Warren. 1923. Page 257.
43. Rail 371, typed extracts from the Director's Minutes of the Liverpool and Manchester Railway. National Archives, Kew.
44. The Hull Packet 14 June 1831.
45. The Leeds Intelligencer 7 July 1831.

Chapter 14.

New Partnerships.

At the dawn of 1832 James Fenton was in his 70th year, and he was still operating as a sole trader. His thoughts must have turned to the succession of the firm because he started changing the structure of the business. Early in the year he approached an employee named Samuel Exley to become a partner in the machine making business [1]. Samuel Exley first surfaced in the family business witnessing Richard Jackson's application for the probate of Matthew Murray's will [2]. Exley was a subscribing member to the Leeds Mechanics Institute in 1826 and a director by 1829 [3]. He must be the foremost candidate for the person described in 'National Education':

> "*A weaver who got a situation as timekeeper at Fenton Murray & Co.'s by availing himself of the benefits to be derived from this institution, qualified himself to superintend its mathematical and drawing classes, which he resigned on being taken in as a partner in the above firm.*" [4].

The Heads of Contract for the partnership between James Fenton and Samuel Exley survive and are probably representative of those that had previously governed the earlier partnerships with David Wood, Matthew Murray and William Lister. In this case the partnership was restricted to the Machine Department, namely the making of machines for preparing, spinning and manufacturing flax, hemp, tow, silk, cotton and woollen goods, and included any future manufactory that may have been set up by the partnership for making cloth from any of these fibres. It specifically excluded James Fenton's other business as iron-founder, millwright and steam engine maker.

The partnership was to be for 14 years and was to be known as Fenton Murray & Co, the name to be subject to change by agreement if any new partners were admitted. Exley was to manage the concern and not to engage in any other business. He was to receive a salary of £150 for the management of the concern, and was to share profits and losses to the extent of one sixth. While his accumulated capital was below £2000, Exley could take out annually his salary, balance of interest and one third of his profits, when his capital exceeded £2000 he could withdraw up to two thirds of his profits.The business was to use the capital assets currently employed by the Machine Department and pay the Foundry an annual rent of £400 for them. An interest account of 5% was to be kept in the Foundry books of all parties' capital. Exley could invest money in the business at 5%. He could draw out of such money £100 at his convenience. He could withdraw larger sums by giving one to three months notice to the Cashier at the Foundry.The Foundry was to act as banker for the Machine Business, which was not to have its own bank accounts. The Foundry was to have priority for all orders for iron, brass and malleable iron unless they could not meet the required delivery, or if better quality could be obtained elsewhere for the same price.

The Machine Business was to pay 25% of the rates and taxes of the site until a proper valuation was done. It was to have its own gas meter and pay its bills. They were to pay a fair proportion for all the carts and wherries used. Exley was to pay £10 per year for the watchman's expenses.

All partners were to have access to the partnership books and Exley was to have access to those of the Foundry books affecting him.

Exley was not to make commitments above £150 without the consent of the partners. If he withdrew money from the partnership in excess of that laid down in the agreement he would be charged interest at 20% per annum so long as he was in debt.

If any of the partners did not wish to renew the agreement when it reached its term, they were to give 12 months' notice of their intention.

If the partnership was dissolved at the end of the term, Exley could withdraw £1000 of his capital at the end of the term and quarterly sums of £250 with interest thereafter. If Exley died during the term of the partnership his estate could draw out £250 at death and £200 more within the year. The remainder was to be taken in equal quarterly instalments of £200. No profit would accrue in the year of death, but 10% interest would be paid on capital until the date of death. The partnership was not to be ended by the death of any of the partners in the Foundry. Any dispute that could not be resolved internally was to go to arbitration, one umpire to be chosen by Exley the others by his partners [1].

Two or three months after this agreement between Exley and Fenton, Richard Jackson was admitted to the partnership for the Machine Business on the same terms. However at the same time James Fenton and Richard Jackson agreed to become partners in the Foundry, Millwright and Engine business, which partnership was styled Fenton Murray and Jackson [1]. The partnership terms for Fenton Murray and Jackson have not been found, but with James Fenton as the senior partner, it is unlikely that they differed from those of the Machine Making Business to any great degree.

The cash book for the Leeds and Selby Railway records a payment to Fenton & Co on January 28th 1832 for Iron Castings for a tunnel [5].

At the meeting of the Institute of Civil Engineers on the 31st January 1832 on the subject of whether it was best to use hot water, air or steam for space heating, Mr

Walker made an aside on the subject of the strength of cast iron. In order to prove the quality of some cast iron chairs being supplied for a railway by Messrs Walker and Yates he had had some of the material recast into two straight bars which were meant to be 1 inch square, however one of the bars was about 1 1/16 x 1 inch. The trial was made at Messrs Fenton & Co's Foundry in Leeds. The bars were suspended in the middle of a steelyard with their ends held down, and a weight was moved along the arm of the steelyard. A length of 3 feet (between the fulcra) of the off square bar, bent one inch with 10 cwt and broke with 11 cwt 1 qtr and 16lbs. An 18 inch length of this bar broke with a weight of 21 cwt 3 qtrs. The 1 inch square bar broke with a weight of 10 cwt and 2qtrs [136].

John Watson was doing some work for the Clarence Railway, which had received its first Act of Parliament in May 1828. The railway's initial line, opened in 1833, ran from Port Clarence on the north bank of the River Tees near Billingham to Simpasture junction on the Stockton and Darlington Railway near Newton Aycliffe. A letter survives from Fenton Murray and Co, dated February 21st 1832, quoting a price for Locomotive engine, equivalent to those the firm had furnished to the Liverpool and Manchester Railway. The price exclusive of a tender was £775. They gave the power as 16 to 17 H.P. depending on the pressure of steam and its management; and they thought that a fair speed on a good and level road with a 20 ton load was 15mph, and with a 50 ton load 9 mph. They added that to go up a hill of 1 in 100 at these speeds the engine could only manage about 1/3rd. of the weights given. Fenton and Murray went on to say that they knew of no gradient steeper that the one on the Liverpool and Manchester Railway at 1 in 96, and so could not advise the level of gradient that would bring the engine to a standstill. In the event no order materialised for Fenton and Murray for the Clarence Railway, which used very locally made engines when it introduced regular steam traction in about 1835. The letter concluded with information on the current engine works at Garforth Pit which have already been noted [6]. Fenton and Murray may well have ceased to be considered for the work after the death of John Watson on the 16th May at his home at Picton Place in Newcastle [7].

The Leeds Mercury for the 26th May 1832 carried another letting advertisement for the fireproof mill left in Matthew Murray's will which it now described as "near to the Steam Engine Manufactory of Fenton Murray and Jackson, to whom apply for particulars". This appears to be the first public use of the new partnership name, and the timing supports the details given by Exley in his submission to Chancery.

On the 4th June 1832 the Reform Act was eventually passed after 15 months of tense political activity, with more than one prime minister resigning, and serious rioting in the last quarter of 1831 (the Bishop's palace, prisons and 45 houses were razed to the ground in Bristol alone). The reforms were to change the constituencies that could elect an M.P. to Parliament, and to change the franchise. 56 boroughs (the Rotten Boroughs) which no longer carried enough houses to justify their own M.P. were abolished, a further number which had two M.P.s lost one of them and the number of urban M.P.s was increased and the voting rights of individuals in towns was franchised at adult male occupiers of houses with a rental value of £10. The county franchises were more complicated. It was a reform, but not a really momentous one in terms of what followed; however it was extremely popular as it demonstrated a change to the status quo. Leeds along with many other places held public rejoicings culminating in the 'Great Reform Dinner' on the 15th of June, attended by Mr Macaulay and Mr John Marshall (the M.P.s). Tickets to the dinner were presented to many workforces in Leeds by their employees including Marshall & Co, Fenton Murray & Co, Taylor Wordsworth and many others [8].

Two locomotives from England had been ordered in early 1832 for the first French railway to operate with steam locomotives the Andrezieux-Roanne line. This railway had already used some early Stephenson locomotives (1828) and French locomotives by Marc Seguin (who first defined the use of multi tubed boilers) and possibly one by Hallet. One of the 1832 locomotives was ordered from R. Stephenson & Co, the other from Fenton and Murray. The Leeds engine was delivered about the 5th August 1832, and was named Jackson. The locomotive was recorded as differing little from the Stephenson one except that the cylinders were horizontal rather than slightly inclined. The weight was less than 6000 kilograms, and it was thought she would return a higher maximum speed. The locomotive had only one driving axle [9].

On the 3rd August 1832 the Directors of the Leicester and Swannington Railway discussed the need for more locomotives. They sent enquiries as to price and delivery to Robert Stephenson, Rothwell Hick and Rothwell, Horsley Co. and Fenton Murray and Co. All the companies responded, and the offers were read at a Board Meeting on the 17 August, at which Robert Stephenson was present. At a later date the order was placed with Stephenson & Co [10].

In November the Low Close Mill at Sheepscar in Leeds was advertised for auction at the end of the month along with its 30 H.P. engine by Fenton and Murray [11].

On the 26th of November the Liverpool and Manchester Railway recorded that due to an accident to the Fury locomotive they were short of engines and Mr Dixon (the Resident Engineer) was to go to Leeds where Fenton and Murray had lately had an engine for sale. On the 3rd December the Board had a letter from Mr Dixon advising that he had procured for £780 an engine equivalent to the Vulcan from Fenton and Murray to be delivered in January 1833. Another letter from Dixon, read out to the Board, complained that Simon Fenwick, a driver, had gone from Liverpool to Manchester in one hour and 8 minutes with the 10 o'clock train led by the Vulcan, which he (Dixon) considered a dangerous speed. Fenwick was ordered to appear before the sub-committee the following week [12].

A letter addressed to Richard Jackson dated the 19th December from the South London Water Works referred to an earlier discussion and asked the firm to look into the viability of installing a larger pump or some supplementary pumps to be worked off their existing engine [13]. They quoted some instances of pumps operated at other companies in London.

During 1832 the firm continued to supply marine engines to France. An engine was provided to a major boat builder M. Guibert at Nantes for a boat built to ply on the Loire. This was a single 8 H.P. engine with a cylinder diameter of 0.432 metres and a stroke of 0.584 metres, making 35/38 strokes per minute with a pressure of ¼ of an atmosphere (3.66 lbs per square inch). It consumed a hectolitre of coal an hour [14].

On the 25th January 1833 the South London Water Works wrote again and stated that they had yet to resolve the way forward with the new pump, but were 'particularly obliged' by his letter of the 3rd January. They were in need of a new boiler, and gave the dimensions, and could he let them know the terms they might obtain in the Leeds area for such a boiler [13].

At the board meeting of the Liverpool and Manchester Railway on 4th February the treasurer reported that the new locomotive from Fenton Murray & Co had commenced operations. The new engine was to be called Leeds.

James Garth Marshall wrote to his father, John Marshall snr, on 10 February 1833 and a sizeable part of his letter relates to his position that Marshalls needed to reduce its outside orders for machines. Whether his proposals were fully implemented is unclear. After some other matters he detailed discussions he had had with Fairbairn over a new variation of a machine and then moved on to the general machinery needs and who was making what. They were employing a firm called Gon to make machines for the Ditherington Mill and in Leeds they had orders placed with Fairbairn, Fenton and Murray, Maclea and March, and Mirfin. When these orders

310

were completed that year, he proposed that no further orders went to Gon, Fenton and Murray, or Maclea and March, as Marshall's requirements were nearly fulfilled. Gon was stated to have no claim on Marshalls so the orders would stop. In regard to Fenton and Murray it was seen that they had reasonable prospects as they had Hives' new mill to satisfy and that of Mulholland in Belfast who was building a very large mill with an increased power of 180 H.P. of steam engines. There were also numerous new firms starting up. Fairbairn would have reduced work but it had been agreed with him which machines he should make exclusively for Marshalls and he had been encouraged to sell machines from the rest of his portfolio to other manufacturers. The fate of Mirfin is not disclosed [16].

On the 23rd February Fenton Murray and Jackson were paid £66-10-5d by the Leeds and Selby Railway for some more iron castings for a tunnel [5].

Fenton Murray and Jackson had replied to the South London Waterworks with a quote, which the latter accepted by a letter dated 22 February 1833 addressed to Mr Dickinson. John Mavin Dickinson was the accountant and cashier for the firm until the closure of business. He also started to undertake a fair amount of the correspondence. He was obviously a trusted member of the management. The boiler in the accepted tender had a 3 foot diameter centre tube of 3/8 inch thick iron, a bottom 7/16 inch thick, sides 3/8 inch and a top of 5/16 inch. It was stipulated that it should not weigh less than 6 ½ tons and was to have a manhole. The price was £175 delivered to the waterworks [13].

On March 7th Dickinson wrote to the Stockton and Darlington Railway concerning their order for suction pipes for the Force Pumps at Middlesbrough, and advised that they were being made and it was better to wait until they were finished before sending any fitters. He stated that Mr Jackson was in London [17].

The Leeds Times for March 28th celebrated the first instance of an export direct from Leeds through Goole. The Iris, Captain Burkenshaw commanding, took a pair of 20 H.P. marine engines by Fenton Murray and Jackson to Havre de Grace (Le Havre). They were for a steam vessel to ply on the Rhone (both the Leeds Times and Intelligencer called it the Rhine). A drawing in the Arts et Metier archives in Paris probably identifies these as ones supplied to a M. Theodore Paul at Chalons-sur-Saone, the drawing states that the engines were imported into France in March 1833. Paul was a cousin of M. Galline whose boatyard built the Hirondelles. It is therefore likely that the engines worked another boat on the Chalons to Lyon run. Though the more powerful engines may well have taken such a boat further afield; to Arles for example.

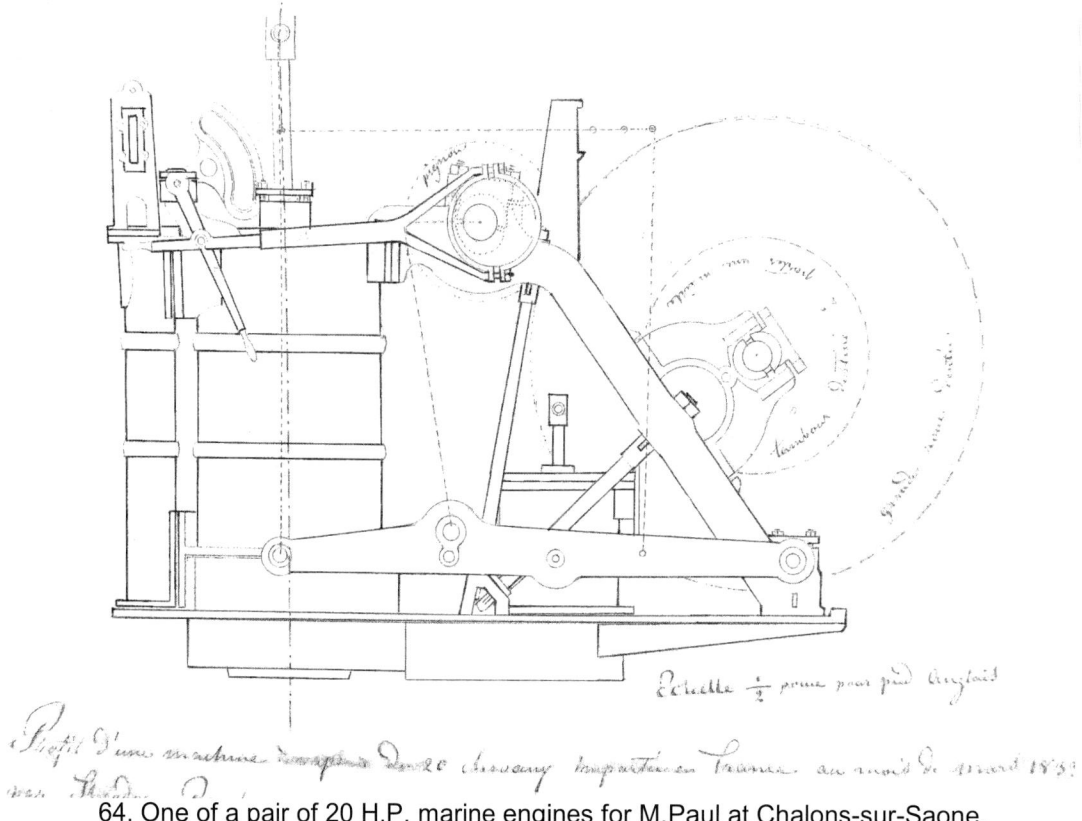

64. One of a pair of 20 H.P. marine engines for M.Paul at Chalons-sur-Saone.

In April the results of the Leeds elections for the Overseers (of the Poor) were reported and included Peter Fairbairn for the Lower North West Division and Samuel Exley for Holbeck, both stated to be machine-makers.

The outcome of an experiment on a railway in France was reported in the French and the Leeds papers in May, and widely picked up by various publications including Niles Register in the United States. The experiment was upon the Loire Railway (Andrezieux – Roanne) using a goods locomotive engine manufactured by Fenton Murray and Jackson. The engine travelled with a load of 15,000 kilograms (nearly 14 tons), inclusive of engine, tender water and fuel. The engine climbed a gradient of 4.5 in 100, using steam pressures no higher than 38lbs per square inch. The gradient was around 2184 yards long and the engine ascended it in 6 minutes and descended it immediately afterwards with great ease, regulating its speed on the descent [18]. It is a reminder of how experimental the early railways were, and this appears to be the first improvement over a trial carried out on the Liverpool and Manchester Railway on the 1 or 2 in 100 gradient that Fenton and Murray had used to support their position in their letter to John Watson in early 1832. An illustration of one of these Fenton Murray and Jackson engines is given in a French book (and reproduced below) [19], however it appears to be a copy of the standard illustration of a Planet locomotive in goods configuration.

312

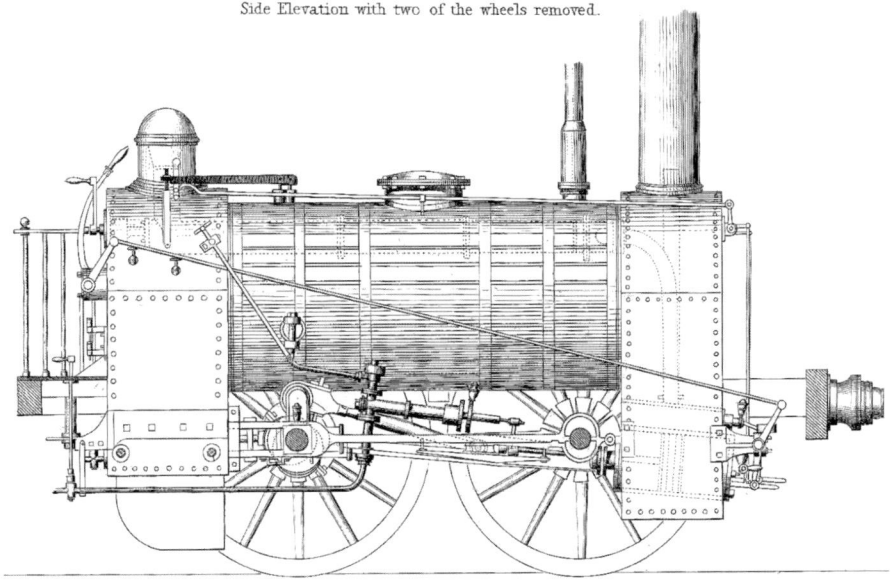

JACKSON'S LOCOMOTIVE ENGINE HAVING FOUR WHEELS & TWO LOOSE ECCENTRICS.

Side Elevation with two of the wheels removed.

65. Fenton Murray & Co 0-4-0 locomotive as supplied Andrezieux-Roanne Rlwy.

At the end of June the South London Water company wrote that they had received the boiler, but that there was no manhole cover and they would be grateful to know when this essential item would be delivered [13].

A 20 H.P. Fenton and Murray engine, a 30 H.P. engine by Boulton and Watt and a 10 H.P high pressure engine were advertised for sale in many papers including the Morning Chronicle of July 17th, being now surplus to requirements at the London Docks. Particulars could be had and viewing given by attending the office of the Superintendent at the London Docks.

The Charleston and Hamburg Railway in South Carolina, which had commenced operations in 1831 with indigenous locomotives, reached Hamburg in 1833 and by 1834 the line had so much freight traffic that it was in need of more locomotives. By the time it reached Hamburg the line was 136 miles long, and was the longest railroad in the world at the time. A Mr William C. Molyneux of Liverpool was employed to purchase some English locomotives. Molyneux appears to have been a merchant and agent trading with the Eastern Seaboard of the United States, and an importer of cotton bales from Charleston. A Memorandum of Agreement between Molyneux and Fenton Murray and Jackson was agreed on the 1st August

1833 for the supply of a locomotive steam engine. The price for the engine was £800 and £200 more for spare parts. The agreement specified that

- The flanges of the wheels were to agree with the drawing, the track being 5 feet with ¾ inch play allowed. The touching parts of the flange surfaces to be 4 foot 11 ¼ inches apart
- The fuel was to be pitch pine with a large proportion of pitch/ turpentine. The flues to be adapted to cope, and soot to be readily cleared out.
- The empty weight of the engine was not to exceed 6 tons.
- The centre of gravity was not to be higher than 4 feet above the rails.
- The engine was to climb gradients of 30 feet per mile (1 in 176) and pass curves of 500 feet radius.
- The power of the engine was to enable it to move five times its weight at 12 mph.
- An engineer to be sent out to run the engine for a year at not exceeding £3 sterling a week.
- The spares were to be made up of an iron fire box the same as on the engine, brass bushes for all the wearing parts, 12 brass pipes and ferrules, two 4 ½ feet wheels complete, 1 crank axle complete, 1 eccentric tumbler with two brass hoops, taps and dies suitable for the bolts of the engine [20].

At this time, presumably also through Mr Molyneux, the Charleston and Hamburg ordered 3 locomotives from E Bury and 3 from R Stephenson [21].

The Norfolk Chronicle carried a sale notice for equipment no longer needed on the Navigation Works at Lowestoft amongst which were two capital 6 H.P. engines by Fenton and Murray, which had been used for pumping on the navigation and were as good as new. Particulars could be had of Mr William Cubitt, and others [22].

On November 12th 1833 Fenton Murray and Jackson submitted a quote to the French Navy for marine engines [23]. The technology in the French steam engine builders at the time was lagging behind that available in the United Kingdom, and there was a deliberate policy to buy from the U.K. to acquire the technology.

The Leeds Times covered a farewell dinner for Benjamin Cubitt held on November 21st at the Woodman Inn Holbeck. The dinner was held by the workmen. Benjamin Cubitt was presented with a silver snuff box inscribed;

"Presented, November 21 1833 to Mr Benjamin Cubitt, by the workmen in the employ of Messrs Fenton Murray and Jackson, engineers Leeds as a token of the high estimation in which he was held during the seven years he was their manager."

On December 21st The Leeds Intelligencer stated that the machine makers of Fenton Murray & Co had become annual subscribers (for 2 guineas) to the Leeds General Infirmary. They hoped the example would be followed by the workforces of other large establishments. This was therefore presumably unusual in being made directly by the workforce as opposed to the firm.

The Benyon's flax mill at Shrewsbury had presumable ceased to be profitable as it was advertised to be let with immediate possession in January 1834 along with its 75 H.P. steam engine by Fenton and Murray [24].

On the 28th of January 1834, just 3 ½ years after the previous one, a fire broke out at Jonas Brown's sawmill next door to Fenton and Murray. The fire was spotted by some women who were retiring for the night, and raised the alarm. However there was a great stock of timber being held on the premises and the fire raged throughout the night not being really got under control until the morning. Had there not been prompt assistance from the fire engines from Marshalls, Fenton Murray and Jackson, Titley Tathams and Walker, and the public Fire Offices, the fire would have spread to Fenton and Murray's premises and some adjoining cottages [25].

The Liverpool and Manchester Railway Board meeting of the 3rd February 1834 heard a letter from Fenton Murray and Jackson read out offering to supply a locomotive for £1000 to be delivered in 6 or 7 months. A letter from Edward Bury also offered an engine to the same specification for £1140. However no further orders were placed by the Liverpool and Manchester with Fenton Murray and Jackson [26].

The Sheffield Independent dated February 8th carried notice of the sale, due to bankruptcy of Samuel and Thomas Darwin, of the Soho Works in Sheffield, an iron and steel rolling mill powered by a 30 H.P. Fenton Murray and Wood engine.

The 20th of that month saw a letter from the secretary of The South London Water Company acknowledging Fenton Murray and Jackson's letter and account for a new boiler which he said he had had to refer to the previous secretary for clarification, possibly due to an additional charge for the manhole cover of £1-12-7d [13].

In March 1834 Messrs Fenton and Murray were refused permission to export flax spinning machinery to Antwerp by the Board of Trade. While there were quite a few requests by other Leeds firms to export various articles this seems to have been the only appearance in the lists of either firm now working at the Round Foundry.

There are plenty of requests from Brownlow and Pearson, the shippers, however, so permission may usually have been sought through that route [27].

On the 10th May 1834 Richard Jackson wrote to Beckett's bank authorising John M Dickinson to sign on the account of Fenton Murray & Co [135].

In May the York papers carried a description of the works to be undertaken by Thomas Rhodes, an engineer of local origin, to improve navigation on the Yorkshire Ouse. It advised that a steam dredger was being built by a Mr Teal with all the machinery being made by Fenton Murray and Jackson, and was to be ready by August [28].

At the end of May James Dufton of Leeds was offering for sale an Hydraulic Bale Press made by Fenton Murray and Wood. And in the middle of June the particulars of a 2 H.P. steam engine with boiler and governor, currently working, by Fenton and Murray, could be had on application to the Leeds Mercury [29].

Fenton Murray and Jackson wrote out an introduction for the French Engineer Marc Seguin and his companion on June 24th to Messrs Pease & Co of Darlington, of the Stockton and Darlington Railway. The letter explained that Seguin was visiting England to gather information on railways to assist him in his work on the Lyon – St Etienne line in France (this line was connected with Andrezieux – Roanne line) [30]. It may well have been this opportunity when Seguin ordered a locomotive from Fenton Murray and Jackson for the Lyon – St Etienne, which was delivered in 1835.

July 1834 saw adverts for a "Valuable freehold property in Sweden". The property was about 3 miles outside Gothenburg and included an Ale and Porter Brewery, which was worked by an 8 H.P. engine built by Fenton and Murray [31].

July also saw Fenton Murray and Jackson receive payment from the Leeds and Selby Railway of £257-2-11d for fencing plates for bridges over the railway [5].

During the same month Robert and John Campion were offering for sale a 12 H.P. cylinder with side valve, throttle valve and side pipes in excellent working order, made by Murray & Co. It was used in their flax mill in Whitby, and the engine must have been upgraded as when the mill went up for sale some years later it had a 16 H.P. engine by Murray & Co [32].

Within many of the local papers of the realm, there appeared adverts for the patent of one J. McDowall (number 6606), which had been granted on the 12th May 1834. The patent was for a new form of metallic piston and air pump bucket. As usual with such adverts at the time the firms that were adopting the patent were given in order to support the usefulness of the product. In this case among those listed were Fairbairn of Manchester, Sherrats of Manchester, Marshalls of Leeds, Maclea and March of Leeds, Fenton Murray and Jackson, and Benyon & Co of Leeds [33].

316

The minutes of the directors meeting of the South London Water Works on the 5th August 1834 stated that it was necessary to order an engine, of similar power to that currently in use, with as little delay as possible. The secretary was minuted to write to Fenton Murray and Jackson, Boulton and Watt, Hicks of Bolton and Mr Fairbairn at Manchester to request their prices and timescales. A letter was duly sent to all four parties:

"I will thank you to let me know, for the information of the directors, of this Company, what would be your price for a Steam Engine of 40 Horses power, two boiler &c &c, to correspond with the one of that power at present in line at these works. Also the shortest time in which you would undertake to deliver, and fix, the same at these works. Our present Engine may be seen here, by your Agent any day between 8 o'clock in the morning and 5 in the afternoon." [34].

On the 15th August some new competition to the Round Foundry announced itself. Newton Taylor & Co of the Globe Foundry were in business to make flax and tow machinery and were set up as brass and iron founders. The proprietors were ex-employees of Fenton and Murray [35].

On Thursday the 21st of August the dredging vessel was launched (on time) on the Ouse Navigation. This was the vessel referred to earlier, which had now been built by Mr Teall to an improved design by Thomas Rhodes. The engine and machinery were by Fenton Murray and Jackson [36].

James Garth Marshall's letter of February 1833 implied that Fenton Murray & Co were and would continue to supply machinery to mills in Ireland (Mulholland and others). Further evidence is supplied by an advert appearing in The Belfast News Letter of September 2nd 1834. Mr Thomas Evans had been granted Letters Patent (No 6361) on 10 January 1833 for machinery for dressing flax and other fibrous materials. He had now taken space in this Belfast paper to state that his machines were now being made by Fenton Murray & Co and Maclea and March. Both companies were authorised to take orders for the patent machines one of which could be seen at work at Mr Sugden's manufactory in Dewsbury Road Leeds. So while the evidence is tenuous, it would appear that the firm were active in the Irish market, despite the indications that most of the machinery for Ireland was copied by local firms.

On the 4th of September Fenton Murray and Jackson had delivered a locomotive, which was named Nelson, to the Leeds and Selby railway. The engine was 'launched' on the 5th and was rated in the press at 18 or 20 H.P. according to the Leeds Intelligencer. It was taken on a run from Leeds on the 6th with 25 people being carried in an open carriage, this may have been a trial by the directors .The Bradford Observer tells that it was tried out several times from the depot at Leeds before

September 11th, and even had a slight accident with being 'run too far before it was stopped' [37].

The Directors of the South London Water Works unanimously resolved on the 17th September to give the engine order to Fenton Murray and Jackson in line with their tender of the 15th August. A letter to this effect was sent to the firm on the 18th September, with letters being sent to Fairbairn and Hicks to inform them they had not won the business [34].

The Norfolk Chronicle advised that an elegant steam pleasure boat, the Salamander, was for sale on the 24th September. She had a length of 30 feet, a breadth over the paddle case of 10 feet and drew 18 inches. She had a speed of 6 mph, a boiler by Fenton and Murray of Leeds and an engine on the most improved and simple construction (but the maker is not given).

By the 18th of September the Leeds and Selby Railway was so far completed that the Directors travelled to Selby and back. The train of four carriages was pulled by the Nelson. Everything went so smoothly that the Directors met on September 19th, a Friday, and decided to commence operations on Monday the 22nd. The first train was to leave the depot at Marsh Lane Leeds at 6 a.m., with the return journey to commence from Selby after the steam boats had arrived from Hull (normally about 11a.m.). The line was duly opened with a train that comprised between seven and nine carriages (according to different newspapers), carrying around 150 to 180 passengers. At 6.30 a.m. the locomotive Nelson was attached. The train made heavy weather along a wet track towards Selby, with the locomotive 'labouring extremely', such that the first 4 ½ miles took about an hour to cover. The remaining 15 miles were done in about 45 minutes. The return journey left Selby at 11.14 a.m. and reached Leeds at 12.27 p.m. The woeful performance on the outbound trip from Leeds was only explained by the Leeds Times. The axles were of too great a diameter for the bushes in the wheels, which stopped rotation and caused the wheels to slide on the rail. At Selby the wheels were removed, the axles were filed and all was well. The Leeds Times also calculated the average speeds for the outward and inbound trips, the outward being done at 8 ½ mph and the return at 16mph [38].

On Tuesday morning a return trip to Selby was without incident, as was the outbound journey on Tuesday afternoon. However on the return journey the throttle valve rod was found broken at Garforth, which resulted in the train having to be drawn back to Leeds by horses, not arriving until 9 p.m. The Nelson was repaired overnight and resumed service on Wednesday morning.

It would appear that the Nelson was the only locomotive which had been delivered to the Leeds and Selby at the time service commenced. Nelson was the only locomotive ordered from Fenton Murray and Jackson at that stage, the remainder were placed with E Bury. By the 3rd of October Bury's report to the board indicated that his first engine would be delivered on the 5th November against a contract date of March! Bury promised two more in November and one in December [39]. Mr Gott and Mr Marshall, after being appointed by the Board, promptly went out and bought a second hand engine called the North Star along with a tender from William McKenzie for £354-17-6d [40].

Horsforth Mills, a scribbling mill, near Leeds, went up for sale towards the end of October along with its 10 H.P. Fenton and Murray engine [41].

Towards the end of 1834 the Fenton and Murray 2-2-0 locomotive was delivered to the Charleston and Hamburg Railway in the U.S.A. and went into service under the name Columbia [21].

At a meeting on the 11th January 1834 the Works Council for the French Navy had decided to ask Fenton Murray and Jackson for their engine plans; as they recognised their engines advantages in terms of weight, strength, compactness and price. It was also decided to send the French naval engineer M Faveau to England to make a survey of the engines and factories. On his return M. Faveau produced a report dated 30 September 1834 in which he recorded his visit to Hull to see the Transit which had a 140/160 H.P. engine apparatus by Fenton Murray and Jackson. He also stated that he could see the advantages of the machine. The Works Council decided on the 2nd October to adopt Fenton Murray and Jackson's engines, as progress in design demanded that such English engines must be trialled. The offer was for a price of 170,000 francs for a 160 H.P. apparatus, 240,000 for a 200 H.P. one and 262,000 francs for 240 H.P. An order for a 160 H.P. apparatus was placed on the 15th November 1834 [23].

Another nationwide series of adverts appeared in the press and technical magazines in November, this time in regard to Samuel Hall's patent improvements to the steam engine (patent no 6556 of 13 February 1834). Owners of engines were invited to contact a list of licenced manufacturers for either new engines or to have their old ones altered. The list included most of the major steam engine manufacturers in England including Fenton Murray and Jackson [42].

In November the Whigs called a meeting in Leeds, which the mayor would not sanction, but it went ahead anyway at 12 mid-day on the 24th of November. The main speaker was Mr Baines, a prominent Leeds citizen, and proprietor of the Leeds Mercury. Reports reached The Morning Post in London. The meeting seems to have been singularly ineffective, and was timed to allow the employees of various manufactories to attend in their dinner time, among whom where those of Fenton and Murray. It was reported that one of Fenton Murray's men said to a person in the crowd:

"There are a hundred of us here, but not one in ten of us, if we had votes, would give them to Baines, if we could do as we liked" [43].

The implication being that if there were secret ballots, Mr Baines would not get elected. The secondary point was that if workers had a vote, there was often pressure to vote as their employers demanded.

The report in The Leeds Intelligencer, Baines main rival, spurred a letter from Fenton Murray and Jackson, which the Intelligencer alluded to in their issue of December 13th:

"Messrs Fenton Murray and Jackson deny that they ever intimidated their workmen to vote contrary to their will. We never said they did. These gentlemen need not be so "outrageously virtuous". It is but very rarely, they say, that they read the Intelligencer. That is no proof of their wisdom or good taste."

On the 17th of December 1834 James Fenton died in his 73rd year [44], never having married.

On December the 30th the Leeds and Selby Railway paid Fenton Murray and Jackson £900 on account for the Nelson and tender [5].

James Fenton's will left the bulk of his estate to William his brother, after four legacies to nieces, nephews and friends [45]. William was living at Peckham Rye in Surrey at the time. Samuel Exley informs us that new partnership arrangements were drawn up for the machine business and signed by the 10th January 1835. The only other change for Fenton Murray & Co was that Exley was now to receive a salary of £125 a year for managing the business and was to share profits and losses to the extent of 1/5th. His capital for interest purposes was to be limited to £3000 [1]. No doubt the foundry and steam engine business were as rapidly put on a new footing.

Fleet Mills, near Leeds, was selling off equipment in January 1835, including a 5 H.P. engine and a 4 H.P. engine by Fenton and Murray [46].

Mr Dean, a wharfinger of Bingley, had three boilers for sale in February: a 26 H.P. in very good condition by Fenton and Murray, and two 18H.P. boilers in excellent condition by Boulton and Watt [47].

On the 2nd March Fenton Murray and Jackson delivered the drawings for the marine engines for the French Navy to the consulate (presumably in London) [23].

In March various adverts appeared for a Fenton Murray & Co 10 H.P. double stroke engine with pipes, pumps and all apparatus, suitable for drainage operations.

It had been used in the excavation of the North Level Drain in the Fens for over 8 miles near Wisbech and was currently standing at Tid Gote [48].

On the 25th March 1835 Fenton Murray and Jackson wrote to Mr Otley the secretary of the Stockton and Darlington Railway to express their regrets that due to the current workload Mr Jackson would not be able to conduct the matter they wished him to. This was probably to do with valuing the railway's locomotives as this is the subject of most of the letters in this file [49].

One of the matters occupying Richard Jackson's time was that he had been summoned as a witness in a civil action over a paddle wheel patent. The case was 'Morgan and another v Seaward and Others'. A Mr Morgan, who had acquired rights to a patent on paddle wheels taken out on 1829 by Elijah Galloway, a well-known steam engine engineer, claimed that the firm Messrs Seawards had built paddle wheels for a boat called Levant, that infringed the patent. Many well-known engineers testified on both sides, including the Brunels and the Donkins. Those in support of Seaward claimed that Galloway's wheels were unsafe in that the axle did not pass completely through the wheel, making it weak. Another defence was that paddle wheels to the design used by Seaward had been built in France by M. Cave (one of the premier engine builders in France) prior to Galloway's patent. Further defence was offered by evidence that three vessels were in operation on the Humber, and had been prior to April 1833, with paddle wheels to a similar design to Seawards which had all been made by Fenton Murray and Co. The vessels were described as two steam tugs belonging to the Aire and Calder Navigation (presumably the Leeds and the Calder) and the Liberal belonging to Bromley and Co of Goole. Richard Jackson was required to attest to a drawing from Fenton Murray and Jackson, that it was a correct representation of the paddle wheels fitted to all three boats. The learned argument never seemed to quite resolve whether the Fenton and Murray wheels were relevant, as under patent law they would need to have preceded the date of the patent to invalidate it, and there was no evidence they did. M. Caves did precede the patent. This evidence seems to have been more about whose veracity could be trusted in regards to the reports of a conversation between Galloway and John Seaward. On appeal and after much legal debate (including ruling that prior invention and manufacture outside England did not invalidate an English patent!), it would appear that Morgan and Galloway got the better of the judgement. This fortunately did not seem to encourage them to take a similar case against Fenton Murray and Jackson, it had probably all been much closer legal argument than they had expected. So far it is the only source of information on the apparatus for the Liberal [50].

On the 3rd of July the William Darley was launched from the ship yard of E Gibson & Co. into the Port of Hull. She was the longest vessel built at Hull to that time, being 156 feet long at the keel and 164 feet long on deck. Her registered tonnage was 436 tons. "She had two very superior engines of 140 H.P. each (actually they were each of 120 H.P.)" made by Fenton Murray and Jackson. The launch attracted a crowd of between 5 and 8 thousand people, and was skilfully accomplished at a point where the river was only 170 feet wide, but helped by the old dock basin, which was 240 feet long, being opposite the site. The William Darley was built for trading with Hamburg, to the order of Brownlow and Pearson, and seems to have established a very good reputation. She did her first trip to Hamburg in September 1835, and her return journey was made in 49 hours against a strong head wind all the way (she later achieved a record of 42 hours) and on this leg carried the M.P. of North Derbyshire. He had twice tried to leave Cuxhaven on London bound packets which had been forced to return to port by the weather [51].

On the 22nd of July Matthew Murray junior died in Moscow, at the age of 42 years. It took until the 5th of September for the information to reach the Leeds papers [52].

At about the same time that the French Navy was trying to improve the performance of its steam powered fleet, the French Government was also planning to establish a fleet of steam powered sloops to provide a service around the Mediterranean. In support of these aims they agreed a contract with Fenton Murray and Jackson on the 30th July 1835 for a pair of marine engines to power one of these ships. The contract was stated not to become effective until the 10th August, and if ratified shipment was to by 1st March 1836. The engines were again of 80 H.P. each giving the apparatus a power of 160 H.P. The French Commission executing the contracts In England consisted of Engineer Faveau and M Gay-Lussac, an eminent French scientist [53]. During the same month Gay-Lussac and Faveau signed contracts with Miller and Ravenhill of London for 2 sets of apparatus, with Maudslays for 2 sets, and with Bury of Liverpool for one set. The other engine apparatus were to be supplied in France by Hallette, Cave and the government factory at Indret near Nantes [54].

The Nautical Magazine in 1835 carried a public letter, dated 6th August Liverpool, from Richard Jackson to Samuel Hall in praise of his patent apparatus.

"I was much pleased in witnessing the working of the engines on board the "Windermere" this morning, with your patent condensing apparatus. Although I only saw them working a very short time, yet I consider you have accomplished a more perfect mode of condensation, as was shewn by the barometers attached to the condensers indicating a real vacuum of 29 inches of mercury. I consider this mode of

condensation of great value in marine engines especially those for sea voyages, where we have witnessed so much destruction in the boilers by the use of salt water. I am, dear sir, for Fenton Murray and self etc. N.B. Although the vessel was plying in salt water, I was pleased to find the water supplied to the boiler from the air-pumps of the engines perfectly fresh."

In August Cadman and Co. of Lady Lane Leeds were advertising for sale a 4 H.P. condensing engine complete by Fenton and Murray. It was in good repair and only being sold as a larger engine was now needed [53].

Fenton Murray and Jackson supplied a locomotive to the Lyon – St Etienne railway, again it was a form of the Planet engine, a drawing of the locomotive, signed by Richard Jackson on the 4th August, survives in the archives of the Arts et Metiers museum in Paris.

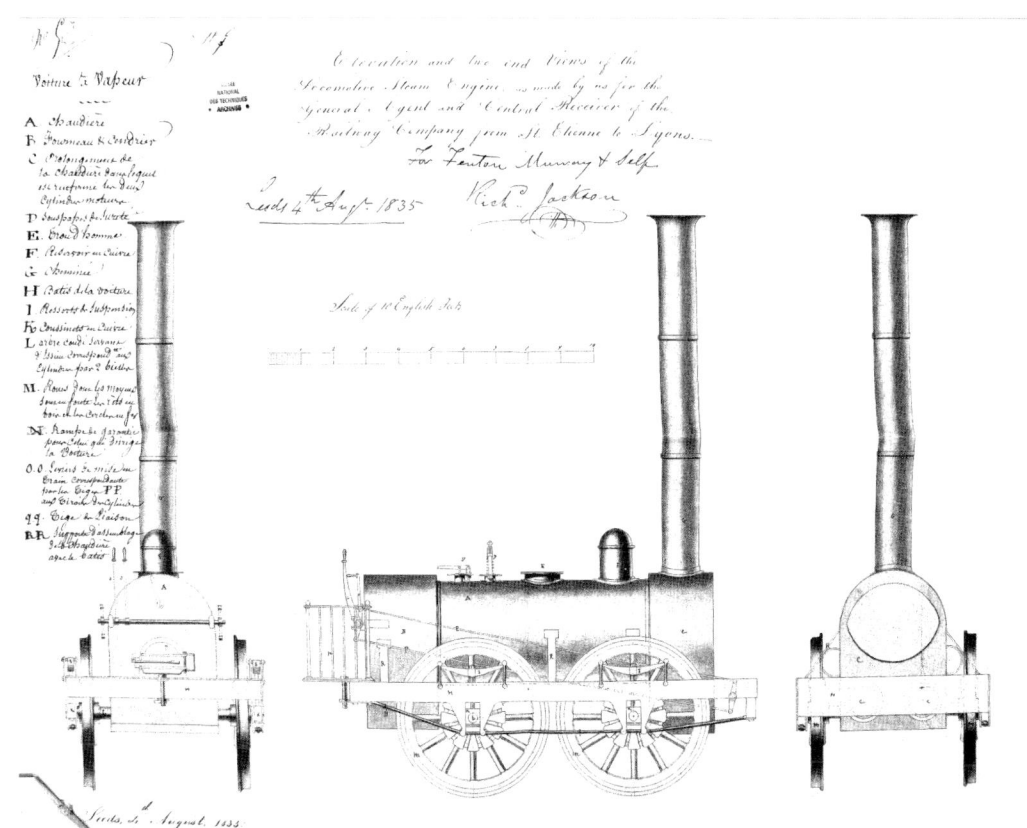

66. Fenton Murray & Co locomotive for the St Etienne-Lyon Rlwy.

Drawings dated September 1835 survive again in the Archives of the arts et metier in Paris of a twin 9 H.P. apparatus supplied to M. Guibert at Chalons-sur-Saone, which would appear to be very similar to the engines provided for the 'Hirondelles' (see Chapter 18) [56].

On October the 15th the Leeds and Selby Railway cash book recorded a payment of £10-18-3d to Fenton Murray and Jackson for work done on the Nelson and Rodney apparently at the instruction of E Bury. Rodney was supplied by Bury, so it was easier for the work to be done in Leeds, rather than Bury sending the necessary from Liverpool, and the work on Nelson may have been some harmonisation [5].

Also on the 15th the Directors of the South London Water Works approved an action taken by their Chairman. He reported that he had retained the engineer sent by Fenton Murray and Jackson to erect the new engine to carry out repairs to the old engine. The old engine had not been stopped still for three entire days in the last 13 years for maintenance and this was the opportunity to examine it and carry out any necessary work [34].

In the week ending the 9th October the Laetitia had sailed from Hull with 137 packages of machinery being parts of two marine engines. They were duly delivered around the 24th of October at Nantes, the port for Indret the home of French marine steam engines. The Fenton Murray and Jackson engines for the French naval sloop 'Papin' had arrived in France [57].

In October a house in Holbeck that was used to manufacture fireworks blew up killing five people and injuring several more. A subscription was raised and a committee appointed to relieve the sufferers. Fenton Murray and Jackson gave the average subscription but had no representation on the committee. Both Richard Jackson's bothers in law worked on the committee [58].

At a Director's meeting on the 5th of November The South London Water Works agreed to give a sum of 10 guineas to Mr Peary the engineer from Fenton Murray and Jackson for his good work on both the old and the new engines [34].

Richard Jackson did serve on the committee to try and erect a toll bridge at School Close to connect Water Lane, Hunslett Lane and Meadow Lane to the north bank and West of Leeds. It was proposed at the November meeting that a committee was formed to obtain an Act of Parliament to give authority to raise such a bridge. Among the committee were Jackson, Maclea, Thos Benyon, 2 Marshalls, Anthony Titley, Joshua Wordsworth and others [59]. The bridge was completed about 2 years later.

On December 11th Fenton Murray and Jackson were advertising for one or two brass moulders [60].

The Benyons were still trying to dispose of their Shrewsbury Mill, which was to be offered for sale at auction on the 21st December. The mill was advertised as having a 75 H.P. engine by Fenton and Murray [61].

December also saw some interesting correspondence in Berrow's Worcester Journal. There was a proposal at that time to improve the navigation on the River

Severn and it was suggested that Thomas Rhodes, who had been employed on a number of Telford's works, should be employed as engineer. There were doubts cast on the scheme, and a correspondent who had worked with Telford and Rhodes corrected some statements made about Rhodes. The ability of the required dredgers to shift as much gravel and mud as claimed, was also called into question. This prompted a response from a Mr George Willoughby, the Dredging Contractor on the River Ouse (see above), and Mr James Rose, the Resident Engineer. Willoughby stated that the Ouse Dredger frequently raised 50 tons of gravel in 20 minutes. The offer was made that the doubter would be shown the dredger in action. He confirmed that the power of the dredger was 10 H.P. and that the equipment was made by Fenton Murray and Jackson to a plan of Mr Rhodes [62].

On Saturday the 26th December 1835 the Leeds Municipal elections took place as required by the King's order in Council of 11th September of that year. Following which the Alderman were elected, and these included 12 of the people just elected as councillors. At the swearing in of the new councillors a John Wilson, who had been elected to Hunslet Ward and who was a member of the Society of Friends, refused to make the declarations required by Act of Parliament. He therefore could not take up his position. Thus on the 11 January 1836 further elections were held for 13 posts. Thomas Benyon had been elected as a councillor in Holbeck Ward, and then as an Alderman, which created one vacancy at Holbeck. There was only one candidate standing for Holbeck in the elections of the 11th January 1836, due to a very hasty meeting of the Ward Committees, and he was returned unopposed. Richard Jackson became a councillor for Holbeck [63].

One of the subjects before the London Committee of the London and Birmingham Railway when it sat on the 13th January 1836 was the acquisition of ballast locomotive engines. These engines were used to support the building of the railway, being used to move supplies and ballast etc. The committee instructed the secretary to write to Messrs Fenton Murray and Jackson for a price for delivering engines and tenders to the River Thames, which met Mr Stephenson's specifications [64]. At a meeting of the Birmingham Committee on 15th January, having decided on 5 of the 6 ballast engines required, the committee faced a choice of: an offer from Rothwell & Co of the Hercules engine for £760 including the tender, buying the Northumbrian from the Liverpool and Manchester Railway for £450, neither of which were to Stephenson's specification, or to order a new compliant engine as offered by Fenton and Murray for delivery in June 1836. They resolved to accept the offer from Fenton and Murray [65]. At the end of January the London Committee read letters from

Fenton Murray and Co and Sharp Roberts and Co which advise them that neither company was in a position to supply any further locomotives in the year 1836 [64].

In France the Papin having been built at Indret was launched on February 3rd 1836 and moved to Lorient in Brittany to be armed.

On the 26th February the Birmingham Committee of the London and Birmingham Railway read a further letter from Fenton Murray and Jackson, dated the 13th February, which also stated that they could not supply any more locomotives that year. By this time the company was considering where to place its passenger and goods locomotive orders [65].

In early March Fenton Murray and Jackson were advertising for a pipe moulder [66].

Provisional navigation permits were issued for two tug boats in March 1836 at Macon on the Saone, Macon lies between Chalons–sur-Soane and Lyon. The boats were named 'Le Jackson' and 'Le James Fenton' (recorded as Frenton) and the constructor was given as M Guibert. The engine manufacturers, unusually were not recorded, but the boat names and the delivery of engines by Fenton and Murray to Guibert at Chalons in September of the previous year, are all good indications of the provenance for the engines for these boats [67].

On April 2nd the auction was advertised of two recently erected power-loom weaving mills situated at Southowram Bank Bottom near Halifax, called Bank Foot Mills. The concern had been that of Messrs Buck and Kershaws and made Worsted Stuff. The mills shared a 20 H.P. engine built by Fenton Murray & Co [68].

The engine apparatus for the French Government Ship for service in the Mediterranean was delivered by Laetitia to the French naval port of Lorient in April 1836. The landing was dated April 17 but not signed off until May 16th [53]. The engines were installed in a ship named Lycurgue, which appears to have been built at Lorient. There should have been a contract amendment as the engines were supposed to be delivered under contract to Rochefort, but in the end the two ships built at Rochefort had engines by Maudslay. Some Government records stated that Lycurgue was built at Rochefort, but the consensus is for Lorient, and the customs record shows the engines were delivered to Lorient [54].

May 21st saw the Sheffield Independent advising of another auction for newly erected premises, this time at Washington Place in Sheffield consisting of a house, warehouses, workshops suitable for the cutlery trade, and an excellent steam engine by Fenton Murray and Jackson.

In May and June Forbes and Low, with a cotton mill at Poynernook Aberdeen, were looking to sell two steam engines. One was a 20 H.P. engine by Boulton and

Watt, the other a 12 H.P. by Fenton and Murray. The Boulton and Watt engine was supplied to Forbes and Low in 1802, and had a 23 ¾ inch cylinder and a 5 foot stroke. Both engines were in place by 1811 and the sale was because they had been replaced by a modern 55 H.P. engine [69].

Fenton Murray & Co was a recipient of I K Brunel's letter of June 1836, wherein he requested tenders for the first tranche of locomotive engines for the Great Western Railway. Brunel stated that the order was hoped to be completed by March 1837. The companies contacted were: Murray & Co Engineers Leeds, Stephenson & Co Newcastle, Sharp Roberts & Co, Manchester, Forrester & Co Liverpool, Mather Dixon & Co Liverpool, Jessop & Co Butterley Ironworks, Tayleur & Co Vulcan Foundry and Maudslay & Co Lambeth [113]. The Great Western Railway had a letter in reply from Fenton Murray and Jackson declining the business due to their current engagements, but advising that they wished to make some locomotives for the G.W.R.

In Newcastle upon Tyne the River Committee had decided to buy a dredger for the Tyne, and had requested estimates to be sent in. In their deliberations in June it was noted that the Newcastle engineers had not submitted any estimates and the Committee gave preference to an estimate by Fenton Murray and Jackson. A Mr Plummer, a member of the Committee, who had business in Yorkshire offered to make inquiries at their factory [70].

The London and Birmingham Railway had at this point decided to use stationary engines to haul its trains from Euston to Camden and was looking to let the tender for the supply of the engines. A tender document dated the 8th June 1836 survives calling for two combined 60 horse engines to be equal to the most improved marine engine arrangement. The names of the companies to which the tender was sent were Miller and Ravenhill, Seaward & Co, Maudslay & Co, Braithwaite & Co, and Bramah & sons. The net though was cast wider as the London Committee sitting on 6th July heard that Mr Jackson of Fenton & Co had called that morning and told the secretary and engineer that with their current order book they could not undertake to complete such engines within 12 months and for not less than £6500 (against the engineers estimate of £5000), and these terms would need to be agreed by his partner [71]. Maudslays were awarded the contract.

On the 13th of July, Fenton Murray and Jackson wrote to the London and Birmingham Railway to state that the locomotive due in June would not be available for at least two more months and suggested that the railway made its own tender. They were informed that their tender had been accepted and they would be held to their contract [72].

Fenton Murray and Jackson advertised a second hand 40 H.P. engine for sale complete with boilers in working condition in that July [73].

Samuel Exley married Miss Pickard of Ossett on the 6th October at Dewsbury [74].

A letter dated 8th October was read at the Birmingham Committee of the London and Birmingham Railway on the 21st October, which requested a deviation to specification to allow them to supply pipes of 1 ¾ inches in place of 2 inch ones. As Mr Stephenson did not consider this desirable, the company replied that the contract had been broken and they would procure the locomotive as best they could and hold Fenton Murray & Co liable for any losses to the railway company [72].

On the 22 October Fenton Murray and Jackson placed an advert for 6 millwrights, who were wanted immediately. The advert stated that "Men who are not in the club will be preferred", meaning members of a union should not apply. Immediately underneath their advert was another one for Smiths and Moulders, accustomed to engine work. This advert had been placed by John Seaward & Co of the Canal Ironworks London, and offered 2, 3 or 5 year contracts with first rate London wages. On the 5th November Fenton Murray and Jackson advertised again for 6 millwrights and 2 or 3 model-makers. They again stated their preference for non-union men [75].

The London and Birmingham Railway read a further letter from Fenton Murray and Jackson on the 11th November that stated they could have the locomotive ready in five weeks and would this be deliverable. The reply said that as they were that far advanced, and provided they met the new date, the company would accept the locomotive [72].

I K Brunel had obviously been disappointed by the answer from Fenton Murray and Jackson. It should be remembered that the firm had served his father well and that Brunel had met Matthew Murray. Brunel wrote to the firm again on 14th November, acknowledging their position and stated;

"You are possibly aware that several of the most eminent manufacturers are now executing engines for our railway somewhat novel in their construction and calculated for much higher speeds than those made or making for the other railways in this country. I should much regret that a house standing so high as yours should not also enter into the competition and I am anxious that at all events the exception should not arise from any neglect on my part" [113].

A navigation permit (operating licence) was granted on 28th November for 'Le Luxor' a passenger boat to ply on the Loire between Nantes and Orleans. The proprietor was M. Guibert and the 12 H.P. engine was constructed by Fenton Murray and Jackson [67].

A steam excavator that had been supplied in 1833, powered by a 5 H.P. engine with machinery and tackle erected by Fenton Murray and Jackson was offered for sale in the Ipswich Journal of December 3rd 1836. The excavator had been used to clear Southwold Harbour, and with the job completed, the Harbour Commissioners now wished to sell it.

On the 18th of December the widow of Matthew Murray, Mary Murray, died in her 73rd year [76].

On the 31st December the sale was announced for the 5th and 6th of January 1837 of all the household furniture, valuable books, plate, prints, pictures and effects of the late Matthew Murray, as instructed by his surviving executor (Richard Jackson). There was as usual a list of the effects, which included a small library of 300 volumes [77].

Fenton Murray and Jackson wrote to the London and Birmingham Railway on the 5th of January 1837 to advise them that the locomotive and tender were being shipped via Hull to the Thames [72].

In January 1837, a further auction notice was made in respect of Matthew Murray's four splendid marble vases. The notice stated that they had been presented by a foreign nobleman of distinction and that they were formed of a hard and curious marble and exquisitely made [78].

The Leeds and Selby Railway paid Fenton Murray and Jackson the sum of £1040 for a locomotive in February, which had been delivered in 1836 [5].

At a meeting of the Tyne River Committee towards the end of March 1837, it was reported that Mr Cubitt who had been engaged to examine certain aspects of the projected works on the river, had examined the dredging boat being built at Shields and expressed approval. He had also agreed with the whole proposed system of dredging. It was then reported that while the boat was well advanced there was a delay in engine and machinery. The reason was that Mr Jackson, the acting partner of Fenton Murray and Jackson, had met with an accident. This had resulted in the works falling behind schedule, and the men had taken advantage of the situation and gone on strike for an increase in wages, which the firm were unwilling to pay. The delay was estimated at one month, of which two weeks had passed [79].

Whether due to this accident or work reasons it is noticeable that Richard Jackson seems to have been absent from council meetings in February and at the end of March. However he did attend a working breakfast at the Mayor's house at the beginning of April to meet the new Recorder of the Borough. He also attended the council meeting at the end of April [80].

On the 29th of April, and on other dates and in various papers, a flax mill was advertised for auction in Darlington with a:

"*capital steam engine, 30 horse power, (fire proof) made by Messrs Fenton Murray and Wood of Leeds, with two cylinder boilers*" [81].

The mill was in the occupation of Charles and Robert Parker. The Parkers had succeeded John Kendrew at Haughton Le Skerne near Darlington. A Charles Parker had been Kendrew's father in law, so this may have been the developed version of Kendrew's old mill.

It was reported on May 20th that the River Tyne Committee had received the inventory of the machinery for the Tyne Dredging Boat from Fenton Murray and Jackson, and notification that it had been shipped from Hull [82].

About the same time the Leeds papers carried a death notice for John North Patrick who had been the foreman to the late Matthew Murray and had worked for the firm for upwards of 30 years [83]. Also in May a collection was made in aid of the destitute inhabitants of the Highlands and Islands of Scotland. Mr Maclea, one of the collectors, received 10 guineas from Fenton Murray & Co. Whether this was from the machine company or the combined companies is unclear, as so often the naming is unclear [84].

At the end of May a case came to court in Leeds that reflected the unfortunate position of the trade in the town, and in much of the country. The machine makers had 'turned-out', or in modern parlance gone on strike. As often the case, not all the men obeyed the call of the 'union' to strike and some remained in work. This resulted in a number of cases of intimidation and this case was one of these. At around a quarter to two in the morning on Sunday 28th May Joseph Swales a mechanic employed by Fenton Murray & Co had two panes of glass broken in one of the windows of his house by a thrown brick. The men at Fenton and Murray had 'turned-out' and Swales was described in the newspaper report as a 'black sheep'. Swales however recognised his assailant as one John Stead a machine maker, who was in court on the charge. Stead's witnesses could not account for him after midnight and he was found guilty of damage and charged in the sum of £4 10s. It appeared that the aim had been damage and not assault [85]. Other cases reported in Leeds were not so mild, and involved death threats.

In France another operating permit was granted on the 3rd July for a ship to ply by sea between Nantes and Bordeaux, carrying both passengers and goods. The hull was made by M. Guibert and the engine by Fenton Murray and Jackson. The ship was named 'Nantes et Bordeaux' [67].

In August a subscription was made In Leeds for William Hirst and family, the wool cloth manufacturer who was in dispute with his creditors, which raised £5 from Fenton Murray and Jackson. Hirst was the gentleman who had been one of the first customers for Murray's hydraulic press [87].

On the 16th September 1837 there was an advert under 'Sales by Private Contract' in the Leeds Mercury. It was for a 12 H.P. steam engine made by Fenton Murray and Wood in 1814 (see 1814 above) and since used for drainage. It was for sale with its boiler, which was calculated for an engine of 24 H.P. The engine and boiler were at Sutton Saint Edmunds in the County of Lincoln, in a good state of preservation. Sutton Saint Edmund was the first permanent steam pumping station erected in the Fens, and the date given here implies that it was put up 2 or 3 years earlier than has previously been thought. The station had been made redundant by the digging of the North Level Main Drain, which had been excavated using a Fenton and Murray engine (see 1835).

At the end of September a 30 H.P. engine with equivalent boilers and patent feeders by Firmstone of Manchester was for sale as part of the extensive scribbling and fulling mill called Elmwood Mill, which was in Camp Road Leeds. Earlier adverts had omitted the details of the engine, but a new one at the end of the month stated that the steam engine was new and made by Fenton Murray and Co, perhaps an added inducement to buy [88].

September 30th saw the announcement of the annual elections for Leeds Town Councillors. Among the vacancies to be filled was one at Holbeck for:
 "*Mr Richard Jackson, (late of Marshall Street Holbeck, but now of) Park Place Leeds (sic) iron-founder*" [89].
Presumably he and his wife had been living with her mother in Matthew Murray's old house, but they had now moved to the prestigious address of 35 Park Square Leeds.

The elections of the new officers for the Leeds Mechanics Institute also were published on September 30th and included Mr J.O.March as a vice president and Samuel Exley as a committee member [89].

Mr Hirst's situation was becoming more complicated as a number of the creditors, including Fenton Murray and Jackson, petitioned for a meeting with his assignees. They did not agree with the way the affairs were being handled [90].

In December the death of another long serving employee of Fenton Murray & Co took place. At the age of 78 Benjamin Hinsliffe, a mechanic, died having served the firm for 42 years during which period he only lost 2 hours of time, when a snowfall delayed his journey from Beeston where he was living [91].

On the 30th of December it was announced that the Leeds and Selby Railway had taken delivery of a locomotive called Hawke made by Fenton Murray and Jackson. She was one of the first 2-2-2 engines (Patentee type) by the firm and described as swift and powerful, having great strength and steadiness having added two small wheels to the other four [92].

A snippet from the Journal du Commerce de Lyon, probably dated August 1837, gives further insight into the volume of the firm's exports to France of marine engines and the dire industrial relations issues in the manufacturing districts. 'Steam tugs are definitely planned to be tried out on the Upper Rhone between Lyon and Montout in April of next year. A case of force majeure has prevented the company from achieving its plans. The hull, made in Lyon, will be stored over winter, due to industrial action which has closed all factories for four months. Messrs Jackson of Leeds, from whom all the engines for the Lyon Geneva service were ordered has had to postpone delivery until early next year'. It would appear that the company was named the 'Compagnie du Haut-Rhone' and operated three iron hulled paddle boats with 48 or 20 H.P. until about 1842, when the service stopped due to insufficient demand [93].

On the 6th of February 1838 the directors of the North Midland Railway sat down to discuss their tenders for locomotives. They had 9 bids and 5 no bids, and the bids ranged between £1150 and £1500 for the locomotives and £150 to £265 for tenders. Fenton Murray and Jackson and Hicks & Co had submitted 'no bids' for tenders, although Fenton Murray and Jackson said that they could buy them in. The Board resolved to place the following orders Tayleur & Co 2 off, Fenton Murray & Co 3 off, Miller & Co 2 off, Stephenson & Co 3 off, Mather and Dixon 3 off and Todd Kitson & Co 2 off, making 15 locomotives in all [94].

There was an address made to the Institute of Civil Engineers on the subject of Cornish Steam Engines by a well-known Yorkshire Engineer Thomas Wicksteed on the 9th January. Wicksteed was a great proponent of the Cornish Engine and while working for the East London Waterworks, where he installed one such engine, published some telling tables in regard to fuel consumption. His January paper was the subject of discussion at many of the subsequent weekly meetings in January and February. The last major discussion recorded on the 20th February 1838 included the following comment.

"Mr Jackson, of Leeds, had surprised him by saying that by cutting off the steam in a low pressure engine, even at [4/8ths] of the stroke, the effect of the irregularity was sensible, by breaking the very fine flax thread manufactured by Messrs Marshall. He thought it right to make these remarks in expressing his thanks, which he was sure were due to Mr Wicksteed for his valuable communications, and he hoped that his

example would be remembered by those engineers who were employed in similar professional inquiries".

It was stated underneath that it had been found in Lancashire that this issue could be overcome by using heavier fly wheels or two connected engines [95].

In March the new engine installed at the South London Water Works had an accident which broke the main rod and crank, and their Secretary was instructed to write to Fenton Murray and Jackson for their replacement [13].

At the beginning of April the firm were advertising for an angle iron smith, a fitter and 3 or 4 riveters all for boiler making [96].

On the 12th of May the Leeds Intelligencer carried notice of an auction, to be held on the 16th May for the materials of Matthew Murray's old house. Kilburn Scott's surmise that Matthew Murray lived in Holbeck Lodge was wrong [97]. Murray's house was situated at the south west end of the Round Foundry plot, on Marshall Street, and faced by the time of its demolition, the site of Marshall's new Egypt Mill and the C, D and E mills. The houses demolition was to make way for a new factory, which was specifically built and tooled to manufacture railway locomotive engines. Taking the deliveries as stated in Lowe's British Steam Locomotive Builders, the annual deliveries of locomotives from Fenton and Murray since they re-entered the business were:

1831	2
1832	2
1833	1
1834	2
1835	1
1836	5
1837	2
1838	7
1839	6
1840	26
1841	7
1842	10.

If Matthew Murray's old statement to the Stockton and Darlington railway about locomotives disrupting his other work still held good, 1836 may have been a difficult year, and 1840 would have been impossible, without the new shop. Railway mania had arrived and new lines were being built and even more being discussed or put to parliament for approval. The decision had been made to invest in the new product in a big way, and like all investments it carried a risk.

At the end of May a mechanic working for Fenton and Murray appeared in court on a charge resulting from getting drunk and abusing passers-by in Water Lane. He then assaulted the policemen who arrested him causing that policeman to be bruised all over and have a serious injury to his arm. The man's foreman swore that he had never known him drunk before or to miss time; and gave him a good character. However the man, William McHugh, was fined £5 with a penalty of one month's imprisonment if he failed to pay. This was a far cry from earlier days, when Round Foundry men had helped the forces of Law and Order [99].

In June the firm were recorded as subscribing 10/- to the Leeds Royal Humane Society for providing 'apparatus for restoring persons apparently drowned or dead' [100].

Queen Victoria's coronation took place on the 28th June 1838. There was considerable rejoicing throughout the country with descriptions of the festivities appearing nationwide by town. Most large factories seem to have given the workforce paid holidays, and many provided dinners and even tea drinking and balls! The recorded happenings in Holbeck included a holiday at Maclea and March along with a dinner, the manufactories of Marshall & Co, Benyon & Co, Titley, Tatham & Co and Fenton Murray & Co were closed (hopefully for a paid holiday) [101].

In June and July Fenton Murray and Jackson were again advertising for an angle iron smith and two or three fitters. They quoted the wages to be paid at five to six shillings a day. The applicants were to be sober steady and first rate workmen. The word sober was missing from the April advertisement, and may give some indication of the cause for re-advertisement [102].

August 1838 saw the announcement, far and wide, of the trial, on the 6th of the month, of the largest locomotive made in Leeds, on the Leeds and Selby Railway, by Fenton Murray and Jackson [103]. The locomotive was for the Paris and Versailles Railway and according to Lowe this 1838 delivery was the 'Matthew Murray'. The locomotive was probably very similar to the illustration below of one of its sister locomotives, which was supplied to the competing railway, although some of the descriptions lead to the belief that it was 'La Versailles' a six wheeler (see chapter 17).

There is no proper physical description given in the press, so it is not stated how many wheels the locomotive had. It did state that the locomotive travelled at 60 mph with one carriage and a tender, and at 20mph with a load of 140 tons.

"In consequence of some improvement in the fire-box, this engine seems to do its work with ease, as, during the trials, it produced more steam than was required, and with the fire-door kept open. By good judges, this locomotive engine is considered of first rate workmanship and finish, and it does great credit to the makers" [103].

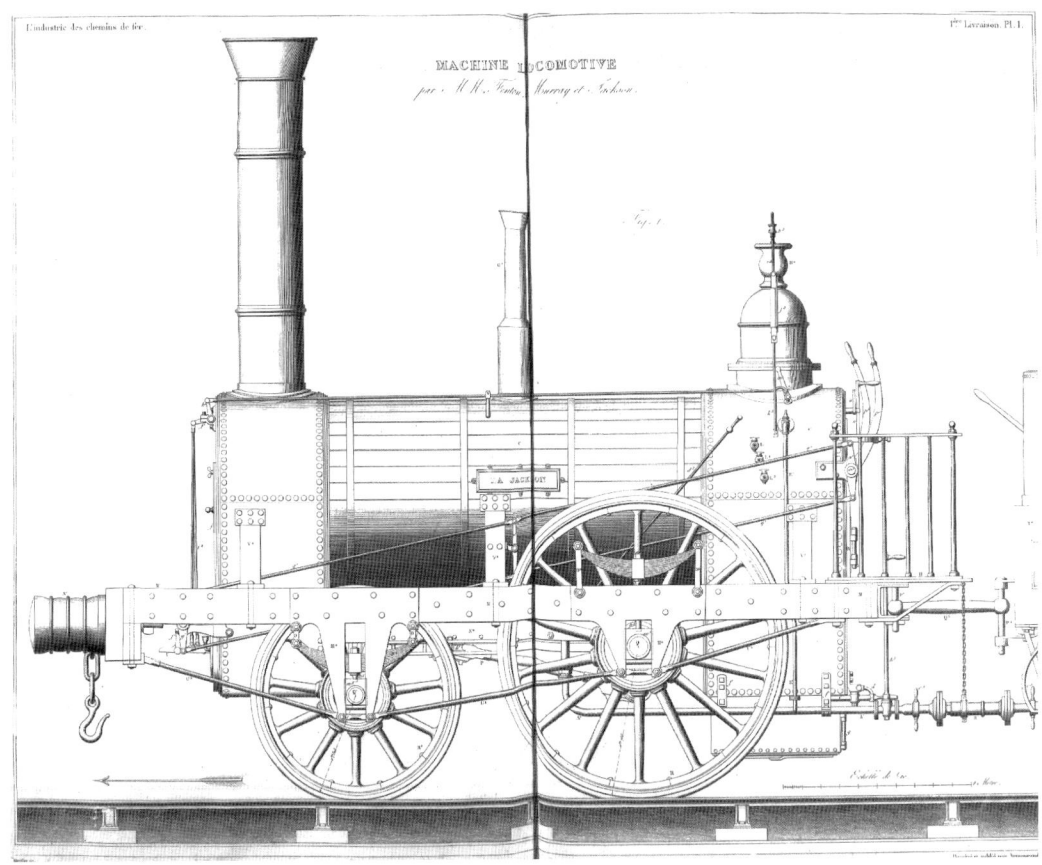

67. La Jackson by Fenton Murray and Jackson for the Paris/ St Germain-en-Laye Rlwy [104]

A model maker from Fenton Murray and Jackson visited Nasmyth Gaskell probably in September 1838, presumably to view their capital equipment to see if any would suit for the new Locomotive Works. The visit, of a type that was encouraged by Nasmyth Gaskell to boost machine sales, was acknowledged in a letter sent to Fenton Murray and Jackson on October 3rd [105].

The Newcastle Courant carried a sale advert for a 12 H.P. Fenton Murray & Co engine with a 16 H.P. boiler for £220, in October. The engine was at Wisbech [106].

And in November an 8 H.P. engine with a 12 H.P. boiler by the firm; was offered for sale by Josh Batty a builder in Leeds [107].

Among the new members elected to the Leeds Philosophical and Literary Society in November 1838 were Richard Jackson, Peter Fairbairn and Joseph Ogdin March [108].

On the 20th November the eldest son of Richard and Margaret Jackson, William Murray Jackson, described as a Civil Engineer of Hunslett married at Salem Chapel Leeds one Anne Pearce of Holbeck [109]. Anne Pearce is thought to be the daughter of Charles Worth Pearce who after managing an iron works near Stourbridge came to be a manager at Fenton Murray and Jackson. Her brother, John

Charles Pearce, served part of his apprenticeship at the Round Foundry, but he finished it elsewhere, and became a leading engineer at the Bowling Ironworks, where he stayed for the rest of his working life. During his time there he took out a number of patents on steam engines and boilers.

In early December the Sheffield Independent advertised the sale of the Washington Works with extensive workshops, etc., with 40 H.P. of steam power provided by two engines, one by Fenton Murray and Jackson and the other by Davy Brothers of Sheffield, both working the same shaft [110].

In 1838/9 the firm delivered a large order of five locomotives for the Montpellier and Cette [modern usage Sete] Railway in the south of France. The first three engines were shipped to Le Havre for onward delivery to the south of France arriving in October 1838; the second two seem to have been shipped direct to Cette in early 1839. They were named 'Notre-Dames-des-Tables', after the patron saint of Montpellier, 'Herault', 'Montpellierenne', 'Cettoise', and 'Rosine'. They weighed 8 tons each and could operate at 20 mph [111].

The Newcastle City Accounts for 1838 record the payment of £1343/2/9d to Fenton Murray & Jackson for their machinery for the 10H.P. dredger for the River Tyne.

An article appeared in The Engineer on 1st April 1927 about Castleton Mill in Armley Road Leeds. The first stone of the mill was laid in 1837, and the engine was built by Fenton Murray and Jackson in 1838. The engine's cylinder had a diameter of 33 5/8ths inches, a stroke of 6 foot, the length of the beam was 26 feet and the flywheel had a diameter of 25 feet. It is believed the engine was supplied as a 40 H.P. one. As the mill was expanded, by the eventual owners [in 1927] the engine was loaded to 120 H.P. and then further to 180 H.P. At this time a small horizontal engine was added to take some of the strain. However further expansions led to a decision in 1920 to replace both these engines with a 450 H.P. Corliss. The Fenton Murray & Jackson engine was running at around 23 revolutions per minute on a boiler pressure of 30lbs from around 1900. The highest power recorded was 181 H.P. but this was reduced to 90 in 1913. It used around 25 -28 tons of coal in a 56 hour week running at 140 H.P. After the horizontal engine was added their combined consumption for a 56 hour week was 27-31 tons of coal for 150-170 H.P. The Corliss running at 80lb / square inch used 20-22 tons of coal for a 51 hour week.

William Mann Crosland began an apprenticeship at Fenton Murray and Jackson in 1838 which he finished in Manchester. In 1845 he went to work for Maudslay, Sons and Fields, where he stayed for the remainder of his working life.

336

1839 did not start well. On the 7th of January a hurricane force storm swept through Leeds bringing down a number of chimneys. Peter Fairbairn's foundry chimney was blown down and through the roof. So was the chimney of the new foundry at Fenton Murray & Co's, which also penetrated the roof [112]. This must have been the foundry for the new locomotive works.

The outlook improved later in the month with a letter from the Great Western Railway (GWR) dated January 30th which presaged the largest locomotive order the firm would ever receive. It was written by Daniel Gooch, who was at this time the Locomotive Superintendent of the GWR. It appears from the letter that Richard Jackson and Daniel Gooch had already had a meeting during which Jackson had expressed the wish to build some locomotives for the GWR. He also described a new form of locomotive wheel that Fenton Murray and Jackson made. The letter advised Fenton Murray and Jackson that the GWR were about to order 20 new locomotives, 10 off with 7 foot driving wheels and cylinders with a 15 inch diameter and 18 inch stroke (Fire Fly class locomotives) and 10 off with 6 foot driving wheels and slightly smaller cylinders. Gooch wanted to know if the firm could make 5 or 6 locomotives by the end of 1839, "not later". He advised that the GWR would provide all the drawings. Gooch then went on to ask about the new wheel, which he rather liked, and as he wanted to adopt a particular type of wheel could he have details. Would the firm provide drawings and a cost estimate for wheels of 7ft, 6ft, 4ft, and 8ft 6inches? The GWR needed wheels with tyres not less than 1 ¾ inches thick from best hammered iron. He also wanted an early answer [114]. Jackson replied on the 3rd February, but this letter does not appear to have survived.

In February the papers carried an item on the Dredging Machine in the Tweed. It had been constructed to the plans of the civil engineer Thomas Rhodes, who hailed from near Leeds and had worked with Thomas Telford. The engine and machinery had been made by Fenton Murray and Jackson. The dredger could excavate 100 tons of material in an hour. The piece was confident that Berwick would soon have one of the finest ports on the east coast [115].

February also brought a further letter from Gooch dated 7th acknowledging a response. He stated that the wheels would be 5 ¼ inches wide, information for Jackson to estimate the wheel costs. He also asks if there would be any objection to the GWR adopting their wheel plan and having them made by any manufacture constructing their locomotives. Fenton Murray would be asked to make any spare wheels. Gooch also states that the detailed specification should be with them the following week and that it was sufficiently detailed for a tender to be made. The drawings were also nearly ready [114].

On the 4th of March I K Brunel formally wrote to all the potential contractors for locomotives for the GWR. In accordance with the Board's deliberations the order was to be for a total of 16 locomotives with 7 feet wheels (Firefly class) and 4 with 6 feet wheels (Leo class). Tenders would have to be for a minimum of two locomotives. This time around the locomotives were to be built to GWR (Gooch) specifications. The letter was sent to Stephenson & Co, Longridge Starbuck & Co, Hawthorn, Murray and Jackson, Jones Viaduct Foundry, Haigh Foundry, Sharp Roberts, Mather Dixon, Rennies, Maudslay, Ravenhill and Miller, Seawards, Bramah, and Stothert [127].

In March Fenton Murray and Jackson advertised for 2 pipe moulders [116].

The Sheffield and Rotheram Railway were planning to open their lines to goods traffic fully and therefore announced the purchase of two more locomotives: *"one of them, Messrs Fenton Murray & Co of Leeds, have contracted to supply on the very best principle"* [117].

On the 17th of March 1838 William Fenton died [118]. William Fenton and his wife Isabella (Bella) had one child, a daughter Mary, born in Leeds on the 30th September 1799 [119]. While the family was living in London, she had married a Richard Yale, a hosier, of the parish of St Bartholomew by the Royal Exchange, on the 13th November 1819. Richard and Mary Yale had a son, Richard, born on September 25th 1820[120]. Richard Yale senior is presumed to be the gentleman who died in Southwark in 1823 as he is not heard of anymore. William Fenton's will after some small bequests including £5 to Margaret Jackson 'the wife of my partner' stated:

"I direct that my wife shall carry on the trade until my grandson attains the age of 21 years, then I give and bequeath all my estate and effects of what kind quality or nature soever the same unto my grandson Richard Yale....he paying thereout the sum of £1000 per year unto my said wife for her life and the sum of £250 per year unto my daughter Mary Yale for her life" [121].

Thus at the time of William Fenton's death, Richard Yale was two and half years away from his majority. Therefore Isabella Fenton and Richard Jackson had the management of the Foundry Parnership, and were joined by Samuel Exley in the Machine Business. Although Exley stated in his submission to Chancery that the partnership deeds were all changed to replace William Fenton with Richard Yale [1].

On the 28th of March the Bristol Committee of the GWR discussed the tenders and recommended a purchase of 10 locomotives at £1850 from Sharp Roberts, 4 at £1835 from Newton Viaduct, 2 at £1835 from Hawthorn, 2 with tenders from Bramah at £2000 and the engineer was to contract with Stothert for 2 or 4 locomotives if they would agree a price of £2240 each [128].

In April a 20 H.P. engine by Fenton and Murray was advertised along with a powerful shaft at the Tobacco and Stuff Factory of Messrs Lundy and Foot in the Minories in the City of London [122].

A 10 H.P. engine by Murray and Co was also available in April, along with a 16 H.P. boiler by the Bowling Company, at the Wheatley Wire Mill at Halifax, lately occupied by the late firm of William Morris & Co [123].

On the 22 April 1839 the London Committee of the GWR, having on the 1 April endorsed the Bristol Committees recommendation, but having received further offers from manufacturers, confirmed an order of two locomotives to be awarded to Fenton Murray & Co [129]. At some point the two locomotives ordered on Bramah were cancelled.

The Sheffield Independent of April 27th describes the delivery of the Agilis locomotive, made by Fenton Murray and Jackson, to the Sheffield and Rotheram Railway on a truck drawn by 16 horses, on the 23rd of April. The Leeds Mercury describes the Agilis as having flanges on all six of its wheels, and that if its eccentrics became broken or damaged, the driver would remain in control of the locomotive [124].

The GWR wrote on the 16th of May that a parcel of drawings had been sent by railway and the remainder should follow shortly. Gooch also wanted to know how the firm would tackle making the plates for the firebox and joining them together. He made a suggestion but indicated that any acceptable way would do [114].

May also saw the launch of a subscription for a school in Holbeck for the established church. Lady W. Gordon donated the land and £100. Fenton Murray and Jackson, and Maclea and March both donated £20 [125].

A petition was made to the mayor of Leeds in May to summon a meeting in order to consider making an address to the Queen to demonstrate strong attachment to Her Majesty, her government, and the principle of liberal principles of government. Richard Jackson was among those who signed the petition to the mayor. This was a reaction to the so-called Bedchamber Crisis, initiated by the West Indies planter's refusal to accept the Slave Emancipation Act, which resulted in the Prime Minister, Melbourne, nearly losing a vote in the Commons and resigning. Wellington and Peel both refused Queen Victoria's request to form administrations and Melbourne was left with no alternative but to resume office [125].

On June 10th the GWR wrote to acknowledge receipt of a model of the fire box. However Gooch did not like the method of joining the plates, indicating the method (jump joints) had proved troublesome. Cap joints were needed, and Gooch enclosed an iron model, which he said was close to the one Fenton Murray and Jackson had sent. He also wanted the top finished as round as possible, not all angles. Gooch explained that he had not sent any more drawings as the copier was late, but they would all be sent together when complete, which would be about Wednesday. He explained that the GWR would supply certain brass work (as listed). 6 days later he wrote again to say the remainder of the drawings had been sent on Friday and please could their receipt be acknowledged and any questions sent in at the same time [114].

A notice dated the 15th of June appeared in the Leeds papers advising that John Milner of Leeds, a glover and hosier had assigned all his assets to Richard Jackson, engineer of Leeds and Peter Greenough of Manchester. His stock was all put up for auction [126].

On July 5th the GWR advised of an error in the drawing containing the reversing gear, and sent a correcting drawing [114].

Fenton Murray and Jackson advertised on 23rd July the sale of a 24 H.P. steam cylinder with piston, D valve and case, valve rod etc. and air pump and condenser etc., all from John Howard's carpet manufactory at Low Fold Mill, Bank, in Leeds [130].

In 1839 the Mechanic's Institute in conjunction with the Philosophical and Literary Society decided to hold a Leeds Public Exhibition, after the success of similar events at Manchester, Sheffield and elsewhere. Requests for loans of exhibits began to appear in the press in May 1839. The subject range for the exhibition was far more varied that one would expect today and ranged from paintings, sculpture, drawings etc; hydraulic pumps and presses, fire engines, steam engines, locomotive engines; models of buildings, millwork, textile machines; natural history; antiquities; coins; manuscripts and all that could be of interest. Among those sitting on the committee were to be found Messrs Fairbairn, Exley, Marshall's, March, and Maclea. The exhibition ran from the 10th July (opening 1 day late) until October 5th 1839, and was deemed a great success. It was open from Monday to Saturday from 10a.m to 10p.m.

Within the exhibition was a section of machines, smaller than factory size, which demonstrated the processing of flax. It consisted of a 'first drawing frame', a 'second drawing frame', a 'roving machine', and a 'spinning frame' all driven by a 4 H.P. steam engine. The machines had been especially constructed for the exhibition. Fenton Murray & Co supplied the two drawing frames and Mr Fairbairn the roving machine and the spinning frame. The steam engine was supplied by Turner Ogden & Co of School Close. Messrs Marshall & Co supplied flax and some of their employees operated the machines and process [131].

340

Additional contributions were added, including a portrait of Queen Victoria which she had sent after a request, when space permitted and the objects were of sufficient interest. The Leeds Mercury records on the 27th of July that the fresh contributions that week included:

"a *model of a sawing frame in operation at the Government works at Chatham offered by Messrs Fenton, Murray and Jackson*"

On 10th of August the GWR advised those firms manufacturing locomotives for it that it had forwarded the gauges and templates for the wheels and also the templates for the axle boxes, and expressed the hope that they were progressing with the engines [114].

In November a James Oldham of Hull was advertising for sale a double hydraulic press in excellent working condition, together with pumps and driving gear made by Fenton Murray and Jackson and fitted with Blundell's patent boxes for eight cakes, all for seed crushing [132].

In that month the son of John Millar, John, died in Leeds. John Millar senior was one of the senior engineer/erectors for Fenton Murray and Jackson, being entrusted with erecting some of the GWR locomotives and some of the French marine engines [133].

Fenton Murray and Jackson advertised for a steady active persevering man of good character to take charge of a 20 H.P steam engine and hydraulic presses in an oil mill in the country. Applications were to be made personally to the firm. Later in December the firm were advertising for an angle iron smith[134].

Notes

1. S. Exley's submission to the Court of Chancery, dated 20 February 1844 in Exley v Jackson. C14/218/E13. National Archives. Kew (offsite).
2. See Chapter 12 note 61.
3. Records of the Leeds Mechanics Institute, Annual Reports 1826, 1827 & 1829. WYL 368. WYAS Leeds.
4. National Education by Frederic Hill Vol. 2 page 196. 1836. London.
5. Rail 351/10 Records of Leeds and Selby Railway, Cash Book. National Archives. Kew.
6. Watson Papers Wat 3/19. N.E.I.M.M.E. see also Chapter 13.
7. Newcastle Courant 19 May 1832.
8. The Leeds Mercury 16 June 1832.
9. La machine locomotive en France. J Payen with B Escudie, J-M Combe. 1988 Lyon and Paris. Also Chemin de Fer D'Andrezieux a Roanne, Annales des Ponts et Chaussees. Paris 1832.
10. Rail 359/2. Leicester and Swannington Railway, Minutes of Directors Meetings 2 and 17 August 1832. National Archives. Kew.
11. The Leeds Intelligencer 22 November 1832.
12. Rail 371. Liverpool and Manchester Railway, typed extracts from Director's minutes. National Archives .Kew.
13. SV/1/165 South London Water Works. London Metropolitan Archives.

14. Annales de la Societe Academique de Nantes et du departement de la Loire-Inferieure. Vol. 9. Nantes 1838.Pages 92/3.

15. Rail 371. Liverpool and Manchester Railway, typed extracts from Director's minutes, 4 February 1833. National Archives .Kew.

16. Marshall Papers. MS200 folio 17 10 February 1833. Brotherton Library.

17. Rail 667/1048. Stockton and Darlington Railway. National Archives. Kew.

18. The Leeds Intelligencer 18 May 1833,

19. The Students Guide to the Locomotive Engine. Flachet and Petiet. English Edition, John Williams & Co London 1849. Plate 5.

20. Manuscript MS 852B. Smithsonian Institute Libraries.

21. The Charleston and Hamburg. Thomas Fetters. The History Press 2008.

22. The Norfolk Chronicle 10 May 1833.

23. Accueil, introduction et developpement de l'energie vapeur dans la marine militaire francaise au XIX siècle. C.V.Dominique Brisou. Service historique de la marine / Universite de Paris 2001. Vol. 2, pages 439/40.

24. The Leeds Mercury 18 January 1834.

25. The Leeds Times 1 February 1834.

26. Rail 371. Liverpool and Manchester Railway, typed extracts from Director's minutes, 3 February 1834. National Archives .Kew.

27. BT 6/151. Board of Trade Export Requests 1825 – 1842. National Archives. Kew.

28. The Yorkshire Gazette 3 May1834.

29. The Leeds Mercury 31 May 1834 and 14 June 1834.

30. Rail 667/1056. Stockton and Darlington Railway. National Archives. Kew.

31. E.g. The Morning Chronicle 1 July 1834.

32. The Leeds Mercury 12 July 1834.

33. E.g. The Sheffield Independent 2 August 1834.

34. SV/1/5 South London Water Works. London Metropolitan Archives.

35. The Leeds Mercury 23 August 1834.

36. The Preston Chronicle 30 August 1834.

37. The Leeds Intelligencer and The Leeds Times of 6 September 1834, The Bradford Observer 11 September 1834 and The Yorkshire Gazette 13 September 1834.

38. The Leeds Mercury 20 September 1834. The Leeds Mercury, The Leeds Times, The Leeds Intelligencer and The York Herald all of 27 September 1834.

39. Rail 351/1. Leeds and Selby Railway, Directors minutes. National Archives. Kew.

40. Rail 351/1 and 351/10. Leeds and Selby Railway. Also William Mackenzie, David Brooke. Newcomen Society 2004, page11. Mackenzie had bought North Star from Liverpool & Manchester Rlwy(delivered to them by Stephensons 1830) for £250. North Star travelled to Leeds by canal.

41. E.g. The Leeds Times 25 October 1834.

42. E.g. Woolmers Exeter and Plymouth Gazette 22 November 1834.

43. The Leeds Intelligencer 29 November 1834, The Morning Post 27 November 1834.

44. The Leeds Mercury 20 December 1834.

45. Wills of Yorkshire Vol. 191, no 63 microfilm 1090. Borthwick Institute, University of York.

46. The Leeds Times 31 January 1835.

47. The Leeds Mercury 14 February 1835.

48. E.g. The Lincoln, Rutland and Stamford Mercury 6 March 1835.

49. Rail 667/657. Stockton and Darlington Railway. National Archives. Kew.

50. The Repertory of Patent Inventions 1835. Law reports on Patent Cases, Morgan and another v Seaward and others. 2 parts.

51. The Hull Packet 3 July 1835. Home tour through the Manufacturing Districts of England, Sir George Head. 1836 reprinted Frank Cass 1968. The Early History of Hull Steam Shipping, F H Pearson 1896 reprinted Mr Pye Books 1984.

52. E.g. The Leeds Times 5 September 1835.

53. Archives du Musee des arts et metiers. CNAM. S145.
54. See note 23, but page 444.
55. The Leeds Times 15 August 1835.
56. Archives du Musee des arts et metiers. CNAM. S156.
57. The Hull Packet 16 October 1835 and note 23.
58. The Leeds Times 7 November 1835.
59. The Leeds Times 28 November 1835.
60. The Leeds Mercury 12 December 1835.
61. Salopian Journal 16 December 1835.
62. Berrow's Worcester Journal 31 December 1835.
63. The Leeds Intelligencer 16 January 1836.
64. Rail 384/32 London and Birmingham Railway. London Committee Minutes. National Archives Kew.
65. Rail 384/67 London and Birmingham Railway. Birmingham Committee Minutes. National Archives Kew.
66. The Leeds Mercury 5 March 1836.
67. Ministry of Public Works, navigation Permits, 1827 to 1868, F/14/4229 to 4232. Archives Nationale de France, Paris. Also note 53 above.
68. The Leeds Times 2 April 1836.
69. Aberdeen Journal 1 June 1836, 28 June 1843.
70. The Newcastle Journal 11 June 1836.
71. Rail 384/42 London and Birmingham Railway. London Committee Minutes. National Archives Kew.
72. Rail 384/68 London and Birmingham Railway. Birmingham Committee Minutes. National Archives Kew.
73. The Leeds Mercury 30 July 1836.
74. The Leeds Intelligencer 15 October 1836.
75. The Leeds Mercury 22 October and 5 November 1836.
76. The Leeds Intelligencer 24 December 1836.
77. The Leeds Times 31 December 1836.
78. The Leeds Intelligencer 21 January 1837.
79. The Newcastle Journal 20 May 1837.
80. The Leeds Times 18 and 25 February, 4,15 and 29 April 1837.
81. The Leeds Intelligencer 29 April 1837.
82. The Newcastle Journal 20 May 1837.
83. The Bradford Observer 25 May 1837.
84. The Leeds Mercury 27 May 1837.
85. The Leeds Times 3 June 1837.
86. The Leeds Mercury 27 July 1837.
87. The Leeds Intelligencer 19 August 1837.
88. The Leeds Mercury 30 September 1837.
89. The Leeds Intelligencer 30 September 1837.
90. The Leeds Mercury 9 December 1837.
91. The Leeds Times 23 December 1837.
92. The Leeds Mercury 30 December 1837.
93. Navigation sur Le Haut-Rhone Francais. S Aubert.
94. Rail 530/41 North midland Railway. London Committee minutes. National Archives. Kew.
95. Transaction of the Institute of Civil Engineers 20 February 1838.
96. The Leeds Mercury 7 April 1838.
97. Matthew Murray Pioneer Engineer, E Kilburn-Scott. Jowett Leeds 1928.
98. British Steam Locomotive Builders, James W Lowe. Goose & Son 1975.
99. The Leeds Times 11 August 1838.
100. The Northern Star and Leeds General Advertiser 9 June 1838.
101. The Morning Post 26 June 1838.
102. The Leeds Mercury 30 June, 21 July, The Leeds Times 30 June, 7 July, 21 July 1838.
103. The Leeds Times11 August 1838.
104. L'Industrie Des Chemin de Fer, Jacques-Eugene Armengaud. Paris 1839.

105. James Nasmyth and the Bridgewater Foundry, J H Cantrell. Manchester 1984. Page 43.
106. Newcastle Courant 5 October 1838.
107. The Leeds Mercury 3 November 1838.
108. The Leeds Mercury 17 November 1838.
109. The Leeds Mercury 24 November 1838.
110. The Sheffield Independent 8 December 1838.
111. Histoire de la locomotion terreste. Les Chemins de Fer. Dollfus 1835.
112. The Leeds Mercury 12 January 1839.
113. Rail 1149/2 I K Brunel Letter Book. National Archives Kew.
114. Rail 1008/26 Great Western Railway. National Archives Kew.
115. The Leeds Intelligencer 2 February 1839.
116. The Leeds Mercury 9 March 1839.
117. The Sheffield Independent 2 March 1839.
118. The Leeds Intelligencer 23 March 1839.
119. Parish Records Leeds Parish Church. Baptisms January 1800. Mary Fenton born 30 December 1799.
120. Parish Records United Parish of Christ Church and St Leonard, Foster Lane, City of London. Baptisms November 1820 Richard Yale born 25 September 1820.
121. Wills of Yorkshire Vol. 199, no 324 microfilm 1096.William Fenton. Borthwick Institute, University of York.
122. Liverpool Mercury 12 April 1839.
123. The Bradford Observer 18 April 1839.
124. The Leeds Mercury 4 May 1839 and Sheffield and Rotheram Independent 27 April 1839.
125. The Leeds Intelligencer 18 May 1839.
126. The Leeds Mercury 15 June 1839.
127. Rail 1149/5 I K Brunel letter Book. National Archives Kew.
128. Rail 250/99 GWR Bristol Committee. National Archives Kew.
129. Rail 250/83 GWR London Committee. National Archives Kew.
130. The Leeds Mercury 27 July 1839.
131. The Leeds Mercury 10 August 1839.
132. The Lincoln Rutland and Stamford Mercury 15 November 1839.
133. The Leeds Mercury 16 November 1839.
134. The Leeds Mercury 7,14 and 21 December 1839.
135. The Royal Bank of Scotland Group Archives, BEL 48/2 Signature Book.
136. The Athenaeum Journal 1832 page 131.

Chapter 15.

The Final Years

At the beginning of 1840 Fenton Murray and Jackson had a good quantity of work, so much in fact that they nearly lost an additional order from the GWR. On the 20 January the GWR in the person of Charles Saunders, the secretary, wrote to the firm to say that having been informed by Gooch, after his visit to the Round Foundry, that they were making very satisfactory progress with the current order, and that with a bit of effort some further locomotives might be managed; they would like Fenton Murray and Jackson to produce two further locomotives by April or at the latest May 10[1] of the current year. If the firm could undertake this without impacting their current delivery schedule, the directors would be happy to confirm the order. Fenton Murray and Jackson replied promptly on January 24th, and expressed their great regret that it was totally out of their power to deliver any more locomotives in the timescale required. They stated that they were pushed very hard in locomotive engines, plus they had a pair of marine engines for France and a couple of steam dredging engines and machinery for the Government in Ireland; all to be completed by the 30th of June [2]. Their locomotive order book on top of those for the GWR, included some for the Hull and Selby, one for the Belgian State, the North Midland, the London and Southampton Railways and probably their last one for the Paris and Versailles. The new locomotive department was up and running. The firm advised in a postscript that they would be most happy to contract for any requirements in the second half of the year.

However it was soon apparent that the GWR wanted to place the order with Fenton Murray and Jackson, as on the 28th January they (Saunders) wrote to confirm an order for two locomotives on the basis that the directors were confident that the effort would be made to deliver them as early as possible in the next 6 months. Additionally they enquired if the firm could undertake an order for two goods locomotives to be finished in the autumn [3]. A similar letter was sent at the same time to R W Hawthorn of Newcastle. This letter to Fenton Murray & Co from Saunders bore the same date as one from Gooch. Gooch was talking of even more work as he wanted the firm to offer for as many as they could make in 1840 as he thought the directors would extend the current orders considerably. He also wanted the firm to supply its own tenders, at as low a price as possible. In one of the two new orders, Gooch requested that the cylinders should be bored out an extra ½ inch to 15 ½ rather than 15 inches, but if this would cause even one day's delay it was not to be done. Gooch was willing to pay "*any little additional expense*", and please to let him know if it was possible by return of post. Gooch also wanted an estimate of the cost for 6 goods engines, which the GWR were ready to order. The goods engines were to have 4 coupled 5 foot wheels and a pair of leading wheels of 3 foot 6 inches. The cylinders were to be 15 inches with an 18 inch stroke and all working parts as for the current passenger locomotives and the boiler to be smaller being only 3 foot 6 inches with a fire box 4 foot 4 inches by 4 foot 2 inches and containing 96 tubes. These were all wanted in 1840 and some as early as possible [4].

On March 6th 1840 the Hull Packet published a report on the annual meeting of the Hull and Selby Railway. The report stated that in addition to the 6 locomotives ordered from Fenton Murray and Jackson, 6 additional locomotives upon an improved plan had been agreed by the directors. The improvement was the use of John Gray's patent, Gray being the principal engineer in the locomotive department of the Hull and Selby. The improved locomotives were built by Shepherd and Todd of Leeds (better known as an early form of the Railway Foundry in Hunslett). Charles Todd had been an apprentice to Matthew Murray. The Fenton Murray and Jackson order was scheduled for 3 deliveries in March 1840 and 3 in June. The Shepherd and Todd deliveries were for 2 in June and one a month from July to October 1840. The report also mentioned that Fenton Murray and Jackson had been paid £1450 on account for locomotive work.

Dated the same day, in the Leeds papers, was an advert by Fenton Murray and Jackson aimed at Boiler makers, saying that the firm required two or three platers or fitters-on [5].

The GWR wrote on March 27th, this time the letter was from T R Crampton, at that time one of Gooch's assistants. The letter advised that the drawings and specifications of the Goods Locomotive had been forwarded and gave some extra clarifications [4].

In April a Flax Mill at Kilbirnie, between Glasgow and the coast near Largs, was offered for sale. It had among its machinery 2036 wet spinning spindles, and advised prospective buyers that the textile machinery was chiefly made by P. Fairbairn, Fenton Murray & Co and Maclea and March Leeds [6]. This advert provides one of the few clear indications that the firm's forays into Scotland were not confined to the east.

After a further visit to the GWR locomotive suppliers, from which he had just returned, Gooch wrote on the 18th April to Mr Osler, the secretary of the Bristol Committee of the GWR, and opined that deliveries now seemed less satisfactory. Osler was about to set out on his own tour of the suppliers. Gooch reported that: Stephenson 8 to deliver, 1 possible by end of May and 1 three weeks afterwards, other 6 had not much progress.

346

Hawthorne 1 delivered, 9 to be delivered, 1 ready, 1 in around 3 weeks, 1 about a month after, and 1 a month after that, 5 not started.

Fenton Murray & Jackson 8 to deliver, 1 ready, 1 in around 6 weeks, 6 not started.(so the firm had received orders for another 4 locomotives).

Sharp Roberts & Co 10 to deliver, 1 ready, 1 in around a month, rest at 3 week intervals thereafter. Their locomotives for the Bristol and Exeter were in a very backward state.

Jones Turner & Evans, Newton delivered 2, 1 more is on its way, 3 more at intervals of 2 to 3 weeks.

Rennie 2 to deliver, so little progress hardly know when they would be delivered.

Stothert & Co not visited recently so cannot estimate their delivery on their 2 engines.

Gooch summarised the position as 3 delivered, with monthly estimates as April 4, May 7, June 5, July 2, August 2, Sept 2, giving a total of 25 by September.

The figures are a mixture of all orders then in place; so the Stephenson ones relate to Star Class locomotives, and the rest to a mix of Fire Fly and Leo class [4].

In April the death was reported of one Thomas Lawson who had worked for Fenton Murray and Jackson for 16 years [7].

On the 20th April Fenton Murray and Jackson delivered the first locomotive for the Hull and Selby Railway. The locomotive was named Kingston. It was given a trial on the line on the 22nd by the directors between Hessle and Hull and was reported as fully answering the expectations held for it. The locomotive was used for hauling materials to complete the line [8].

Locomotive Department. Marshall Street.
Land, Marine & General Work. Water Lane.
Leeds, Yorkshire 23 July 1841

Fenton, Murray & Jackson,
Engineers
Manufacturers of Land, Marine & Locomotive Steam Engines, also
Boilers, Hydraulic Presses, Flax & Tow Machinery & of Machinery in general.

68. Fenton Murray and Jackson's new letterhead after the opening of the Locomotive Department.

The first Fire Fly class locomotive from Leeds, the Charon, was also ready and was tried on the 29th April on the machine erected for that purpose at the works. Although on this first test it was decided that due to the weight of the locomotive it was unsafe to test either the power or speed. The machine was erected by Mr F Bennet [9]. This early rolling-road was described, testing a subsequent delivery, by David Joy (then an apprentice at the works) in his diary under August 1842:

"After all the work of a day, I well remember staying over hours to see one of these engines (the last, (so it must have been Argos)) tried in steam. It was placed on special rails with struts in front and behind, and the middle wheels resting on underground pillars (rollers), which they drove round, and to which was attached a counter to show the revolutions per minute or the miles per hour. After that test a dynamometer was yoked on at the trailing buffer, and the hauling power noted." [10].

Joy, who was college educated, also recorded his first day at the works:

"So one fine Monday morning I found myself at 8 a.m. at the foundry in fustian clothes like a working man. (Many a time after I met my father in the streets, and he did not know me). That Monday I shall never forget, it quite disillusioned me. I had to help a smith at the anvil holding things, and standing, standing, oh! Standing for ever. I got home somehow, never before so tired, but had to begin it again next morning at the place at six, after half an hour's walk." [10].

Joy also gave his conditions:

"There were three of us apprentices at the works, and, being too far from home to return for meals, we fed in one of the workmen's houses close by. Breakfast 8 to 8.30; dinner 12 to 1; afternoon tea 4 to 4.30; works closing at 6 and opening at 6 a.m. Catering thus together we did it for about 6s 6d per week. Saturday closing time was 4 p.m."
"I was only a shop apprentice, as they took no others at Fenton, Murray & Jackson's. I got four shillings per week." [10].

On the 1st May the North Midland Railway passed a payment for £1500 for a locomotive [11].

The GWR wrote, in reply to a letter from Fenton Murray and Jackson, on the 4th May saying they were glad the locomotive was so near to delivery. Mr Jackson's note had been shown to the directors, who had agreed to a payment as soon as the firm wrote to say the engine was ready to ship. They pointed out that any delay was now in the firm's hands. In reply to a query Gooch explained that apart from the first two, all wheels for tenders and locomotives were to have strutted spokes and a welded rim and all the springs were to be elliptical [4]. A further letter followed on the 7th May from Crampton stating that Drawing no. 2 for the Goods Locomotives showed the coupled wheels cutting into the frame and so to increase the distance from the wheel centre to the end of the frame by 6 inches [4]. On May 9th Fenton

Murray and Jackson wrote to say the locomotive was on board ship and would sail on Monday morning. The Bill of Lading and List of Materials were attached. They expressed the hope that the amount of payment would be at least the £1500 mentioned by Mr Jackson as they had some large bills that had to be settled on Thursday next, and they would therefore be particularly obliged by a remittance before then [2]. On May 12th Crampton wrote to clarify which drawings for the Goods Locomotive had been sent when, and emphasised that all the spokes were strutted [4]. Gooch wrote on the same day, in which the copy letter surviving states that he is anxious to try the Fenton Murray Jackson locomotive with 11 ½ inch cylinders, which must be copyist error, as it can only mean 15 ½ inches as Gooch requested in his letter of 28th January. He also advised that all improvements in the Goods Locomotives were to be applied to all the locomotives and the drawings for the Goods Locomotives were to be used for the Passenger Locomotives if applicable. He said that Mr Sanders had promised to credit their account with £1500 that day [4].

The North Midland Railway opened from Derby to Rotheram on the 11th May, and gave access to Sheffield via the Sheffield and Rotheram Railway [12].

On the 14th May the second locomotive for the Hull and Selby, by Fenton Murray and Jackson, the Exley was delivered. The locomotive was described as a very beautiful one with six wheels, those in the centre being much larger than the others [12].

There is an entry in the Securities Book of Beckett's Bank dated May 18th 1840, which recorded the receipt of the deeds for premises at Holbeck from Fenton Murray & Co to secure their account [110]. In all probability the money had been used to finalise the Locomotive fitting shops and install the new machinery.

On the 19th May the North Midland Railway passed a further payment to Fenton Murray and Jackson after receiving a certificate from Mr R Stephenson, who managed their locomotive requirements [11].

The 23rd May saw another missive from Gooch on the subject of steeling locomotive tyres, and asking for a list of their requirements so that they could be ordered from the Haigh Foundry. He also asked the firm to market the product to local railways and use them on all their locomotives. In addition he reminded them to order their brass work for the GWR locomotives either through him or direct with Gordon of Manchester. He invited Jackson to the launch of the Charon, hopefully before the end of the week, as she was wanted for the opening on the 1st June [4]. The date of 1 June referred to the opening of the GWR railway between Reading and Steventon, in Oxfordshire. Charon was the locomotive used to haul the director's train on its trial run from London to Steventon on May 30[13], performing the journey in one hour and ten minutes [13].

The Newcastle Courant provided an update concerning locomotives for the North Midland Railway on the 29th May. The firms contracted to supply them with locomotives were listed as Robert Stephenson, Charles Tayleur, Mather Dixon, Fenton Murray and Jackson, Kitson and Laird, Thompson Cole of Bolton, Longridge , and R & W Hawthorn. The locomotives were to have numbers and not names.

Fenton Murray and Jackson wrote to the GWR on the 25th June asking for a further remittance of £400 relating to the Charon locomotive [4].

On July 1st 1840 the Hull and Selby Railway was opened. One of the locomotives used for the opening was the Exley. At the time of the opening, it was reported, that there were 8 locomotives available, 6 from Fenton Murray and Jackson, the other two from Shepherd and Todd with Gray's patent motion. It was stated that the locomotives with Gray's motion were for light traffic only (passenger), and were designed to use less water and fuel to perform a given quantity of work than any method previously used [14].

On the same day the North Midland Railway opened to Leeds, although the first train had operated the day before.

July saw a bid by the Radical Whigs, led by the younger Marshalls to form an association to push for a new Reform Act to further extend suffrage, secrecy of ballot, rebalancing of representation, limiting parliaments to three years, and the abolition of a property qualification for voting. The proposal was indorsed in the Leeds Intelligencer by some 60 citizens and these included Peter Fairbairn, Richard Jackson (assumed to be the member of the firm) and Samuel Smiles [15]. At the inaugural meeting of the new association at the end of August, which was chaired by J G Marshall, a committee was elected which included Samuel Smiles and Richard Jackson [16].

At the Leeds Borough Sessions on July 11th two individuals were found not guilty of stealing copper from Fenton, Murray and Wood (sic) [17].

The Traffic Committee of the Great Western Railway met on the 26th August 1840 and one of the subjects under discussion was the adequacy of the current orders for locomotives. A proposal from Nasmyth Gaskell for 20 locomotives was approved. An offer from Sharp Roberts was read, but it was considered 'inexpedient' to accept it, as there was an ongoing issue with them on quality. An offer from Fenton Murray and Jackson was also considered to construct 8 locomotives at one a month from March 1841 until January 1842, but without deliveries in May and November. It was resolved to place an order with them for 5 locomotives in the first 5 delivery slots offered. An offer from Longridge Starbuck for 6 engines was also considered, and Brunel was authorised to place an order with them subject to price, quality and workmanship [18].

On the 5th September Brunel ordered 6 locomotives from Longridge Starbuck, the letter included the following paragraph:

"The engines to be made according to the specification and drawings furnished by the Company of the Class A with 7 feet driving wheels and 15 inch cylinders, to be in every respect of the best material and workmanship and at least equal in these respects to the engine Charon now running on the line." [19].

On September 10th Brunel ordered 6 further locomotives from Fenton Murray and Jackson, one to be delivered in each of March, April, June, July, September and the last by October 1841. These engines to be in every respect equal to the Charon [19].

On the 3rd of October H C Marshall and one John Forster were nominated as candidates for the municipal elections by the Whig Radicals of Holbeck. One of the retiring councillors was quoted as stating in the meeting:

"He thought that if such great firms as Marshalls, Benyons, Fenton Murray & Co, Maclea and March, Browns etc would come forward at the ensuing election, in the same manner as they got other firms in Wortley to do last year, there was no doubt they would prove conquerors, for the workmen could not get off voting their way, if the masters only used the great influence they possessed". [20].

On October 12th the GWR in the person of Crampton wrote to Hawthorns and to Fenton Murray and Jackson making some further correction to the drawings for the Goods Engine [4].

A 20 H.P. engine by Fenton and Murray with two boilers, all in excellent condition were advertised for sale for £400 by Messrs Pigou of the Gunpowder Mills at Dartford in Kent, in the Newcastle Courant of 16th October.

Margaret Jackson died on the 23rd October in Southampton after a long illness, the cause of death being given as liver disease [21].

In November the Hull and Selby Railway carried out trials between the Fenton Murray and Jackson locomotives and those made by Shepherd and Todd to Mr Gray's patent. The trials were very widely reported at the time [22]. The reports began by stating the improvements included in Gray's design:

1. by building the locomotive on a larger frame with 6 wheels to improve stability

2. reducing the number of sub-frames, so that the crank-shaft has fewer bearings and reducing the amount of machinery by simplifying the valve motion

3 using the steam expansively with the new valve motion, and allowing steam from 25 to 90 per cent of the stroke as required by adverse gradients or wind.

4. by using the best combination of dimensions from experience.

All the above combined to give safety, simplicity, accessibility, and economy.

In initial operations the Fenton Murray and Jackson engines produced a lot of sparks which caused fires to the sides of the track and the fire boxes did not last, resulting in three of them being altered. The results with the altered locomotives were an improvement and Mr Gray called for a comparative trial.

All three types of locomotive were tested, Gray's locomotives (Star and Vesta), the Fenton Murray and Jackson altered locomotives (Exley, Kingston and Selby), and the Fenton Murray and Jackson unaltered locomotives (Collingwood, Andrew Marvel and Willington). The results were summarised in a table:

Class of locomotive	Gray's patent	FM&J altered	FM&J unalt.
Load in tons conveyed over one mile in lbs	1649.400	1649.400	1649.400
Elsecar coke used per trip of 31 miles in lbs	446.980	686.150	1007.780
Coke used per mile in lbs	14.410	22.130	32.590
Coke used per ton per mile in lbs	0.271	0.416	0.614
Water used per trip of 31 miles in lbs	2672.000	4618.000	6432.000
Water per mile in lbs	86.190	148.400	207.500
Water per ton per mile in lbs	1.620	2.790	3.900

It was also calculated that for the usage on the Hull and Selby the annual cost of the coke and boiler costs £2000 £3250 £4500.

The rules applied appeared to be fair on the surface in that all locomotives were brought up to a pressure of 55 to 66 lbs per square inch before starting, and the fires and boilers filled to a given point. At the end of running the pressure in the boilers was about half that at commencement and the fuel and water were replenished and the usage recorded. There were 16 trips made by Gray's patent locomotives, 24 by the unaltered Fenton Murray locomotives and 10 by the altered ones. The trials were witnessed by representatives of all the parties, Messrs J Millar and T Lindsley for Fenton Murray and Jackson; Messrs J Craven and J Barrons for Shepherd and Todd and Messrs E Fletcher, W Bray, J Lynde jun, J Farnell, and J Gray for the railway company. Mr Gray appeared to have made his point, as the report stated that all the witnesses attesting to the results were satisfied with the manner in which the trials were conducted [22].

However the report generated a response in the Leeds Mercury from one of the managers of Fenton Murray and Jackson. Matthew Murray Jackson was the second son of Richard Jackson and was to have a fairly illustrious career at the Swiss engineering company Escher Wyss, and later as Chief Engineer of the Austrian Danube Navigation Company (some 2000 vessels, many of them steam boats). In his letter to the paper, he drew out some facts that had not been made clear in the Trials Report.

1. That the locomotives from Fenton Murray and Jackson had been ordered in 1838 to be to the same specification as the Hawke and Anson on the Leeds and Selby Railway. These had been delivered in 1836/7 and as pointed out locomotive technology was progressing very quickly.

2. The locomotives were not specified to work expansively as they were for working full loads (goods), not the quarter loads used in the trial. A fairer trial may have been to load the firm's engines to their maximum load at a given speed, and then loaded Gray's locomotives with the same load proportionate to their cylinder sizes.(It is interesting to note that in the report made when Gray's engines were ordered, they were stated to be for light i.e. passenger traffic only (see July 1st above)).

3. M.M.Jackson claimed that the water in Gray's locomotive tenders was heated before the start and that in the firm's locomotives was always cold.

4. He wondered that - as the method of working expansively had been patented by James Watt 60 or 70 years ago, what was new in Gray's patent.

5. He pointed out that in the data provided to Fenton Murray and Jackson the coke used by the unaltered locomotives ranged from 0.436 lbs per ton per mile to 0.85, in the altered locomotives from 0.306 to 0.622, and the patent locomotives on some trips had used 0.358 [23].

But then such variations do occur within averages.

At this length of time it is impossible to judge the veracity of the trials, but Mr Gray had an axe to grind, and some doubt over the truth must be allowed over the loading issue.

The following week saw a reply from Shepherd and Todd which appeared in the late editions of the Leeds Mercury, and was repeated the week after. Shepherd and Todd repeated the facts regarding the trails and that they had been accepted by all the representatives. They said great care had been taken to ensure that nothing other than absolute facts were stated in the report. Richard Jackson and Charles Todd had stood back from the trials quite correctly. Shepherd and Todd could only conclude, if Fenton Murray and Jackson were dissatisfied, that either the witnesses had been "shockingly imposed upon" or the premise of the experiments had not been correct. If so, they argued, the matter should be reopened as none of the parties wished to impose a fallacy upon the public [24]. M. M. Jackson was clearly unhappy with the basic premise of the trial but it does not seem to have been revisited, at least in public.

In December 1840 a 30 H.P. cylinder with pipes, valves air pump and condenser etc were up for sale at James Brown and Company's Bagby Mill at Woodhouse. The parts were suitable for an engine with a six foot stroke and were being taken out because more power was needed. The prices and details could be had from Fenton Murray and Jackson, who were presumably performing the upgrade [25].

Mr Bruce of Park Lane Leeds was attempting to sell a Fenton Murray & Co hydraulic press on the 2nd of January 1841.

Meanwhile additional correspondence had been exchanged with the GWR over further orders. On the 15th January Isambard Brunel wrote to Fenton Murray and Jackson in response to their letter earlier in the month. He advised them that after discussions with the Directors he was authorised to order 8 more engines from them to the existing drawings and specifications. Brunel then went into some detail on the delivery schedule, which was now to be as follows, the three Goods locomotives at one a month, February, March and April 1841 and the balance of 18 Passenger locomotives, all to be delivered in 1841, April to July at 2 a month, August and September at 3 a month, October 2, and November and December 1 each. The 8 additional locomotives to be assumed as the 2nd, 4th, 6th, 8th, 10th, 12th and 16 in order of delivery! The price was to be £2000, with a progress payment of £1000 when any 2 boilers for the programme were in the firm's yard completed and ready for inspection. The other parts to be delivered within 6 months and if the schedule was missed the progress payments would be used to substitute for the next ones due. He also pointed out that if the firm chose to procure their boilers from another maker, that was not for the GWR to consider, but Fenton Murray and Jackson would be responsible for such boilers as if they had made them. The firm were to supply tenders for each of the additional 8 locomotives ordered at £500. The copy letter is annotated at the end with the phrase "Tenders at £800". On the surviving invoices the tenders were all charged at £500, so whether this was a mandate from the Directors to negotiate which I Brunel did not have to use, or Fenton Murray and Jackson's original quotation, remains unclear [27].

The comment made by the firm to the GWR in 1840 concerning their inability to take more work due to their current commitments which included a "couple of dredging steam engines and machinery for our government in Ireland" is explained in an article which appeared in the Leeds Times on February 6th 1841. The piece was headed Shannon Improvement Commission. It stated that two steam dredgers had been started in operation on the shoals near Banagher. The vessels which contained the machinery were 80 feet long and had a 19 foot beam and were made from wrought iron. The engines were on the marine principle and of 15 H.P. each and these along with the machinery had been made by Fenton Murray and Jackson, whose agent Michael Peary had fitted them up. The trials were monitored by Col.

354

Jones, one of the Commissioners (for the Shannon) and Thomas Rhodes, the principal engineer. They had expressed their approval after one of the dredgers had removed 38 tons of hard clay and gravel in 20 minutes. The dredgers were named Victoria and Albert [28].

On the 18th of February Matthew Murray Jackson married Mary Shann, the third daughter of Thomas Shann of Hilary House, Woodhouse Lane [29].

In March, after a public meeting on the indebtedness of the Leeds Dispensary, Fenton Murray & Co was one of the firms and individuals who agreed to increase their subscription, in their case from 2 to 6 guineas [30].

On the 24th of March John Dickinson wrote on behalf of Fenton Murray and Jackson to Gooch at the GWR to acknowledge receipt of their account in respect to the locomotives Charon and Cyclops, which was agreed and enclosing theirs with the GWR. He also sent the material list for a tender, advised that their engineer would be with the GWR the following day, and that they would pay the carrier. Dickinson also stated that they did not like the price given for the Glass Gauge and whistle but would discuss it with Gooch on his next visit [31]. They had also invoiced the GWR for another locomotive, most probably the Cerberus on March 22nd.

On the 24th March the Mayor of Leeds was requested to call a public meeting to consider petitioning parliament for the repeal of the corn laws, which request was endorsed by, among others, J G Marshall, P Fairbairn, H C Marshall, Maclea and March, Taylor Wordsworth, Fenton Murray and Jackson and Laird Kitson & Co [32].

During April an advert was placed in Leeds by Mr Lovatt a portrait and miniature painter, in which he said that he had specimens of his paintings including a half-length portrait of Richard Jackson of Park Square, and these were on view at Mr Abraham's Repository of Arts [33].

On the 12th April Fenton Murray and Jackson wrote to the GWR with the invoice for Hecla, the first of their 3 Goods Locomotives to be delivered. The covering letter also describes a potential issue over the delivery of the boiler, as the London and Birmingham Railway had advised that to deliver boilers they would have to be carried on special trucks such as those used by Sharp Roberts & Co. They were in the process of trying to borrow these trucks as all was ready for delivery. The letter also advised that Mr Jackson had made the clash pipe at the reduced diameter of 3 ¾ inches, the let off cocks had been moved and the screw plugs inserted. The description of the locomotive on the invoice includes the following information: Two cylinders of 15 inches diameter and 18 inch stroke, wrought iron connecting rods, 4 eccentrics, wrought iron axles, 4 driving wheels of 5 foot diameter coupled together with outside cranks and connecting rods, 2 trailing wheels of 3 foot 6 inches diameter, ash framing, steel springs; wrought iron boiler with 94 brass tubes of 2 inch diameter and 2 of 1 5/8ths inches diameter, copper fire box, copper steam pipes, 2 brass safety valves, 2 feed pumps and pipes, polished brass cover for the fire- box ; 2 handrails, splashers, buffers, protectors, alarm whistle etc.

The locomotive was of the best materials and workmanship and was to be completely assembled, delivered in London for £1950 in accordance with the contract.

For the steel tyres for the driving wheels there was an extra charge of £42 and for the steel tyres for the trailing wheels one of £16/10/0 [31].

Mr Jackson junior, either William or Matthew, gave evidence on the Employment of Children at the Round Foundry, before J.C. Symons on the 23rd of April 1841, and the following is what was reported;

"That they employ few boys before they are 14 years of age in their business, but several above that age; they are employed by the masters, and paid by them, and not by the workmen; they are not apprenticed. The only exception is in the Boiler making department. In this, a man contracts to do the whole work, and he employs a few lads himself. The hours of work are 59 ½ per week; they begin at 6 in the morning and leave off at 6 in the evening, except on Saturdays when they leave off at 5 in the afternoon; 2 hours are allowed for meals every day out of the 12 hours; these are regularly observed; 1 hour for dinner and ½ an hour for breakfast and tea each. Wages are very good: they vary for adults according to their skill, from 24s to 32s a week; and the total wages paid to men and lads will average between £1 and £1-1s per week. The lads are well treated, and they are not overworked. They have a great variety of employment: filing, fitting, turning, and a variety of jobs. Some of the machinery might perhaps be boarded off more, but there are so few accidents it is not required. The only process at all harmful is that of brass-founding, which is supposed to emit a sulphurous exhalation occasionally producing asthma". [34].

The London Gazette for the 26th of April 1841 listed among the bankrupts Peter Carr, John James Robinson and Christopher Bell flax spinners of Leeds. The petitioning creditors were Isabella Fenton (Richard Yale being still a few months off his majority), Richard Jackson and Samuel Exley described as machine manufacturers [35].

In May a 12 H.P. engine by Fenton Murray & Co with a patent piston, boiler etc. all complete was to be had of William and John Bairstow of Steeton or Keighley [35].

On the 13th of May Richard Jackson wrote to Gooch asking for money. He advised that the first boiler from Tayleurs, for the new 8 locomotive order, was in their shops, which entitled the firm to 1/8th of £9000, as agreed or £1125 all the other parts being on schedule. He then commented that he had hoped that the last payment from the GWR would have been for £2500 not the £2000 received. He mentioned this as they had great expenses and outlays and an extra payment of £500 would ensure the firm's maximum effort to achieve fulfilment of the orders. In about a week they would be despatching a passenger locomotive and a goods tender and two weeks after a goods locomotive and another tender. Jackson then mentioned he had started work on a brass fire box ¼ full size and would let Gooch know the results of the experiment [4]. The letter has two interesting points. Jackson's claim does not coincide with Brunel's letter of January 1840, which clearly states that a progress claim of £1000 would be payable when there were two boilers available. This appears however to be the result of subsequent negotiations and agreement. Two payments are shown in the accounts paid on May 17th, one of £500 and the second of £1000 [36]. Jackson received the payments he needed for it is apparent that the firm was in great need of cash. The second point is the last item in the letter, a number of the letters between GWR and Fenton Murray and Jackson mention changes, and revised parts or experiments, so it is clear that Gooch was looking at alternatives in his specifications and that Fenton Murray and Jackson, in the pre Swindon days, were one of the companies entrusted to build and try them out.

In May the machinery of Messrs Carr Robinson and Bell was up for auction at the mill in York Street Leeds. It consisted of 9 spinning frames, 3 gill roving frames, 4 hackling machines, 1 breaking machine, a switching machine etc. The advert stated that the machinery was new and had been made in the last few months by Fenton Murray & Co and other respectable makers. Further details could be obtained from Fenton Murray & Co or the auctioneer [37].

The GWR was planning the establishment of its own locomotive works at Swindon at this time and had started obtaining quotations for various pieces of equipment. They wrote to various companies for quotes for stationary engines. On June 14th Fenton Murray and Jackson submitted their quote which summarised was: one pair of condensing engines £1000, another the same £1000, 6 boilers and grates for these £720, a 4 H.P. high pressure engine £340 and the boilers and grates for the 4 H.P. at £260 in all £3220. Each pair of condensing engines was to have one fly wheel and one governor, 4 double acting pumps and 3 safety valves. The boilers were to contain 27 tubes each of 4 inches diameter. Quotes were also received from Nasmyth Gaskell & Co at the Bridgewater Foundry, the Vulcan Foundry, Harrington and Stothert and Slaughter at the Avonside Ironworks. Repeat quotes from Nasmyth and Stothert are recorded in October and November and are much lower (£1500 to 1700) than the Fenton Murray and Jackson figures, so these were probably the front runners to win the business [38].

A further letter from Fenton Murray and Jackson was written on the 23rd of July in response to one of Gooch's of the 19th. The letter addresses the layout of smith's shops and locomotive matters. In regards to smith's shops they sent Gooch three plans showing the arrangements for the smith's shops at the Round Foundry, and gave prices for jewel irons ranging from £6/6/0 a set for the largest to £5/10/0 for the smallest size. They described using a fan to distribute the blast up to 150 yards with main light cast iron pipes of 10 inch diameter and branch pipes of 5 inch diameter. They piled the anvil block, and drove the piles in with a large striking hammer. They could supply green sand castings, loam castings and anvils and they would have been glad to have the order for the fan. On the locomotive front they said they would be careful to alter the next set of pistons as Gooch requested. Whether this was because they had not made the last ones to specification or Gooch wanted to try something different was not clear. They were loading the Harpy on the 24th and testing the Pluto in the works the following morning, and had ordered the ships to transport them both to London. A few more locomotive names were requested in order that they could start making the nameplates [39].

On July 26th Fenton Murray and Jackson sent the invoice for the passenger locomotive Harpy with a covering note that she had been shipped that day on board the 'Coke' commanded by Captain Smith. The description is very similar to that of Hecla with the necessary changes as; the cylinders were 15 ½ inches diameter with an 18 inch stroke, 2 driving wheels of 7 foot diameter with 4 supporting wheels of 4 foot diameter, the boiler had 127 brass tubes of 2 inches diameter and 4 of 1 5/8ths inches diameter. The price was £2000 with the additional price for the steel tyres at £74. The tender was described as having 6 iron wheels of 4 foot diameter, a wrought iron tank, ash framing, 6 elliptical steel springs, wrought iron axles and brakes apparatus etc etc. the price of the tender was £500 plus £57 for the steel tyres [31].

The invoice for Pluto followed on July 29th [31].

On August the 10th Jackson wrote to Gooch from Toulouse. The main concern again was for payment. He hoped that the GWR had two more of their locomotives in full operation since they had last met. He would not be able to get any letters from Leeds until he reached Marseilles in a few days. However, "*as our House in Leeds is poor*" I know you will, as in the past do what you can to expedite our payments. He closed by hoping to see Gooch in London at the back end of the month [40].

On the 17th of August Fenton Murray and Jackson wrote to Saunders at the GWR, again on the subject of payment. Having hoped that that the last two locomotives delivered were giving satisfaction, they went on to say that they had some major payments due the next Friday and needed another remittance. Tayleur had delivered another boiler and wanted payment. They would be despatching the next locomotive, Minos, in the next week. They enclosed an account for £5159/10/91/4 d, which included the progress payment due on the second boiler. Minos was the locomotive invoiced on September 2nd [31].

Matthew Murray was not forgotten within the Marshall family, and when writing, on 10th September 1841, to J G Marshall, Henry Marshall, in discussing a proposal for a hackling machine received from Taylor and Wordsworth, said (in précis) 'their plan of heckling both sides at once by two flexible gill sheets is a very old one: - it is as Girards patent of the same date with that for gill preparing & wet spinning. Matthew Murray had a plan of the same sort. Whether the heckles be connected into a chain by links or by leather straps' [41].

On the 27th of September Fenton Murray and Jackson advised the GWR of the despatch of Ixion on board the 'Coke', captained by John Smith, sailing on the morning of the 28th. They also notified the despatch of 2 Jewel Irons, which had presumably been ordered by Gooch after receiving the firm's letter of the 23rd July. The descriptions on the invoice for the locomotive and tender do not add to those already recorded, but there are additional items. A wrought iron cranked locomotive axle turned, fitted up and case hardened is included and charged at £90. The 2 wrought iron water jewels for smith's fires were charged at £3/10s for the pair. And Israel Jackson's freight for the Pluto and Tender were deducted after being paid by the GWR [31].

The finance report of Newcastle upon Tyne Town Council dated 29 September 1841 noted that a need for a new dredger had been raised despite the fact that they already had one. A financial provision of £2000 was budgeted [42].

In October 1841, Britain was in the grip of a long recession, and there were thousands of people without work, and many were living in total poverty. A few of the working population of Leeds started a movement to properly record the distress and facts of the unemployed in the borough. They formed a committee which was called the 'Unemployed Operatives Enumeration Committee', which spread the message through much of the employed population of Leeds, who agreed to fund the work of the committee out of their wages. The funding started coming in in September 1841 and carried on until January 1842. The Committee employed 20 literate members of the unemployed, to interview the households in each ward, 2 per ward, and a further person who collated and performed the statistical calculations. They published their report in October which was summarised in the press in the following statistics: Leeds borough contained 747 families or 3960 individuals with an income of 1s.4d per head (under 7p in today's money without inflation adjustment); 214 families or 1294 individuals with an income of 4d 3 farthings (about 2p); and 1946 families or

5776 individuals with no income at all. There were 4752 families or 19937 individuals subsisting on 11d and a farthing per head (under 5p). The committee accepted some of the data given to them was probably inaccurate, but said even if only 50% was accurate the picture was still wholly unacceptable. Mr Beckett, one of the MPs for Leeds ensured that a copy of the report was given to the Queen, and others went to the Prime Minister Peel. There was strong feeling that the politicians in London were completely 'effectually excluded from a knowledge of common life'. The Duke of Wellington had asserted 'that every Englishman can obtain a competency by industry', something the press would not allow him to forget. Holbeck was unsurprisingly the worst hit ward in Leeds and the statistics compiled for it were:

				Holbeck Ward						
Number of families	4	3	18	52	907	137	162	42	25	766
Number of persons	21	15	84	270	484	811	932	238	912	3767
Number of employed	8	5	35	105	173	248	219	43	0	836
Unemployed + dependents	13	10	49	165	311	563	713	195	912	2931
Total Weekly Income	£4/2/2	£2/11/0	£11/18/0	£30/9/2	£43/11/4	£53/0/9	£39/12/3	£4/14/3	£0	£189/19/1
Av. Per head per week	3s/11 1/2d	3s/4 3/4d	2s/10 d	2s /3d	1s/91/2d	1s/3d	10 1/2 d	4 3/4d	0d	1s0d

Some groups of operatives from most of the main employers in Holbeck gave donations, maybe not every week, but most. There were three groups collecting at the Round Foundry, William Hesketh collected at Fenton Murray & Co, a collection came from the boiler makers at Fenton Murray and Jackson and another from the Old Side (Foundry?), with various people making the collections [43].

On the 22nd of October Fenton Murray and Jackson submitted the invoice for Hecate to the GWR, advising that she would be despatched the following Monday. They stated that Mr Parkinson had been late supplying the Dome and Splashers, so he was presumably Gooch's nominated supplier for these parts. The next locomotive the Gorgon was to be tested the following week. They said they were setting up a boiler from Tayleur, expected another in a day or two and a third next week. They could not get the hoops (steel tyres) from the Haigh Foundry fast enough. And of course they would be particularly obliged by another payment! The otherwise standard invoice includes a charge of £6/6s for extra lengths of necks on the leading and trailing axles for Ixion and Hecate[31].

Following the bankruptcy of the partners earlier in the year, October also saw the auction of the 16 H.P. engine by Murray of Leeds and the oil presses that had been used by George Popple & Sons Merchants (oil) from Hull [45].

On the 11th November, Fenton Murray and Jackson submitted to the GWR the invoice for the Gorgon and its tender. The invoice covered the standard charges with the modifications for steel tyres and long neck axles. They referred to an issue with a wheel on the Minos, which Mr Jackson had investigated. Unfortunately the word describing the matter is blotched out [31].

The Bristol Mercury of November 13th carried an advert by one Harman Visger to sell among other items a 6 H.P. condensing steam engine by Fenton and Murray of Leeds with boiler complete.

On the 7th of December the invoice for the Vesta was submitted to the GWR which was to leave Leeds the following day with Captain John Smith and the Coke. They were also sending a new 7 foot hoop for the Minos wheel. Payment for at least two locomotives and tenders was requested. On the 9th of December they acknowledged receipt of £2000, being the value of one locomotive without a tender [31].

On the 18th of December Brunel responded to a query from Maudslay and Field, who were investigating possible locomotive manufacturers for a foreign client:

"To recommend one manufacturer is to a certain extent to condemn others & therefore it is rather an invidious task to reply to your enquiry – at all events you must consider it private and you must not quote me. Fenton Murray & Co have turned out the best work decidedly – their workmanship is perfect – Stephenson of course is very good – both of these makers get high prices. Hawthorns turn out good work & moderate prices – Nasmyth's Gaskell of Manchester and Stothert of Bristol have made excellent engines for us and at low prices – but they all require close looking after – and [?] should be employed in England to look after a foreign order". [46].

High praise indeed to be ranked for quality above Stephenson, and the general tribute was repeated by Daniel Gooch in his diaries:

"all the makers did their work well: but the best were built by Fenton Murray and Jackson of Leeds, but the great durability of the engines has proved that all did their work well. I very frequently visited the various works where they were built." [47].

On the 23rd of December the firm invoiced the GWR for the locomotive Proserpine, according to the surviving list [31].

In the early morning of the 24th December Hecla (one of the Leo class goods engines) left Paddington hauling a mixed train of passengers and goods. At Sonning cutting, between Twyford and Reading, the embankment had slipped, causing earth to cover the rails. Due to the poor visibility the driver failed to notice this and the locomotive ran directly into the earth at around 7 a.m. The sudden stop from a speed of between 16 to 18 mph caused the locomotive to jump the tracks and the carriages to violently impact together and turn over. 8 persons were killed on the spot, and a number of injured, some seriously, were admitted to the Royal Berkshire Hospital in Reading [48].

Towards the close of the year 1841, on December 27th Fenton Murray and Jackson provided a statement to the GWR, showing that they were owed £5880/16/5 farthing, which either crossed or prompted a payment by the GWR of £5000, which the firm acknowledged in a letter to Saunders at the GWR on December 31st [31].

January 6th 1842 saw Fenton Murray and Jackson invoicing the GWR for the next locomotive, Acheron, and its tender which were entrusted to Captain Smith and the 'Coke' [31].

Acheron is probably the most looked at of the Fenton Murray and Jackson locomotives as she appeared as the locomotive shown emerging from a tunnel in John Bourne's famous drawing used as the frontispiece of his tome on The Great Western Railway.

69. Acheron from the frontispiece of Bourne's GWR.

January 1842 saw a meeting in Leeds of engineers and businesses on the subject of Smoke Nuisance. The object being that there should be agreement that smoke could be consumed without injuring boilers and with reduced coal consumption, and agreement on the best methods. Richard Jackson was in attendance, as was Peter Fairbairn, Charles Maclea and a wide selection of others. William Beckett MP was in the chair. Makers of Smoke Consuming apparatus had been invited to attend and speak for 15 minutes on their methodology and inventions [49].

An inquest on the 18th of January was held into the death of George Howden 32 years of age. Howden was a smith working for Fenton Murray and Jackson. On Saturday the 15th of January he had been working on a large locomotive wheel which weighed a ton. It had needed to be placed in the fire, so he had suspended it from a

362

crane, some part of which had failed, and the wheel and crane fell on Howden crushing him to the workbench. He was taken to the Infirmary, but died on Monday. The verdict was 'Accidental Death' [50].

On the 5th of February the invoice for the Erebus and its tender was made out, the covering letter also advised that a tender would be sent off on the Wednesday by railway. The firm asked that the GWR should supply 4 more names to complete those required for the order, and could they receive a payment. On the 10th a tender was duly despatched and invoiced. The carriers were Staley and Sherman. Fenton Murray and Jackson had been asked to quote for a hydraulic press for the GWR and been advised that their price was too high, and said they would have another look at it. There is no record that they ever delivered a press to the GWR. The payment for the locomotives offered, they said, would be very acceptable [31].

The papers in the middle of February announced the Holbeck subscription for the relief of the distressed poor. At a meeting in January it had been decided to raise a subscription due to the "severe and extensive distress existing in the Township, caused by the long depression of trade". Thomas Benyon was the chairman of the committee and H. C. Marshall the treasurer. Fenton Murray and Jackson contributed £20, among the top seven contributions that were recorded [51].

On the 28th of February the GWR were told that the Styx and tender were being shipped on board the 'Young Man's Endeavour', Captain Thomas Earnshaw commanding. The letter requested a meeting between Gooch and Jackson. Jackson was leaving for London and a note could be left at the Queens Arms, Newgate Street as to whether a meeting was possible the following Wednesday either in Paddington or the City. Jackson was looking for more orders [31].

Saturday 5th March saw the adverts for the sale of Campion's Flax Mill near Whitby, Mr Campion having been declared bankrupt. The mill was described as having the carding room on the ground floor, spinning room and workshop on the first floor, the preparing room on the second floor with a Heckling Garret above and having an excellent 16 H.P. engine by Murray [52].

March 30th was the day upon which the GWR was invoiced for the 'Hydra' by Fenton Murray and Jackson [31].

On April the 5th the firm followed up with the invoice to the GWR for the 'Lethe' [31].

In April a 10 H.P. engine with the fly wheel and boiler by Fenton Murray & Co could be had at auction at Kidacre Mill near the North Midland Railway Station in Leeds [53].

The Leeds Times published a letter on the 16th of April which was part of the correspondence on Smoke Prevention. The letter was from a company called Dircks & Co in Manchester, and brought to the attention of the readers the Patent Argand Furnace of C. Wye Williams, which they made. The following paragraph related to Leeds

"At Messrs Fenton, Murray, and Jackson's, Leeds, the Argand Furnace is applied to a waggon boiler, the grate is considerably lessened, small slack is used as formerly, there is an abundance of steam, and no smoke is made. When a little round coal, for experiment, was mixed with the slack, the generation of steam was so much increased, as to oblige the fireman to cool the furnace by opening the door." [54].

Five men were arrested for highway robbery and committed for trial in mid-April. A John Motley, landlord of the Rising sun in Camp Field, and a grinder at Fenton Murray & Co was returning home about 1 o'clock on Sunday morning. He was attacked by the five men, dragged along the ground until nearly senseless by his neckerchief and robbed. He hung on to one of the villains by the leg until help came [55].

On the 30th of April Marshall & Co advertised for sale a 30 H.P. condensing steam engine with two boilers and all their apparatus complete. The engine could be seen working on the premises in Water Lane, and had a 30 ½ inch diameter cylinder and a 6 feet stroke and was made by Fenton and Murray. The sale was in consequence of the decision to close Mill A in Water Lane as the new mills in Marshall Street could spin much finer yarns due to the new wet spinning machinery. This was the engine supplied to Marshall before the end of Watt's patent, so had been at work for over 30 years, but the new mills had modern 70 H.P. engines, so this one was surplus [56].

On the 3rd of May Fenton Murray and Jackson prepared the invoice to the GWR for the locomotive 'Medea' and tender. The ship was again the 'Coke' under Captain John Smith. The Letter asked for a remittance that week [31].

On Sunday May 8th an appalling accident happened on one of the railways in Paris. On that day had been celebrated the King's fete at Versailles with great waterworks and a firework display. The train in question, run by the Paris Versailles Left Bank railway, left Versailles around 5.30 in the evening and consisted of 16 carriages pulled by two locomotives. The first locomotive, a four-wheeler, was called 'Mathieu Murray' and made by Fenton Murray and Jackson. The second locomotive was a six- wheeler, called 'L' Éclair' and had been manufactured by Sharp Roberts. On descending into Meudon, in the suburbs of south west Paris, the 'Mathieu Murray' began to oscillate across the tracks to the extent that the front wheels left the rails and were jumping from sleeper to sleeper. At some point the front axle broke and fell and was left 90 to a hundred yards behind the place where the 'Mathieu Murray' suddenly stopped and overturned. The second engine ploughed straight into the wreck causing extensive damage and both engines dropped their fires. Unfortunately the impetus carried them over the fires and some of the wooden, freshly painted carriages came to rest on the remains of the fires and immediately ignited and blazed. As was a common practice at the time the carriage doors were locked. There were around 168 casualties of whom 54 died, some at the scene, some later. The discussions over the sequence of events, the precise causation lasted months. The axle was determined to be of excellent metal. There was much discussion over the prudence of putting a more powerful 6-wheeled locomotive behind a 4-wheeled one of about 12 horsepower [57].

Sometime earlier there had been a disastrous fire in the city of Hamburg and there was a collection taken up in Leeds for the relief of its citizens, to which Fenton Murray and Jackson contributed £20 [58].

On the 24th May Fenton Murray and Jackson invoiced the GWR for the 'Medusa' and tender and said that they had been loaded on the 'Mary Ann', commanded by Captain Horn [31].

On Monday the 13th of June the first rail trip by a British reigning monarch took place, when Queen Victoria travelled from Slough to Paddington on the Great Western Railway. The journey took 25 minutes. The train was drawn by the locomotive 'Phlegethon', which with its tender had been manufactured by Fenton Murray and Jackson [59]. The locomotive had been delivered under the name 'Styx', but someone in the GWR had had second thoughts about the name, and it was changed to 'Phlegethon'. Whether it had temporary name plates or Fenton Murray and Jackson were just slow in invoicing is unclear, but the firm did produce *"two extra name plates with brass letters 'Phlegethon'"* and they were invoiced on September 2nd [31].

The 'Medusa' was followed by the 'Ganymede' and tender on the 17th of June and the covering letter included a sentence on the recent royal trip

"We were much pleased to see in the papers that Her Majesty had honoured you with her first trip on your railway, and we were also glad that all went off so well".

The letter concluded by requesting a remittance as the firm had heavy payments to meet at the end of the week [31].

At this period many manufactories and foundries operated a custom known as 'footing', which required the principal person to open a 'pot' for drinking. Typically for a wedding the groom would have to stump up around 10/- (a huge sum) and the others in the shop would put in 6d each. This practice was not acceptable to many of the Victorian reformers and so changes tended to be noticed. Who was behind or why the following change was implemented in Fenton Murray and Jackson remains a mystery. However on June 25th 1842 the Leeds Times announced "A Noteworthy Example" in that Thomas Long of the firm was presented with an elegant bible on the celebration of his marriage instead of *"footing him"* as would have happened before. Footing was the custom on the completion of an apprenticeship, or another milestone in a man's life such as marriage, and on July 21st it was noted that the smiths and stokers at Fenton Murray and Jackson presented a bible to the young men completing their apprenticeships [60].

On the 8th of August 1842 Fenton Murray and Jackson invoiced the GWR for the 'Argus' and tender which were consigned with Captain John Smith on the 'Coke'. The invoice contained a few extra items in terms of copper pipes and screws and bolts. The firm also advised the GWR that they were estimating the cost for the new regulator and would let the GWR know the price. The 'Argus' was the last locomotive to be built at the firm of Fenton Murray and Jackson, and the delivery heralded the beginning of the end of the firm [31].

August 1842 saw severe industrial action in many areas including Leeds. The action was aligned to the Chartist Movement (for electoral reform and male suffrage) but the root causes were more immediate than political. Many operatives were being faced with wage reductions, shorter hours and lay-offs. At the same time the price of bread and other food was rising. Attempts to even up the political divide by obtaining electoral reform, in the guise of a petition to Parliament signed by over 3 million people, had recently been rejected by 287 votes to 49. The striker's plan was to visit the industry in each region on a particular day and attempt to stop all the works mainly by knocking the plug from the steam engine boilers, hence the common name for the actions 'the plug riots'. The aim was to do this relatively peacefully and if the mill-owners co-operated there was no violence, but the works were to be stopped. The unrest spread from Lancashire and Yorkshire through Staffordshire, Derbyshire and the rest of the Midlands, with the coal miners striking even in Scotland. The Governments reaction was firm in that a Royal Proclamation was issued offering a reward of £50 for successful prosecutions. Special Constables were signed up, and the police were armed with cutlasses. The army was called out, and military law enforced under the Riot Act.

The army force for Leeds was the 17th Lancers commanded by Prince George of Cambridge, 4 troops of the Yorkshire Hussars commanded by Col. W Beckett M.P. and 48 Horse Artillerymen with cannon. Fortunately matters passed off relatively peacefully in Leeds town (Manchester (policemen killed) and Halifax were

366

far more violent); there were more serious disturbances at Pudsey. The main event, around Leeds, was in Holbeck on Wednesday 17th August. In the morning a large body of demonstrators were dispersed by the military and the police; however in the afternoon further protesters marched around Water Lane and Meadow lane. The mills at Marshall's, Benyon,s, Titley Tatham and Walker's, and Maclea and March's all had their engines stopped, after various degrees of scuffling and threats mainly at the two latter establishments. At Fenton Murray and Jackson's one report said all the men were hanging out of the windows watching, and so were told they could leave, and presumably the engine was stopped. The major disturbance was at Maclea and March's as this was where the police caught up with the demonstrators and most of the arrests were made. Some 40 men appeared in court on Thursday and Friday mainly for unlawful assembly. One police officer was stabbed, but in the confusion there was insufficient evidence to convict on this offence. The men were all fairly local and 3 were reprimanded and discharged, 10 were fined 5/- or 2 months imprisonment and bound over for 6 months, and 27 were committed to York Castle to be tried at the next assizes. This item has been dealt with in some detail in order to highlight the state of the manufacturing districts, and the intense feelings of the operatives. The Leeds Times while stating that they thought the operatives had gone about their protest in the wrong way made it clear that they could not be blamed as they had to do something to try and gain alleviation of their plight [61].

On September 2nd Fenton Murray and Jackson wrote to the GWR apologising that their erector J Millar had had to leave so suddenly, but that he was urgently needed to fix a steam engine in Lincolnshire. They enclosed their account for £4615/5/1 ½ d and requested a remittance for as much of the balance as possible in the next week. They were hard pressed by their copper and brass tube suppliers. The letter also covers an additional charge for some modified valves on the Argus [31].

At the time the legal system moved swiftly (certainly by today's standards) and by the 10th September the papers were reporting the sentences handed down for the Leeds plug rioters, which varied from being released on their own surety only to be sentenced if there was a future offence, with the majority receiving sentences of from 3 weeks to 6 months in prison. The one offender thought to have stabbed a policemen, who had anyway been in the forefront of trying to pull the plug at both Titley Tatham and Walker and Maclea and March received18 month's hard labour.

On the 1st December Mary Anne Jackson, Richard Jackson's eldest daughter, married Andrew Faulds junior, a coal agent from Darley near Barnsley [62].

In Newcastle upon Tyne the river Committee reported that 84,000 tons of ballast had been dredged up from the Tyne, and that new equipment had been purchased including a dredger with equipment from Fenton Murray and Jackson which cost £1500 [63].

The records of the Charleston and Hamburg Railway record that locomotive no 12 the Columbia built by Fenton and Murray was rebuilt [64].

A case before the Leeds Borough Sessions was reported in the Leeds Intelligencer of December 31st, between the parish of South Crossland and the Parish of Holbeck. The Parish of Holbeck had sent a pauper, one Thomas Brayshaw, back home to South Crossland to be on their charge. South Crossland argued that Thomas Brayshaw had been contracted under an agreement for 4 years to Fenton and Murray of Holbeck, the contract being sound had resettled the man in Holbeck parish. Holbeck argued that the contract was such that it did not constitute a resettlement of Brayshaw into their Parish. The judgement went in favour of South Crossland subject to a ruling in the Court of the Queen's Bench [65].

In January 1843 the Wellington Street Mills in Leeds went up for sale, including a 40 H.P. steam engine with 3 boilers of 33 H.P. each, all by Fenton Murray & Co [66].

February 1843 saw Marshalls and most of the other flax spinners shorten their hours again. The Leeds papers reported a large company failure in the flax business in Doncaster/Rotheram. They also reported that the value of a factory erected several years ago for £60,000, had now been valued at just £13,000 [67].

At the end of December 1842, Richard Jackson's brothers in law had cancelled their partnership, Charles G Maclea leaving the firm for health reasons [68]. The firm continued to operate as Maclea and March under the control of J.O. March. In January Charles G Maclea was treated to dinner by his former workmen and presented with a gold snuff box [69].

Maclea continued to do some work for the firm as evidenced by his presence as part of a deputation of flax and tow machine makers from Leeds who had an interview with the Earl of Ripon, whose deputy was W.E.Gladstone, on February 21st 1843 at the Board of Trade. The other members of the deputation were P. Fairbairn, J. Wordsworth (Taylor Wordsworth), S. Exley (Fenton Murray & Co), M. Johnson and J Hetherington. The purpose of the visit was to represent to the Government the depressed state of the machine making business, and to obtain permission to export flax and tow machinery to the continent. The export of such machinery, at the time, was illegal. Although the Government was setting up a select committee to reconsider the matter, the machine makers were aware that there was a large value of orders waiting to be placed by continental manufacturers, and unless assurances could be given in the very near future, the orders would be given to continental firms [70].

A further extract from David Joy's diaries indicates that the order book for Fenton Murray and Jackson was bad:

"*Spring came and times were slacker. Finally Mr Jackson sent for me to tell me that they were going to close the works, so I should not be wanted*" [10].

On March 7th 1843 the Securities Book for Beckett's Bank records securities on the Foundry Estates at Holbeck in the name of Fenton Murray & Co and also under the partner's names Jackson and Yale [110].

In April the Steam Mills at Spalding, Lincolnshire, used for Corn, Cake and Bone were put up for sale by Henry Hawks. They were operated by a 20 H.P. steam engine by Fenton Murray and Jackson fitted with Stanley's patent fire-feeder [71].

In May the case of Thomas Brayshaw [see December 1842] came before the Queen's Bench for a ruling. The case is of interest because it describes Brayshaw's terms of employment. On May 31st 1824 a written agreement was made between T Brayshaw, on the one part and James Fenton and Matthew Murray, whereby Brayshaw agreed to serve Fenton and Murray for 4 years (he had already worked for them for 2 years by this time). He was to turn iron or any other artisan work they set him and he promised to give his whole time to their work between their usual working hours from 6 in the morning to 6 in the evening whether in the shop or working off site. He agreed to direct and assist those he worked with and those under his care, and he would account for his time when asked. He would not communicate with or make known to or show any of Fenton and Murray's plans or methods of working to any person in the trade. In return he would receive a weekly wage, which would be 8/- in the first year, 9/- in the second year, 10/- in the third year and 11/- a week in the fourth year. A few years earlier the hours had been 6 a.m. until 7 p.m., but the workmen had 'turned-out' and since then it had been 6 until 6. The evidence then showed that outside this time the workmen were free to do other work for other masters and while it was unusual for a man to refuse overtime from Fenton and Murray, which was usually on offer when times were busy, they could. If times were slack they might work short hours, but would have their wages cut pro rata. The point for the case was Brayshaw had gone to live and work in Holbeck and obtained 'settlement rights', and should not be sent back to his original parish when he became a pauper [72].

Mr and Mrs Land's mill in Water Lane with a 20 H.P. steam engine by Fenton and Murray was advertised as available for letting in May [73]. Later in the month the Flax and Tow Mill at Priestgate in Darlington was for sale with its 30 H.P. steam engine by Fenton Murray and Wood [74]. Both these mills made regular appearances in the 'for sale' or 'to let' columns of the press.

In June adverts for the sale in July of Flax and Tow machinery at the Castle Mills in Knaresborough appeared. The machinery was stated to be nearly new and from the best makers. The analysis shows a split in the manufacture as follows: Fenton & Co – drawing and roving frames, Lawson & Co – strong drawing and roving frames, Taylor Wordsworth & Co – tow cards and heckling machine and Hyde & Co tow drawing and roving frames [75].

In July the announcement of the Grand Jury for the Leeds Borough Sessions also stated that 5 individuals had been summoned and not answered their names. As a result they each incurred a £5 fine. One of these latter was Richard Jackson, who was no doubt away trying to find business [76].

An excellent 6 H.P. condensing steam engine by Fenton Murray & Co was offered for sale by W & J Slingsby of Bell Busk Mill on July 15th [77]. A week later a Hydraulic Press by the same was available for £120 on application to the offices of the Leeds Intelligencer [78].

In the House of Commons on the 10th August the second reading of the bill on Exportation of Machinery took place. In a long speech in support of the removal of export restrictions Mr Gladstone made some interesting remarks. Gladstone was at this time the deputy to the Earl of Ripon with whom the February meeting had been held, and at which Gladstone was present. Gladstone said that within the last 3 days he had been approached by a most respectable machine maker of Leeds, who had also visited him the previous autumn. There were a number of orders waiting to be placed by continental firms for textile machinery, which could be delivered to the Leeds firms if the customers could be assured that they would be able to import the machines without duty and legally. As no such assurances could be given, this gentleman had advised in the last few days that the orders had gone to the Belgian machine makers. Mr Gladstone quoted the firms who had been approached to supply these extensive orders in the flax field; they were Messrs Taylor Wordsworth, Fenton Murray and Jackson, and Mr P Fairbairn [79].

On August the 23rd Beckett's Bank recorded in the Securities Book against Jackson and Yale and the firm that a memorandum had been written not to exceed the £4000 security [110].

On the 28th of August I K Brunel wrote to Richard Jackson saying that unfortunately he was leaving town for 10 days and therefore could not meet him as suggested. The locomotive engines required for the Bristol and Gloucester Railway were to have a longer stroke and smaller driving wheels than those of the GWR.

370

They should also be lighter with less ornamental brass work. He said that he would be very happy to respond to any queries.

Letters went to other suppliers as well which said that the position was not the same as for the firefly class, as it was down to the suppliers to propose the specification they would supply [80].

Various newspapers at this time highlighted the woes of the copper industry, with widespread reductions in wages.

In October, after a public meeting, an organisation called 'The Leeds Tradesmen's Benevolent Association' was set up. There were Ward representatives appointed to manage the local affairs of the society. In Holbeck ward, one of the 4 appointees was Samuel Exley [81].

At the end of October Clarke Plummer & Co's Northumberland Flax Mill in Newcastle upon Tyne offered for sale parts of a 40 H.P. condensing Steam Engine, now working but about to be taken out, and made by Fenton Murray & Co. The components included the cylinder, piston and piston rod, air pump etc. A selection of flax and tow machinery was also up for sale, but the makers were not stated [82].

On the 7th Of December the Bradford Observer contained an advert offering a Hydraulic Packing Press by Fenton Murray & Co for £120, applications were to be made to John Clapham a general commission agent in Mill Hill Leeds.

On the 18th December 1843, as required under the partnership agreement, Samuel Exley wrote to Richard Jackson and Richard Yale in the following terms;

"Gentlemen, I hereby give you notice pursuant to the 19th Clause of the Minutes of Partnership under which we have carried on business, that it is not my intention to form a fresh Partnership with you or either of you on the expiration of the term of our present Partnership carried on under the firm of Fenton Murray & Co but that it is my intention to retire from and determine the co-partnership now existing between us under the said firm on the 31st December 1844." [83].

Under the terms of the partnership agreement Fenton Murray and Jackson served as bankers to Fenton Murray & Co and during 1843 Exley had been having more and more difficulty and delay in securing the monies needed from Fenton Murray and Jackson in order to run Fenton Murray & Co, he had to draw on them for all his cash including wages.

During 1843 Fenton Murray and Jackson carried out some further work for The Shannon Commissioners and produced some lock sluices for the lock at Meelick near Bannagher. These still exist and have the only known name plate ever produced by the firm that has survived.

70. maker's plates from Meelick Lock on the Shannon Navigation.

On the 26th January 1844 Exley was served with a writ or summons in an action by William Foster & Co against him and his partners. It was an action to recover a debt of £158. William Foster & Co was a large copper company involved in mining and supplying its product. In their letter to the GWR on September 2nd 1842 Fenton Murray and Jackson had used as a reason for asking for another payment that they were hard pressed by their copper and brass tube suppliers. The sum was owed by Fenton Murray and Jackson (partners Richard Jackson and Richard Yale). While Exley knew that he could in no way be liable for this debt, the fact that Fenton Murray and Jackson had allowed a debt to be unpaid for so long that it resulted in such action, aroused Exley's suspicions. Exley immediately went to Fenton Murray and Jackson's counting house, and finding neither Jackson nor Yale present he spoke to their cashier John Mavin Dickinson. Dickinson told Exley that W Foster were not the only creditor chasing Fenton Murray and Jackson for payment, which he substantiated by producing a statement of liabilities and a letter from W. Foster. Dickinson added that their liabilities greatly exceeded their assets and in his opinion they could not continue in business beyond February. He also advised Exley that the assets of Fenton Murray & Co were being used to discharge the liabilities of Fenton Murray and Jackson, and that any further receipts from the Machine Business, or a large proportion of them would go the same way [83].

The Barnsley Union Calender was to be auctioned on February 1st with its two excellent steam engines one of 12 H.P. and one of 17H.P. The advert stated that one was made by 'Watt & Bolton of Birmingham' and the other by Fenton and Murray of Leeds [84].

Exley decided Fenton Murray & Co's position could not continue and went and collected £450 of money owed to the firm, which he did not pass to Fenton Murray and Jackson. He did inform Jackson of his actions and explained that he was merely trying to ensure that the money was used for its proper purpose. Jackson admitted that Fenton Murray and Jackson was in the hands of certain of its creditors who had been directing the business for some time. About a year previously W Foster had been owed over £600, which had been reduced to the £158 'given in the writ' using monies from Fenton Murray & Co.

Jackson however demanded that Exley hand over the £450 he had collected, which Exley refused. Jackson then advised the remaining debtors of Fenton Murray and Co that they should not pay their dues to Exley [83].

On the 10th of February 1844, Exley's solicitors, Snowdon and Preston received a letter from Jackson's solicitor Mr Shaw:

"Re Fenton Murray & Co
Mr Jackson and two of the principal creditors of this firm [Fenton Murray & Jackson] have called to consult me as to the course to be pursued in consequence of the recent proceedings of Mr Exley and from the statements made to me I have felt it right to advise Mr Jackson not to make any payment for either debts or wages connected with the Machine Business so long as Mr Exley retains the money he has collected in breach of the terms of their partnership articles. The consequences of this will no doubt be that the workmen will give up their work and that the present action against Messrs Jackson and Exley will proceed and if a verdict be obtained against Mr Exley all the other creditors who have furnished goods for the Machine Business will proceed against Messrs Jackson and Exley jointly for their debts. That this will be very disastrous if not ruinous to both parties is too clear for dispute but it is the only course Mr Jackson can take in justice to the creditors and Mr Exley will have brought down all the mischief on his own head by his violation of his partnership articles. I have also advised a meeting of the creditors in order to lay the case before them as I think they ought to be informed of what is taking place. If there yet remain any chance of averting the calamities that seem to be impending I should be very happy to contribute to so desirable an end and with that view I should be ready along with Mr Jackson to meet you and Mr Exley to discuss the subject if you think it will be of any service to do so
I am etc"

Snowdon and Preston responded immediately for their client:

"In reply to your note just received we beg to inform you that if Mr Jackson or the firm of Fenton Murray and Jackson do not pay the wages this evening due from Messrs Fenton Murray &Co on the Machine Department in which concern only Mr Samuel Exley is a partner Mr Exley will incur no liability in carrying on the business further. With regard to the collection of the debt by Mr Exley due to the concern in which he is a partner, we think any court or body of creditors will justify the proceeding. We beg you however most distinctly to understand that Mr Exley has had no other motive in collecting the money than that of procuring its proper application. Mr Jackson has

allowed Mr Exley to be sued for a debt which he must be well aware is owing from the concern in which Mr Exley is not a partner; and what reliance could Mr Exley have on Mr Jackson, paying the debts owing from Mr Exley's concern when he neglected to pay the debts owing from Fenton Murray & Jackson. You will also bear in mind that Mr Exley is a large creditor of Fenton Murray & Jackson he wants we believe about £3000 and it is high time Mr Exley was protected. As we understand from Mr Exley the Machine Department is solvent, however embarrassed the concern of Fenton Murray & Jackson may be, and it is hard on him that he should be subjected to the harass of being sued for the debt of a concern in which he has no more interest (except as being unfortunately a very heavy creditor) than either you or we have. We have appeared to the Action [acts] of Williams and others for Mr Exley and we by this post send up instructions for Plea. As far as our information goes and we believe we are in possession of the facts the concern in which Mr Exley is a partner has been kept perfectly distinct from that in which he is not a partner and Mr Exley has a right to insist that the debts owing to the concern in which he is a partner shall be applied to the debts owing from it and that he shall come in rateably along with other creditors of Messrs Fenton Murray & Jackson for the large debt owing to him from the last mentioned firm. We have not had an opportunity of conferring with Mr Exley since the receipt of your note but we cannot doubt he will adopt our advice and give Mr Jackson a meeting as you proposed and that as soon as convenient. We are etc." [83].

On the 10th of February an advert for the sale of a capital hydraulic packing press, made by Fenton Murray & Co, appeared in the Leeds Intelligencer. It also noted that without the masonry work the press had cost £120.

On the morning of Monday the 12th of February, Jackson had a notice read to the workmen of Fenton Murray & Co, and subsequently delivered to Exley, which stated:

"We Fenton Murray & Jackson do hereby give notice that from this time and until an affair on the part of Mr Samuel Exley is settled we will not make any payments for either debts or wages connected with the Machine Business of Fenton Murray & Co Leeds. Fenton Murray & Self Richard Jackson February 12th 1844." [83].

The wages had been due the previous Saturday. Exley faced with this statement and being unwilling to risk any more of his money, over and above his inaccessible capital, informed the workmen that he personally would not meet their wages either. By the 24th of February, due to the actions of the workforce, the dispute was picked up by Leeds papers and reported in some detail in both the Mercury and the Times, both of which however initially confused the two firms involved. The Leeds Times reported:

"Summonses for wages – Messrs Fenton Murray & Jackson – A good deal of surprise has been occasioned by the number of workmen employed at the extensive establishment of Messrs Fenton Murray & Jackson, engineers and machine makers, of this town, applying for summonses to compel their employers to pay them their wages; the firm having been considered one of the oldest, most extensive, and most eminent, in the engineering business, existing in the country. The first case came on for hearing before Richard Bramley and Griffith Wright Esq's at the Court House, yesterday, on a summons obtained by one of the workmen, named Smith, for £10-7-

6d wages alleged to be due to himself and three other workmen, for work which they had taken by contract or "piece" as it is called. It appeared that the work, however, was unfinished and the amount due for the portion completed was estimated by Smith at £5, he claiming one-fourth as his share. But he contended that they ought to be paid the whole amount (£10-7-6d) as the contract had been violated by their Employers; both Mr Jackson and Mr Exley, two of the partners, having refused to allow them to finish the work. The engine, it appeared, had been stopped, and the portion of the works called "the Round Building" in which the machine department was carried on, had ceased working. Mr J H Hill, barrister at law, appeared on behalf of Mr Exley to answer the summons. Mr Exley himself, also, was present, but none of the other partners. Mr Hill contended that the case was placed out of the jurisdiction of the magistrates, in consequence of the claim amounting to more than £10, but this difficulty was got over by the magistrates hearing the case, only so far as it concerned Smith, and the claim which he had to make. The reason stated by Mr Hill, for the non-payment of the wages, was a misunderstanding betwixt Mr Exley and Mr Jackson. Several suggestions were made by Mr Hill, with a view to obtaining a postponement of the cases, but they were all overruled by the magistrates, who thought it extremely cruel that the men should be kept out of their wages. A good deal of difficulty was found in deciding the amount due to Smith, in consequence of his forming one of "a set" who had taken the work by contract, and also of part of the work being unfinished but ultimately the magistrates settled the amount due to him at 25s, and ordered that sum to be paid. Immediately after this decision, about fifty other workmen, who were in court applied for summonses against Messrs Fenton Murray & Jackson for wages amounting altogether to about £80 or £90. Mr Exley however, in order to avoid the expense of so many summonses, agreed to come to some arrangement with the men, either that afternoon or the following day, and on this understanding, the granting of the summonses was postponed." [84].

It would appear that a corrective statement was issued as all the main Leeds papers printed a follow up on March 2nd and the wording in the Leeds Times and the Leeds Mercury was almost identical and hardly different to that contained in the Intelligencer. The paragraph from the Times/ Mercury was as follows:

Case of Wages.
We last week noticed a case of wages as between Fenton, Murray and Jackson and their workmen, which wages, we understand, in consequence of a warrant of distress in the case reported, and about 50 summonses in the other cases having been issued, have been paid by Fenton Murray and Jackson. We are now informed that the case was not between the workmen of Fenton Murray and Jackson, Engineers, Founders &c but between Fenton Murray and Company, Machine Makers, and their workmen, Mr Exley being a partner only in the latter firm. We understand that the cause of the non-payment of wages was a dispute between the firms of Fenton Murray and Jackson, and Fenton Murray & Company, and that the matter is now before a higher tribunal, Mr Exley having filed a bill in Chancery.

Exley had indeed filed a bill in Chancery, dated the 20th February 1844. His submission contains most of the data given in describing the events leading up to the demise of Fenton Murray & Co. Only Exley's submission seems to have survived, but it must be reliable, as under the way the law worked in those days, if Exley had submitted an inaccurate fact within his whole submission, and the defendants Jackson and Yale could prove him wrong the whole submission would have failed.

Exley's request to Chancery was for the partnership of Fenton Murray & Co to be dissolved and its affairs wound up and that the value of Exley's capital in the hands of Jackson and Yale was to be ascertained. Any surplus in the liquidation accounts of the firm should go to pay Exley for his capital. Exley asked the Court to appoint a proper person to conduct the winding up and collect all monies due and make all due payments. He also asked for an injunction on Jackson and Yale ordering them not to take any action in receiving or paying monies related to Fenton Murray & Co [83].

On the 7th of March the Court of Chancery issued a restraining order against Yale and his agents from collecting or disposing of the assets of Fenton Murray & Co. On March the 29th the Court followed this up by appointing one John Young to be the receiver and manager of Fenton Murray & Co in its winding up [85].

On the 23rd April the London gazette announced that bankruptcy proceedings had been instituted against Richard Jackson on the 16th of that month. The petioning creditor was Benjamin Hallewell, a Wine and Spirit Seller of Duncan Street (between Briggate and Call Lane) Leeds. A meeting of creditors was called on the 10th May to discuss the actions to be taken by the assignees (creditors appointed to manage the estate of the bankrupt along with the Official Assignee appointed by the Court), the ways forward in regard to the two businesses and defending a suit in Chancery (Exley's).

The Flax spinning Factory called Rawcliffe, between Goole and Selby, was advertised to be auctioned in the papers of 11th May and subsequently. The auction to be held on the 24th May along with a 20 H.P. steam engine and boiler by Fenton Murray & Wood. On the 30th of May adverts offered a 30H.P. engine by the firm with boilers in Manchester [86].

On the 1st of June Exley submitted a further document to Chancery in order to bring Jackson's assignees, Thomas Clayton, William Collier and the official assignee Charles Fearne, into the case, and ensure that they could not ignore the Chancery proceedings in settling the estate of Jackson in his bankruptcy [85].

Consequent on his bankruptcy, the contents of Richard Jackson's house at 35 Park Square Leeds were advertised for auction in the Leeds papers of the 8th of June [87]. The auction was to be held from the 17th to the19th of June. The summary was written as follows

"Sale of unusually valuable household furniture, capital Britzka [a type of carriage] nearly new, Phaeton with head complete, cabinet piano in Rosewood by Ellwart, 250 ounces of silver plate, oil paintings, books and prints, mirror, and chimney glasses, chandeliers etc. etc."

And this was followed on June the 22nd with an advert for the lease of the house itself [88].

The Standard in London of June 19th carried the notice of an auction on the 12th of July for a Tobacco and Snuff manufacturer, Lundy Foot and Co, situated in Vine Street and New Square, Minories. The equipment was worked by a 20 H.P. steam engine by Fenton and Murray.

Despite all the actions pertaining to the case in Chancery in regards to Fenton Murray & Co and Jackson's bankruptcy, no definitive comment on the affairs of Fenton Murray and Jackson appears until 23rd July. At this point the bankruptcy of Richard Yale, in partnership with Richard Jackson and trading as Fenton Murray and Jackson, was announced in the London Gazette. The petitioning creditor was John Millar of Leeds, Engineer. This was probably the John Millar who worked for Fenton Murray and Jackson, and had recently erected the locomotives for the GWR. If this was so, it would seem likely that the firm was still stumbling along but after Jackson's bankruptcy, when any men left turned to Yale for any payments due, he could not oblige, so Millar, maybe on his own, or acting for his colleagues took the action to petition for Yale's bankruptcy.

Exley had also obtained the ruling he wanted in Chancery, this is borne out by the appearance in the Leeds papers, on the 27th July 1844, of the announcement of the auction on the 5th to 7th of August of all the valuable materials of Fenton Murray & Co. The adverts inform us that the sale was:

"pursuant to an order of the High Court of Chancery, made in a cause of 'Exley v Jackson', with the approbation of Sir William Horne, Knight, one of the masters of the said court"

The materials comprised various castings, metals, a small fluting machine, an extensive range of files and rasps, assorted timber including pine, cedar, birch, plane, alder, and box. There were also a number of unfinished machines : a 116 spindle twisting frame, a double transfer hackling machine, two third drawing carriages, two second drawing carriages, an experimental spinning frame with 12 spindles, 16 cylinder hackling machines, a first drawing gill machine, spindles, and various models [89].

There was further action to come, this time from very close to home. On August 6th the London Gazette announced a further fiat in bankruptcy against Jackson and Yale trading as Fenton Murray and Jackson. The petitioners were Andrew Faulds and Mary Anne his wife; that is Jackson's daughter and his son in law. By this time matters were getting too complicated, Jackson and Yale each had a separate fiat against them and now a joint fiat in bankruptcy. The matter was referred to the Bankruptcy Court of Review, and the issue was heard on August 20th. The ruling was that the individual bankruptcies were impounded and all the debts and proof of debts and all the proceedings of the two individual fiats were to be transferred to the joint case in bankruptcy awarded on August 6th [90]. On the 6th of September Clayton and Collier were agreed as the assignees for the joint bankruptcy as well, with George Young as the official assignee [91].

The most tragic event of the whole collapse of Fenton Murray occurred with the death of the 24 year old Richard Yale on the 21st September. The cause of death was given as asthma and delirium tremens.

At a meeting on the 23rd October many of the issues over the effective amalgamation of the different fiats in bankruptcy were ironed out. The assignees also sought authority to sell any part of the land of the ex-partnership, to letting any part of the land, mills, tools, fixtures and fittings of the ex-partnership, to insure the assets, to mortgage the assets, to keep them in good condition [92]. With the difficulty of disposing of industrial property in this time of great slump, these rights appear to set the necessary background for the legend of the 'Forty Thieves'. Production at the Round Foundry did not appear to cease, with the demise of Fenton and Murray, and until Smith Beacock and Tannent (all ex-employees of Fenton, Murray & Co) acquired the site in 1847, it was run by various of the workmen, no doubt in agreement with the assignees. Smith, Beacock and Tennant are unlikely to have been involved, unless they used the workforce on sub-contract, up until their purchase, as they had been in business at the Victoria Foundry in Victoria Street Camp Fields, since 1837 [93]. When they purchased the Round Foundry site in 1847 they transferred their works name with them [94].

On the 31st October Exley made his third surviving submission to Chancery, to ensure his rights under the final joint bankruptcy proceedings, dictated by the court in August, and to bring Clayton, Collier and Young into the proceedings as the assignees of the final joint bankruptcy [91].

On the 5th November it was announced in the London Gazette that Richard Jackson's Certificate of Conformity would be allowed and confirmed on the 26th of that month unless due cause be shown to the contrary. So by the end of the month Jackson was discharged from Bankruptcy

On the 2nd November Joshua Buckton of Meadow lane Leeds was advertising for sale two 8 inch ram hydraulic hot presses by Maclea and March and a six inch ram hydraulic packing press by Fenton Murray & Co [95].

On the 9th of November the first advert of many for the sale or letting of the premises and stock of Fenton and Murray appeared in the Leeds press. These adverts appeared on and off with changes in wording until January 1847. The earlier adverts ran along the following lines but they were much, much shorter by 1846/7. The initial adverts for the stock in trade and tools, for which the last abbreviated advert was in January 1846, read:

"*Fenton Murray and Jackson's Foundry, Marshall Street and Water Lane Leeds. Large sale of Smith's, Founder's, and Boiler Maker's tools; moulding boxes; core barrels; castings and bar iron; locomotive boilers; 10 horse steam cylinder and case; carts; wherries; and other effects, free from all auction duty.*
Messrs T and W Hardwick beg to announce that they are instructed by the Assignees of Richard Jackson and Richard Yale, lately carrying on business under the firm of Fenton Murray and Jackson, to sell by auction, on Monday next, upon the premises in Marshall Street and Water lane Leeds,
All the extensive and valuable Smiths', Founders', & Boiler makers' tools, stock in trade, and other effects, comprising upwards of 50 tons moulding boxes, 70 tons scrap iron, Pig and Bar Iron, Boiler plates, cast steel, shear and blister steel, copper and brass sheets, block tin and spelter, old copper and brass, 300 dozens new and old files, iron blocks and ropes.

Two Locomotive Boilers, with Copper Fire Boxes and Stays
for 12 inch engines; cylinders for ditto, pistons and lids, ball and socket brass joints, stop valves, brass slide bars, wrought iron connecting rods, levers and joints, 4 brass eccentric circles, wrought iron arms, steel springs, 2 brass oil boxes, with 6 taps each, fire doors and frames, tender tank, and large Locomotive wherry, with cast iron wheels.

One Ten Horse Steam Cylinder and case
Air pump and condenser; one 20 horse parallel motion, complete; one 20 horse governor, 2 centre gudgeon plummer blocks, 2 cross shaft ditto, cross bar and slide rods, brass foot and delivering valves, air pump bucket and rod, manhole lids, flue frames and doors, standards and wheels for boiler work, wrought iron heavy shafting, with cast iron drums and cones; foundry tools and implements, portable cupola, rings and core barrels, a large assortment of smiths' tools, cast steel chisels, maundrills, turning tools, drills, hand hammers, bolts, nuts, screws, screw keys, screw jacks, large steelyards, circular and common bellows, scales and weights, weighing machines, metals for bending boiler plates, triangle crane, iron jennies, straightening plates, iron and wood hand wherries; an immense stock of iron and wood models for land, marine, and locomotive engines, hydraulic presses and pumps, and general millwright work; a small stock of Ash timber, crab tree, mahogany and deal;

Valuable Counting House Furniture
Including two large mahogany desks, with drawers, brass railings, and curtains; mahogany and painted drawing tables, plan tables with drawers and cupboards; chairs, stools, fenders and fire irons. Also three large wherries, with patent arms, one broad and two narrow wheeled carts, with patent arms, cart gearing, gig harness, and other truly valuable and important effects.
The property may be viewed one week previous to the sale by making application at the offices of the auctioneers. Catalogues will be ready ten days previous to the sale, and any further information may be had of George Young Esq, Official Assignee or of Mr John Blackburn solicitor Leeds." [96].

The early adverts for the buildings, for which the last short advert was in January 1847, read:

Valuable and Eligible Buildings, steam engines, machines, and tools, shares in the Zoological Gardens and Victoria Bridge for sale.

To be sold by private contract, either altogether or in lots, or to be let from year to year, all these important and valuable freehold estates, situate in Marshall Street and Water Lane, Leeds, lately in the occupation of Messrs Fenton Murray and Jackson, including all those extensive newly-erected buildings used in manufacturing Locomotive engines; with large erecting shop, 72 feet square, and turning and fitting up chamber over; Carpenter's shop, 120 feet by 18 feet; Boring Mill, 75 feet by 34 feet with fitting chamber over; Smith's shop, 132 feet by 22 feet; Grinding and Glazing House, 58 feet by 12 feet, with Engine House, Watch House and small iron warehouse, and an extensive range of Counting Houses, and Drawing Offices, fitted up with Water Closets, Iron Safes, and cupboards. Also a 12 Horse condensing steam engine and a 20 horse boiler, with suitable drums, shafting and going gear.

Also all that Circular Building recently used a machine shop, four stories high, and 27 yards diameter, with a 6 horse steam engine and two boilers, and the drums, shafting, and going gear, with the Engine and Boiler House. Grinding Sheds, Store Room, and Smiths' Shop, and an extensive area of vacant land, suitable for building purposes.

Also all those other buildings used as Boiler Shops, Brass Foundry, Smiths' Shops, Model Sheds, Gas Works, Lime House, Two gas meters, Painters' Shop, Mess House, Charcoal House, Model Shed, 3 stall stable, with Gig House attached, and sawpit, and a large quantity of vacant ground.

To be sold by private contract all that convenient dwelling house fronting into Water Lane, lately occupied by Mr Matthew Murray Jackson.

Also all that plot of building land, comprising 200 square yards, with the stables and buildings thereon, situate on the north side of Water Lane, and adjoining the Beck. And also that extensive Maltkiln, with drying kiln, warehouse, Smiths' Shop, shed, and two stall stable, with loft over, in the occupation of Messrs Skelton and Co.

Also 10 shares in the Leeds Zoological gardens.

Also 3 shares in the Leeds Victoria Bridge Company.

The valuable machines and tools, consisting of planing machines, drills, lathes. boring engines &c &c used in the Boring Mill, Fitting up Shop, and Smiths' Shops, may be taken by a tenant at a valuation.

Although a large proportion of the buildings have been occupied by Fenton Murray and Jackson, for general engineering purposes, and machine manufacturing, they are admirably adapted both in extent, convenience and strength, for any business requiring room, light and power.

The whole of the property is in a locality favourable to trade, and is situated midway between the North Midland Railway station, and the contemplated stations of the Leeds and Thirsk, the Leeds and Dewsbury, the West Riding Junction, and the West Yorkshire Railways.

For particulars as to price and rentals, application may be made to George Young Esq Official Assignee; or to

John Blackburn Solicitor Leeds

Leeds 8th Nov 1844." [97].

Although the situation at Fenton Murray and Jackson was obviously dire at the beginning of 1844, they had still a few orders. The Reports of the Commissioners for the Caledonian Canal in 1844 included one from a contractor stating that a new dredger had been ordered with a new steam engine and was being made in Aberdeen. It is apparent that the steam engine and dredging equipment were either planned to go to Fenton Murray and Jackson or had been ordered. The following year in the 40th report of the Commissioners there was included report number 2 from Messrs Walker and Burges, one of the contractors, dated 19 February 1845. This included a statement that the excavation at Bona had been delayed. They awaited the new steam dredger which was also delayed due to:

"the execution of the machinery being transferred from Messrs Murray & Jackson of Leeds, on account of their stoppage, to Messrs T Vernon & Co of Aberdeen."

The fate of the buildings, some of which still survive and some of which were demolished in the 1980s can be briefly summarised. The majority of the property became the Victoria Foundry of Smith Beacock and Tannent, who successfully manufactured machine tools, on an extensive scale, at the site for a number of decades until about the end of the 19th century. The site survived a horrendous boiler explosion in October 1847, shortly after the new firm took residence [98]. It also suffered a very destructive fire in September 1875 when the round building, that had given the site its popular name of the Round Foundry, was destroyed [99]. In the 20th century the site was occupied by various engineering firms (eg Herbert Alexander) Cotton Warehouses, a mineral water factory, etc [100]. It continued in multiple occupation until the renewal and refurbishment carried out in the first decade of the 21st century, and is now again in mixed occupation.

Jackson's new locomotive works were sold off separately and became known as Campfield Mill. They were offered up for lease or sale by G E Donisthorpe & Co, machine top and noil (short fibre wool) manufacturers, oil refiners and wool merchants of Larchfield Mill Hunslett in 1859 [101]. They were being used as textile shops by a Mr Glover from 1860 to 1865, and he also let powered rooms [102]. In 1865 Campfield Mills were up for sale again [103] and were purchased by Messrs Emmanuel & Co, the trading name of a German called Worms, who moved from Providence Mill Bingley. He undertook extensive alterations [104] and shortly after moving in suffered a disastrous fire, which destroyed the mill and threatened the Leeds Industrial Co-Operative mill next door and the adjoining premises of the Victoria Foundry [105]. The mill was rebuilt by insurance and continued in textiles until it again came up for sale in 1882 [106]. The new proprietors were Messrs Wilson Crosby & Co who used it as a flour mill [107]. At some time towards the end of the 19th century the premises were taken over by the Leeds Industrial Co-Operative next door and eventually became part of their extensive operation [108].

The mill erected by Matthew Murray, next to his dwelling in Marshall Street, and left to Richard Jackson in his will had been extended before the firm crashed and it was this mill that was purchased by the then Leeds District Flour Mill Society in

1848, as a flour mill to produce high quality flour for its members [109]. This organisation developed into the Leeds Industrial Co-Operative Society. The mill was extensively enlarged and itself suffered a large fire (October 1881), but its construction was sufficient to limit the damage. The premises were therefore re-united in the early 20th century, but were both demolished in the 1980s when the site was no longer appropriate for the society.

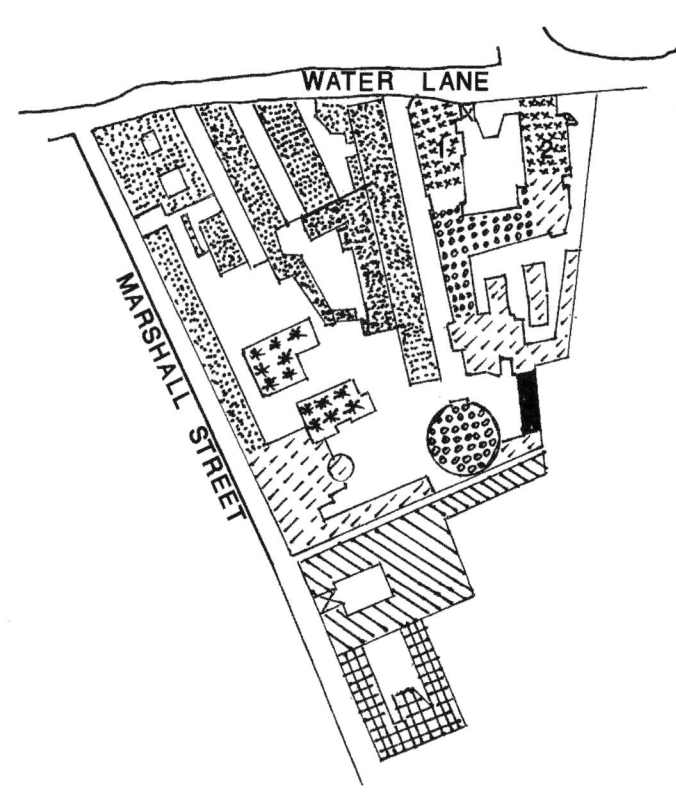

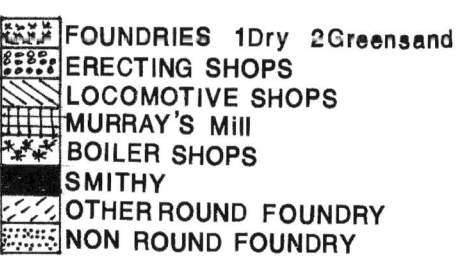

FOUNDRIES 1Dry 2Greensand
ERECTING SHOPS
LOCOMOTIVE SHOPS
MURRAY'S Mill
BOILER SHOPS
SMITHY
OTHER ROUND FOUNDRY
NON ROUND FOUNDRY

71. Plan of the Round Foundry early 1840s

While the extent of the damage to Campfield Mill under Messrs Emmanuel was extensive it probably did not destroy the façade onto Marshall Street which bears the hallmarks of a building put up in the 1840s. Some photographs survive of the premises from the early 20th century showing the frontage of Jackson's Locomotive Shops and the Murray Mill incorporated into the Co-Operative premises.

72a. Marshall Street, Murray's Mill with locomotive works behind.

72b. Frontage of Fenton Murray & Jackson's locomotive shops

Notes.

1. Rail 253/91. GWR. C S Saunders to R W Hawthorn 22 January 1840, annotated with, similar letter sent to Fenton Murray & Jackson. National Archives. Kew.
2. Rail 1008/37. GWR. National Archives. Kew.
3. Rail 253/91. GWR. National Archives. Kew.
4. Rail 1008/26. GWR. National Archives. Kew.
5. The Leeds Mercury 7 March 1840.
6. The Belfast News Letter 7 April 1840.
7. The Leeds Mercury 18 April 1840.
8. The Hull Packet 24 April 1840, Lincolnshire Chronicle 1 May 1840.
9. The Leeds Mercury 2 May 1840.
10. Some Links in the Evolution of the Locomotive. The Railway Magazine May 1908.
11. Rail 530/4. North Midland Railway, London Committee Minutes. National Archives. Kew.
12. The Hull Packet 15 May 1840, Yorkshire Gazette 23 May 1840.
13. The Reading Mercury 6 June 1840.
14. The Hull Packet 3July 1840.
15. The Leeds Intelligencer 11 July 1840.
16. The Leeds Mercury 5 September 1840.
17. The Leeds Mercury 18 July 1840.
18. Rail 250/116. GWR. Traffic Committee Minutes 26 August 1840. National Archives. Kew.
19. Rail 1149/6. I K Brunel Letter Book. National Archives. Kew.
20. The Leeds Intelligencer 3 October 1840.
21. Hampshire Advertiser 24 October 1840, copy death certificate.
22. E.g. The Hull Packet 27 November 1840.
23. The Leeds Intelligencer 5 December 1840.
24. The Leeds Mercury 19 December 1840.
25. The Leeds Mercury 12 December 1840.
26. The Leeds Mercury 2 January 1841.
27. Rail 1008/37. National Archives. Kew.
28. The Shannon Navigation, Ruth Delaney. Lilliput/Waterways Ireland 2008. Page 162.
29. The Leeds Mercury 20 February 1841.
30. The Leeds Mercury 20 March 1841.
31. Rail 254/549. GWR. National Archives .Kew.
32. The Leeds Intelligencer 27 March 1841.
33. The Leeds Mercury 10 April 1841.
34. Children's Employment Commission. Appendix to the Second Report of the Commissioners. Trades and Manufacturers Part 1 Reports and Evidence from Sub-Commissioners.
35. The Leeds Mercury 1 May 1841.
36. Rail 254/459. GWR. FM&J to GWR statement of account, 24 March 1841. National Archives. Kew.
37. The Leeds Intelligencer 15 May 1841.
38. DM 1713/3. FM&J to D Gooch. 14 June 1841. Brunel Archives, Brunel Institute Bristol.
39. DM 1713/4. FM&J to Gooch. 23 July 1841. Brunel Archives.
40. Rail 1008/17. National Archives. Kew.
41. Marshall Papers. MS200 folio 17. Brotherton Library.
42. Newcastle Town Council Minutes (bound). Newcastle upon Tyne Central Library.
43. The Northern Star 23 October 1841. For operative collections see The Northern Star 2 October 1841 to 1 January 1842 and The Leeds Times.
44. The 5th column in the number of families, whilst correct to that published, does not appear to be valid. However the mathematics works for the number of persons to the average per head.
45. The Leeds Mercury 23 October 1841.

46. DM 162/10/2b folio 237. Brunel to Maudslay and Field. Brunel Archive.
47. Sir Daniel Gooch Memoirs & Diary. Ed. R Burdett Wilson. David and Charles 1972. Page 35.
48. The Reading Mercury 24 December 1841. The Morning Post 25 December 1841.
49. The Leeds Times 15 January 1842.
50. The Leeds Intelligencer 22 January 1842.
51. The Leeds Intelligencer 19 February 1842.
52. The Leeds Intelligencer 5 March 1842.
53. The Leeds Intelligencer 9 April 1842.
54. The Leeds Times 16 April 1842.
55. The Leeds Times 23 April 1842.
56. The Leeds Times 30 April 1842.
57. The Leeds Intelligencer, The Leeds Times, Morning Chronicle (London) all of 14 May 1842, plus extensive coverage elsewhere including the Mechanics Magazine (many issues).
58. The Leeds Intelligencer 21 May 1842.
59. The Leeds Intelligencer 18 June 1842.
60. Berrow's Worcester Journal.
61. The Leeds Times, The Leeds Intelligencer 20 August 1842, The Northern Star 27 August 1842, The Hull Packet 19 August 1842.
62. The Leeds Times 10 December 1842.
63. The Improvement of the River Tyne 1815-1914, R W Rennison. The Newcomen Society vol 62 1990-1991. Page 118.
64. The Charleston and Hamburg, Thomas fetters. The History Press 2008.
65. The Leeds Intelligencer 31 December 1842.
66. The Leeds Intelligencer 14 January 1843.
67. The Leeds Times 11 Febraury 1843.
68. The London Gazette 14 December 1849.
69. The Leeds Times 4 February 1843.
70. The Morning Chronicle 22 February 1843, The Leeds Mercury 25 February 1843.
71. Lincoln Rutland and Stamford Mercury 21 April 1843.
72. The Queen v The Inhabitants of Holbeck, Cases in the Queen's Bench 3 May 1843.
73. The Leeds Mercury 13 May 1843.
74. Newcastle Courant 19 May 1843.
75. The Leeds Mercury 15 June 1843.
76. The Leeds Mercury 8 July 1843.
77. The Leeds Intelligencer 15 July 1843.
78. The Leeds Intelligencer 22 July 1843.
79. House of Commons reports, The Standard (London) 11 August 1843.
80. DM 162/10/2c item 198. Brunel Archives.
81. The Leeds Intelligencer 28 October 1843.
82. The Newcastle Journal 28 October 1843.
83. S. Exley's submission to the Court of Chancery, dated 20 February 1844 in Exley v Jackson. C14/218/E13. National Archives. Kew (offsite).
84. The Leeds Times 24 February 1844.
85. S. Exley's submission to the Court of Chancery, dated 1 June 1844 in Exley v Jackson. C14/218/E13. National Archives. Kew (offsite).
86. The Leeds Mercury 11 May 1844, Bradford Observer 30 May 1844.
87. The Leeds Mercury 15 June 1844.
88. The Leeds Intelligencer 22 June 1844.
89. The Leeds Times, The Leeds Mercury of 27 July 1844.
90. B/1/203 Bankruptcy Records. National Archives. Kew.
91. S. Exley's submission to the Court of Chancery, dated 31 October 1844 in Exley v Jackson. C14/218/E13. National Archives. Kew (offsite).
92. The London Gazette 1 October 1844.

93. The Leeds Mercury 25 November 1837.

94.   The name remains to be seen in Foundry Street at the time of writing.
95.   The Leeds Mercury 2 November 1844.
96.   E.g. The Leeds Intelligencer 9 November 1844.
97.   E.g. The Leeds Intelligencer 30 November 1844.
98.   The Leeds Mercury 6 October 1847.
99.   The Leeds Mercury 29 September 1875.
100.   Charles Goade's Insurance Plans for Leeds 1901.
101.   The Leeds Times 16 April 1859 to 6 August 1859.
102.   E.g. The Leeds Mercury 15 March 1862.
103.   The Leeds Mercury 22 June 1865.
104.   The Leeds mercury 6 July 1865.
105.   The Leeds Mercury 28 May 1866.
106.   The Leeds Mercury 7 January 1882.
107.   The Yorkshire Post 22 October 1884.
108.   Industrial Development 1780-1914 by Connell and Ward, in A History of
          Modern Leeds Ed. Derek Fraser. Manchester University Press 1980.
109.   The Leeds Industrial Co-Operative Society Ltd, G J Holyoake. 1897.
110.   The Royal Bank of Scotland Group Archives, BEL 45 Securities Book.

Chapter 16.
Millwork, Machine Tools, Mills and Stationary Engines.

Matthew Murray has a solid reputation as an engineer or mechanist, and it is probably in his abilities as an expert in all the aspects of mills that his reputation is founded, even if that is not what he is known for today. The partnership in which he was involved constituted the toughest competition that the firm of Boulton and Watt had in the decade after the end of James Watt's patent, for the supply of stationary steam engines. However the engines provided by Fenton Murray and Co were basically built to Watt's patent. Murray used different details, such as his 'D' type valve, and he ensured that his engines were manufactured to the highest standard, but he made no earth shattering change to the fundamental design of the condensing engine. While Murray was alive it is noticeable that the prices of the Leeds engines were always lower than those from the Soho stable. In the first published account of Matthew Murray, by Samuel Smiles in 1863, Smiles wrote

"Mr Murray's faculty for organising work, perfected by experience, enabled him also to introduce many valuable improvements in the mechanics of manufacturing. His pre-eminent skill in mill-gearing became generally acknowledged, and the effects of his labours are felt to this day in the extensive and still thriving branches of industry which his ingenuity and ability mainly contributed to establish. All the machine tools used in his establishment were designed by himself, and he was most careful in the personal superintendence of all the details of their construction"[1].

While Smiles has been shown sometimes to be selective, there is some justification for taking his brief writing on Murray's reputation, if not the early myths, seriously. Smiles was the editor of the Leeds Times from 1838 to 1845, and lived in Leeds from 1838 to 1854. Latterly he lived in Blenheim Square where he was in the same neighbourhood as Mr March and Mr Maclea. He had certainly discussed Murray with March as stated in his book.

John Farey writing in Rees's Cyclopaedia, earlier than Smiles, describes the works of Fenton Murray and Wood:

"The engines they send out cannot be excelled in beauty and perfection of workmanship, and they perform as well as any others. Their factory at Leeds is very extensive, and provided with every convenience for making all the parts of the engine in the best manner, and with the least labour. They have three steam engines in the works, one for boring cylinders, and turning large lathes; a second for turning small lathes, grinding, drilling the centres of wheels, tapping screws, etc, and for blowing the furnaces of the foundry; and a third engine for working a great forge hammer, by which the heavy wrought iron work is forged. The boring machines for cylinders, of which they have three in number, are very capital, as by an ingenious movement, invented by Mr Murray, for drawing the borer through the cylinder, it is made to advance regularly from one end to the other without any interruption. These machines are worked by a separate steam engine, which is never stopped during the operation of boring a cylinder through, as it is found to make a sensible mark or ring if the motion is stopped. The best means are also taken to prevent the cylinder from

changing its figure by its weight, or by the pressure of the parts which hold it in its position. The whole of the factory is lighted by gas lights in winter time. The boilers are manufactured by the aid of several machines to cut out the plate, pierce the holes, and bend the joints. Before any of the smaller engines are sent away, all the parts are put together in a building on purpose, where there are boilers fixed and they are actually tried, to insure that every part is perfect: they are then taken to pieces, with marks and directions for putting them together, and packed up for carriage, which is easy as there is a canal at the gates, which has communication by water to every part of England. For such engines as are too large to be put to work at the factory, workmen are sent out with them, to assist and direct in setting them to work"[2].

John Farey had been employed by Marshalls for a period from 1822 and he would have had the opportunity to visit the Round Foundry over the road, as his brief had included the loading and economy of Marshall's steam engines [3].

Murray's name is often quoted in connection with the manufacture of machine tools, but the known records of sales are few and far between. There were sales to Russia (the best documented being those in 1806)[4] and the well-known sale, presumed by Fenton Murray & Co, of a boring machine to the French works at Chaillot [5]. A rereading of Murray's response to the request to manufacture locomotive engines for the Stockton and Darlington Railway:

"excepting that it does not suit with the present arrangement of our business to take orders for high pressure or locomotive engines, we have not made any this 8 years, and if we do begin of it, it must be where they may become a regular article of sale" [6],

fits extremely well into the context of what is known today as production engineering. Thirteen years earlier than the above comments, Murray, in writing to John Watson about the Kenton and Coxlodge locomotives, stated that:

"we hope we shall be able to complete it in 2 months, but those engines are by no means yet a straight forward job and are attended with many difficulties in making them and perhaps will be the case for some time yet "[7].

Add to this evidence Simon Goodrich's notes on his first visit to Fenton Murray and Co in 1803 in which he described Murray's boring machine and his virtually automatic machine for fluting rollers. In the same visit Goodrich noted a machine for cutting wheels up to four feet in diameter with either straight or bevelled gears and that Mr Murray would supply one of these machines for £60 [8].Goodrich's journals over time also record other details of Murray's machines such as his lathe construction. Murray, no doubt in conjunction with David Wood, was one of the first engineers to fully recognise that as well as designing the product; quality and economical manufacture required that the process and the machine tools to produce the items had to have the same attention to their design, and all facets were of equal importance. The evidence above supports that all these were combined in the works of Fenton Murray and Wood. The firm also appears to have been one of the first to assemble and test its products before delivering them to the customer. This ability to achieve cost effective and quality manufacturing enabled the firm to achieve production costs that drove their prices to undercut the competition.

All the information available on wage rates, researched to date, indicate that even in Leeds in the early days the firm had to pay a near national wage to its key employees, or at least the good ones, else they would decamp to another employer. Competitive pricing could not therefore have been achieved at the expense of the employee's wages. Alongside the paucity of evidence of the sale of machine tools there is unsurprisingly not much evidence remaining of Murray's mill-work (as opposed to mill building). However some written evidence has already been seen. Two records of his mill gearing wheels still exist, the first from 1795 in the Marshall Papers described in Chapter 4 pages 57, and the second is the illustrated section from Buchanan's Essay on the Teeth of Wheels published in a revised form in 1808, see Chapter 8 page 175. Marc Brunel seems to have been happy to ask Fenton Murray & Co to quote for millwork in support of his sawmills and indeed it is very likely that the firm supplied Brunel with millwork for Messrs Borthwick's sawmill at Leith (Maudslay was moving shop at the time), and indeed some of the millwork at Chatham seems likely.

Another area where Murray seems to have been in advance of his contemporaries was in the area of mill shaft connection. He came close to adopting a universal joint, and seems to have come up with the next best thing. This connector, along with a much simpler one, were still being described in technical books and encyclopaedias in the 1840s. The simpler shaft required one bearing per length of shaft and is represented below:

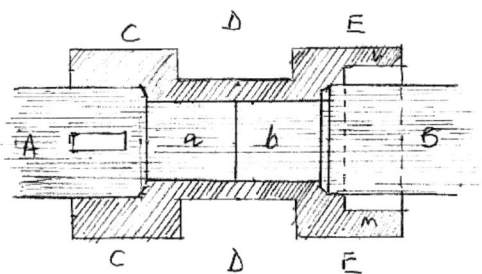

73. One of Murray's shaft couplings from Encyclopaedia Metropolitana.

'A' and 'B' are the two shafts to be joined and each ends in a pivot 'a' and 'b'. The coupling box 'CDE' is bored inside to take the ends of the shafts and turned at 'DD' to give an outside neck to take the bearing. The shafts are fixed by keys 'lm' slotted into each shaft end and fitting into slots machined inside the coupling. The pivots are held tightly in the coupling box but the shafts are not which allows a little freedom of movement without straining the bearing and a slight deviation from a straight line does not produce an unacceptable strain.

The major coupling for a long lines of shafting carrying a heavy load has two bearings and is given below:

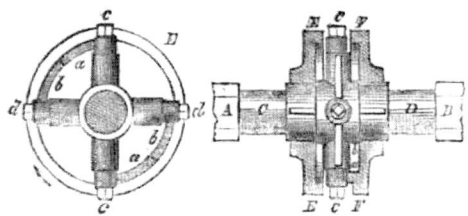

74. Murray's advanced shaft coupling Encyclopaedia Metropolitana and others.

The two shafts to be united are 'A' and 'B'. They have collars 'C' and 'D' which lie in the bearings. The end of each shaft has fixed upon it, by a square with wedges, a box 'E' and 'F'. Each box has two pieces which, being offset, project into the other box at 'aa' and 'bb'. Within the box is an iron cross 'cc,dd' with screws fixed in each end. The projecting pieces 'aa,bb' act upon the iron cross to pass the motion from one shaft to the other. The cross is quite detached within the box and acts as a universal joint communicating the motion between the shafts 'A' and 'B' [9].

A further method of shaft coupling used by Murray was given by Goodrich in 1819 and is shown in Chapter 11 page 246/7.

The machine tool sector offers some more detailed information. The 1803 visit of Bentham and Goodrich provides the drawing of the boring machine (which is the one that J Watt jnr saw), a mention of an engine for cutting wheels (gears) up to 4 feet in diameter with either straight or bevelled teeth, and the drawing and description of the automatic fluting machine at the Round Foundry [8]. From Goodrich it is known that Fenton Murray and Co supplied to Russia in 1806: a complete forge mill, a boring machine, a screw cutting machine with attendant steam engines shafting etc. [4]. In his paper 'The Early History of the Cylinder Boring Machine'[5], E. A. Forward demonstrated that the basic development of these machines to enable acceptable cylinder boring had been accomplished by John Wilkinson with his Bersham Boring Mill circa 1775. It was with this mill that the early Watt cylinders were bored. The issue that Forward left open was who had been the first to introduce to one of these types of mills a positive automatic feed of the cutters through the cylinder. Forward recognised that Michael Billingsley of Bowling Ironworks near Bradford patented a vertical boring mill with a form of positive automatic feed on the 22 December 1802. Mr Rhys Jenkins, in discussion said that James Watt jnr had reported a screw feed in

Murray's works in 1802. This can be confirmed, and the report was in a letter to Rob Boulton written on 15th June 1802. The full quote provides the additional information that the boring mill at Boulton and Watt's Soho Works did not have an automatic feed at that time:

"*His* (Murray's) *cutter block is pushed forward upon the boring rod by an endless screw, which, or some similar contrivance we must adopt, both to guard against the negligence of the borer and to save part of his wages*" [10].

So the first automatic feed could well have come from Murray or Billingsley. Forward also describes Murray's boring machine of 1817 which was fully described by John Farey in the Encyclopaedia Britannica [11]:

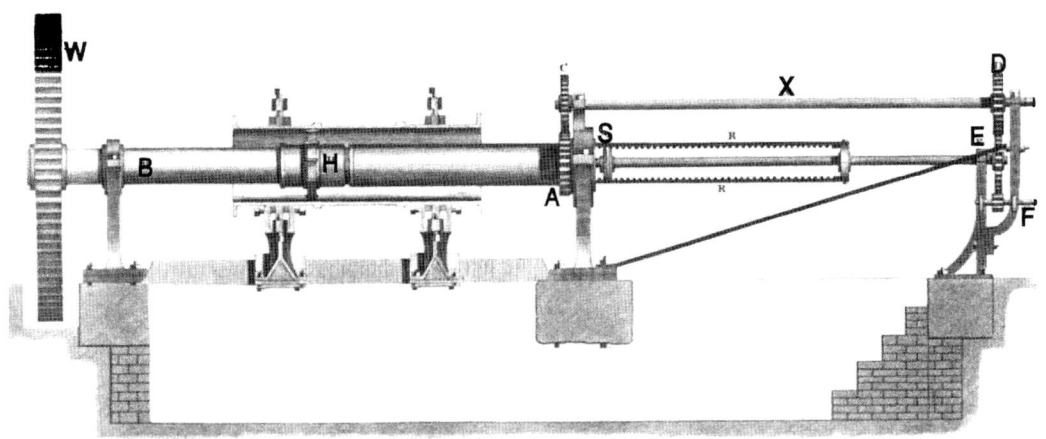

75a. Murray's 1817 boring machine side view

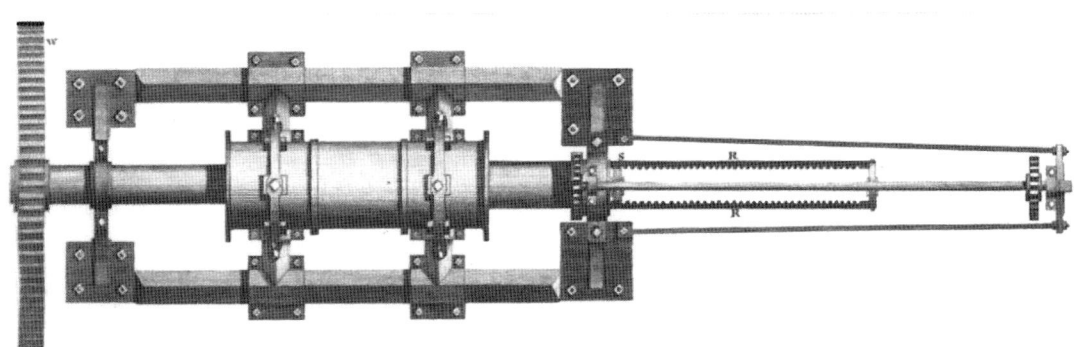

75b. Murray's 1817 boring machine plan

W in the plan and elevation is the spur wheel, deriving its motion from water or steam, and communicating a revolving motion to the boring-bar. The toothed wheel *A* (elevation) moves round with the boring-bar *B* on which it is fixed; it gives motion through the wheels *D* and *E*, and to the screw *S*, whose threads act on the two racks, which racks are fixed to the cutter-head *H*, and revolve with it. The velocity with which the cutter-head is impelled along the cylinder, depends upon the number of threads of the screw in a given length, and on the proportions of the wheels *A*, *C*, *D*, and *E* to each other. By varying the velocity of the screw, the cutter-head may be made to move in either direction, up or down the cylinder. *F* is a pinion, whose axis ends in a square, which may be wrought by a key, so as to bring the cutter-head out of the cylinder, or push it home by the hand when that is required.

The cylinder is fixed in its bed by screws passing through two iron rings, as represented below; in this way the cylinder is equally pressed in the different parts of its circumference.

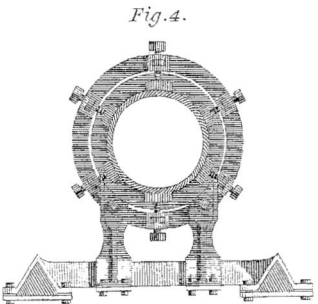

75c. Murray's 1817 boring machine detail of cylinder fixing to bed.

The drawing below is a traverse elevation of the collar in which the end of the bar at *A*, elevation, turns; *X* is the gudgeon in which the spindle *X*, turns. It also shows the two apertures through which the two racks pass.

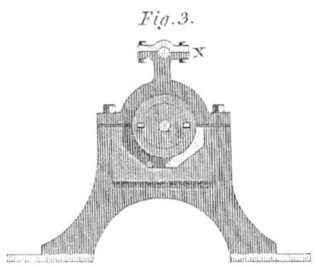

75d. Murray's 1817 boring machine detail of collar at end of bar.

By this machine also, the flanges are turned truly plane, so that the lid of the cylinder may fit on exactly [11].

Two boring machines, of the same build, were described together in France in the early 1820s one at St Quentin and the better known one at M Perier's works at Chaillot, these are ascribed to Murray, but there does not seem to be any hard evidence other than the fact that the firm was known to make machine tools for others and for export. The most detailed description of the Chaillot machine is in French [12], but it is covered by Forward, and also in Gill's Technical Repository. Forward states that the machine gave good work but was clumsy being 22 foot long for a travel of only 5 foot, making it only suitable for the smaller range of cylinders.

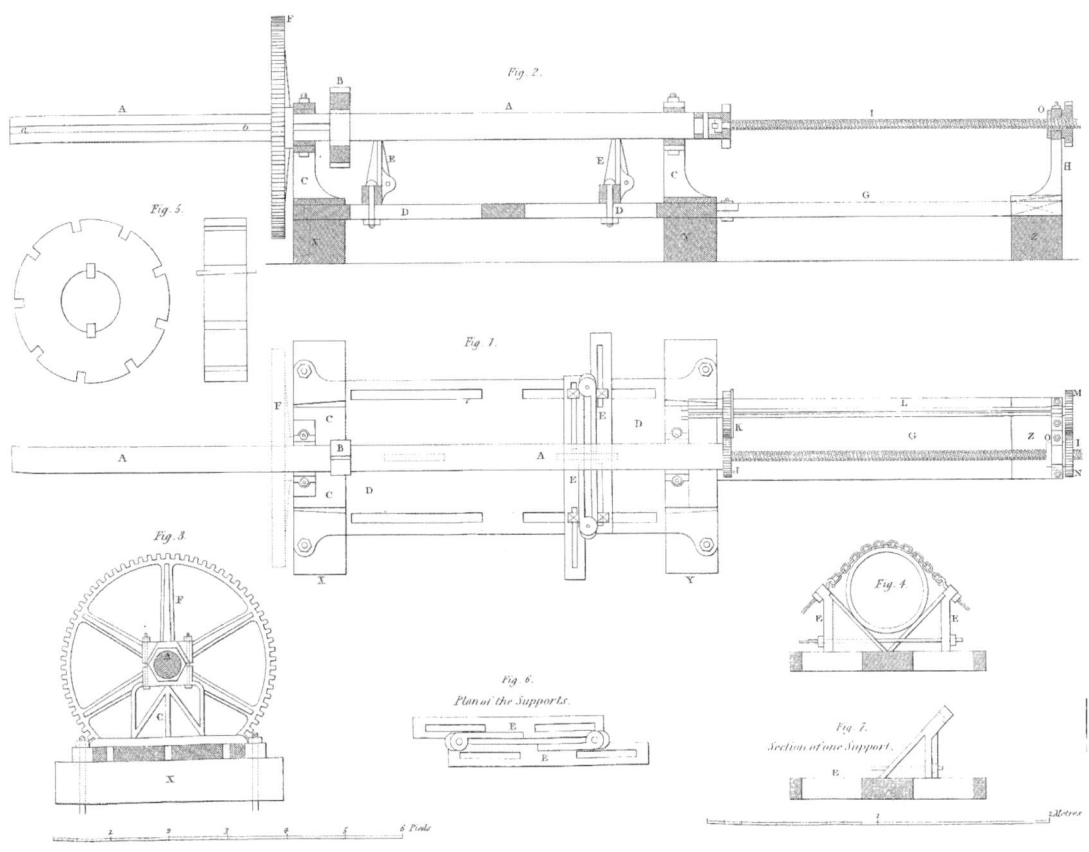

76. Chaillot boring machine [13]

In a book published between 1816 and 1818, the author described a proposal by Matthew Murray for grinding flat surfaces by machinery, a parallel grinder.

"two grindstones must be fitted up, to run with their flat surfaces horizontal; the under surface of one of them must lie over the upper surface of the other, and the circumference of the one extend to the centre of the other, as nearly as the axes will permit. If these stones be sufficiently thick to prevent their bending; if one of them be constructed to sink, or the other to rise, as the stones wear away, and the whole machine is steady, the surfaces thus working together will form each other into planes; and will communicate their flatness to any material ground upon them" [14].

However this was only a plan and it is not known when or where such a machine was first built.

The origins of the planing machine can be traced back to France, and while they were on a small scale, no doubt certain principles were laid down and published. A Nicolas Focq built a planing machine as long ago as 1751. In 1805 a Parisian locksmith, named Caillon built a small one to assist his manufacturing and developed it over the next decade and a half. The development on this side of the channel is not easy to establish. A patent by Samuel Bentham in 1793 may have tried to develop his patent of 1791 for planing wood into a machine for planning iron, but there is no evidence that one was ever built. Bramah took out a patent in 1802 'Machine for producing straight, smooth, and parallel surfaces on wood and other materials'. He supplied one to the Royal Carriage Works at Woolwich in 1805 for wood but there seems no clear evidence that he applied it to a machine for metal. Murray patented his famous 'D' slide valve at the turn of the century and needed some device for machining the necessary flat surfaces. It is clear that he had some form of planing machine in operation by 1814. There is on record for that year a clear remembrance of it by Mr J O March, Murray's son in law, and the head of the very successful firm of machine manufacturers Maclea and March:

"I recollect it very distinctly, and even the sort of framing on which it stood. The machine was not patented, and like many inventions in those days, it was kept as much a secret as possible, being locked up in a small room by itself, to which the ordinary workman could not gain access. The year in which I remember it being in use was, so far as I am aware, long before, any planing machine of a similar kind had been invented" [1].

There is a curious corroboration of March's testimony, to be found in a book published in 1874 by a Yorkshire man William Mortimer Baines. Unfortunately the date of the item is not given but relates to an interview in New Zealand with an old Leeds man Mr Fearnley, in about 1851:

"I was apprenticed to the famous Mr Matthew Murray, of Leeds, who had probably done as much as James Watt himself to perfect the steam engine. I believe that the planing machine we had in use whilst I was at the works at Holbeck, which was of his invention and Make, was the first ever constructed" [15].

It is known that in 1814 Fox of Derby was working on his planing machine and that it was built that year. March's description above is one of a machine that had been erected for some time and in place for a while.

Descriptions of the firm's cranes have also survived.

" The application of a column of water to lift weights was made, many years ago, by the late Mr Matthew Murray of Leeds, who employed it to raise the heavy boilers he manufactured for the spinning-mills in that district.

His mode of using it was very simple and effective. From a cistern placed upon a lofty building, a water-pipe communicated with a cylinder set upright upon the top of 'a triangle' formed of three stout trees, and fitted with a piston and rod, which passed through a collar of leather or stuffing box in the bottom of the cylinder; on the end of the piston rod was a loop or shackle, to which strong chains were attached, for suspending large boilers, engine beams etc.

By admitting the water between the cylinder bottom and the piston, the load was lifted sufficiently high to allow a waggon to pass under it, and, by allowing the water to escape, the weight descended upon the carriage. [60]"

The popular name for the works in Water Lane was the 'Round Foundry, after the round building erected early on but for a great deal of the time it was occupied by Fenton Murray & Co they referred to it simply as the 'Steam Engine Manufactory

Leeds', in deference no doubt to their most popular product. At this distance of time it is difficult to define the firm's absolute contribution to the steam engine.

It has been shown that Fenton Murray & Co had produced engines prior to the expiration of Watt's patent; but that it was done in such a way to avoid infringement of the patent in that the condenser was not fitted until after the patent expiry. Murray had also agreed with at least two clients Nixon and Fisher and Marshall and Benyon that they should approach Boulton and Watt for a license to operate before the patent expiry. In both cases the granting of a license was categorically refused, and it has to be assumed that the engines were operated atmospherically until after the patent expired in 1800. There is little evidence that the quality of the early engines was other than the standard later set by the firm even though many of the components had to be out sourced until the partnership had established its own premises and invested in the necessary manufacturing plant, as the pre 1800 engine supplied to Marshall's Mill A was not put up for sale until that mill was being closed in 1842. From the correspondence within the Boulton and Watt collection there are numerous references to the basic engine structures being identical to those of Boulton and Watt. Marshall probably encouraged and invested in Murray and Wood to produce steam engines suitable for textile machinery at lower prices, better quality and without the restrictions placed by Boulton and Watt in regard to such factors as moving the engines. Most of the restrictions lapsed along with the patent, but the relationship between Boulton and Watt and Marshall and his various partners was never good.

The first contribution of the firm to the steam engine has already been described above, which was the introduction of precision standardised manufacture leading to a high quality affordable product which was also, apart from the larger engines, tested on site before despatch. The second, which is the second item for which Murray is most closely connected, was the introduction of the 'D' valve. Murray patented the idea in his 1802 patent and used it in the manufactory's products. Its fame however is due to its widespread adoption by other manufacturers particularly for locomotive engines. The valve was ideal for operating expansive steam, where the supply was cut off at a pre-determined point and left to drive the piston on by its own expansion. It has been claimed that the 'D' valve was invented by Murdoch, who did indeed use a 'D' valve but the two were very different. Murdoch's valve used up less steam in its passages, but was of a far more complex construction which inevitably would lead to maintenance actions. Murray's valve was of simple robust construction which in the end offset its higher waste of steam, and led to its wide adoption.

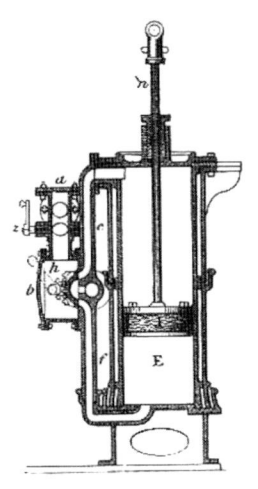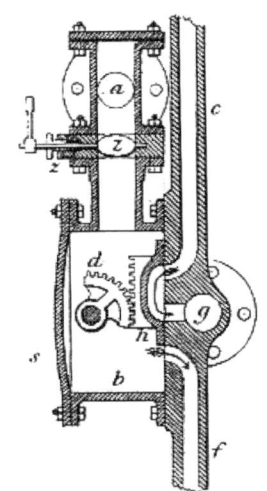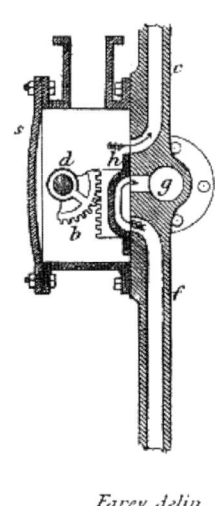

Section of the Cylinder

Farey. delin.

77. Murray's 'D' valve, 3 views from Farey.

Murray did display his fairly radical thinking when he patented and then produced hypercycloidal steam engines. The principles were known and their use in an engine had been discussed and published by others [16], but it was Murray who put it all into practice, and although he recognised the weakness inherent in the design and appears to have stopped their manufacture, other firms were still using the design in the 1830s [17].

Possibly one of Murray's most influential engines was his portable engine of 1805, where he moved the beam under the cylinder. The basic layout was used widely in early steam boats, and extensively developed. Murray did not patent the idea, but had it published by writing with a full description to Mr Nicholson for publication in his Philosophical Journal, see Chapter 7 pages 151/2.
Murray seems to have developed the engine for the firm to sell also, as the engine supplied for the Deptford Victualling Yard Brewery, as drawn by Joshua Fields, was to this layout, Chapter 8 page 182. It is also probable that some of the engines the firm supplied to saw mills were of similar layout. Once adopted for marine engines the type became quite common in mills and forges as for example the one supplied by Fenton Murray and Jackson to Kirkstall Forge and listed in their stock book for 1839. The description was for:

"Steam Engine for Tilt
One eight horse marine condensing engine, flywheel with one twelve horse boiler all complete in good working condition, Fenton Murray & Jackson cost £400, deduct for 4 years wear £80, balance £320"[18].

Murray applied himself at various times to the issue of controlling the steam raised, the usage of fuel and the reduction of smoke. The first item was a damping device as part of Patent No 2327 of 1799 and the last was his paper of 1821 on a smoke consumer which was published in the London Journal of Arts and Sciences (see Chapter 11 pages 251/2. The 1799 self-acting damper was used for years [19], see the drawing of such a device on a Swedish engine from Fenton Murray & Co on page 160. It was described as follows:

"A, a small cylinder upon the boiler, in which is fitted the piston and rack B,B, which are made to move freely up and down. C is a small wheel upon the shaft or spindle D, and works with the rack B. E is a damper fixed to the end of the spindle D in the chimney F, where it has free liberty to turn round. G is a circular cone fixed upon the spindle d, from which the weight H is suspended by a small chain. I is a pointe upon the spindle, and is made to go round the dividing scale K. Now, as the steam increases in the boiler beyond what is necessary, it will press upon the piston and rack B, which will turn the wheel C and shut the damper E, at the same time it will wind up the weight H, by which means the draught of the chimney is suspended. And the further consumption of the burning coal is stopped till the superfluous steam is wrought out of the boiler, while the divided scale shows the density of the steam and regulates the attention of the fireman" [20].

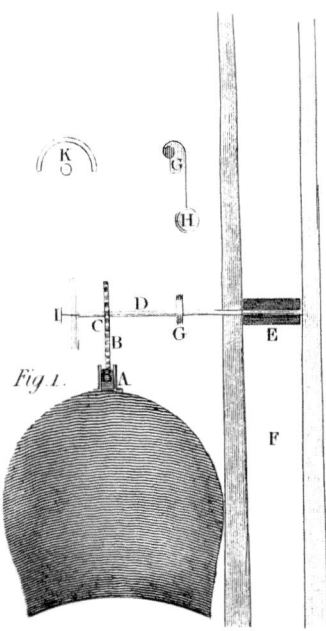

78. Murray's damper from his 1799 patent. [20].

This system was used as an example of the development of feedback control in a book published in 1975 [21]. The author wrote that the pressure measure would have been inaccurate due to the friction of the piston in the cylinder. However for the era in which it was developed, when the rules were only just surfacing and being formulated it will have worked as a practical solution to reducing fuel costs. Fenton Murray & Co as well as developing their own control systems for fuel and smoke always seem to have been happy to use other reputable systems that were available from John Roberton's system in 1802 to Brunton's fireplace around 1823. This policy continued in the life of the firm as the legislation on smoke nuisance became tougher and tougher.

An additional area where the firm may well have influenced an advance in engine design is the substitution of iron beams for wooden ones. Again it is very difficult to establish which firm was first, and who very quickly followed on. This matter has already been raised in Chapter 8. Boulton and Watt were producing metal beam proposals in 1800, so it is reasonable to assume that Fenton Murray & Co and Bateman and Sherrat were too; and certainly the Boulton and Watt drawings of these competitor beams are among their own earliest [22]. The evidence is yet to be found as to who made the breakthrough. Many of the problems in the early engines were derived from the wooden beams, which contracted and expended, and distorted after erection. This led in many cases to rough and uneven running, which would have been unacceptable to a textile maker, where it would lead to threads breaking and machinery running untrue. With this in mind it is certainly an area that would have received early consideration from Murray.

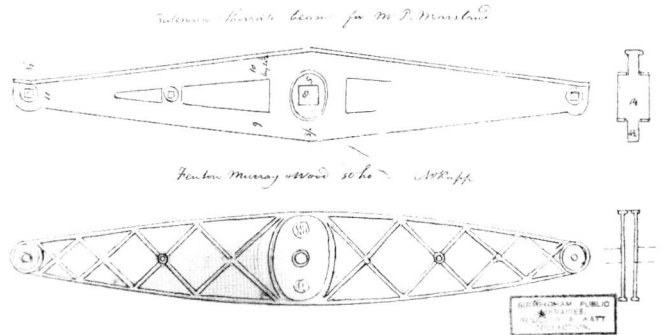

79. Early iron beams Bateman and Sherrat & Fenton Murray & Co.

Another innovation was the use of iron bedding plates for engines; which came into use about the same time. Fenton Murray & Co would have found these of great benefit for engines that were assembled tested and disassembled prior to despatch, but they also would have increased the stability of the engines.

It has to be said that not all Murray's ideas worked and the wondrous new pump which was part of his patent of 1801, the one that was subject to scire facias, did not work (see Chapter 6) and Fenton Murray & Co had to retrofit a few pumps. Which is probably another reason why Murray did not fight Boulton and Watt on the day the patent went to court.

The extent of the firms' dominance of the local market is shown in a survey conducted in 1824 [23]. The survey identified that out of the 129 steam engines recorded in the Leeds area 77 (60%) were by Fenton Murray & Co, 7 (5%) by Boulton and Watt, 37 (29%) by other large local suppliers and the balance of 6% small makers. Pullan and Sons were the second largest local supplier, followed by

Zebulon Stirk, with Low Moor and Bowling only accounting for 6 engines. If expressed in terms of horsepower supplied the figures become Fenton Murray & Co 64%, Boulton and Watt 7%, local suppliers 23%, and others 5%.

The following series of illustrations covers some of the engines proposed or delivered between 1800 and the 1830s. Many of these are reproduced earlier in the book, but here are shown in larger size to give the detail.

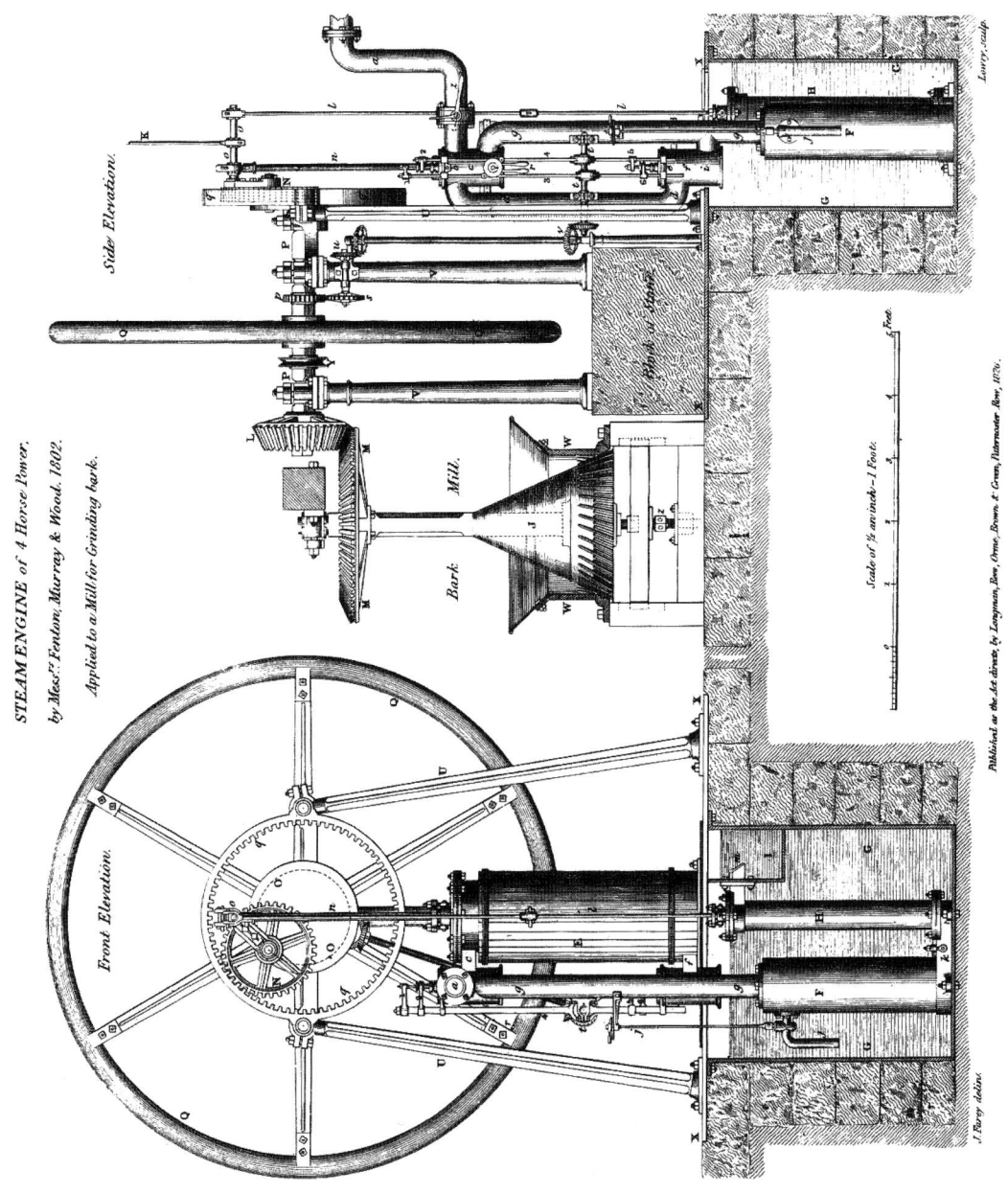

80. Hypercycloidal engine as supplied to a London Bark Mill in 1802

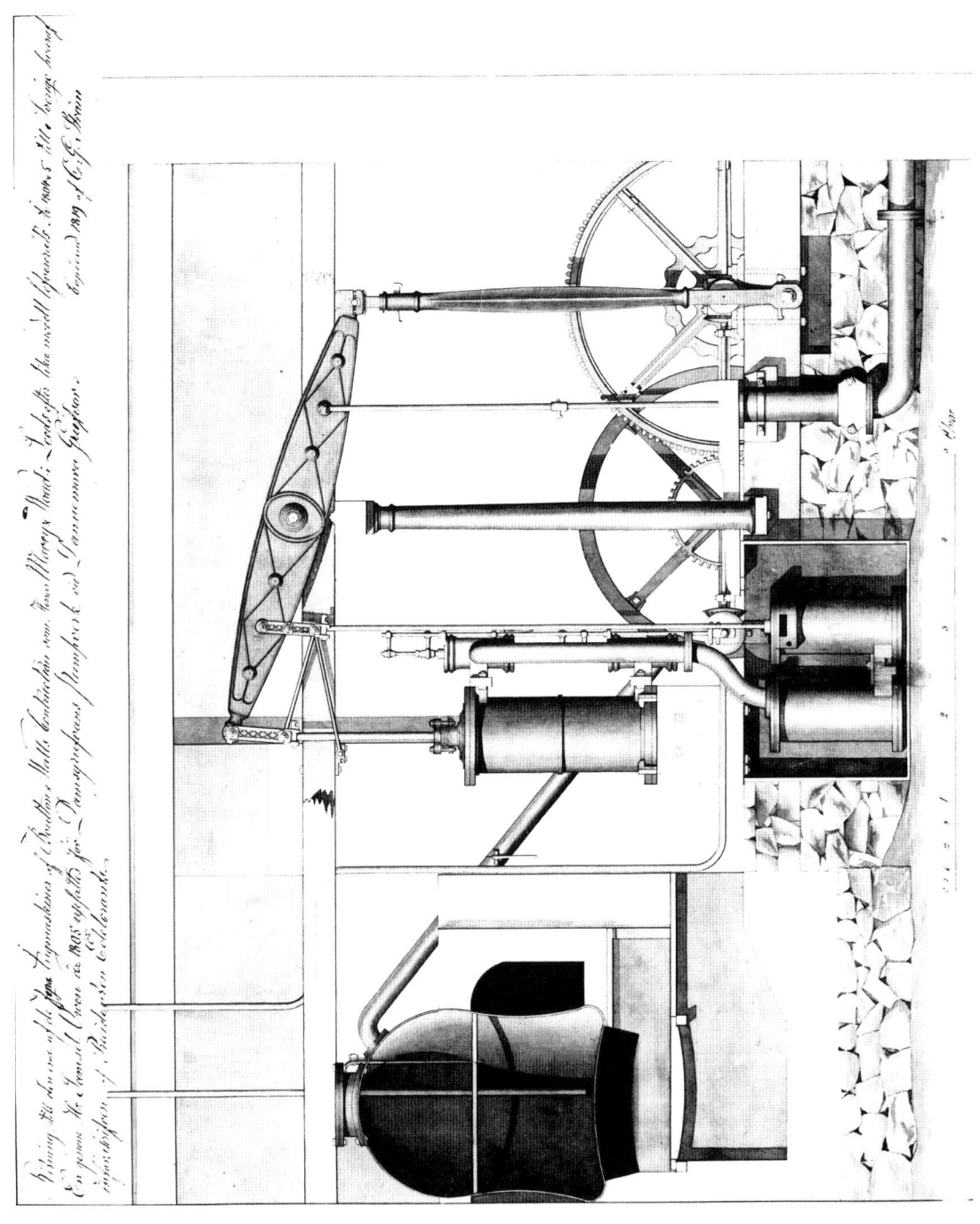

81. Engine supplied to Sweden, erected in 1805, used at the Dannemora Mine (see item 28).

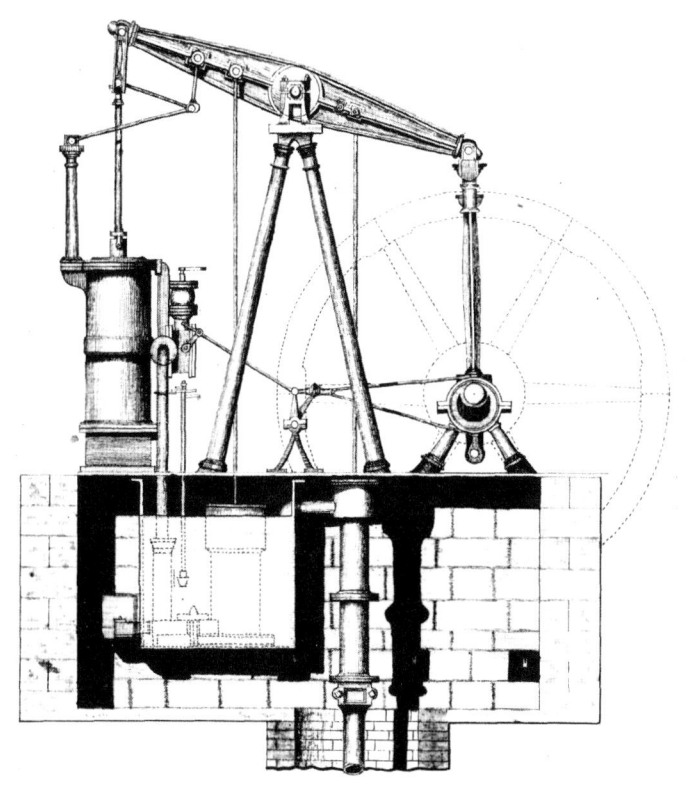

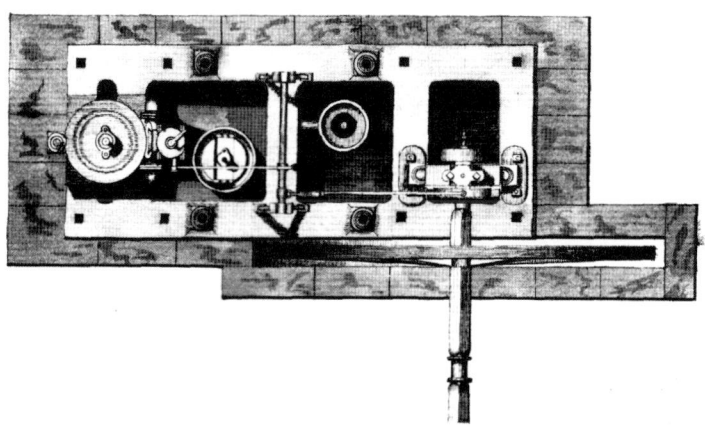

82. Engine drawing of 1810 sent to John Watson (item 43).

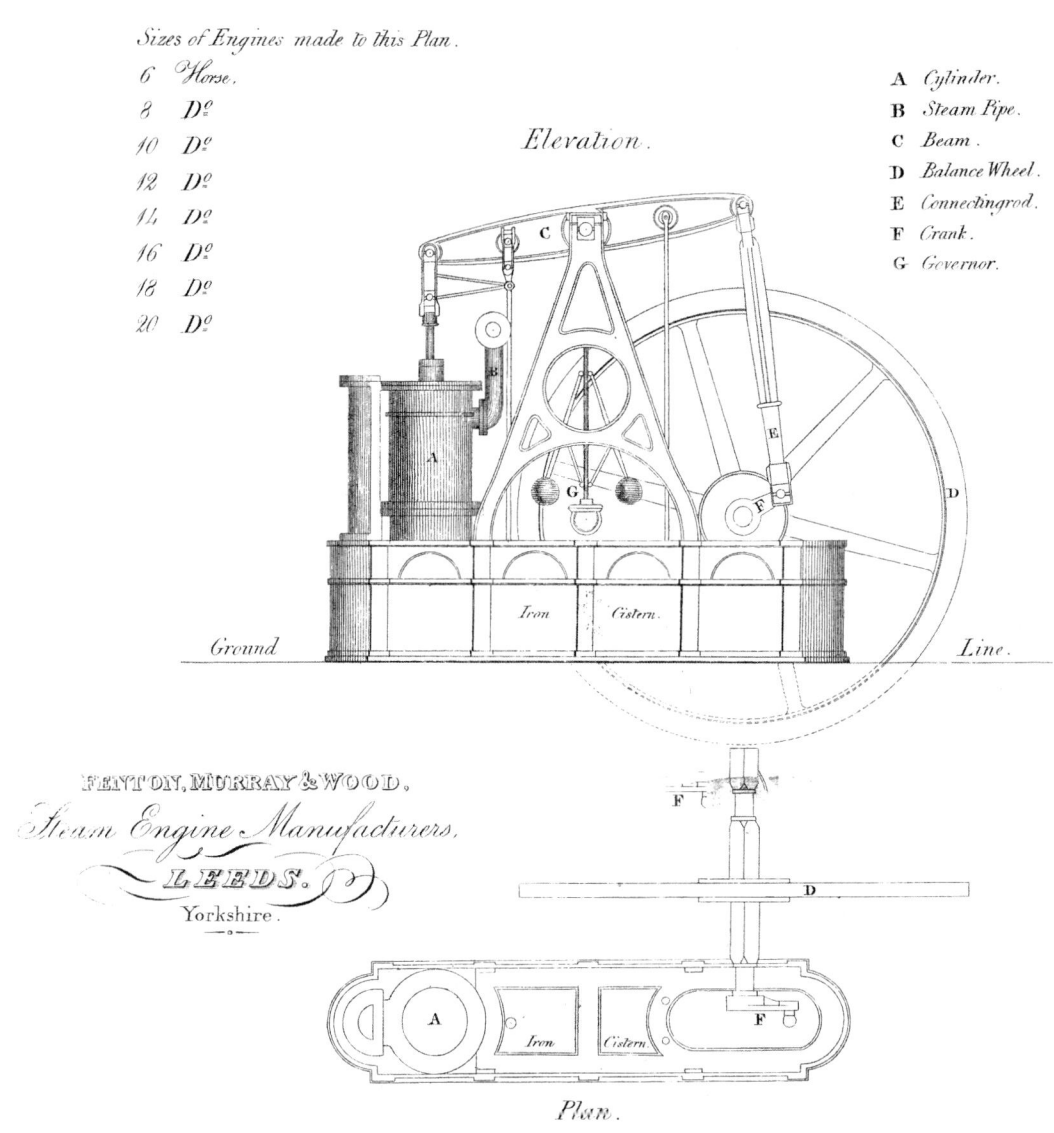

83. Drawing sent to Col Heth 1819 (item 53).

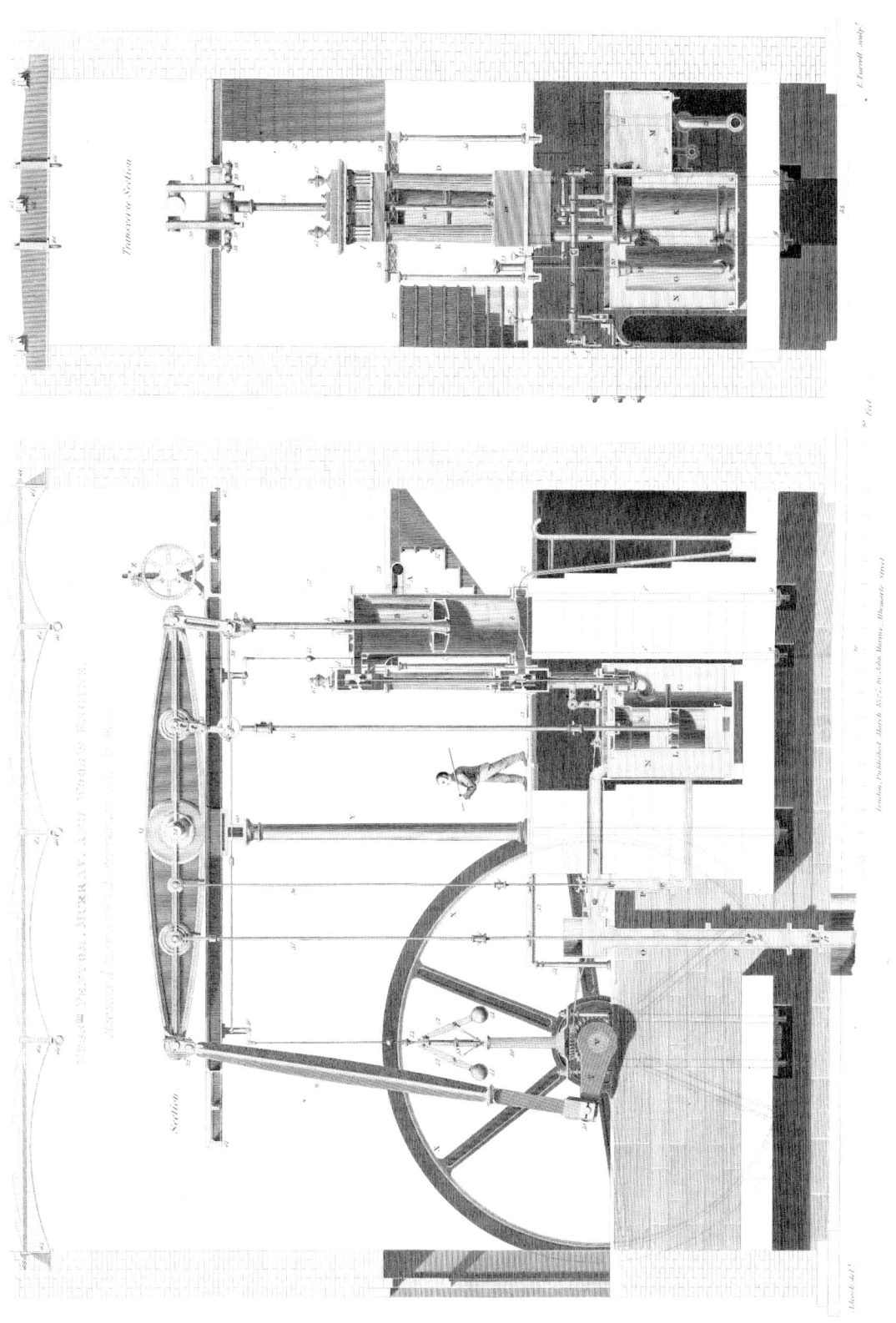

84. Messrs Fenton Murray and Woods Engine published by Birkbeck and Adcock 1827.

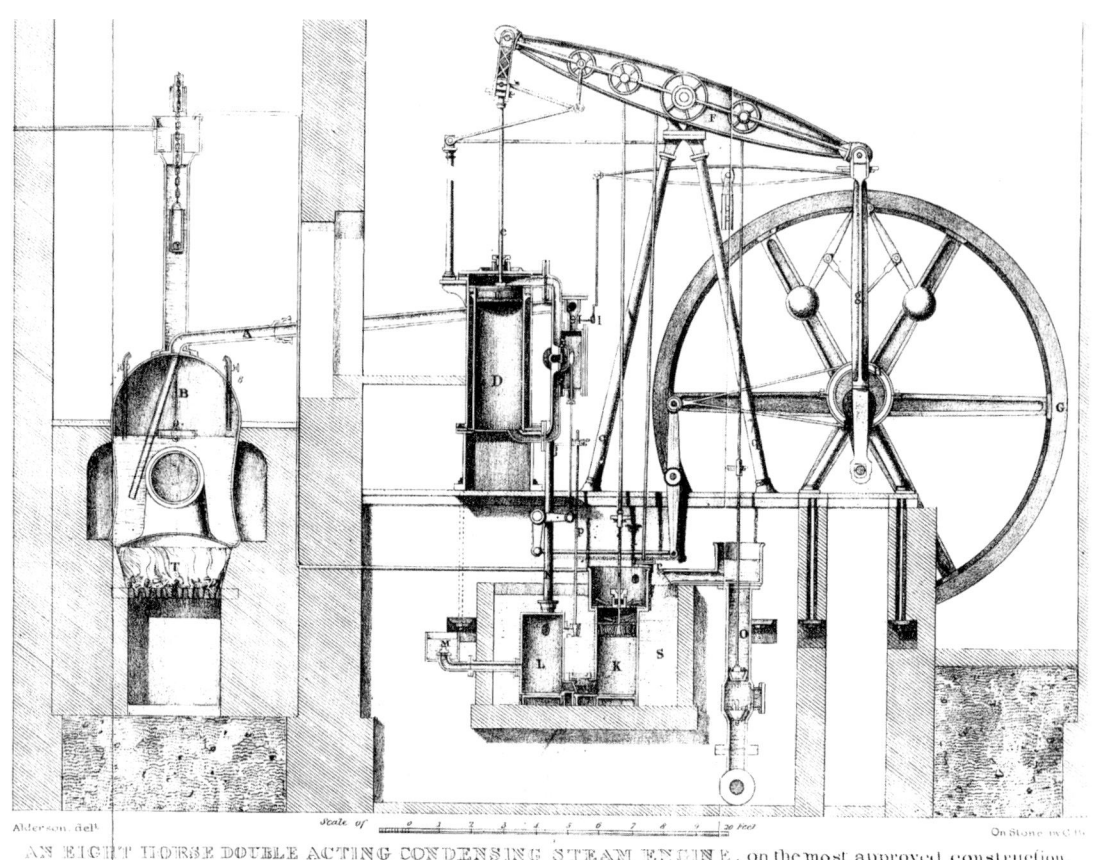

AN EIGHT HORSE DOUBLE ACTING CONDENSING STEAM ENGINE, on the most approved construction.

85a. 8 HP engine, elevation, by Messrs Fenton & Murray published in 1834 [29].

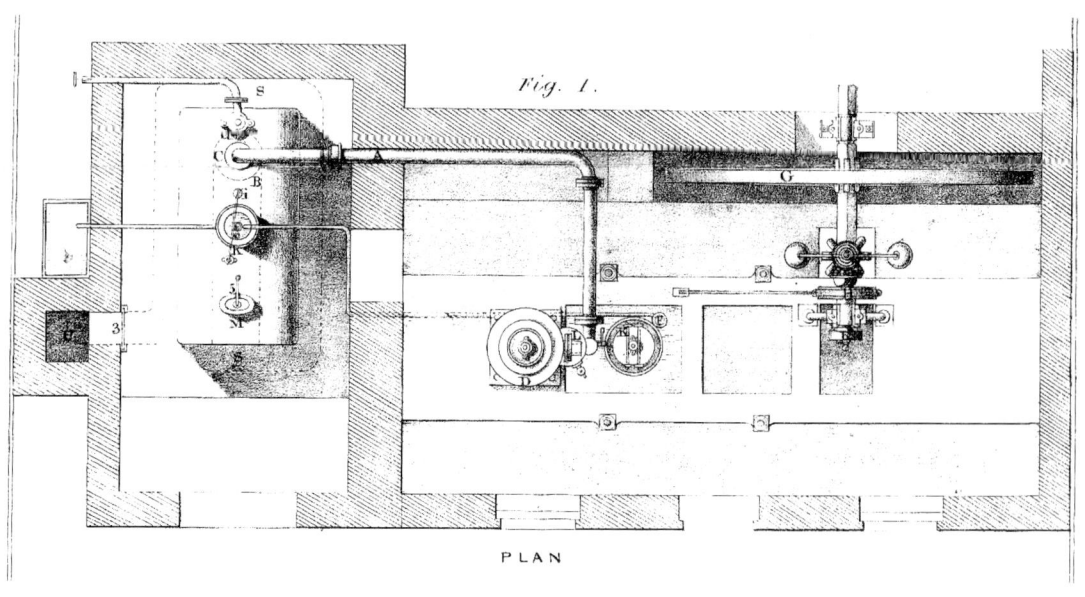

Fig. 1.

PLAN

85b. 8 HP engine, plan, by Messrs Fenton & Murray published in 1834 [29].

In 1817 the entry for Fenton Murray & Co in Baines Directory of Leeds read as follows:

Fenton Murray & Wood; manuf of steam engines, flax spinning and mill machinery, constructors of fire-proof buildings, water presses, & gas light apparatus. Water Lane.*

This was one of the most extensive lists they published, when they were probably at the height of their business. It also includes the least well known aspect of their business, at least until the last couple of decades, that of being erectors of fire-proof mills. The firm had been advertising this aspect of its offer since at least 1811, when it had appeared on their letter heading, see Chapter 8 page 191. They continued to advertise themselves as fire proof mill constructors until at least 1834.

Since the destruction by fire of Marshall and Benyon's new Mill B on the 2nd February 1796 (see Chapter 4), Matthew Murray had been acquiring the technology for fire-proof buildings, at first alongside John Marshall and then applying the techniques to the advantage of Fenton Murray & Co. The early range of buildings at the Round Foundry that occupy the east side of Foundry Street, demonstrate the evolution of these techniques. The remaining three floors of the structure immediately north of the lower roofed building on the end of the row are built of oak beams with brick arches sprung from them, following the principles developed by Strutt in his mills and warehouses between 1792 1nd 1795.

86. Brick arches springing from oak beams at the Round Foundry [30].

This building was erected prior to 1800, and probably dates to about 1796/7. The Round Foundry was under construction by those years as attested by the 1796 Fire Insurance policy. The most southerly block in this range is built to the succeeding fire proof technology and is of iron frame construction, which Murray will have seen when fitting out the Marshall and Benyon Mill (1796) at Shrewsbury (Ditherington). The design for the Shrewsbury mill for the partnership was by Charles Bage (earning him partnership status), who had corresponded with the Strutt over the development of the iron framing. The iron work for the Shrewsbury mill was made in Shropshire (see Chapter 4).

87. Brick arches sprung from cast iron beams, Round Foundry.

It is interesting to note that the columns at the Round Foundry are round, unlike those at Ditherington. Many of the mills attributed to Fenton Murray and Wood have a cruciform cross section similar to the original Ditherington ones with the addition of the more decorated capital shown in Chapter 10 page 224. It is also highly probable that a number of mills of early fire proof construction were by Fenton, Murray & Co but due to their construction with standard hollow round columns, and the fact that Fenton, Murray & Co did not appear to take credit for the mill construction, their design and build cannot be clearly assigned. For that group with a closely aligned plan and distinctly patterned cast iron structural parts, whilst there is no direct documentary evidence that provides incontrovertible proof of the designer and builder, for what may be termed the Bage Succession Mills, there is a wealth of evidence that ties Fenton Murray & Co to this group of mills.

The opportunity for Fenton Murray & Co to become involved and understand fire-proof mill building as practised by Charles Bage came with the breakdown of the relationship between John Marshall and the Benyon brothers. This resulted in the formal termination of the partnership on 30th June 1804 [32]. Negotiations over the share of the business had already started. Marshall had in his share the mill at Ditherington, and the Benyon's had some of the machinery and a cash settlement. The Benyons had no intention of leaving the flax spinning business, and formed a new partnership with Charles Bage. The first act was to build a new fire proof mill off Meadow Lane in Leeds to the design and direction of Charles Bage.

The main part of this spinning mill was well under way by August 1804, when Simon Goodrich visited Fenton Murray & Co. Goodrich recorded on the 4th August that Murray had sent him along with Murray's foremen to view Benyon and Bage's new flax mill. Goodrich noted that the mill was upon the same plan as Mr Marshalls. . Goodrich was probably repeating part of a discussion with Murray or his foreman and the reference was to Ditherington, which came to Marshall & Co under the partnership settlement. Goodrich also made some of his sketches of the mill which are shown above in Chapter 7. The fact that Murray sent Goodrich to see the mill does not necessarily mean that Fenton Murray & Co were the constructors, the steam engine was noted by Goodrich as of their build however. Two further pieces of information tie Fenton, Murray & Co into the mill. The majority of the mill, not just the spinning mill was probably complete by the end of 1806 or thereabouts, as in April 1807 building land containing a valuable bed of clay (bricks) near to the mill was advertised for sale; with interested parties being directed to Fenton Murray & Co [33]. Later in the same year Lord Milton one of the Yorkshire MP's visited Leeds and was shown the mill. The structure of the mill would still have been considered very innovative at the time, only being matched in Leeds by Gott's Armley Mills and maybe one or two others. Lord Milton was conducted over the works by Mr Benyon and Mr Matthew Murray, the latter explained the process of the factory as far as time permitted [34].

This pattern of association exists between the group of mills under discussion and Fenton Murray & Co. Equally important is the lack of any evidence linking these mills in their formative period to any other major engineering company. The buildings erected prior to Marshall's C Mill are all of similar dimensions to Benyons Leeds Mill or of a double unit, but the width of the mills does not vary by any significant amount. Bage's structural calculations for the loading of the beams across the mill width would not have required revision and length was a matter of adding what was in effect another standard bay.

Before addressing the individual mills, the other mill built by Benyon and Bage at Castlefields Shrewsbury is worthy of a short consideration. The mill was erected between 1806 and 1809, but all the buildings apart from the flax warehouse were demolished in 1835. The architect and designer would have been Bage, who visited Boulton and Watt in May 1803 to discuss an estimate for steam engines. James Watt jnr wrote to Benjamin Gott on the 8th June 1803 and in mentioning the visit stated:

"Mr Bage called about a fortnight ago for estimates for 20 & 56 Horse engines which we gave him, but although there seems to be strong schism between him and Marshall, I suspect he cannot prepare his machinery without the aid of Matthew

(Murray), & the latter may not choose to execute it without he has the engines also [35].

The Castlefields mill had a 20 HP by Boulton and Watt and a 75 HP by Fenton Murray & Co [36]. James Watt jnr was obviously under the impression that Benyon and Bage would need to procure their flax machinery from Fenton Murray & Co. The columns in the surviving warehouse resemble those on the ground floor of the Ditherington Cross Mill, and are close to the ones in the main mill. They would also appear to resemble some used in a flax hackling shop at Broadford Aberdeen, built circa 1808 and photographed on demolition in the 1980s. Unfortunately the photographs do not seem to have survived [37]. This situation coupled with evidence below might indicate that Fenton Murray & Co were heavily involved in the construction of the Castlefields Mill as well as populating it with machinery and a steam engine and presumably the mill work. Fenton Murray & Co may well have done a deal on the two Benyon and Bage mills, and if so Murray's chances of technology acquisition on cast iron frames would have been doubled.

The next candidate was further afield, and was constructed for a Mr Ford of Montrose, Scotland in 1805. The first engine man was George Stephenson who was employed to look after the Boulton and Watt engine [38]. No such engine is listed in the Boulton and Watt archives and it is safe to assume that the description referred to a condensing steam engine upon Boulton and Watt's plan. The mill was enlarged in 1814/5 (see note 31 Chapter 7). James Ford went bankrupt shortly afterwards owing money to Fenton Murray & Co, who pursued their debt through the Scottish Courts (see note 27 Chapter 11). The mill was advertised for sale in 1817 (see note 32 Chapter 7), and was described as of fire proof construction with two steam engines by Fenton Murray & Co, presumably the original one and one for the extended works. By 1814/5 indigenous fire proof mill builders were appearing in Dundee (30 miles away), and elsewhere; it would therefore seem that employing a mill constructer from Leeds at that time would probably only arise where the original 1805 mill had been constructed by the same firm. The mill was purchased from Ford's assignees by John Maberley an English MP, banker and government contractor. On his bankruptcy (1832) the concern was purchased by Richards & Co. At a time of decline for the linen trade and industrial unrest, the mill closed in 1877 when Richards & Co consolidated their business at Broadford in Aberdeen. The buildings and machinery were advertised for sale in December 1881 and the buildings afterwards demolished. This advert indicated that the fire proofing was of the kind used in the family of mills under consideration, as upwards of 300 tons of cast metal columns and beams were offered for sale along with a condensing engine with a 42 inch cylinder and 7 foot stroke, the latter possibly being the surviving larger Fenton Murray & Co engine [39]. However no drawings of the mill or its columns are known to survive, but the 1817 sale notice is remarkably similar to those of other mills in the family.

Following on from Montrose the next Mill was constructed 30 miles away in Dundee, in 1806/7. The Bell or West Ward Mill, for which again there are unfortunately no known depictions of the columns of beams on the fire proof floors, was constructed for James Brown of Canonsyth (a village about 5 and a half miles north west of Arbroath, or 11 miles south west of Montrose). The plans for the mill and the chief engineer came from Leeds and most of the flax machinery and the steam engine were supplied by Fenton Murray & Co [40]. A description of the original construction, which was noted during the demolition, described the fire proof floors as brick arched on cast iron beams and cruciform columns [41]. The mill was demolished in 1959, but external photographs survive that show that inverted arches were used in the wall construction to spread the loading. This technique which was not common at this period was a feature on two of the other mills in the family, Bell and Bragg's in Whitehaven and Marshall's Mill C in Holbeck.

The original mill at Broadford in Aberdeen was the next construction in sequence. At some point between February and September 1807 land at Broadford was acquired by a firm called Scott and Brown [42]. A case has been made that the Brown of Scott and Brown was James Brown of Canonsyth. He had married a Mary Scott at St Vigeans in Arbroath in December 1783. The two families had both been involved with Trottick Mill on the Dighty Water to the north of Dundee [43]. In 1810 Trottick mill was managed by John Scott [44] who was training James Brown's son William at the mill. A John Scott was advertising a lint mill for rent giving his address as Broadford in February 1810 [45]. There are a number of other indications of the realtionship.

Scott and Brown had built the original mill at Broadford, which still stands albeit altered at each end by the addition of subsequent mills. The Mill is typical of the family of Bage Succession Mills (see picture Chapter 8), even to the extent of having originally had an iron roof. This roof may have been succeeded when the first addition was made around 1820 but was there when the mill was surveyed for gas lighting by Boulton and Watt in 1815.

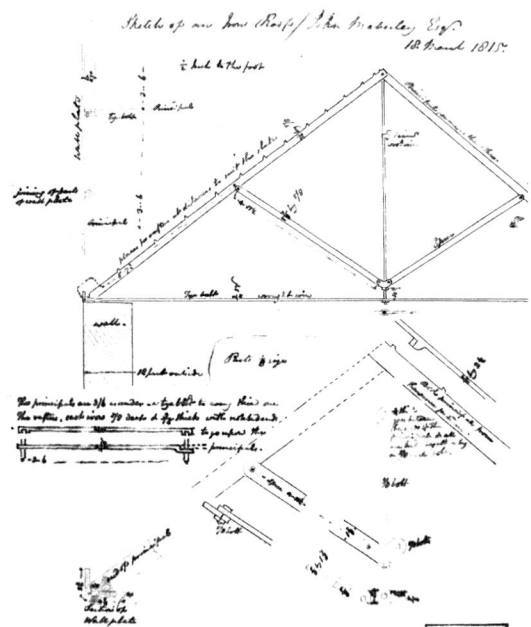

88 Broadford Mill Aberdeen, drawing of the roof from gas light survey 1815.

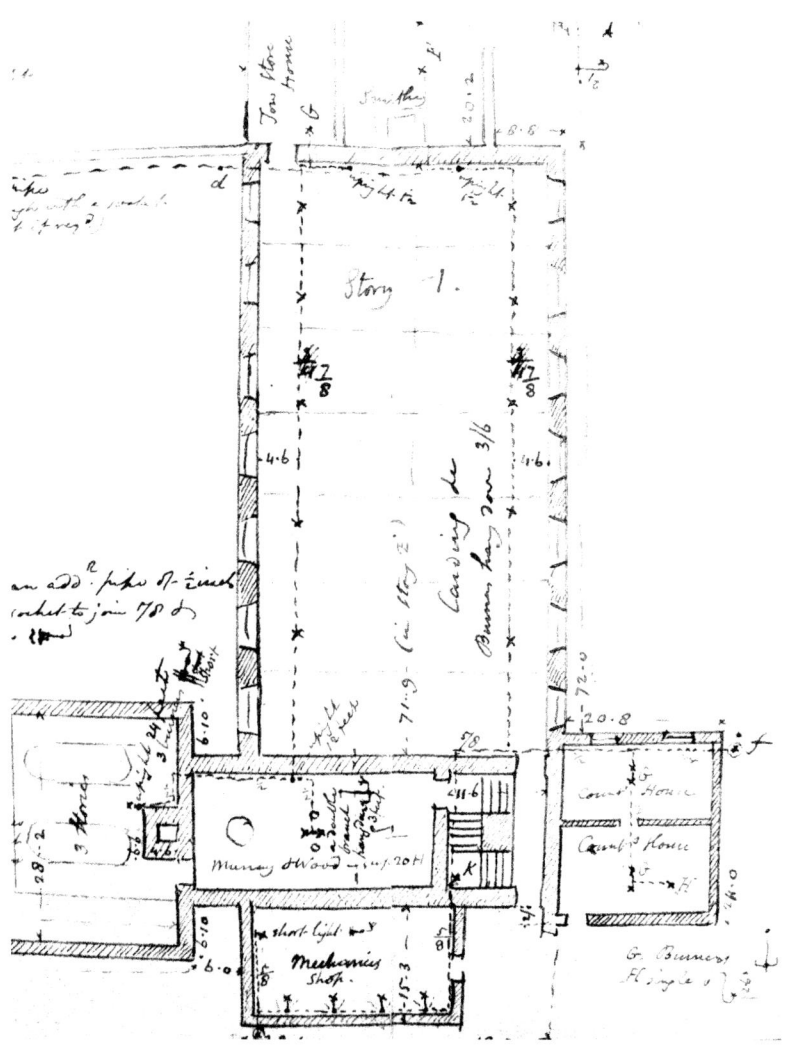

89 Ground Floor of Broadford Mill Aberdeen from gas light survey 1815.

The plans tell us that at this time the carding was done on floor 1, spinning on floors 2 and 3 and preparation on the tables on the top floor [46]. This data is confirmed by some old plans that were retained by Richards Plc, which show the machine layout for the ground floor carding engines and the first floor spinning machines [47]. By the middle of 1810 Scott and Brown were in sequestration, unsurprising after a major investment being followed by a rapid downturn in the market due to trading restrictions imposed by Napoleon on large areas of Europe with Britain. James Brown's death was recorded in the Aberdeen Journal on the 16 January 1811, and shortly afterwards Broadford Mill was advertised for sale on the 27th February 1811, followed by the Bell Mill in Dundee on the 6th March [48]. The sale notice for Broadford stated that the mill was fireproof and that the 20HP steam engine and the greater part of the machinery was constructed by Fenton Murray Wood & Co of Leeds. The mill was purchased by John Maberley (as Fords Mill in Montrose was later), and also passed to Richards & Co, who continued to operate the increasingly large works until 2002, when financial pressures forced closure.

At about the same time as the last two of these Scottish Mills were being built Marshall & Co had started to expand along what would become Marshall Street in Holbeck Leeds. They had a warehouse / counting house and a drying house constructed. The first building which still exists showed all the traits of the Bage Succession Buildings [49].

90. 1806 Warehouse and Counting House Marshall Street.

The drying house has been demolished and was of a different construction. It used open chequer plate flooring supported by a different arrangement of cast iron beams and solid cylindrical columns [50]. However as the Round Foundry is on the other side of the road it would have made sense that all the ironwork was made there.

In 1809 a flax mill was erected in Whitehaven for Messrs Bell and Bragg, and conforms to the features of the mill family. It is however double length with the engine house in the middle [51]. The mill still stands and at one time served as a military barracks and today is more commonly known as Barracks Mill or Catherine Street Mill. Documentary evidence links Fenton Murray & Co to the mill in regards to both the steam engine and some flax machinery. The mill was put up for sale by Joseph Bell & Co in 1853, and the advert included a 30 HP steam engine by Fenton Murray & Co [52]. In 1809 Matthew Murray was awarded one of the Royal Society's gold medals for a flax hackling machine (Chapter 8). In his letter of submission to the society dated 10 April 1809 Murray stated:

"Messrs Bell and Bragg, proprietors of a large flax mill at Whitehaven have also given orders for machinery to be begun upon this principle [53]."

The placing of the columns relates Whitehaven very closely to Broadford in that they both had a double row of columns on the ground floor with lugs for carrying the drive shafting for the machines. The Whitehaven Mill also uses inverted arches, clearly seen below underneath each window, in its construction which are also found at Bell Mill in Dundee.

91. Bell & Bragg's Mill Whitehaven showing inverted arches.

1811 to 1812 are the dates indicated for the building of the new Cross Mill and Flax Warehouse at Marshall & Co's Ditherington Mill near

Shrewsbury, by the entries in John Marshall's accounts [54]. The Cross Mill replaced the earlier one which had burnt down. The columns on the ground floor of the Cross Mill are more akin to the main mill and the surviving columns from the flax warehouse at Castlefields, while those on the other floors of the Cross Mill and in the warehouse are typical of the Bage Succession Buildings. In the same period the original Boulton and Watt steam engine in the south engine house was augmented with a more powerful 60 HP steam engine. It is unlikely that the engine was made by a firm other than Fenton Murray & Co. The Lindley list shows that by 1824 Marshall & Co had 5 steam engines in their Leeds mills and they were all by Fenton Murray & Co (i.e. the two original engines by Boulton and Watt had been replaced by Fenton Murray & Co products) [23]. The firm was also making regular quantities of flax machinery for Ditherington as shown by the Marshall Papers [55]. It is therefore quite logical that the structural cast iron for the Cross Mill and warehouse came from the same source.

In 1814 Robert Campion of Whitby replaced his old spinning mill at Bagdale with one on a larger scale and improved plan, with a steam engine of 12 HP by Fenton Murray & Co [56]. Subsequent to Campion's bankruptcy the flax spinning mill and the newly erected sail cloth manufactory were put up for sale with a 15 HP steam engine by "Murray" [57]. The mill still survives today and a portion of it is typical of the family of mills under consideration. Another section however is constructed using cylindrical (presumably hollow) columns. Some areas upstairs use slender cylindrical columns but as the whole roof is of a later date, these may well be contemporary with that rebuilding. It is tempting to consider that the area with cruciform columns may well be the earlier mill referred to by the Rev Young, and the large cylindrical columns may be a Fenton Murray & Co extension.

In 1816 the building of the great Mills for Marshall & Co along today's Marshall Street began with Mill C. The size of the mill is different and the technology to adapt Bage's theories had become understood either in Marshall's or Fenton Murray & Co, who by this time were attracting some skilled people. The 70 HP engine came from Fenton Murray & Co as did much of the machinery. The structural iron work conforms to the Bage Succession Mill type with solid cruciform columns although they were by this time outdated, and the neighbouring D and E mills used the standard hollow cylindrical columns when they were built between 1826 and 1830. The iron work most probably was again supplied from Fenton Murray & Co, the builder also used inverted arches. By the time of the two later buildings Fenton Murray & Co were probably standardising on the more usual hollow cylindrical columns and the Bage Succession Mill ceased to be distinctive.

Apart from the mills in Shrewsbury and Leeds, the others all bear a common location, in that they were built at seaports. One of the other discussions has to be over whether even if Fenton Murray & Co were responsible for the Bage Succession Mills whether they produced plans and patterns for the iron work or produced it themselves and shipped it to the location. Assuming the Montrose mill was a double mill by the time it had been extended in 1814/5, then the iron work was about 300 tons for a double and say 150 tons for Dundee and Aberdeen. In 1804 Fenton Murray & Co had quoted Mr Goodrich shipping from Leeds to London at £2 a ton, so if this was increased by 50% for Dundee and Aberdeen a shipping cost of £450 could be expected. If as has been projected the firm were offering customers turn-key flax mills, there would also have been the cost of shipping the steam engines and flax machinery. It is known that the Dundee mill cost its proprietor £7000, which if inclusive of shipping would have made the charge just under 7%, probably an acceptable level. It is unlikely that Fenton Murray & Co would have entrusted their steam engine drawings and patterns, or those for their machines to a prospective competitor, and the same probably applied to the structural iron work.

An interesting paragraph appeared in Saunders's News-Letter of Dublin in January 1819. It concerned premises to be sold or let at 52 Cork Street Dublin. They were described as a

"Large building with three floors, about 50 by 32, containing an eight horse power steam engine made by Fenton Murray Wood and Co. of Leeds; and a row of lying shafts in two floors, supported by metal pillars".

Was it the shafts or the two floors that were supported by the metal pillars? The advertisement bears great similarity to the sale of other properties in which the firm had been involved in the construction. However the plot now seems to support a Lidl supermarket, no doubt of modern steel frame construction.

It is worth repeating while on the subject of structural ironwork that there is some, most probably made by the firm on view in Chatham Dockyard, albeit to the design of Marc Brunel. As described in Chapter 9, due to the fact that budget delays caused Brunel to be required to build both the Woolwich Arsenal and the Chatham saw mills in overlapping timescales, he probably could not expect Maudslay to produce both at the same time. Fenton Murray & Co had bid for the Woolwich work in 1808, but by the time it was ordered Maudslay was ensconced in his new factory and ready to make the machinery. Brunel's request to Fenton Murray & Co for Woolwich specifically asked for Iron Columns 12 inches diameter, 16 feet long, hollow in the middle [58]. With the second order Brunel had to spread the load, and the engine at Chatham was certainly ordered from Fenton Murray & Co. The strongest clue is the Navy Board Letter to the Chatham Officers advising that some of the structural iron work was delayed at Selby [59], which was a trans-shipment port for Leeds. Certainly at this time the Leeds firm would have had more experience in structural ironwork than Maudslay.

92. Chatham Saw Mill – columns probably by Fenton Murray & Co.

As constructers of mills Fenton Murray & Co also installed stoves and ventilating systems into the mills, one of which has been described in 1815 as having been in a flax-mill for at least 7 years [61].

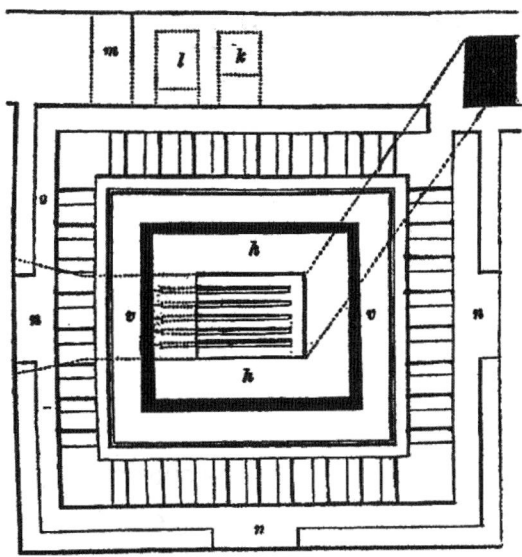

93a. Horizontal Section of a stove by Murray [62]

415

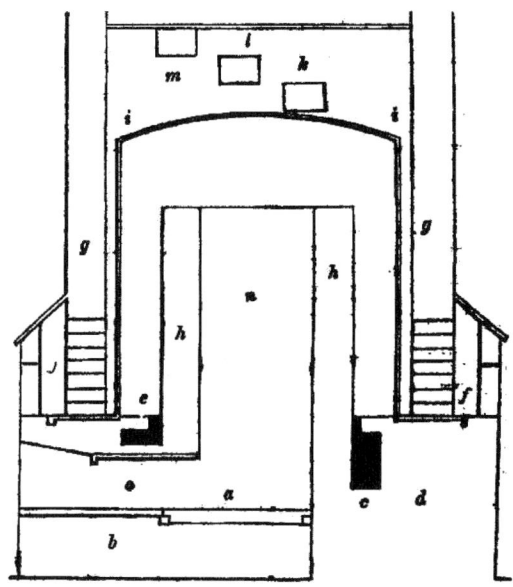

93b. Vertical section of above [62]

The fireplace (vertical section) was placed in a brick tunnel and was 19 inches long, 15 inches wide and 45 inches high above the furnace bars. The hood or cockle was placed 9 inches above the tunnel and was made 3/8 inch plate iron riveted together. The solid black areas represent the smoke flues and the chimney. K, l and m are the apertures whereby the hot air travelled to the floors for heating. N are the apertures to intake the external air and were 16 x 10 inches on three sides of the stove [62].

Notes.

1. Industrial Biography: Iron Workers and Tool Makers. Samuel Smiles. John Murray 1863.
2. Rees's Cyclopaedia. See volume 5 of David and Charles reprint 1972.
3. Marshall's of Leeds. W G Rimmer. Cambridge University Press 1960.
4. See chapter 7 page [173/4].
5. Early History of the Cylinder Boring Machine. Paper by E .A. Forward, given to the Newcomen Society, 17 December 1924. Also in The Engineer 19th and 26th December 1924.
6. See chapter 12 page [315].
7. See chapter 9 page [234].
8. See chapter 5 page [131-133].
9. The Encyclopaedia Metropolitana, volume 8. Smedley, Rose and Rose. London 1845. Also Rees's Manufacturing Industry, volume 3. David and Charles 1972.
10. Boulton and Watt Papers MS3147/3/51/8.
11. Supplement to the 4th, 5th, and 6th Editions of the Encyclopaedia Britannica, 1824.
12. Bulletin de la Societe d' Encouragement pour L' Industrie Nationale No CCXXIII January 1823.
13. Gill's Technical Repository.
14. The Mechanic or Compendium of Practical Inventions by James Smith, Liverpool.
15. The Narrative of Edward Crewe by W.M.B. London 1874. Page 59.

15. The Narrative of Edward Crewe by W.M.B. London 1874. Page 59.
16. A New Century of Inventions by James White, Manchester 1822, in which he states that he had a medal from Napoleon Bonaparte for the invention in 1801.
17. Goodrich Journals B68 Science Museum Library – at Rastrick's Engine works as observed by Goodrich in 1830.
18. Photocopy of Stock book no 1 Kirkstall Forge. Now lodged at WYAS reference unknown.
19. 'Matthew Murray, A Centenary Appreciation', by G F Tyas, Newcomen Society, February 1926.
20. Patent no 2327 of 1799 Specification of Matthew Murray, Steam Engines.
21. The Origins of Feedback Control by Otto Mayr. MIT press 1975.
22. Boulton and Watt Papers, MS 3147/5/1378.
23. William Lindley list, MSS 18. Special Collections Brotherton Library, Leeds University.
24. A treatise on the Steam Engine by John Farey, 1827.
25. From the collection of the Technical Museum Stockholm.
26. Watson Papers Wat 2/2/308. NEIMME. Note the near copy in Farey's Steam Engine at plate XVIII, presumably a standard type for that time.
27. MS 38-114. University of Virginia Library U.S.A.
28. The Steam Engine Theoretically and Practically Displayed, G Birkbeck and J Adcock. John Murray London 1827.
29. An Essay on the Nature and Application of Steam, M. A. Alderson. London 1834.
30. Photograph by Courtesy of Dr R Fitzgerald.
31. Author's photograph.
32. London Gazette announcement dated 11 April 1805.
33. Leeds Mercury 11 April 1807.
34. Leeds Mercury 14 November 1807.
35. Boulton and Watt Papers MS3147/3/99.
36. Salopian Journal November 25 1835 for the 75 HP engine, whether this was the original engine is unclear but the B & W drawing for gas lighting dated April 1811 shows that the main engine was by "Murray".
37. Matthew Murray and Broadford Works Aberdeen: evidence for the Earliest Iron-Framed Flax Mills, Mark Watson. Textile History vol 23 1992. Pages 236-7.
38. Story of the Life of George Stephenson, Samuel Smiles. John Murray 1860. Pages 37/8.
39. The Dundee Advertiser 13 December 1881.
40. Note 8, 9, 9a Chapter 8. Sale notice Aberdeen Journal 6 March 1811.
41. Jute and Flax Mills in Dundee, Mark Watson. Hutton Press 1990. Page 32.
42. Aberdeen Journal, land for sale various adverts 1807.
43. Information from Mark Watson.
44. London Metropolitan Archives. Sun Fire Insurance Policy CS/88/837828.
45. Aberdeen Journal 28 February 1810.
46. Boulton and Watt Papers MS3147 folio 808.
47. These two drawings are now in the care of Historic Scotland and on the Canmore website.
48. Both in the Aberdeen Journal.
49. A reconstructed plan of how the original warehouse/counting house looked can be found in Yorkshire Textile Mills 1770-1930, Giles and Goodall. HMSO 1992. Figure 92.
50. Development of the cast iron frame in Textile Mills to 1850, Ron Fitzgerald. Industrial Archaeology Review vol. X Spring 1988. Page 133 and Plate 4.
51. As per note 37 but page 238.
52. Preston Chronicle 4 June 1853.
53. Journal of the Royal Society for the Arts volume xxvii. Pages 148-153.
54. Marshall Papers MS200 folio 1. Brotherton Library.

55. As for 54 but folio 28.
56. A History of Whitby and Streoneshalh Abbey, Rev. George Young. 1817. Vol. 2 page 558.
57. Leeds Mercury 26 August 1843.
58. M I Brunel's Letter Book. LBK 54. MIB to Messrs Murray and Wood, June 13/14 1808.
59. CHA/E/110. NMM. Navy Board to Chatham Officers, 1 September 1813.
60. Rudimentary Treatise on the Construction of Cranes and Machinery. Joseph Glynn. London 1849.
61. A treatise on the Economy of Fuel and Management of Heat, Robertson Buchanan. Glasgow 1815.
62. On the History and Art of Warming and Ventilating Rooms and Buildings. Walter Bernan London 1845.

Chapter 17.

Locomotive Engines.

The name of Matthew Murray is recognised more for his association with the early rack locomotive than with any of his other achievements. The fact that he was capable of building a locomotive in 1812 that worked and could operate commercially for 23 years remains to his great credit. In many areas and for some decades, the locomotive has been referred to as Blenkinsop's locomotive and as he was the one who patented the rack rail and presumably paid Fenton Murray & Co to develop the engine, he has the right to much of the credit. As was common with many machines at the time, the rack locomotive was commonly referred to as a 'patent locomotive'. The main patents involved were the rack system (Blenkinsop) and the high pressure steam (Trevithick) although the latter had sold most of his rights in the patent and would not have received much benefit from the royalties paid on these rack locomotives.

The firm of Fenton Murray & Co was the first series producer of railway locomotives, and after a long period of not making any locomotives its desire to re-enter the business as a volume manufacturer was one of the factors that led to its demise. The investment required in a state of the art new locomotive shop on Marshall Street over extended the firm's credit, all at a time of great difficulties in the manufacturing industries.

In the first phase of locomotive building, the rack engine phase, Fenton Murray & Co produced 6 or 7 locomotives. 5 engines in all were ordered by the Middleton Colliery, one of which was diverted to the Kenton and Coxlodge Colliery and the Middleton Colliery was paid for it by Messrs Williams and Watson (the proprietors/managers of the Kenton and Coxlodge) and entered as such in the Middleton Colliery receipts on the 24th June 1814 [1]. On the face of it the Kenton and Coxlodge had 3 engines, the one transferred from the Middleton Colliery and 2 further ones ordered by John Watson from Fenton Murray & Co and acknowledged in a letter from Fenton Murray & Co dated 14 October 1813 [2]. Blenkinsop actively tried to sell the concept elsewhere and succeeded in coming to an agreement with John Clark, a banker, one of the proprietors of the Orrell Colliery in Lancashire. His engineer was Robert Daglish, a man of good repute as an engineer, who along with a Mr Holding had direct discussions with Blenkinsop [3], and quite possibly with Murray. The upshot was that Daglish had three locomotives built by the Haig Foundry for the Orrell Colliery. The first locomotive was claimed as built in 1812 for use in January 1813 [4], but in view of the date of the Blenkinsop letter to Clark this seems a trifle early. The locomotives were known locally as Walking or Yorkshire Horses, and there is no reason not to believe that at least the first build was to the Fenton Murray & Co design.

While the Leeds built locomotives were all to the same basic design, they were not all the same dimensions. The main innovation was the use of two cylinders, which did away with the need for a flywheel. The engine also had two safety valves, but they were both readily accessible. There was a single flue of 14 inch diameter, passing through the boiler to the chimney. The pistons drove connecting rods to two cranks on either side which drove axles to the front and rear of the main drive axle to which they were connected by geared wheels meshing with a larger geared wheel on the main drive axle. The models made for the Science Museum and the Leeds Museum in the 20th century follow the French drawings which had the main gear wheel in the centre of the drive axle, which would have been logical if there had been two rack wheels on the system as these drawings show. However the contemporary Embleton model, currently on display in the Leeds Museum, has the main gear wheel positioned just the other side of the main frame member to the single driving rack wheel. The French drawings also show a locomotive with a water pump (the box on the front of the frame).

In a letter to Watson in April 1813 Murray had commented, in regard to the Middleton build engine being transferred to Kenton and Coxlodge, that he would not have recommended the swap as the engine was not sized to undertake the Kenton and Coxlodge journey [5]. Earlier in the month Blenkinsop had observed in one of his letters that the boiler needed filling every 5 miles [6]. The design no doubt evolved for purely technical reasons as well as sizing the boilers for greater distances. The first experiment which was briefly described by Farey in Rees Encyclopaedia, stated that a small condensing engine was used but that the water became too hot and they had to use a high pressure engine [7]. The experiment must have had very poor results to persuade Murray, with all his bias against it, to select high pressure steam. Perhaps the most surprising aspect of the Middleton and Orrell engines was the oval boiler, which is shown extremely clearly in the drawings reproduced in an 1815 Bulletin de la Societe d'Encouragement pour L'lindustrie Nationale [8].

The description of the Orrell locomotives handed down by Daglish make it clear that his boilers were also oval [9]. However dimensions recorded for the Kenton and Coxlodge locomotives merely give a diameter, and the boiler is some 2 feet shorter than the Middleton and Orrell locomotives [10]. The following table shows the surviving dimensions of some of the known engines, all given in inches:

420

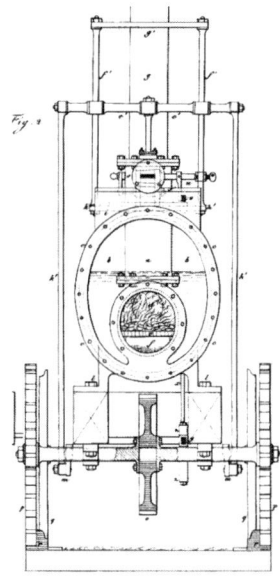

94. Front view of Murray/Blenkinsop locomotive from the 1815 French drawing.

Source	NRM Leeds	Wat 6/61 K & Cox	QAR 5/39 Orrell
Boiler Diameter		56	
Boiler oval	37 x 32		51 x 38
Oval to round approx.	34.5		44.5
Boiler length	114.0	85	108
Diam. of cylinders	9	8	8
Stroke	22	24	24
Carriage wheels diameter	35	36	36
Wheel base	88		86
Carriage length		140	
Cog wheel diameter	38.2	38	
Chimney height		144	168

A rough calculation of the boiler sizes shows that the Kenton and Coxlodge boilers were almost twice the size of the Middleton ones and that the boilers of the Orrell locomotives were about midway in size between the other two. In support of the contention that the Kenton and Coxlodge boilers were round it can be noted that Nicholas Wood in describing Brunton's locomotive stated that the boiler was nearly similar to Blenkinsop's being cylindrical with a tube passing through it [11]. Additionally the gauge of the Kenton and Coxlodge way was 4 foot 7 ½ inches [12] as compared to the Middleton at 4 foot 1 inch and the Orrell at 4 foot [9], and therefore the locomotive frame must have been wider, which would have made a more stable base for a broader boiler. Interestingly the Embleton model also shows a cylindrical, rather than an oval, boiler. This may have been for the ease of the model maker, but does beg the question as to whether it is actually a model of a Kenton and Coxlodge locomotive, rather than a Middleton one.

The correspondence between Leeds and Newcastle on the Kenton and Coxlodge locomotives mentions a further engine which has already been referred to in Chapter 9. This was the engine described as:

"of larger dimensions but of nearly the same description in hands these 6 months for Mr Buddle, it is now nearly finished, we believe it is for drawing coal upon an inclined plane ." [13]

An engine was described as a Trevithick engine underground in the Percy Main by Mathias Dunn [14]. It remains unclear whether it was a locomotive or stationery engine, but it is hard to believe it to be anything other than a locomotive from Matthew Murray's letter. Murray's general lack of enthusiasm for high pressure engines would also argue against the engine he provided to the Percy Main at this time being a high pressure engine, unless it was a locomotive. There were other firms making high pressure engines locally and the engine mentioned by Dunn may be another engine altogether. If it was a Murray locomotive it may have been a direct traction locomotive, and as it was sold to Mr Buddle in 1813 it may have been used as something of a model for the early Steam Elephants that Buddle and Chapman collaborated on. If so it might explain the similarities in construction.

The life of the Kenton and Coxlodge rack railway does not appear to have been long, as the locomotives were withdrawn after a couple of years due to a dispute between the colliery owners and with a neighbouring colliery, and while calculations exist for the economics of their re-instatement, there is no hard evidence they were. Surely with their cost however they will have been used in some form or another. At Middleton however the system survived the fatal explosion of a locomotive in 1818 (driver error) and continued until 1834 when a second locomotive blew up again killing the driver. The engines ceased to operate in the following year. A locomotive survived until 1860 when it appears to have been scrapped. The Orrell locomotives enjoyed similar longevity, and it was claimed that they worked for 36 years until the 1840s. One of the locomotives was in the colliery stables cutting feed for the horses and was only scrapped in the 1920s [4].

It has been mooted in various journals and books, including Lowe's British Steam Locomotive Builders, that the firm sent a small Blenkinsop type locomotive to Russia. The end source of these reports appears to be an article published in the Railway Magazine [15]. The article 'Matthew Murray and the Blenkinsop Locomotives' by R. E. M. Bleasdale quotes a letter from Murray to the effect that he had sent a small locomotive engine to Russia several years earlier. This letter was written in reply to one Murray had received from Simon Goodrich. At the beginning of Murray's letter it is clear that he was responding to a request from Goodrich for the loan of models and information on steam engines and locomotives that Goodrich needed because he was lecturing to a "Philol. Society". In the first paragraph Murray stated that he had no models available. The reference in full later in the letter reads

"I am sorry I have not any models of engines – I had a small loco-motive engine, but it was sent several years ago to Russia, or you should have had it.[16]"

This context makes it fairly clear that the subject under discussion was a model, and as there is no other evidence for a proper locomotive being sent to Russia, the matter can be put to rest.

Matthew Murray produced a further design in 1825/6, which has been covered in Chapter 12, but the firm declined all enquiries to build locomotives after the initial burst, until 1831, a period of 15 years.

After the Rainhill Trials and the ascendancy of the Rocket locomotive, its makers Robert Stephenson & Co began to be inundated with orders for locomotives from 1830. One of the first results was that Robert Stephenson became associated with Charles Tayleur of the Vulcan Foundry near Warrington in order to build more locomotives. To do this he had to satisfy his partners at Newcastle that the Forth Road Works there would always have his main attention. R Stephenson & Co supplied drawings to other companies capable of manufacturing locomotives, one of which was Fenton Murray & Co. At the time Fenton Murray & Co was run by James Fenton, the sole remaining long standing partner. The desire to re-enter the locomotive business must have seemed an attractive proposition, and there were no doubt sufficient managers in the firm capable of undertaking the work. Richard Jackson and Benjamin Cubitt are two names that spring to mind. Their first customer was the Liverpool and Manchester Railway, to whom they delivered locomotives number 19 Vulcan (illustration Chapter 13) and 21 Fury in 1831 and number 30 Leeds in 1833. A French engineer F M G de Pambour wrote a book on locomotives published in 1835, which described experiments conducted on the locomotives of the Liverpool and Manchester and Stockton and Darlington Railways [17]. In the book, although Fenton Murray & Co are not mentioned, their three locomotives featured in most of the experiments concerning the Liverpool and Manchester Railway. In the list of the 12 best locomotives on the railway all three were listed along with the Liver by Bury and 8 locomotives by R Stephenson & Co. The table also gives the dimensions of the fire-boxes and boilers, and while the Fenton Murray & Co locomotives were all Planets, there were differences in the detail to the Stephenson ones. As an example no two Stephenson locomotives in the list had the same number of boiler tubes, whereas the Fenton Murray & Co ones standardised on 107 tubes. The arrangement of the firebox and grates differed between them however, the Vulcan having a larger area than the Fury. The maintenance records show that by 1833, the Vulcan had covered 36750 miles and was still going strong, while the Fury had managed 21330 miles and needed a new fire-box. The Vulcan was one of the exceptions as in the sample of 15 given only three were still fit for work. Vulcan was joined by Planet (Stephenson) and Liver (Bury), while other locomotives needed fixing after 8000 miles [18]! Vulcan was sold in 1841 to T Pearson for £399, Fury was broken up circa 1842, and Leeds was sold in 1840 to the Chester & Birkenhead Railway, for £414 [19]. Some of the other locomotives only lasted 5 years.

Prior to the delivery of 'Leeds' to the Liverpool and Manchester Railway, Fenton Murray & Co made their first two deliveries into their most lucrative export market for locomotives. These were to France, where the firm had already established a market for marine engines. The first railways in France had been established in the region bounded by the Upper Loire and Upper Rhone in the Lyon area. The area was developing as one of the heavy industry centres of France due to its coal mines, for which the first railway was developed between St Etienne and Andrezieux. Although this was the first railway it did not adopt steam traction until after its two close neighbours, and maintained horse traction until 1844. From 1825 a separate company, under the technical direction of Marc Seguin, started building a line from St Etienne to Lyon. The third line was built from Andrezieux to Roanne (Le Coteau) by Messrs Mellet and Henry. The three lines were amalgamated after various financial re-organisations into the Railway of the Loire, but this happened after the period involving Fenton Murray & Co [20].

The Andrezieux – Roanne obtained final permission to import two English locomotives in 1831. One locomotive was ordered from Stephenson's and the other from Fenton Murray & Co. They were expected to haul 60 tons at 12 m.p.h and have a power of 12 horses and had a configuration of 0-4-0 (Sampson type). The price was 25,000 francs including transport, which was equivalent to the prices in Britain. The Stephenson locomotive was delivered first in May or June 1832, and named 'La Loire'. The Fenton Murray & Co locomotive was commissioned on the railway in August 1832 and named 'Constance'. 'Constance' was used to undertake trials in September pulling a train of 110 tons, and in climbing the inclined plane at Biesse (45mm/m), with the wheels uncoupled. This is probably the same trial that was reported internationally in the middle of 1833 [21] and described as: with a weight of 15000 kilograms (c.14 tons), including the locomotive, tender, water and fuel, she ascended an inclined plain 2000 metres (2184 yards) long with a rise of 45mm per metre, with a steam pressure not exceeding 38 lbs per square inch. The climb was made in 6 minutes, and was made with great ease both up and down. The wheels were not coupled for the trial. She was driven by an Englishman called Georges, who went on to become the Chief Engineer of the Paris - Versailles Railway and met his death in the infamous Meudon disaster of 1842. Constance was lent to the St Etienne –Lyon line at the end of 1833 to make a trial run between Lyon and Roanne. Constance was withdrawn in 1842 and rebuilt and was recorded as having six wheels in 1844, 2-4-0. Locomotive No 3 was also built by Fenton Murray & Co, but appears to have been ordered by M.Edwards at the Chaillot works who had received an order for two locomotives from the Andrezieux – Roanne that were to be built to the same specification as the first two. Locomotive No 3 was delivered to Chaillot and used as a model, prior to being delivered to the railway as the first part of the order. This locomotive was called 'Jackson' and commenced operations on the line in the first two months of 1834. 'Jackson' was lent to the St Etienne –Lyon line at the end of 1833 to make a trial run between Lyon and Roanne. Chaillot's copy does not appear to have been delivered until 1836, but the French copies were about half the price of the English locomotives which were also normally subject to customs duty. 'Jackson' appears to have been withdrawn from service by 1846.

Gras [20] states that the machine from Fenton Murray & Co (No 2) had slightly inclined cylinders whereas those of the Stephenson engine were horizontal; this information may not be accurate as if the locomotives were all four-coupled Planet type they could all have been the same. Therefore although these drawings are attributed to a Fenton Murray & Co locomotive in the publication they can be found in other publications [22]. As has been seen with the locomotives on the Liverpool and Manchester, external similarity hid detailed differences in matters such as the number of boiler tubes used.

95. 3 views of Fenton Murray & Co's locomotives for the Andrezieux/Roanne Rlwy.

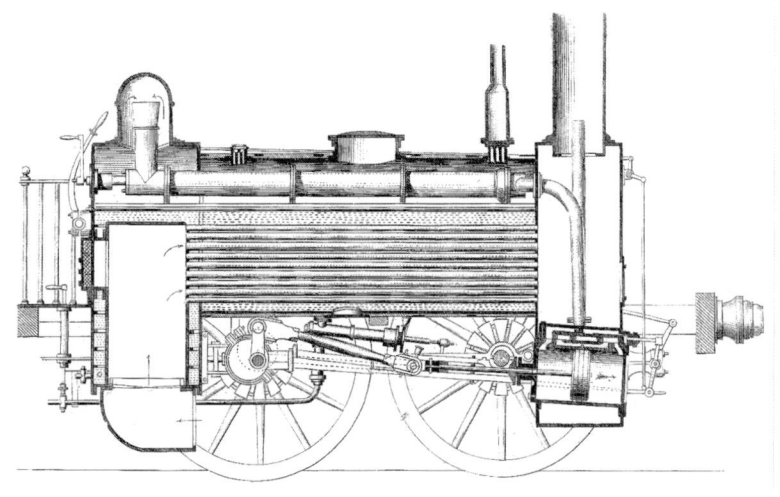

95a. longitudinal section

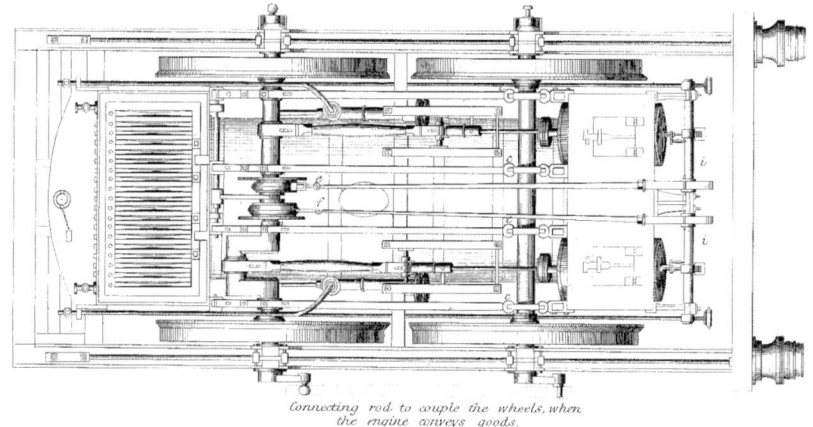

Connecting rod to couple the wheels, when
the engine conveys goods.

95b. plan of underside.

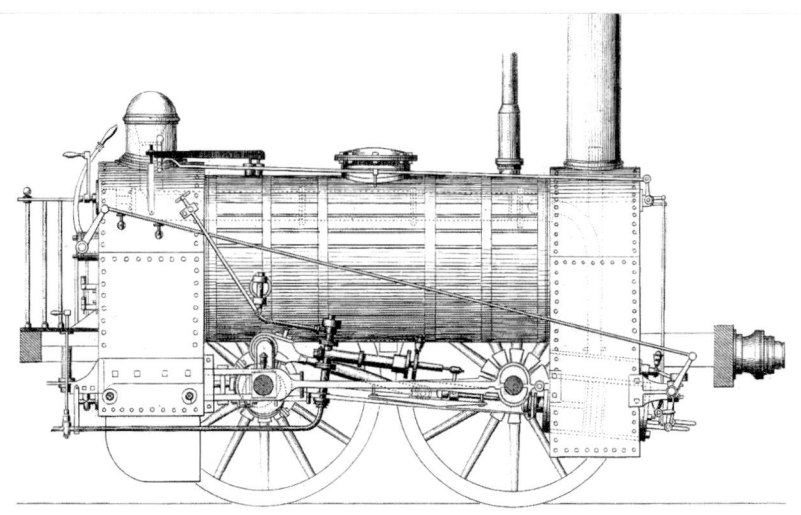

95c. side elevation with two wheels removed.

The line from St Etienne to Lyon, under the direction of M Seguin until February 1835, was the first to use steam locomotives of the three railways. Two Stephenson locomotives had been imported into France in 1828, one went to the workshops of Hallette at Arras and the other to Lyon for Seguin. They were of the same type that Stephensons had built for the Stockton and Darlington Railway. The next 12 locomotives for the St Etienne – Lyon were built to Seguin's designs which to some extent used the Stephenson one as a model. Seguin had however patented the use of multi-tube boilers earlier than any in England, and this major improvement was incorporated into his locomotives. Seguin's multi-tubular boilers were of the return type with the chimney behind them and large blowers mounted in front to provide the draught. The system worked out in England for the Rocket was more satisfactory, and probably indicates that very little discussion took place on the subject between Stephenson and Seguin.

426

In the 1834 trials between the 'Jackson' of the Andrezieux – Roanne line (on loan) and Seguin's no 8, the Fenton Murray & Co locomotive proved of interest. Seguin noted that she had more power and better stability due to the almost horizontal positioning of the cylinders. The St Etienne – Lyon decided to buy two English locomotives, one of which was ordered from Fenton Murray & Co by Seguin when he visited England in 1834. She was delivered in June 1835 as locomotive no. 13 and named 'Jackson'. This second 'Jackson' was rebuilt in 1843, but was sufficiently useful that the company ordered 5 more machines of the type from French manufacturers. A drawing undertaken by Fenton Murray & Co survives in the Archives du Musee des arts et metiers Paris (S183).

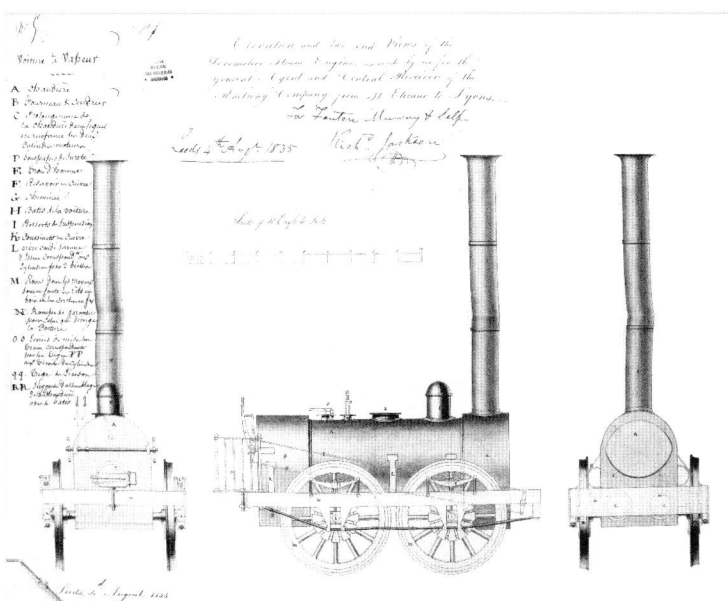

96. Fenton Murray & Co locomotive for St Etienne/Lyon Rlwy (see item 66).

Of the 5 French versions, 2 were supplied by Edwards at Chaillot, one by Tourasse, and a further 2 by Creusot after Edwards moved there subsequent to the failure of Chaillot.

In 1834 the firm delivered its only locomotive for America to the Charleston and Hamburg Railway of South Carolina. The locomotive was adapted for wood burning and the outlines of the agreement between Fenton Murray & Co and Mr Molyneux of Liverpool, the railroads agent, are given on page 352 (Chapter 14). Two 0-4-0 locomotives were ordered from Bury, and 3 from Stephenson at the same time, but the Fenton Murray & Co locomotive, Columbia, was a 2-2-0. Columbia was rebuilt in 1842, while one of the Stephenson locomotives did not survive beyond 1838 and another was scrapped in 1842 [23].

The locomotives of the Leeds and Selby Railway present a challenge, as many of the sources for data that has been published remain obscure. The Directors minutes cover the early Bury locomotives clearly and the first locomotive from Fenton

Murray & Co. The Engineers Report of 7 August 1834 states that the two firms had been selected because their locomotives had been proved upon the Liverpool and Manchester Railway [24]. The initial orders were for 5 locomotives, 4 from Bury and one from Fenton Murray and Co. The Bury locomotives are well described and consisted of the Rodney and Howe which had cylinders with a 12 inch diameter and 18 inch stroke and 4 wheels of 5 foot diameter each (0-4-0); and the Duncan and St Vincent which had 11 inch diameter cylinders and an 18 inch stroke and 5 foot diameter driving wheels with 3 ½ foot leading wheels (2-2-0). The specification of the Nelson, from Fenton Murray and Co, was left to the Engineer.

Bury was extremely late in delivering his locomotives, the first had been ordered for delivery in March 1834, however the only locomotive available to open the railway on the 22nd September 1834 was the Nelson, and the only gleaning from the papers was that she had 18 horsepower. It is very likely that she was closely based on the locomotives for the Liverpool and Manchester, and the three Fenton Murray and Co locomotives for that line had 11 inch diameter cylinders with a 16 inch stroke. This could make it the locomotive advertised for sale by the Leeds and Selby in the Railway Times in March 1839, a locomotive with 11inch by 16 inch cylinders and four 5 foot wheels, the other contender being the North Star (see below).

The Leeds and Selby also purchased an ex Liverpool and Manchester early Stephenson locomotive (1830) called the North Star from the contractor William McKenzie in late 1834 to cover for the late delivery of Bury's locomotives [24.] She was probably only used for a short while.

A further locomotive was purchased from Bury in 1836 (with possibly another in 1837), payment being made in September 1836 [25]. This may have been Exmouth, although she is generally regarded as a Fenton Murray and Co locomotive. Exmouth was operating in 1836 [26]. The delivery from Fenton Murray & Co in December 1836 may have been the Anson, which was followed a year later by the Hawke. Both locomotives were by Fenton Murray & Co as they are so described by Matthew Murray Jackson in a letter in December 1840 [27] and Hawke had a short piece about her in the papers [28]. Matthew Murray Jackson describes these locomotives as the basis of the order received by the firm for the Hull and Selby Railway, which order was for all 6 wheeled locomotives. Hawke is so described and it is reasonable to assume Anson was the same and that they were both 2-4-0 coupled locomotives as the Hawke was described.

"Yesterday, a splendid engine called the "Hawk" left the foundry of Messrs Fenton Murray and Jackson, of this town, for the Leeds and Selby Railroad. This machine has cylinders cylinders 12 inches in diameter, 18 in. stroke, six wheels, (four 5 ft

diameter and two 3 ft diameter) on the most improved principle, and is finished in the first style of workmanship. The weight is nearly 10 ½ tons"[28].

They had been ordered to contend with the increasing freight traffic on the line.

By 1839 the Leeds and Selby was looking to replace its early locomotives and made a purchase of a 2-2-2 from Kirtley of Warrington in April/May 1839[29]. In the shareholders report for the second half of 1839, the Directors reported that 3 of the old engines had been sold and 9 new engines had been ordered, of which 6 had been delivered; and the report showed an expenditure on locomotives of £2587[30].

The report sent to the Board of Trade by the York and North Midland Railway for the return of 1841 showed that they had taken over from the Leeds and Selby 6 single locomotives by Kirtley of Warrington and one single and 4 coupled locomotives by Fenton Murray and Jackson. This report also states that all the locomotives had 6 wheels[31]. In June 1840 the Leeds and Selby advertised two new Fenton Murray & Co locomotives for sale; one that had been working since August 1839 and one since October 1839[32]. A further advertisement in September 1840 only included one 6 wheel Fenton Murray and Co locomotive, stating that it was too heavy for their track, for sale which may indicate that one had been sold[33]. A locomotive was sold to the Great North of England Railway in 1840 by the Leeds and Selby Railway[34]. If the second advert was unsuccessful this would have left a further single Fenton Murray & Co locomotive, which was the one taken over by the York and North Midland. Hawke was and Anson probably was a coupled locomotive which along with two later purchases made the 4 coupled locomotives handed on to the York and North Midland.

Taking the 5 locomotives handed to the York and North Midland plus the one sold to the Great North of England along with Fenton Murray and Co's original Nelson, the firm's supply to the Leeds and Selby was 7 locomotives. If the delivery in 1836/7 was a four wheeler (Exmouth?), then the total was 8.

The Fenton Murray & Co coupled 2-4-0 locomotives were probably broadly similar to a Stephenson locomotive supplied to the Paris railways, La Victorieuse.

Two or three of the Fenton Murray & Co locomotives supplied to the Leeds and Selby appear to have survived rebuilds and joined the North Eastern Railway with one possibly surviving until the 1870s[36].

The next railways that were supplied with locomotives by Fenton Murray & Co were those around Paris, and around 7 locomotives were made for 4 different railways and delivered between 1836/7 and 1840.

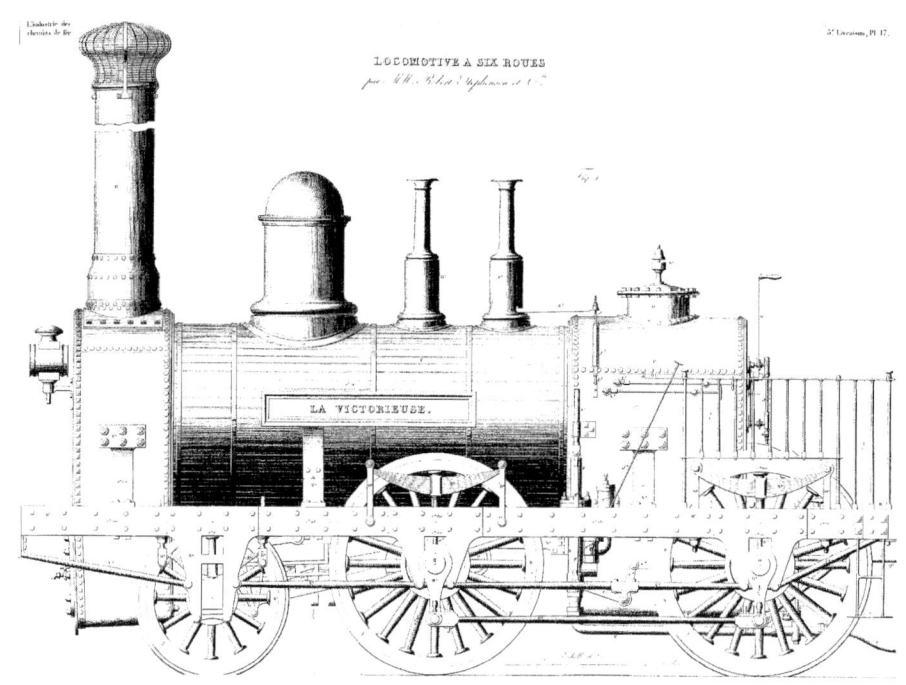

97. La Victorieuse, likely template for Fenton Murray & Co's coupled locomotives for the Leeds and Selby [35]

The earliest railway to commence operations in the Paris area was the Paris – St Germain, which opened on 26 August 1837. Fenton Murray and Co supplied two 2-2-0 locomotives to this line which were named 'Jackson' and 'Denys Papin' [37]. Delivery dates in 1836 are provided by Lowe, which seem to be about right, although they were for the Paris - St Germain, not the Paris – Versailles as stated [38]. The Jackson was still in service in 1849, in a modified state as a fixed tender engine, which in effect made her a 6 wheeled locomotive [39].

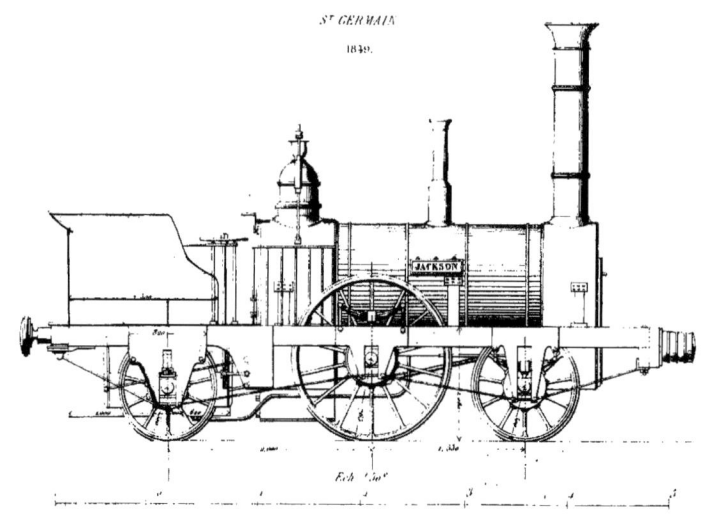

98. Paris- St Germain Rlwy Jackson as later converted to a 2-2-2.

The Paris – St Germain line shared its track near Paris and its St Lazare terminus with two other early railways. The Paris – Versailles Right Bank railway, which opened on the 2nd of August 1839, and departed from the St Germain line at Asnieres to follow its own track to Versailles [40]. In March 1840 Fenton Murray and Co supplied a six wheeled locomotive to this line which was named 'La Versailles'. These 2-2-2 locomotives were called 'Large Jacksons' or 'New Jacksons' by the French to distinguish them from the 'Small Jacksons' or 2-2-0s [41].

The Paris – Rouen Railway departed at the next station after Asnieres, which was Colombes, but did not open until 1843. However it is worth noting that the original company that gained the concession was the Chemin de fer de Paris a la mer. This company did not survive the financial hiatus of 1839 and was wound up in that year. In what was presumably one of its final reports to the investors dated 11 August 1839 it stated that it had ordered and paid deposits on 15 locomotives. These orders consisted of 10 locomotives from Messrs Sharp Roberts (£1460, or 36,938 francs each); 4 locomotives and a tender from Messrs Fenton Murray and Jackson (locomotives at £1500, or 37950 francs each) and 1 locomotive and a tender from Messrs Robert Stephenson (locomotive at £1575 or 41202 francs) [42]. The successor, Compagnie du chemin de fer de Paris a Rouen, purchased all its early locomotives from Alcard and Buddicom, and it has to be assumed that the original orders were cancelled on Sharp Roberts, Fenton Murray and Co, and Stephenson, in the liquidation of the former company.

A further line of 17 km was built from Paris (Montparnasse) to Versailles, known as the Paris – Versailles Left Bank. This railway opened for business on 10 September 1840, and had probably within the preceding year accepted delivery of its two 2-2-0 Fenton Murray and Co locomotives 'Fulton' and 'Mathieu Murray' [41]. It is tempting to ponder whether these two 4 wheel locomotives were originally intended for the Paris a la mer line, as the majority of deliveries by this time were 6 wheelers, with their added stability. The Paris – Versailles Left Bank was the line that suffered the appalling accident in May 1842, in which the Mathieu Murray, with a Sharp Roberts 6 wheeler coupled behind it, left the tracks and the front carriages caught fire from the dropped fireboxes causing many deaths (see Chapter 15). The line had a further 4 wheeler named Denis Papin, which also appears to have been of English manufacture. Denis Papin is stated by Mathias [43], along with Fulton to have come from Hick Hargreaves. The surviving records and drawings of Hick Hargreaves do contain 2 locomotives for Paris - Versailles, but they were 6 wheelers. By this time the only major English manufacturers offering 4 wheelers were Bury and Fenton

Murray and Co, and while it is possible that these two locomotives were supplied by Bury or were Bury designs built by Hick Hargreaves, one or both could have been from the Leeds stable. Again it may have been part of the original order from the Paris a la mer line.

The final Paris railway to receive locomotives by Fenton Murray and Co was the Paris – Orleans of which, the first section to Corbeil, was completed in September 1840. The Engineer for Materiel for the company J-G-L Clarke visited England and Belgium at the end of 1838 to view the different locomotive manufacturer's factories and view their products. He visited Liverpool (Bury, Mather-Dixon), Warrington (Tayleur), Manchester (Sharp Roberts, Fairbairn, Nasmyth and Gaskell), Bolton (Rothwell, Hick), Newcastle (Stephenson, Hawthorn and close by Longridge) and at Leeds, the factory of Fenton Murray and Jackson, 'which enjoyed an old and excellent reputation and who were building at that time a magnificent workshop especially for locomotives'. Also at Leeds he visited Todd Kitson and Laird. London (he noted) had the factories of Rennie, Maudslay and Field and Miller and Ravenhill [44].

The Paris Orleans Railway ordered and received two locomotives from Fenton Murray and Co, both being 2-2-2s, which carried the names 'Jackson' (number 9) and 'Newton' (number 10). It is recorded that these two locomotives (which were to the same design as La Versaille on the Left Bank railway) served until 1852/4 before being sold off. The other English manufacturers picked up quite few of the early orders for this railway; Stephenson 8, with more of the long boiler type to follow, Hawthorn 2, Todd Kitson and Laird 2, Mather Dixon 2 and Sharp Roberts 5 [45].

A contractor's or ballast locomotive 0-4-0, ordered in 1836 by the London and Birmingham Railway was eventually delivered 7 months late in January 1837, by Fenton Murray and Co. In Harry Jacks book [47], the London and Birmingham ballast engines are well covered. He believes that the Fenton Murray & Co locomotive was called Leeds, which others of their engines certainly were. The known dimensions were cylinders of 12 x 18 inches and wheels of 4 foot 6 inches, and she was built to a standard (Stephenson) specification. She seems to have worked the Rugby contract first. The locomotive was listed as being in working order in 1841 and again in 1847; and in 1851 as one of the engines being used by Richard Madigan a contractor. In November 1853 Leeds was sold along with another locomotive to the Rhymney Railway for £600 each. However in February 1854 the railway declined the locomotives due to their poor condition, and they were eventually sold to the English Copper Company for £1100 the pair in March of that year.

Table 1. EARLY LOCOMOTIVE DIMENSIONS.

Manufacturer	Locomotive	Railway	class	Del'y Year	Diam. Driving Wheels metres	Pistons diam. metres	stroke metres	Direct area sq. m.	Tubes Qty	Heating areas area sq. m.	Total sq. m.
Stephenson	Victory	L'pool & Manchester	2-2-0	1831	1.524	0.279	0.406		97	25.87	
Fenton Murray & Jackson	Vulcan	L'pool & Manchester	2-2-0	1831	1.524	0.279	0.406		107	27.49	
Fenton Murray & Jackson	Fury	L'pool & Manchester	2-2-0	1831	1.524	0.279	0.406		107	27.49	
Stephenson	Ajax	L'pool & Manchester	2-2-0	1832	1.524	0.279	0.457		63	20.41	
Fenton Murray & Jackson	Leeds	L'pool & Manchester	2-2-0	1833	1.524	0.279	0.406		107	27.49	
Fenton Murray & Jackson	Denys Papin	Paris - St Germain	2-2-0	1837/8	1.540	0.280	0.410	3.320	82	22.16	25.48
Fenton Murray & Jackson	Jackson	Paris - St Germain	2-2-0	1837/8	1.530	0.282	0.410	3.100	82	21.10	24.20
Bury	La Seine	Paris - St Germain	2-2-0	1837/8	1.546	0.280	0.415	3.070	76	30.00	33.07
Sharp and Roberts	Atlas	Paris-Versailles R.Bank	2-2-2	1838	1.520	0.318	0.348	4.410	117	38.44	42.85
Cave	Gauloise	Paris-Versailles R.Bank	2-2-2	1840	1.670	0.330	0.490	5.830	99	40.49	46.32
Stephenson	Vesta	Paris-Versailles R.Bank	2-2-2	1839	1.680	0.330	0.450	5.690	103,8	46.75	52.11
Shneider Freres	Exposition	Paris-Versailles R.Bank	2-2-2	1840	1.830	0.330	0.460	5.980	138	46.75	52.92
Fenton Murray & Jackson	Versailles	Paris-Versailles R.Bank	2-2-2	1840	1.670	0.330	0.460	5.330	155	52.71	58.04
Sharp and Roberts	Vesuve	Paris-Versailles R.Bank	2-2-2	1840	1.680	0.330	0.460	5.830	162	52.00	57.83
Stephenson	La Victorieuse	Paris-Versailles L.Bank	2-4-0	1837	1.38	0.380	0.460	4.800	145	54.23	59.03
Sharp and Roberts	La Rapide	Paris-Versailles L.Bank	2-2-2	1840	1.67	0.330	0.464	5.032	134	47.99	53.02
Fenton Murray & Jackson	Mathieu Murray	Paris-Versailles L.Bank	2-2-0	1840	1.53	0.282	0.410	4.000	82	22.60	26.60
Fenton Murray & Jackson	Fulton	Paris-Versailles L.Bank	2-2-0	1840	1.36	0.253	0.405	3.400	76	17.60	21.00
F M&J ?	Denis Papin	Paris-Versailles L.Bank	2-2-0	1840							
Shneider Freres	Creusot	Paris-Versailles L.Bank	2-2-2	1841	1.7	0.330	0.464	4.840	120	43.68	48.52
Fenton Murray & Jackson	Jackson	Paris - Orleans	2-2-2	1840	as for Versailles Paris - Versailles L Bank above.						
Fenton Murray & Jackson	Newton	Paris - Orleans	2-2-2	1840	as for Versailles Paris - Versailles L Bank above.						

433

Table 2.
EARLY PARIS RAILWAY LOCOMOTIVES PULLING POWER.

Manufacturer	Locomotive	Railway	class	Gradient 1/1000 Load in tonnes				Gradient 5/1000 Load in tonnes		
				20	40	80	160	20	60	90
				Speeds for above loads in km.s / hour						
Fenton Murray & Jackson	Denys Papin	Paris - St Germain	2-2-0	42	34	24	-	31	19	-
Fenton Murray & Jackson	Jackson	Paris - St Germain	2-2-0	42	34	24		assumed as above		
Bury	La Seine	Paris - St Germain	2-2-0	51	40	30	-	37	23	-
Sharp and Roberts	Atlas	Paris-Versailles R.Bank	2-2-2	49	42	32	22	38	26	21
Cave	Gauloise	Paris-Versailles R.Bank	2-2-2	53	46	35	24	42	27	23
Stephenson	Vesta	Paris-Versailles R.Bank	2-2-2	61	52	39	27	46	32	26
Shneider Freres	Exposition	Paris-Versailles R.Bank	2-2-2	65	55	41	28	50	33	-
Fenton Murray & Jackson	Versailles	Paris-Versailles R.Bank	2-2-2	65	56	43	29	50	34	27
Sharp and Roberts	Vesuve	Paris-Versailles R.Bank	2-2-2	66	57	43	30	51	35	28
Stephenson	La Victorieuse	Paris-Versailles L.Bank	2-4-0							
Sharp and Roberts	La Rapide	Paris-Versailles L.Bank	2-2-2	66	56	43	24	51	34	26
Fenton Murray & Jackson	Mathieu Murray	Paris-Versailles L.Bank	2-2-0	42	34	24	-	31	19	-
Fenton Murray & Jackson	Fulton	Paris-Versailles L.Bank	2-2-0	similar to above						
F M&J ?	Denis Papin	Paris-Versailles L.Bank	2-2-0	similar to above						
Shneider Freres	Creusot	Paris-Versailles L.Bank	2-2-2	58	50	38	26	44	30	24
Fenton Murray & Jackson	Jackson	Paris - Orleans	2-2-2	65	55	43	29	50	34	27
Fenton Murray & Jackson	Newton	Paris - Orleans	2-2-2	65	55	43	29	50	34	27

For sources for the Tables see note [46].

Between 1838 and 1840 Fenton Murray and Co supplied two 2-2-2 locomotives to the Munich and Augsburg Railway. They were christened Vulkan and Mars. It is probable that they were to a similar build standard as the Versailles, Newton and Jackson supplied at around the same time to the Paris railways. The critical dimensions were slightly different as the cylinders were 11 ¼ inches by 16 and the driving wheels were 5 foot diameter [48].

The concession for a railway between Montpellier and Sete (or Cette) was granted in 1836, a line of 27 kilometres length. The company ordered 5 locomotives and tenders for the start of its operations. These locomotives were all 4 wheeled and manufactured by Fenton Murray & Co. They weighted 8000 kg while consuming 3 cubic metres of water and 200 kg of coke for a round trip between the two towns. The first 3 locomotives were delivered in October 1838 via Le Havre at Cette, from there they travelled by canal and river to Montpellier. The other two locomotives arrived at Cette in April 1839, on board the Amelie, and were towed on the railway to Montpellier. The locomotives all had names related to the area, being Notre-Dame-des-Tables (patron saint of Montpellier), L'Herault, La Montpellieraine, La Cettoise and La Rosine [49].

99. Notre-Dame-des-Tables for the Montpellier et Cette Rlwy [50].

In 1841 a new 6 wheeled locomotive was delivered to the railway. It was not of French manufacture and was therefore probably English [51]. In 1843 two of the Fenton Murray & Co locomotives were converted to 6 wheels in the company's workshops. La Cettoise was adapted to power the workshops in 1848. The other two original locomotives were either upgraded to 6 wheels also or kept for the freight trains. Due to the poor quality of the local water the iron fireboxes had to be replaced with copper ones. By 1852 the railway, having failed in its attempts to expand, decided to accept offers to amalgamate with the Lyon to Avignon Railway. This in turn ended up as part of the Paris Lyon and Mediterranee Railway, the PLM, in 1857 [49].

The railway between Sheffield and Rotherham does not appear in the Government Returns relating to Locomotive engines gathered in 1841. Wishaw [52] advises that there were 6 locomotives mainly of six wheels built by Rt Stephenson (Victory, London & Leeds), Fenton Murray and Co (Agilis), Bingley & Co of Leeds (Rotherham) and Davy Bros of Sheffield (Sheffield). There were references to another couple of locomotives but they were leased or temporary only. Wishaw also relates that the wheels of the Stephenson locomotives were without flanges according to their 'improved construction'. Agilis was delivered on the 23rd of April 1839, upon a 6 wheeled dray hauled by 16 horses. Her configuration was 2-2-2 and she had flanges on all six wheels. Descriptions of her also include her having the advantage that if one of the eccentrics broke, became disordered of fell off; the locomotive could still be controlled by its engineer who could safely conduct it along the railway [53]! Bingley and Co of Leeds delivered the Rotherham in early 1840 and it appears to be the only locomotive they made, although they did advertise themselves as locomotive makers in the 1837 directory. It has been suggested that Rotherham was built as a sub license from Fenton Murray and Co[54], this is partially based on a Sheffield and Rotherham specification of February 1838 for 2 locomotives but having very little detail [55]. The fact that Fenton Murray & Co were in the process of building their new locomotive erecting shops may support the assertion.

Agilis had a short claim to fame in that she helped to carry the news of the results of the 1839 St Leger races at Doncaster in a record time to Sheffield, completing the last 2 ¾ miles into Sheffield Station in the 'astonishing' time of 2 ¼ minutes, the whole trip from Doncaster to Sheffield (including horse to Rotherham) took 44 ½ minutes [56].

Neither locomotive made it into the stock of the North Midland Railway. Baxter states that Rotherham was unserviceable and out of use by 1844 [36]. Agilis was severely damaged in a 3 locomotive collision on the 28 September 1844, and was presumably scrapped. While the two remaining accounts are not compatible, the basic facts appear to have been that Agilis and another locomotive had been replenishing their water down the line from Sheffield station and were returning with the tender of Agilis in front, then Agilis, then the second tender followed by the second locomotive. Bartholomew the manager of the engine house then took a third locomotive to water and collided with the other two. Agilis's tender was crushed and the locomotive was 'greatly injured'. The other vehicles were all damaged but not so severely, and two of the men received severe bruises. In the official report it was added that Bartholomew had instructed the other two engines to move some goods trucks towards Masborough after watering, but that they had been refused a signal and returned on the same line [57].

The London and South Western Railway started out as the London Southampton railway but changed its name and gained the right to build a line to Portsmouth under an act of parliament that received Royal Assent on 4th June 1839. The line to Southampton was fully opened in May 1840.

Fenton Murray & Co provided four locomotives to this railway. Two locomotives were delivered in 1839, one invoiced in June and the other in September. These were known as Leeds and Eclipse. Two further locomotives were delivered in 1840, invoiced in August and September, namely Phoenix and Crescent [58, 59]. They carried the numbers 31 to 34. The cost of each locomotive was £1775 plus delivery. They were 2-2-2's with 13 inch diameter cylinders and probably an 18 inch stroke. The probable diameters of the wheels were: driving 5 foot 6 inches, and leading and trailing 3 foot six inches. The minutes of the Traffic Committee include two references that relate, the first on the 27 March 1839 referred to a visit to be made by Mr Wood to Manchester, Bolton and Leeds to view the locomotives then being built for the railway and the second on the 16 April 1939 to the effect that Rothwell & Co were not to be allowed to vary the specification [60]. If Rothwell are the Bolton firm, Fenton Murray & Co the Leeds firm and Sharp Roberts and Tayleur the Manchester firms, it is reasonable to assume that the engines were to a common specification and reading across the known dimensions of the locomotives delivered by these firms for the period [59] it is probable that the dimensions were as stated above.

The Leeds and Phoenix locomotives nearly had an extremely short career, as the locomotive department at Nine Elms suffered from a severe fire, on March 16th 1841, which destroyed most of the machinery and three locomotives. Thirteen other locomotives were rescued and these included the Leeds and Phoenix [61].

Leeds appeared in the Accident Book after being derailed after an employee deliberately moved a rail out of true in the Winchester sidings on 19th May 1847 [59]. Leeds was valued at £994 in the 30th June 1849 valuation along with a tender adding £190 [62].

Eclipse was involved in an accident at Nine Elms on the 17th October 1840. Eclipse had hauled the fast train up from Southampton leaving that town about 3 o'clock. On arrival at Nine Elms the slow train that had left Southampton at 1.30 p.m. and been delayed was still blocking the line, while its locomotives were uncoupled. In the absence of any warning lights the fast train clipped the end of the slow train, unfortunately killing a passenger in the last carriage. The slow train had had two engines in arriving at Nine Elms one of which was Crescent. Crescent was not involved having been uncoupled and Eclipse had very little damage, and took a train to Southampton and back the following day. The damage to the carriages was also very slight and the injuries to the passenger had been sustained by being thrown forward by the impact [63].

On 11th November 1841 John Herapath as part of his trials between 4 and 6 wheeled locomotives travelled on 3 locomotives on the London and South Western Railway. He was quite impressed by the Queen (Fairbairn) the first locomotive he rode, which had only been delivered the previous month. Eclipse was the second locomotive he tried and he rode her from Winchester to Southampton:

"This was a six-wheel engine with outside bearings and a 13 inch cylinder; but I could not get the length of stroke. From the hind axle to the driving she was 5 feet 1 inch, and to the front axle 10 and a half feet, the same as the Queen: her total length was 15 ½ feet; she had a play of 1.2 inches on the rails. The whole of the way I rode with this engine was downhill, and at a high velocity. After making her what I thought a fair allowance for these circumstances, she appeared to me to be more unsteady than the Queen, and had much more sinuous motion. Her apparent sinuous motion was full three to four inches. The coupling appeared to be one foot above the driving axle, and she was pulling upwards. Probably, the play of the axles on the bearings was partly the cause of her greater unsteadiness" [64].

The third locomotive was the Orion by Sharp Roberts which Herapath found much better than the other two as regards the sinuous motion, but that her general motion was exceedingly rough and it felt as if she had no springs behind.

Eclipse was involved in another fatal accident in January 1842. During morning preparation the Transit engine forced the Raven engine into the Mars which pushed the Eclipse against her tender. The deceased was working on the tender and appears to have been trapped between the locomotive and its tender [65]. At the June 1849 valuation Eclipse had a value of £1105 put on her with her tender put at a further £150 [62].

Phoenix worked a trial train between Southampton and Dorchester on the 25th May 1847 [59]. She was valued at £1076 in June 1849 with her tender adding a further £130 [62].

Crescent carried an excursion train of London and South Western Railway employees from Vauxhall to Southampton in May 1842. At Southampton the passengers embarked on the Princess Victoria and visited the Isle of Wight [66]. In June 1842 she hauled the company's state carriage from London to Southampton, on the 15th, and back on the 17th carrying the Queen Dowager and party on a trip to the Isle of Wight [67]. In 1846 Crescent hauled the Director's special with 16 carriages at the opening of the Richmond Railway. The train left Nine Elms at 2.00 p.m. on the 22July 1846 and travelled for two miles on L&SW railway track before turning onto the new line [59]. She was valued at £1009 with a tender worth a further £190 in the June 1849 valuation [62].

Leeds Phoenix and Crescent were unserviceable at 30 June 1842 [59], and they all received rebuilds in 1843/4 at Nine Elms. Eclipse and Phoenix were withdrawn respectively in October 1852 and September 1851. Leeds spent a time ballast laying until damaged in an accident in 1852, Crescent was rebuilt as a well tank in 1852 and operated on the Bishopstoke to Salisbury service until her boiler wore out in 1856 [59]. Against this scenario, common for many of the locomotives purchased by this railway in 1839/40, of poor performance and short life should be weighed the contents of a report written by Joseph Beattie in 1852. In the report Beattie states that the period of 1845 to June 1850 saw very little proper maintenance of the locomotive stock, with locomotives being laid up while the new ones coming on line were utilised to maintain services. He demonstrates how ,in 18 months he had increased the stock of good engines from 80 to 98, reduced the inefficient locomotives from 29 to 17 and the worn out stock from 9 to 5 [68]. The figure of 29 inefficient and 9 worn out locomotives in 1850 agrees with a number quoted by Bradley. The worn out locomotives included Phoenix as well as some delivered later by Sharp Roberts and Fairbairn. Eclipse was counted among the inefficient locomotives as being not worth repairing along with others of the same era.

The London and South Western had a fair amount of problems with axles produced for various locomotive manufacturers by the Patent Axle Co, it is interesting to note that a trailing axle supplied by Fenton Murray & Co with Crescent in 1840 failed on the Venus by Sharp and Roberts in February 1867 [59].

The Belgian State Railways started receiving locomotives in 1835, and purchased a mixture of Belgium made and English locomotives until 1840 after which no more English locomotives were ordered. A total of 42 English locomotives were delivered from 4 manufacturers: 30 from Stephenson, 10 by Longridge, 1 by Sharp Roberts and 1 from Fenton Murray and Co.

The Fenton Murray & Co locomotive was named Firefly and was locomotive number 100 in the Belgian inventory. She was delivered in late 1839 or early 1840 and went into service on the 25th March 1840 and was based at Antwerp (Anvers).

The Belgian Government records show a works number (i.e. manufacturer's number) of 49, which appears to be the only recorded Fenton Murray & Co works number known today, that is realistic

The locomotive does not seem to have started well as she was withdrawn between 16 June and the 18th of July 1840 for two complete new pistons, and again in November of that year for a new smokebox top and two new steam chests. Not unusual in the locomotives of the day but hardly matching the Round Foundries reputation for excellent workmanship! Perhaps there were teething problems when the new locomotive works were opened on Marshall Street in 1839.

The kilometres run by year also indicate that the locomotive may have been altered shortly after she went into service, or at least that some issues were resolved [69].

Year	Kilometres
1840	1348
1841	1426
1842	2050
1843	1158
1844	25
1845	6
1846	3298
1847	7268

The Hull and Selby Railway opened for business in July 1840, thereby completing a railway link between Leeds and Hull that had been started with the opening of the Leeds and Selby in 1834. At the opening ceremony on July 1st five locomotives are mentioned as being active: Kingston, Exley, Selby and Andrew Marvel which all belonged to the Hull and Selby Railway and had been made for the line by Fenton Murray & Co, and Prince (made By Kirtley & Co) one of the recent locomotive purchases of the Leeds and Selby line who had loaned it for the occasion [70]. As well as the 4 locomotives mentioned above Fenton Murray & Co made a further two for the line, the Collingwood and the Wellington, and all six appear to have been delivered by July 1840. The Locomotive Superintendent of the railway was John Gray, who was one of the early proponents of the use of expansive working in railway locomotives, and had indeed patented a valve gear in 1838 that had been applied to a locomotive called Cyclops on the Liverpool and Manchester Railway. Gray had a further 6 locomotives to his design manufactured for the Hull and Selby Railway by Shepherd and Todd at The Railway Foundry in Leeds. Two of these locomotives, Star and Vesta, were delivered in October 1840 and thereafter used in the trials described in Chapter 15, which received world-wide publication in the technical journals of the day. The results were in favour of the use of expansive working. At the trials three of the Fenton Murray & Co locomotives had been converted by the railway to use Gray's expansive valve gear, Kingston, Selby and Exley. It is probably a good assumption that the other three also were converted later on. The Fenton Murray & Co locomotives had 5 foot 6 in. driving wheels and steam cylinders of 12 by 18 inches, they also had outside bearings on all the wheels.

440

Those by Shepherd and Todd had 6 foot drivers and larger cylinders of 12 by 24 inches, and while they had outside bearings on the front and rear axles, the driving or cranked axle had inside bearings. Both series of locomotives were 2-2-2s.

At some time towards the end of 1841, John Herapath one of the prolific writers on Railways of the time, travelled on some of the trains of the Hull and Selby line. Herapath experienced 4 of the company's locomotives the Wellington and Collingwood of Fenton, Murray & Co and the Star and Liverpool by Shepherd and Todd. His comments were as follows: Wellington – she was a very steady engine; Liverpool – appeared to him to be a good engine and performed well; Star was not commented on: Collingwood – observed she was unsteady and inclined to pitch at low velocities, but lost it all at high, and Herapath had very little doubt this was owing to the position of the coupling link as he had found on other lines. Some of the locomotives (those by Fenton Murray & Co) had the capability of lifting the trailing wheels to give more purchase to the driving wheels, but Gray told Herapath it was seldom used as it created instability [71].

Charles Todd, who had left the Railway Foundry in 1844, did an evaluation of the stock of Locomotive Engines on the Hull and Selby Railway dated February 7th 1846. It was, no doubt, performed as part of the change of management in the line. The railway was leased to the York and North Midland Railway from 1 July 1845, who in turn came to an arrangement with the Manchester and Leeds Railway to undertake that lease jointly. The arrangements were agreed by the shareholders of the 3 companies by the middle of 1846 and the lease and acquisition of the Hull and Selby Railway was enacted in July 1846. The listing of locomotives is as follows:

When delivered	No.	Name	mileage	£ value.
July 1 1840	1	Kingston	31767	900
July 1 1840	2	Selby	27003	900
July 1 1840	3	Andrew Marvel	48774	850
July 1 1840	4	Collingwood	46868	900
July 1 1840	5	Wellington	52774	800
Oct. 6 1840	6	Star	50959	900
Oct. 6 1840	7	Vesta	45818	1000
Nov 26 1840	8	Rainbow	68832	1000
Dec 14 1840	9	Manchester	75885	1000
Mar 15 1841	10	Fire Fly	89277	950
Mar 29 1841	11	Liverpool	80070	1000

Feb 17 1842	12	Leeds	38162	1000
Feb 18 1842	13	Urgent	48040	1200
Dec 27 1842	14	Vulcan	39898	1200
Sep 28 1843?	15	Exley	13961	1200
Nov. 4 1843	16	Broadley	10716	1550
Jun 30 1845	17	Tottie		1550

There was also a list of Fenton Murray & Jackson spares amounting to £317-15-10d, covering some 15 items including connecting rods, piston rings wheels etc . And a small list of their patterns. The tenders are also listed and were made by Waddington, Shepherd and Todd and the railway. The whole signed By Charles Todd of the Sun Foundry [72].

Two points of interest are that the original Exley by Fenton Murray & Co seems to have gone off the books and been replaced by a new one in 1843 (probably made by the railway), and the Shepherd and Todd goods engine Hercules is not listed. Hercules may of course have already been diverted by the new management elsewhere.

There seems to be a consensus on the future numbering of the locomotives as being

Name	York & N Midland		North Eastern	out of service
	1846	1849	1853	
Kingston	67	61	293	1857
Selby	58	43	278?	1854
Collingwood	68	73	305	1857
Andrew Marvel	59	44	279	1868?

With Wellington not being taken Into York and North Midland stock [36].

Lowe and Baxter [38 & 36] both list the Great North of England Railway as having an 0-4-0 Fenton Murray and Co locomotive. The early records of this railway are among those preserved at the National Archives. The records that remain cover the period of their initial locomotive purchases, which started with the supply of two Pattern locomotives from R W Hawthorn & Co, one a 2-2-2, the other a 2-4-0. The subsequent orders and payment to Hawthorn, Stephenson, Tayleur, Jones Turner & Evans and Fairbairn. There is even a minute to purchase an 0-4-0 at £1200 from Turner Ogden and Co of Leeds [73] to be used on Number 2 Contract (line construction). There has not been found to date any minute to purchase a locomotive from Fenton Murray and Co.

442

Two sets of entries may probably be linked to satisfy that there was indeed a Fenton Murray and Co locomotive in stock. In 1839/40 the Leeds and Selby was re-equipping its locomotive stock (see above) and putting up for sale the old locomotives as can be seen from a number of adverts in the Railway Times, and further in June 1840 was advertising two of the recently purchased locomotives from Fenton Murray & Co along with a 2-2-0 by Bury and then in September, shortly before the locomotive stock was passed to the York and North Midland Railway, an advertisement for one of the Fenton Murray & Co locomotives. Various Great North of England railway ledgers refer to the purchase of a locomotive engine from the Leeds and Selby Railway [74], and minutes of the 17 November 1840 referring to the Hull engine stated that:

> "*The repairs to this Engine are left to Mr Storey, to make the best he can for their completion*" [75].

There are then entries of payments on the Locomotive Account to Messrs Fenton Murray & Co for £256 in December 1840 and a further £164 in May 1841 [76]. In their return to the Board of Trade in 1841 [31], the Great North of England Railway referred to having 6 merchandise engines, all of which had 6 wheels, by Tayleur, Hawthorn, Jones Turner & Co and Fenton Murray & Co. The latter is the most likely candidate for the purchase from the Leeds and Selby. The Board of trade return further states in the paragraph on merchandize engines that one engine which was employed in ballasting had 13 inch cylinders and 6 four foot diameter wheels. The new Fenton Murray & Co locomotives advertised for sale by the Leeds and Selby Railway specified that they had 13 inch diameter cylinders and part of the work contracted with Fenton Murray & Co by the Great North of England could well have been to convert the wheel arrangement.

One of the railways that purchased Fenton Murray & Co locomotives was the North Midland. This railway did not name its engines and after an early consecutive numbering plan, changed all the numbers in1843, and grouped the locomotives by cylinder and driving wheel size [77].

The task of executing the Board's decisions on procuring locomotives fell to the London Committee, who in turn relied very heavily on R Stephenson. Stephenson was employed, from February 1839, with a percentage of profits to manage the locomotive power and to procure it. Stephenson was tasked in late 1837 with providing a list of potential locomotive manufacturers. The addition to this list of Nasmyth, Patricroft; J Lindsay & Co (Earl Balcarras) Wigan and Todd Kitson & Laird, Leeds was recommended by the Leeds Committee.

On the 16th of January 1838 the Secretary advised the London Committee that request for tenders had been sent to those on the list. On the 6th of February he presented their answers to that committee. Braithwaite & Co had not responded, Hird Dawson (Low Moor) had advised that they were not in the trade and Haigh Foundry, Edward Bury and Nasmyth & Co had declined tendering at that time. The tenders (abbreviated) were as follows:

	Engine Price	Tender Price	Number offered
Fenton Murray & Jackson	1400	-	15
R Stephenson	1550	230	3
Sharpe Roberts	1450 & 1500	246 265	15
Hawthorn & Co	1400	170 & 200	5
Todd Kitson & Co	1150	150	10
Tayleur & Co	1300	220	15
B Hick	1500	-	10
Miller & Co	1450	170	some.

The Committee decided that the initial orders were to be:

Tayleur & Co	2
Fenton & Co	3
Miller & Co	2
Stephenson & Co	3
Mather & Dixon	3
Todd Kitson & Co	2

Being 15 in all. However either Stephenson exercised some powers of persuasion, or by the time the letters arrived other business had arrived, because Miller & Co appears not to have supplied any early engines, but Hawthorn became a supplier.

In January 1839 the London Committee asked Stephenson to report on the necessity of ordering more locomotives. Presumably he answered in the affirmative and in his new position was responsible for their ordering as the Committee and Board minutes do not reflect the placing of the orders, although the accounts still record the payments. From surviving locomotive lists a further 4 locomotives were ordered from Fenton Murray & Co, making a total of 7 [78].

Not all the locomotives were the same and those from Fenton Murray & Co were split between two of the larger ranges. They produced three off to the cylinder dimension of 13 in. x 18 in. with 5 ft 6 in. driving wheels to a 2-2-2 configuration. On joining the Midland Railway in 1844 these were numbered 40 to 42, and renumbered

101, 105 and 106. They all appear to have been replaced by 1851/2, though some sources believe they ran until 1862.

The second batch, again 2-2-2s, had cylinders of 14in. x 18in. and 6 ft. driving wheels. In the Midland Railway numbering they became 13, 15, 16 and 17. The driving wheels had all been reduced to 5 ft. 6 in. by 1849 shortly after which they were withdrawn one in 1851, the others in 1854 [79]. Again other sources have the locomotives running until 1860.

The amounts paid out to Fenton Murray & Co were as follows:

May 1 1840	£1500 locomotive.
May 19 1840	£1933 being 1813 plus balance of 120.
June 2 1840	£1500 on account.
November 11 1840	£2766 loco engines 2263 repair engines 503
January 26 1841	£ 2001 Bill due 2 April J Dickinson for Fenton Murray & Co
March 3 1841	£3531 loco stock 3000 + 300 + 231.

The prices for the second batch were somewhat higher than for the first which does not accord with the tender. None of the prices recorded as paid agree with the tender amounts which were obviously renegotiated and probably dependant on the point of delivery [80].

The majority of the business conducted between the Great Western Railway and Fenton Murray and Co has been included in the main chapters as this business was crucial to the firm in its final years. Indeed it was the final major piece of business that was executed by the Holbeck factory. This section will therefore summarise that business and give a little more detail on the in-service period of the locomotives supplied.

The firm was one of those invited to tender for the first ever locomotives for the Great Western Railway in 1836 by I K Brunel, but they declined to bid on the basis that they had no spare capacity at the time. This led to a follow up letter from Brunel dated the 14 November 1836 asking whether they were now in a position to submit a tender [81]. To date no letter has been found from Brunel to the other non-bidders asking them to review their position. This indicates that Brunel was interested in having his father's old suppliers as a contractor for the G.W.R. Fenton Murray & Co at that time had not increased their capacity by building the locomotive and marine shops on Marshall Street. The additional facilities were reported in progress by J G L Clarke when he visited Leeds in 1838 for the Paris – Orleans Railway [44]. So by the time the next major 'Request for Tender for Locomotives' was sent out in March 1839, the firm had or was near to having the necessary facilities. They responded accordingly. In dribs and drabs they received orders for a total of 23 locomotives, 3 goods engines of the Leo class and 20 Firefly or Priam class (Priam was one of the last 'Fireflys' to be delivered).

The Goods engines were to a 2-4-0 configuration with 15inch diameter, 18 inch stroke cylinders, 94 boiler tubes, leading wheels of 3 foot 6inches and coupled wheels of 5 foot. There were 18 locomotives in the class 3 from Hawthorns, 12 from Rothwells and 3 from Fenton Murray & Co. The three from Fenton Murray & Co were named Hecla, Stromboli and Etna. The class were somewhat light for their purpose and all were converted to saddle tanks.

Hecla was invoiced in April 1841 [82] and appears to have started service in the same month, and according to Gooch's papers was withdrawn by 1867/8 [83]. As described in Chapter 15 Hecla was hauling a mixed passenger and goods train when derailed in Sonning Cutting on Christmas Eve 1841, resulting in 8 deaths and many injured. Her mileage at the end of 1860 was 284247 [84].

Stromboli went into service at the same time as Hecla (the probable invoice date was March 22 1841 [82]). In September 1841 she was chartered by a large party of mechanics and others with family members to undertake a trip to London [85]. Stromboli worked on the South Devon Railway when it was worked by G.W.R. locomotives, and in June 1851 she was hauling a train between Plymouth and Totnes when an unfortunate individual deliberately threw himself in front of her and was decapitated [86]. Her mileage at the end of 1860 was 284855 [84]. She also worked on the Wilts and Dorset line and the Hammersmith extension of the Metropolitan line running to Kensington from 1864 [87 & 88]. This may well have been her last working as she was also shown as withdrawn on Gooch's list of 1867/8 [83].

Etna appears to have started life 2/3 months after the other two, in June 1841. Her recorded mileage at the end of 1860 was 259596 [84]. She was working out of Wolverhampton, when sent to Swindon for General Repairs in April 1864, having previously had attention in April 1861 [83]. She was still listed as working in 1867/8, but the consensus remains that she was withdrawn by the end of 1870.

The Firefly or Priam class were designed for passenger traffic with the configuration of 2-2-2. There were 62 delivered: 6 by Jones Turner and Evans, 10 by Sharp Roberts, 20 by Fenton Murray & Co, 2 by G & J Rennie, 6 by R B Longridge & Co, 2 by Stothert & Slaughter and 16 by Nasmyth Gaskell & Co. The driving wheels were 7 foot and the carrying wheels 4 foot. The specification called for cylinders of 15 inch diameter by 18 inch stroke [9]. In his letter to Fenton Murray & Co of the 28th January 1840 Gooch requested them to make the cylinders for the next (first?) engine 15 ½ inches diameter by 18 inch stroke [89]. Gooch obviously liked the result as certainly from Harpy onwards (August 1841) all the surviving invoices for the Leeds

446

locomotives specify that they had the larger cylinder [82]. All the locomotives that were refitted as Priam class were given 16 inch diameter cylinders, around 1844 [87].

Charon was the first engine delivered and was invoiced on 9 May 1840. The G.W.R. were very pleased with the locomotive despite the fact that they had to paint both Charon and the second locomotive Cyclops and provide some of the brass work. These items were set against the invoices on Fenton Murray & Co's statement of account [82]. On May 31st 1840 she hauled the G.W.R. directors and guests up to Steventon Station the day before the public opening to that point [90]. On the 15th of August 1840, Charon hauled the train taking the Dowager Queen Adelaide from Wallingford Road to Slough. The Dowager Queen had stayed with the Archbishop of York at Nuneham the previous night [91]. Charon was one of the locomotives sent to South Wales in 1850. She was renewed and restarted in service in October 1864 after an expenditure of £909/4/7d. She was still working in 1876, classified as a Tiger Class but was sold to the Metropolitan Board of Works in June 1878.

Cyclops was invoiced on the 29 September 1840 and in service in October. In December 1840 she was used to haul a special train to Faringdon road station to collect engine drivers stokers and others and take them to London to be addressed for two hours by I Brunel on the subject of a new system of signals, to prevent accidents and warn of obstructions [92]. She does not appear to have been an exceptional locomotive, was not renewed and had been withdrawn by 1867/8.

Cerberus was invoiced on the 28th May1841 and in service by June. Cerberus was officially renewed as a Hawthorn Class in 1866 [93], although the consensus is that those names reused in the Hawthorns were actually complete new builds as opposed to rebuilds. The reasons behind this may have had to do with the ability to classify a rebuild as revenue spend (maintenance) whereas a new locomotives would have had to be classified as capital. It is difficult to be certain which they were. The costs of the new build as opposed to the cheaper renewal were higher by £263 or about 28%. However in an official listing of Broad Gauge locomotives of 1876 [94] all the Hawthorn class were still denominated as under their original maker, Fenton, Sharp, Nasmyth or for the sub-contracted new ones Avonside Co. (the old Stothert and Slaughter).

Harpy was invoiced on 26th July 1841 and in service in August. Harpy was renewed and back in service by November 1864 and still classed as working in 1867/8, but was probably withdrawn soon after the latter date, as she was sold to the Steam Coal Co. Abercarn in August 1873 [87].

Next up was Pluto invoiced three days after Harpy and also entered service in August 1841. In July 1844 the G.W.R. was testing a new form of Telegraph between

Paddington and Slough and it just so happened that Pluto was one of the engines involved [95]. However in August 1845 Pluto was holding up the traffic at Ealing after bursting a pipe (tube?) in the boiler [98]. Pluto was not renewed and had been withdrawn by 1867/8.

Minos was invoiced on 2 September 1841 and entered service the same month. She was not renewed and off the lists by 1867/8.

The next locomotive was Ixion, invoiced 24 or 27 September 1841 and in service in October. Ixion was used in the Gauge Trails between the standard and broad gauge, for which she is well known. She also hauled the royal carriage with the Queen and her family from Slough to Paddington in April 1846 on the return of the court to London [97]. Ixion was renewed in 1863 at a cost of £1065/17/11. Ixion was reclassified as a Tiger and was the last of the Firefly/Priam Tiger class to be withdrawn in July 1879 [94].

The invoice for Hecate was dated 22 nd October 1841 and she entered service in November. She was not renewed and had been withdrawn by 1867/8.

Gorgon was invoiced and entered service in November 1841. She was one of the locomotives despatched to South Wales, and featured in the completion ceremony of the new bridge over the River Wye at Chepstow in July 1852. The passenger train with Brunel on board went first hauled by Stentor (Nasmyth Gaskell) and after she had returned, Gorgon passed onto the bridge with ten trucks each containing 10 tons and went to and fro. All tests were pronounced satisfactory [98]. Gorgon was renewed and back in service by February 1866, going on to be reclassified as Tiger Class. She was sold to the Metropolitan Board of Works in October 1878 [87].

Vesta was invoiced and entered service in December 1841. She was despatched to South Wales in 1850. There was no renewal and she appears to have stopped working about the end of 1864 and had certainly been withdrawn by 1867/8.

The next locomotive to be invoiced was Prosperine on the 23 December 1841, however she is not listed as entering service until June 1842. Prosperine was renewed to return to work in November 1865. She was still on the active list in 1867/8 and was sold to Fielding and Platt in July 1873 [87].

The first delivery in 1842 was Acheron, invoiced on 6th January and starting work that month. Acheron was the locomotive emerging from Box tunnel, drawn by Bourne for the frontispiece of his History of the Great Western Railway. Acheron lasted until 1866 when she went through the same metamorphosis as Cerberus and became a brand new, or otherwise, Hawthorn Class locomotive.

February 1842 saw the invoicing and entry into service of Erebus. Erebus hauled the royal carriage before the execution of this duty was not deemed noteworthy. She performed on the Queen's return from Scotland in September 1842 [99]. Erebus received a renewal putting her back into working in September1865, and continuing to do so through 1867/8. She was sold to Fielding and Platt in July 1873 [87].

Hydra was invoiced in March 1842 and went to work in April. She was involved in an accident at Paddington in March 1846, when some carriages had been left on the line in the way of the mail train which Hydra was hauling. The messenger of warning was too late to warn the driver. In consequence Hydra collided with the carriages and was thrown off the track with serious damage [100]. She was not renewed and had been condemned prior to 1867/8.

Lethe arrived in April and was invoiced and ready for work in that month. On the 28th January 1846 Professor Barlow and a colleague attended at Exeter to trial the express trains between there and Taunton for the Gauge Commission. Mr Gooch was also in attendance. This stretch of line included the 'White Ball' gradients. The Lethe hauled 8 (instead of the usual four) carriages weighing 80 tons. The down express used the Mars (Longridge) locomotive. The results were 'most satisfactory' for the Broad Gauge [101]. Lethe was back at work after her renewal by February 1862 and remained on muster through 1867/8 and seems to have remained available in the Tiger class until sold to the Metropolitan Board of Works in October 1878 [87].

Medea was invoiced on the 3rd May 1842, and the record of going into service appears to be wrong at March 1842. She was renewed and back in service in October 1863.She was available through 1867/8 and reportedly ceased in 1870 and was sold to G W Colliery Co. in March 1874 [87].

A locomotive named Styx was invoiced on the 28th February 1842, but was not shown as entering service until May. She may have been waiting for her new nameplates - Phlegethon, which while they were not invoiced until the 6th August may have been delivered earlier. Under her new name Phlegethon had the privilege of drawing the first train upon which Queen Victoria travelled on the 13th June 1842 [102]. The journey was from Slough to Paddington. The experience was repeated two years later when the Queen travelled to Scotland in September in 1844 when Phlegethon was again driven by Gooch himself [103]. Phlegethon also metamorphosed into a Hawthorn class locomotive in January 1866.

Medusa was invoiced on the 24th May 1842 and available for service in June. Medusa was not renewed and was withdrawn by 1867/8.

Ganymede started work in July 1842 having been invoiced ion 11th June. She took the Queen and her party up to London in January 1847 [104]. Ganymede was renewed and back on station by April 1863 and available through 1867/8; reclassified as a Tiger she worked until August 1878 when she was sold to the Metropolitan Board of works [87].

100. Ganymede, a Fenton Murray & Co 'Firefly' for the GWR.

Argus, the last of the Fenton Murray & Co Fireflys, was invoiced on the 6th August 1842 and went into service the same month. She, as a new locomotive, was soon roped in to haul a royal train, within a month of starting. Argus was again supervised by Gooch on this trip from Slough to Paddington[105]. Argus was also renewed by July 1863 and survived through Gooch's list of 1867/8 but did not make it onto the list of 1876 as a Tiger as she was sold to D Joseph for the Bristol Colliery in November 1873 [87].

101. Argus, the last locomotive produced by Fenton Murray & Co.

The Broad Gauge list from Gooch's files [83] used above to determine withdrawals is dated 1867/8. Many of the locomotives are crossed through such that their names cannot be read. Fortunately another copy of the original list survives enabling those crossed out to be stated. The scored list is annotated

"Note, engines scored out have been condemned and replaced with narrow gauge engines."

By this date the G.W.R. was already becoming a narrow gauge railway, in about 7 years the mixed gauge to Bristol was only maintained for through traffic to the West of England and the rest of the workings were mainly narrow gauge. Therefore the large number of Broad Gauge locomotives disposed of at this time was due to necessity and not a reflection on the numerous makers. The older locomotives that were kept were presumably the best or most historic ones. The Broad Gauge listings of September 1876 [94] show 8 surviving Firefly's then categorised as Tigers: Spitfire, Jones Turner and Evans (April 1840); Saturn, R B Longridge (June 1841); Leopard, Sharp Roberts (May 1840); and Lethe, Ixion, Gorgon and Ganymede by Fenton Murray & Co. This possibly reflects both Gooch and Brunel's opinion that the best Firefly's were made by Fenton, Murray & Co (Chapter 15 p361).

The locomotives produced by the firm followed three basic layout designs; Stephenson's Planet, Stephenson's Patentee and Gooch's Firefly and Leo classes. A question that should be addressed is whether the firm slavishly followed the designs of others or whether their locomotives evolved to contain much that was to their own design.

To cover the easy option first, that of the Gooch designs, these were only produced for the designing railway, one of whose major aims was to have parts interchangeability between the different manufacturers, and therefore there was no scope for the inclusion of anything that deviated from Gooch's specifications. While many of the companies had quite a few issues with the G.W.R. over this, Fenton Murray & Co did not have many problems although Gooch did reject a proposal on the box (fire or smoke is not clear) in June 1839 [106], prior to going into manufacture. Gooch also seems to have used Fenton Murray & Co on a few occasions when he wished to try changes; for example with the 15 ½ inch diameter cylinders.

With the Planet type engines although the general layout is to Stephenson's design and the company sent a number of drawings to Fenton Murray & Co [18], the detail, even in the early Fenton and Murray locomotives for the Liverpool and Manchester Railway, did not necessarily conform to the Stephenson locomotives. An examination of the analyses made by Pambour, as used earlier [17] shows that the boilers of the three locomotives by Fenton Murray & Co were 6.5 feet long and contained 107 tubes, whereas the Stephenson locomotives given had boilers between 6.5 and 7.78 feet and between 63 and 140 tubes (but none with 107).

Fenton Murray & Co also persisted with the manufacture of 4 wheeled locomotives longer than Stephenson and needed to add improvements as they could. Many of these 4 wheeled locomotives were exported to France and it is in the early French writings on locomotives that some detail survives on the differences in the detailed design of components [17, 22, 37 & 39]. This action was not peculiar to the Leeds firm and most of the well-known companies that adopted Stephenson design outlines incorporated their own detail.

102. Fenton Murray & Co locomotive components from Flachat & Petiet 1849:

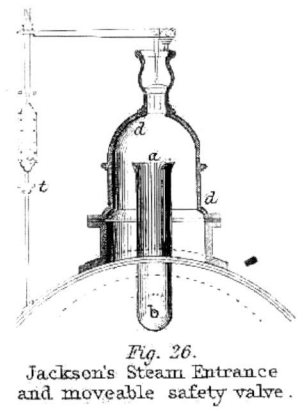

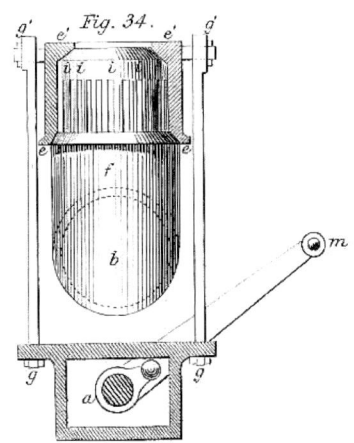

102a.Steam Valve and safety valve.

102b. Regulator valve in 2-2-2's.

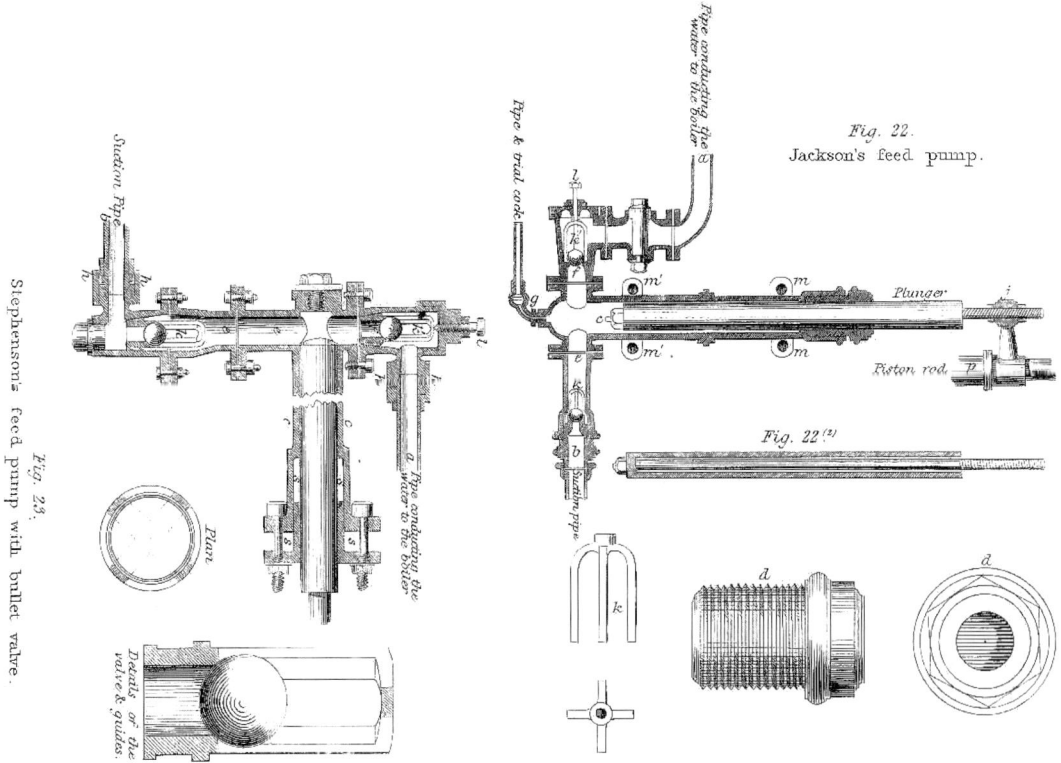

102c. Feed pump alongside one of Stephenson's.

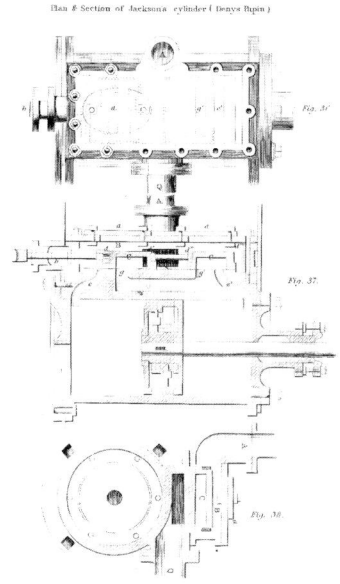

Plan & Section of Jacksons cylinder (Denys Papin)

102d. Cylinder and valves from Denys Papin, 2-2-0, Paris Versailles Right Bank Rlwy.

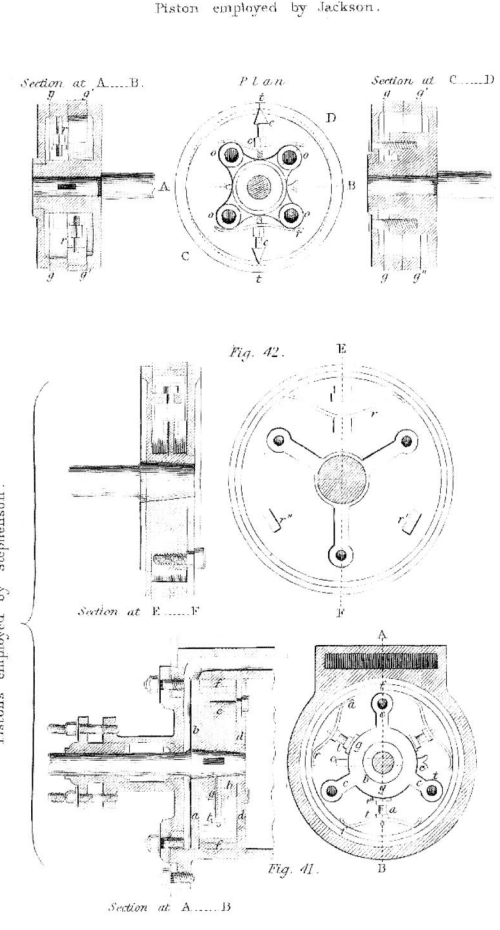

Fig. 43.
Piston employed by Jackson.

102e. Fenton Murray & Co's pistons alongside one of Stephenson's.

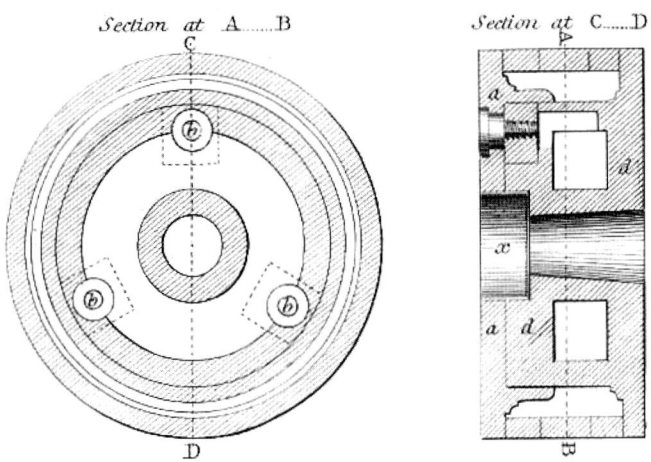

102f. Fenton Murray & Co's metallic packing for pistons.

Fenton Murray & Co. also developed their own valve and eccentric systems, one of which is also included in Daniel Kinnear Clark's Railway Machinery of 1855, but this diagram again comes from the French source:

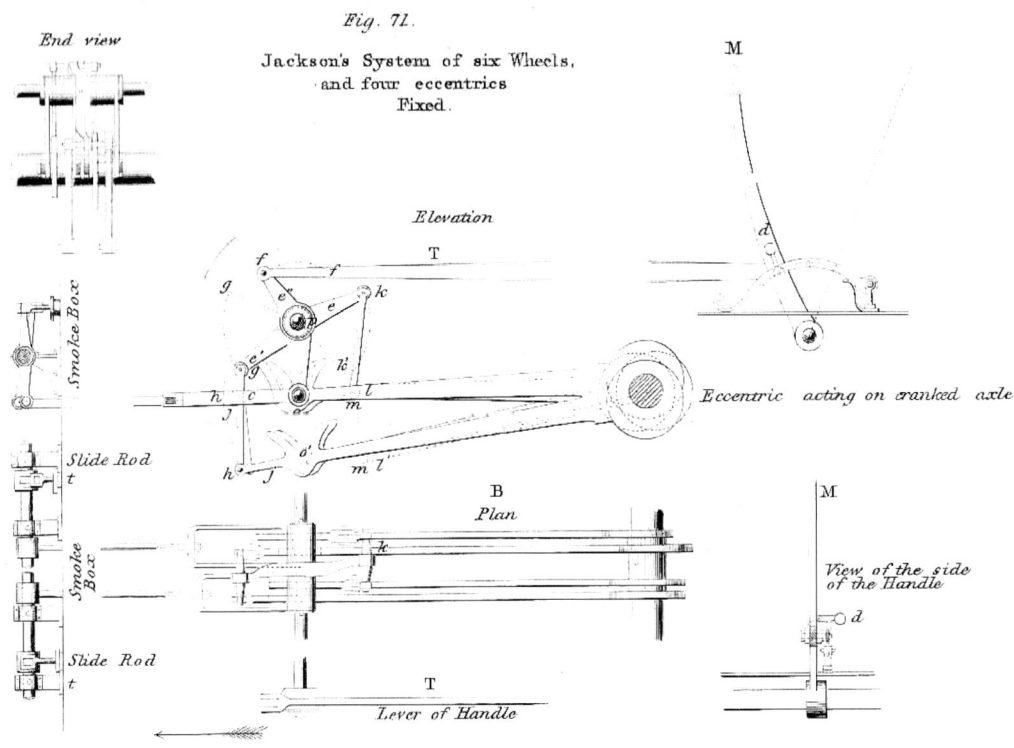

102g. Fenton Murray & Co's eccentrics

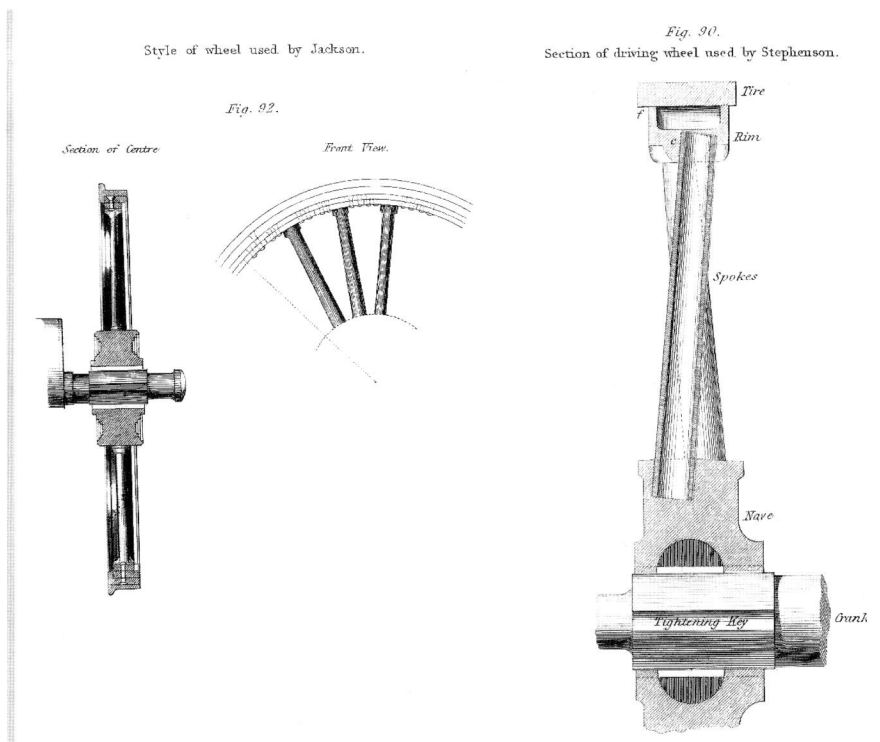

102h. Fenton Murray & Co wheel on left and Stephenson one to right.

Notes:

1. Middleton Colliery Account Books WYL 899 WYAS, Leeds
2. Watson Papers Wat 3/112/17. NEIMME.
3. Photocopy of Blenkinsop Letter Book 22 July 1813. NRO 961/2 Northumberland County Archives, copy also held at WYAS Leeds.
4. Letter from Daglish in Wigan Record Office, for this and other information see A Yorkshire Horse by Richard Daglish, RCHS Journal vol 31 p123-31, and The Orrell Coalfield by D Anderson, Moorland Publishing 1975.
5. Watson Papers Wat 3/13/116. NEIMME.
6. Watson Papers Wat 3/13/115. NEIMME.
7. Ree's Manufacturing Industry, David and Charles 1972, volume 5 page 146.
8. French Technical Journal.
9. QAR 5/39 William Radford's notebook. Lancashire Archives, Preston.
10. Watson Papers Wat 6/61. NEIMME.
11. A Practical Treatise on Rail-Roads, Nicholas Wood, 1831. Also see John Blenkinsop and the Patent Steam Carriages by Sheila Bye, Early Railways 2, 2003.
12. Watson Papers Wat 3/13/117. NEIMME.
13. Watson Papers Wat 3/112/17. NEIMME.
14. Richard Trevithick & Pioneer Locomotives by Jim Rees and Andy Guy pages 213/4 in Early Railways 3. They quote 'View Diary of Mathias Dunn', North of England Open Air Museum, Beamish, 1997-202.
15. Railway Magazine volume 26 January – June 1928.
16. Goodrich Papers A1097. Science Museum Library. See Chapter 12 where the letter is quoted in full.
17. A Practical Treatise on Locomotive Engines upon Railways, Chev. F M G de Pambour. Bachelier France1835.
18. A Century of Locomotive Building, Warren. Newcastle 1923. Page 274.
19. The Liverpool & Manchester Railway R H G Thomas. Batsford 1980. Appendix B.
20. Information on the Railway of the Loire is from: 'La Loire Berceau du Rail Francais' by Fure, Vachez et les 'Amis du Rail du Forez' 2000; and 'Histoire des Premiers Chemins de Fer Francais' by Gras, St Etienne 1924.

21. The Leeds Intelligencer May 18th 1833.
22. All three illustrations from The Students Guide to the Locomotive Engine by Flachat and Petiet English Edition John Williams London 1849. Plates 4, 5 & 6 Jackson's Locomotive having 4 wheels & 2 loose eccentrics, employed in conveying goods upon the St Etienne and Roanne Railway. However see also Grier's Mechanics Pocket Dictionary, Glasgow 1838, which illustrations were reproduced in Warren's A Century of Locomotive Building 1923.
23. Manuscript MS 852B, Smithsonian Institute Libraries and 'The Charleston and Hamburg' by Thomas Fetters. The History Press 2008.
24. RAIL 351/1 Records of the Leeds and Selby Railway. National Archives, Kew.
25. RAIL 351/9 and 10 Leeds and Selby Railway. National Archives, Kew.
26. Leeds Intelligencer 21 July 1838, Vint v Leeds and Selby Railway.
27. Leeds Intelligencer 5 December 1840.
28. Leeds Intelligencer 2 December 1837.
29. RAIL 351/3 Leeds and Selby Railway, Meeting of Railway Committee 26 April 1839. National Archives, Kew.
30. RAIL 1014/17 G.W.R. National Archives, Kew.
31. Parliamentary Accounts and Papers, Railways, Session 3 February to 12 August 1842.
32. Railway Times 12 June 1840.
33. 19 September 1840 Railway Times.
34. Great North of England Railway Rail 232/52. National Archives Kew.
35. L'Industrie Des Chemins De Fer by Armengaud (Jacques-Eugene elder), Paris 1839.
36. British Locomotive Catalogue 1825-1923 Volume 5a, Baxter.
37. Guide du Mecanicien-Conducteur de Machines Locomotives, Flachat et Petiet, Paris 1840. This is a different text to the book at note 22. E. Flachat was the Chief Engineer of the Paris–St Germain Rlwy and his writing reflects his experiences with Fenton Murray & Co locomotives.
38. British Steam Locomotives Builders, Jmaes W Lowe. Goose and Son 1975.
39. Guide Du Mecanicien et Conducteur de Machines Locomotives, Le Chatelier, Flachat, Petiet and Polonceau 1851. Volume of plates number 74.
40. La machine locomotive en France, Jacques Payen with Escudie and Combe. Lyon and Paris 1988.
41. Le materiel moteur et roulant des Chemins de fer de L'Etat du Paris-St Germain au rachat de l'Ouest et a la S.N.C.F., L-M Vilain. 3rd edition Nantes 1972.
42. Rapport de M. Le Directeur-General par interim Messieurs Les Membres du Conseil d'Administration du Chemin de Fer de Paris a la mer. Biblioteque nationale de France. Gallica BnF.
43. Etudes sur les Machines Locomotives de Sharp et Roberts, Felix Mathias. Paris 1844.
44. Rapport au Conseil d'Administration du Chemin de Fer de Paris a Orleans, J G L Clarke. Paris 1839.
45. Un siècle de Materiel et Traction sur le reseau d'Orleans, L-M Vilain. 3rd Edition Paris 1983.
46. The majority of the data in tables 1 & 2 come from the books listed at notes 17, 38, 44 & 36.
47. Locomotives of the LNWR Southern Division, Harry Jacks, 2001. The Railway Correspondents and Travel Society, pages 71-81.
48. Allegemeine Bauzeitung, mid 1841.
49. La Societe du Chemin de Fer de Montpellier a Cette, L Chanuc in; Histoire de Communications dans le Midi de la France, tome XIV no. 52 July-Sept. 1971. Also Histoire de la Locomotive Terreste: Les Chemins de Fer, Charles Dollfus. Paris 1935.
50. Lithograph of Notre-Dame-des-Tables by Laurens in 'Notice sur le chemin de fer de Montpellier a Cette' by Lebrun, Boehm et Cie 1839.
51. Ministere des Travaux Publics, Compte Rendu des Travaux des Ingenieurs des Mines pendant l'annee 1842, page 54. Paris 1843.
52. Railways of Great Britain and Ireland, Wishaw. 1842.
53. Leeds Mercury and Leeds Intelligencer of 4 May 1839. A sketch in the Rotheram Archives under the name Agilis clearly shows flanges on all the wheels, which was

unusual for the time (Rotherham Archives 176/B1/1). Unfortunately the sketch casts no light on the construction of the eccentrics.

54. British Steam Locomotive Builders, J Lowe listed only the one locomotive for Bingley and suggested it was made under license from Fenton Murray & Co. White's History Gazeteer and Directory of the West Riding of Yorkshire for 1837 had Bingley listed as iron and brass founders, and mfrs of locomotives and steam engines, hydraulic presses, flax and tow m/cy, drills lathes and Mill work, Harper Street.

55. Midland Railway Locomotives, Sheffield and Rotherham Section, P C Dewhurst. The Locomotive, Railway Carriage and Waggon Review, May 15 1943 pages 76-77.

56. Sheffield and Rotherham Independent 21 September 1839.

57. Sheffield and Rotherham Independent 5 October 1844, and HMG Accounts and Papers, Railway Dept. Railway Accident Reports, 22 January to 28 August 1846 (pub) Volume 39.

58. RAIL 412/8 London and Southampton Railway. National Archives, Kew.

59. L&SWR Locomotives. The Early Engines 1838-53, D L Bradley. Wild Swan Publications 1989.

60. RAIL 412/4 London and Southampton Railway. National Archives, Kew.

61. London Standard 18 March 1841.

62. RAIL 411/227 London and South Western Railway. National Archives, Kew.

63. Morning Chronicle 21 October 1840.

64. Railway Magazine 1842, page 130.

65. Morning Post 20 January 1842.

66. Northampton Mercury 21 May 1842.

67. Leeds Mercury 25 June 1842.

68. RAIL 411/176 London and South Western Railway. National Archives, Kew.

69. Chemin de Fer, Compte Rendu des Operations de L'Exercise 18xx, Rapport presente aux Chambres Legislatives 1841-1847. Belgian State Papers.

70. Yorkshire Gazette 4 July 1840.

71. The Railway Magazine 1842 page 66, also reported in the Yorkshire Gazette 22 January 1842.

72. RAIL 315/27 Hull and Selby Railway. National Archives, Kew.

73. RAIL 232/6 and 232/52 Great North of England Railway. National Archives, Kew. It has been proposed that Turner Ogden & Co were merely agents for locomotives but it can be established that they were manufacturers in their own right. The firm was set up in Leeds in in April 1836 (Leeds Intelligencer 23 Sept. 1837), but a disastrous fire in 1839 (Leeds Mercury 23 November 1839) coupled with no insurance led to the firms demise in 1841 or 1842. However reports of the fire in the Yorkshire Press clearly describe a locomotive under construction in the shops surviving the fire! The assets were up for sale in April 1841 (Leeds Intelligencer 10 April 1841), although the sale was withdrawn and Turner seems to have still occupied the premises for some time. The firm supplied possibly its final locomotive to the York and North Midland, a four coupled tank, in 1841, which was named Nelson in a Leeds Times report of 28 August 1841. This locomotive may be the reason it has been held in the past that the York and North Midland took over the Fenton Murray & Co Nelson from the Leeds and Selby.

74. E. g. RAIL 232/52 and 232/48 Great North of England Railway. National Archives, Kew.

75. RAIL 232/2 Great North of England Railway. National Archives, Kew.

76. RAIL 232/52 Great North of England Railway. National Archives, Kew.

77. The North Midland Railway, L Wilson. Journal of the Stephenson Locomotive Sociery Vol. 23 1947.

78. RAIL 530/7 & 8 London Committee, RAIL 530/2 & 3 Board of Directors, RAIL 530/4 Leeds Committee, North Midland Railway. National Archives, Kew.

79. The text follows British Locomotive Catalogue Vol. 3a by Baxter but rather different views are to be found in Early Midland Locomotives, Fenton and Co, in the Journal of the Stephenson Locomotive Soc. Vol. 19 1943.

80. RAIL 350/8.

81. RAIL 1149/2 Brunel letter book, contains both letters. National Archives, Kew.

82. RAIL 254/459 G.W.R. National Archives, Kew.

83. RAIL 1008/3 Gooch papers. National Archives, Kew.

84. RAIL 254/35 G.W.R. National Archives, Kew.
85. Reading Mercury 18 September 1841.
86. Royal Cornwall Gazette 27 June 1851.
87. The Locomotives of the Great Western Railway, Part Two, Broad Gauge, The Railway Correspondence and Travel Society 1952.
88. The Railway Magazine 1908.
89. RAIL 1008/26 Gooch Papers. National Archives, Kew.
90. Berkshire Chronicle 6 June 1840.
91. Berkshire Chronicle 22 August 1840.
92. Berkshire Chronicle 12 December 1840.
93. RAIL 254/37 G.W.R. National Archives, Kew.
94. RAIL 254/548 G.W.R. National Archives, Kew.
95. London Standard 18 July 1844.
96. Morning Chronicle 30 August 1845.
97. Evening Mail 27 April 1846.
98. Morning Chronicle 15 July 1852.
99. Reading Mercury 24 September 1842.
100. Lloyds Weekly London Newspaper 20 March 1846.
101. Cornwall Royal Gazette 30 January 1846.
102. Leeds Intelligencer 18 June 1842.
103. Windsor and Eton Journal 14 September 1844.
104. Morning Post 19 January 1847.
105. Bristol Times 3 September1842.
106. RAIL 1008/26 Gooch Papers. National Archives, Kew.

Chapter 18.
The Marine Engines

The first recorded marine use for a Fenton, Murray & Co engine for propelling a boat was by James Linaker, the master millwright at Portsmouth Dock Yard. In the years around 1808 he experimented with a system that can be viewed as a forerunner of today's water jet propulsion. He laid a pipe along a boat's keel with a pump amidships which sucked water in at the bow and expelled it at the stern [1]. Of his two versions one had a hand pump and one was to be operated by a small steam engine. The engine (circa 2 H.P.) was procured from Fenton & Murray [2]. The project was abandoned after the Navy Board declined funding [3]. Linaker knew Fenton & Murray as he was involved with the installation of their 30 hp engine at the Block Mills and had visited their factory [3]. How his boat project advanced is uncertain, but it was described as being 'effectually tried,... but found inferior to paddle wheels' [4]. The engine appears to have ended up being used in the dockyard [5].

The next known event seems to have been a progression from producing Blenkinsop's locomotives and involved fitting a high pressure engine into a boat called *L'Actif* belonging to Mr Wright of Yarmouth in 1813 [6]. The engine was of 8 H.P., had an 8 inch cylinder, 2 feet 6 inches stroke, with a cast iron boiler 6 to 8 feet long . The boat returned to Great Yarmouth and after successful trials began a scheduled service between that town and Norwich under the name *Experiment*. She was later joined by another boat, also with a high pressure engine (not by Fenton & Murray) called *Telegraph*. The engine from *Experiment* was later installed in a new hull and called *Courier* [7]. The service continued until the boiler exploded on the *Telegraph* on Good Friday 1817 with several fatalities, resulting in a House of Commons enquiry [8]. The steam boat service was discontinued by Wright and the fate of the engine from Experiment is not known. The importance of this ferry is that very unusually in Great Britain the engines were high pressure, which while the preferred option in the USA, are not recorded elsewhere in the British listings of the time.

Thirteen days after the explosion at Norwich, an advertisement appeared in the newspapers including the Caledonian Mercury announcing that the Morning Star would be recommencing its services between Alloa (near Stirling) and Newhaven (near Leith) and advised readers that the engine on the boat was of low pressure and had a new boiler constructed of the best malleable scrap iron, riveted together and made by Fenton & Murray [9].

103. Steam Boat off Milmont House, Great Yarmouth - showing either Experiment, Courier or Telegraph.

In July 1815, Matthew Murray, aware as were many others, of the potential for steamboats for naval use wrote to Simon Goodrich at the Navy Board. He proposed a twin cylindered, 40 or 60 H.P. tug with four paddle wheels, without sail power, to manoeuvre ships of war in and out of port in all winds, to move stores around, and to tow fire ships [10]. At this time ships of the line and others could wait for weeks for a favourable wind that would allow them to sail into certain ports.

On 31 December 1813, a patent had been granted in the USA to Mr Francis B Ogden [11]. The patent was the first for using expansive steam in marine engines, and it covered: *a*. combining 2 or more cylinders to form one engine, with a view to cutting off the steam, during the stroke of the piston, and using it expansively for the residue, and *b*. the arrangement of the cranks, to enable them to work without dead points. Neither of these ideas was new to England, but Ogden appears to have been the first to combine them and use them for marine engines [12]. After having a couple of engines built in the USA Ogden visited England and met James Watt. Ogden placed an order with Fenton & Murray [12]. They built him a double acting engine with two cylinders, 30 inches diameter, 4 foot stroke, with the cylinders driving one shaft by cranks at right angles with each other; and with a throttle valve on each cylinder with a cut off at half stroke. It weighed 68 tons [13].

Ogden wrote to Fenton & Murray on 26 November 1818 from Norfolk Virginia, concerning an engine for a friend, but advised them of his entire satisfaction with the engine which they had sent him [14]. He also enquired about an order from New Orleans.

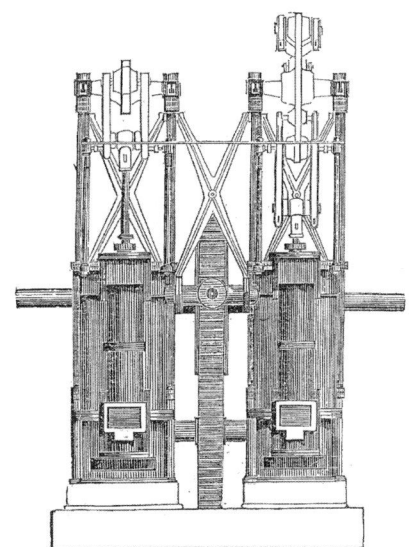

104. Francis Ogden's marine steam engine by Fenton, Murray & Co, circa 1818.

Ogden's boat can probably be identified with a description given by Marestier of the engines of the Roanoke, as comprising two English engines of 30 hp each, operating two cranks at right angles on the same shaft [15].

Ogden's enquiry as to whether Fenton & Murray had received an order from New Orleans ties in with the report that Ogden purchased an engine in 1824 in New Orleans made by Fenton and Murray, from the same patterns as his 1817/8 engine. He had the engine put in tug for the Mississippi, which he described as going upriver against a 3 ½ mile current towing two ships of 300 to 400 tons along with a couple of smaller ones [16]. In certain reports it has been mooted that this was an unauthorised manufacture by Fenton and Murray to Ogden's plans, but the additional information contained in the letter rules this out.

A new steam packet was reported to be building at Selby, Yorkshire by the proprietors of the Favourite which worked between Hull and Selby, in January 1820 [17]. This was probably the steam packet Leeds, which was in service by October 1821 and maybe somewhat earlier, as she appeared in an advertisement by a Leeds firm Fisher Walker and Fletcher who were using the Selby steam packets service, Leeds, Favourite and Humber, as onward conveyance for the goods carried in their fly boats to Selby [18]. She was listed in Battle's Directory of Kingston upon Hull in 1822, White and Parsons Directory of 1826 advised that she had a 30 hp engine, by 1839 she was operating as a luggage packet between Hull and Goole [19] before working between Goole and Gainsborough [20]. The engine was by Fenton Murray & Co [21].

461

In 1821 a new steam packet started on the London Margate route called Hero. She was built at Rochester by Bancham with a tonnage of 427 and two 50 H.P. engines by Fenton & Murray. She had a draught of 6ft 4in; the Paddle Wheels diameter was 14 ft, with a breadth of 8 ft, and a speed at the extremity of15 mph, her depth in the water was 1ft 6in; and her speed was 11 ½ mph (she was the fastest steam boat on the Thames at the time); the coal consumption was 2240lbs per hour; and the engine strokes per minute were 30 [22]. Goodrich noted that the cylinder diameter was 38 ½ inches with a 3 ft stroke, she worked at 33 strokes per minute, and her paddles made 22 revolutions a minute [23].

Captain Percy of the Hero was a witness before the Parliamentary Committee on Holyhead Roads in 1822, where, on being asked if he was familiar with Boulton & Watt engines, replied he was and thought that boats powered by them had fewer accidents. He stated that, on the Hero, the cross heads had gone twice, and some bolts had failed along with the steam pipe. Similar accidents were noted by captains of other boats powered by Boulton & Watt engines, so were presumably fairly frequent across all manufacturer's products [24].

The Committee corresponded with a number of experts on various aspects of its enquiries, including on the subject of steam engines with James Watt, Maudslay & Field, Fenton Murray and Co, William Brunton, and David Napier. Fenton & Murray responded to the standard questions on 27 April 1822 and recommended; double engines with two cylinders, beams to work below, with side rods and cross heads made of wrought iron; a combined boiler being 3 distinct boilers put together as one, fire to pass through each 3 times, with a construction allowing it to be put in or out through the hatchways without impacting the deck; paddle wheels, shafts and arms to be wrought iron with wooden paddles [25].

On the 2nd April 1822 the Yorkshireman was launched at Thorne for Weddle and Brownlow of Hull [26]. She was designed for the Hull to London route and advertisements for her, along with the earlier Kingston, appeared in the press later the same year [27]. For the first 7 years of her life she was powered by a pair of 40 H.P. engines by Fenton Murray & Co, these were however removed in 1829 for some more powerful pieces by Butterley [28].

In 1822 the Navy Board correspondence included the subject of steam engine suppliers for naval vessels [29]. On 3 August the Navy Board received a letter from Fenton Murray and Co asking if the Board required 25 or 50 H.P. engines to which they replied that each vessel was to have two engines of 50 H.P. The Navy Board issued a Request for Quotation to four companies for this requirement; Boulton & Watt, Butterley, Fenton Murray and Co and Maudslay. The abstract notes that

Boulton and Watt were to be told that the Navy Board had had an offer of a 25% lower price than their indication. In August 1822 the bids were received. In October, the Surveyor of Buildings reported the comparative prices to the Navy Board stating that Fenton Murray and Co had made the lowest offer. In response to a request Fenton Murray and Co supplied details of engines supplied, names of their vessels and owners. Coal consumption was requested from the owner of a vessel powered by each of the tenderers. The vessels used in the comparison appear to have been the 'Lord Melville' for Butterley; 'The Engineer' and the 'Sovereign' for Maudslay; and the 'Hero' and the 'Yorkshireman' for Fenton Murray and Co. By the end of the year the Board recorded that supply would be restricted to persons of established reputation as steam engine makers and that they would employ Boulton & Watt and Maudslay [29].

Fenton Murray and Co seem to have been determined to pursue the marine market as in September 1825, when advertising for a factory manager, they specified a preference for experience with steam boat engines [30].

In 1826 a wooden paddle steamer named the Calder appeared in White and Pearson's Directory for Kingston upon Hull described as having a 34 H.P. engine and sailing from Hull to Selby. She was built by Pearson & Co at Thorne and the engine was by Fenton Murray and Co [21]. By 1833 she was working the Hull to Goole route and was still doing so in 1842 [31]. Calder features in two paintings owned by the City of Hull, Three Paddle Steamers in Hull Roads (along with Kingston and Prince Frederick) by Thomas Binks and The Lion and Calder off Goole by William Griffin. (This picture can be viewed on the Hull C.C. collections website using reference KINCM:2005.4989).

In 1829 Fenton Murray and Co constructed an iron paddle steam boat for Mr William Gravatt (who later worked for the Great Western Railway). This was stated to be the first instance of a boat using a tubular boiler [32].

Simon Goodrich visited Fenton Murray and Co in March 1830 and noted that the firm was making a pair of 8 H.P. marine engines which were not to exceed 3 ½ tons each. With boiler and water added at 8 tons the apparatus was not to weigh more than 15 tons. The apparatus was for a boat on the Saone in France. The price was £1150 and it was to be delivered at Le Havre [33]. Before the railway, the main route from Paris to the Marseille area, was via Chalons-sur-Saone, where a boat was taken to Lyon, and from here a larger boat was taken to go down the Rhone. Steam boats were introduced here around 1826. In 1830 Messageries Co. along with Galline & Co built Hirondelle No 1 as competition to the existing fleet, with the Fenton Murray and Co engines. It was a twin (8 H.P. each) coupled low pressure 16 H.P.

one, the internal diameter of the cylinders was 0m.457, the stroke 0m.66, the revolutions were 30 a minute, and the steam pressure was 1 1/3 atmospheres. The boiler had two grates and 30 metres of heating surface. 125 kilos of coals were burnt an hour. The Hirondelle was faster than the competition, and immediately afterwards a Hirondelle 2 was built with the same standard engines and hull [34].

In 1831 S. P. Transit was launched at Hull, and the reports of her launch described her as the second largest steam packet built in Great Britain. Her engines were by Fenton Murray and Co of a power of 70/80 H.P. each with 5 foot stroke. This was the first indication the firm felt ready to build large sea going marine engines. Transit was built by Pearson & Co of Thorne and traded on a regular basis between Hull and Hamburg until the 1850's (see Hull CC website KINCM:1982.518). She also visited London and St Petersburgh [26].

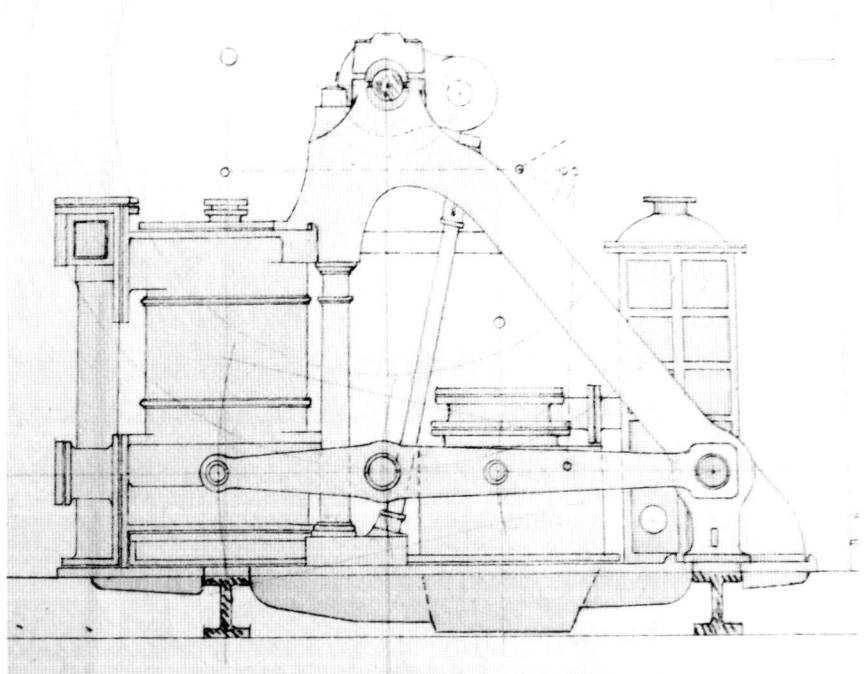

105. Engines for S.P.Transit from a drawing sent by Fenton Murray & Co to the French Government. © SHD – Vincennes, MV 7 DD1.

In the same year an iron steam boat was built on the Humber by Jas Livingstone with a 10 H.P. Fenton & Murray engine capable of 10 mph. She was built for the Castleford Goole trade [26].

There was mention in a patent case, which involved the alleged infringement of a paddle wheel design, of 3 paddle steamers using Fenton Murray and Co paddle wheels and operating in the Goole area. [35]. One is named as the Liberal, owned by Bromley & Co, and the others as Aire & Calder Co tugs. The Aire & Calder Co's minutes of 1831 report the introduction of the first tug on the Castleford Goole run (Mr Livingstone's boat above?), and the ordering of a second [36].

1832 found the firm supplying M. Guibert at Nantes, with a single 8 H.P. engine, with a diameter of cylinder of 0m.432, a stroke 0m.584, which operated at 35/38 strokes a minute. The pressure was ¼ of an atmosphere and the coal consumption was a hectolitre an hour [37], the boat was called Hirondelle 2 (Swallow), a very popular name for river steamers at the time [38]. Also dated 1832 is a drawing of a 20 H.P. engine for Theodore Paul at Chalons [39]. Paul was a cousin of M Galline and involved in building the Saone-Rhone Hirondelle's [40]. The boat for this 20 H.P. engine has not yet been identified, but this may well be the apparatus which was posted in the Leeds Intelligencer on March 30 1833 as the first instance of a cargo being shipped direct from Leeds abroad. This was achieved via the new customs at Goole, and the goods were shipped on the Iris to Le Havre.

At this time the French Navy were buying English engines to improve their indigenous product. Their baseline steam 'frigate' was the Sphinx, which was powered by a 160 H.P. apparatus made by Fawcett and Preston of Liverpool in 1828 [41]. The Sphinx had performed better than any of the ships constructed with French Apparatus, which made the French Navy interested in acquiring further English technology. In November 1833, Fenton Murray and Co submitted a proposal to the French Navy for steam apparatus with a power of 160 H.P. The letter mentioned that Mr Jackson had just returned from France, where the firm had supplied steam engines for boats on the Rhone, Saone and Seine. The engines for Seine boats have not been identified. The letter enclosed a sketch of the Fenton Murray and Co apparatus installed in the Transit.

M.Faveau, a leading engineer in the French Navy, went to Leeds to visit Fenton Murray and Co and then to Hull, where he inspected the Transit, and examined the installation of her apparatus and discussed it with the owners. In July 1834 he submitted his report. He noted that the Transit had been operating for 3 years, including sailing in the winter months, that the machinery had been very satisfactory and had been tried in very rough seas without mishap and that it was considered the best available in Hull (quoting Mr Brownlow, the owner, and another nameless person). Faveau established that the weight of the Fenton Murray and Co apparatus would be about ten of 12 tons lighter than that in the Sphinx (the Transit boilers had not come from Leeds), or even 20 tons if Mr Jackson could achieve the power without adding weight as he claimed.

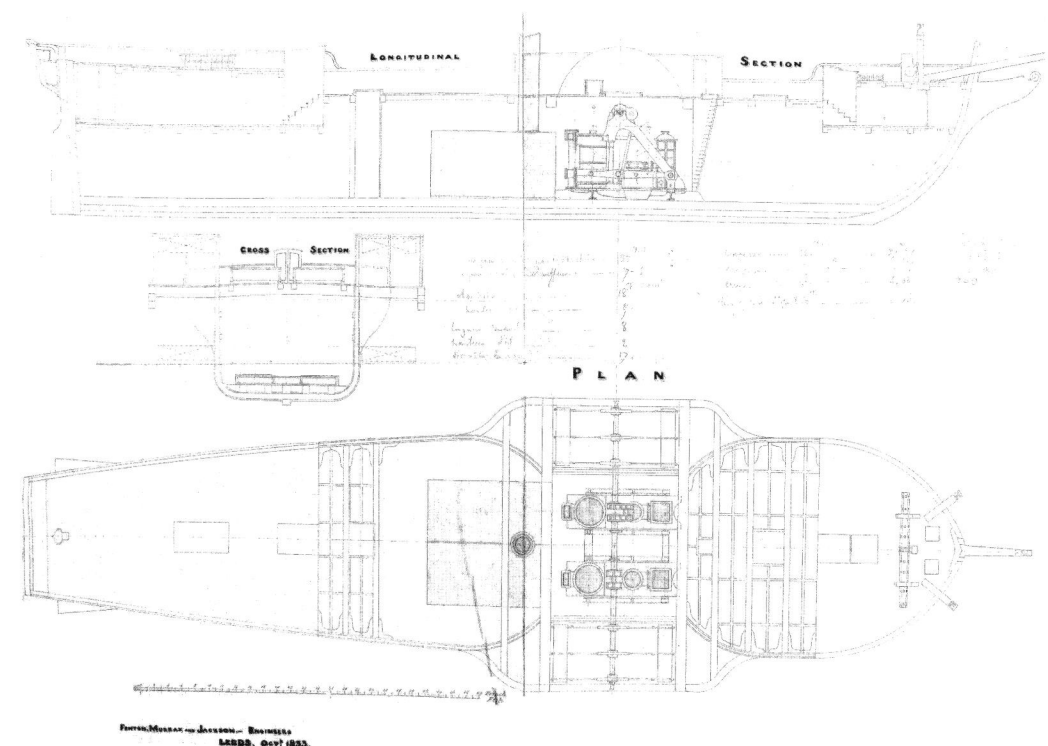

106. Plan of P.S. Transit supplied by Fenton Murray & Co to the French government.
© SHD – Vincennes, MV 7 DD1.

Faveau was supplied with the following data by Jackson:

	2x80=160	2x110=220	2x120=240
H.P. of the engines:			
Diam. of the piston moving at 210 ft/ minute:	48in.	56in.	89in
Stroke of the piston:	5ft 0	5ft 4	5ft 6
Total weight with the boilers full of water:	140 tons	185 tons	200 tons
Price £:	7000	9625	10500
Price Francs:	175000	240625	262500.

He also ascertained in Hull that Fenton Murray and Co had a good reputation as engineers for producing well-made and well finished work, and that they were an agreeable company with which to do business. On the downside Faveau was concerned about the effects of the unusual construction of the apparatus, especially the fact that each engine was unbalanced without its twin. He calculated that on the upstroke the piston had to lift an extra burden of around two tons, reducing the steam action of 12 tons by two, but acknowledged that there was an equal and opposite effect on the down stroke, and that the overall issue was mainly eradicated in the joint operation. However in the light of the fact that the apparatus obviously worked well in the Transit he recommended the purchase of a set of apparatus (engines, boilers and paddle wheels) from Fenton Murray and Co so that at least they could be trialled against apparatus of the same standard as that in Sphinx [42] .

466

On 15th November 1834 the French Navy ordered a 160 H.P. apparatus for the Papin, for which the drawings were delivered in March 1835. A surviving letter from Fenton Murray and Co of March 1835 mentions that they had heard that the French Government were looking to buy some more engines and that they would be 'highly gratified' to receive another order, and that the previous week the William Darley had been launched at Hull and they were fixing a pair of 120 H.P. engines in her. A letter of July 30th mentioned they had had a visit from M. Faveau and M. Guy Lussac, when they had signed a contract for a further supply of engines (see below). On the 12th of September Fenton Murray and Co acknowledged an instruction to send the apparatus to Lorient rather than to Rochefort and advised that they should finish sending the initial order of machinery to Hull for shipment in the next week. The machinery arrived at Indret, near Nantes, from Lorient in October of 1835 [43]. Indret was and is the propulsion centre for the French Navy (and incidentally now provides nuclear power plants for the French Navy).

The William Darley, built by E Gibson & Co of Hull, and launched in July 1835, was a wooden paddle ship with 'two very superior engines, of 140-horse power (see above, they were 120 H.P. each) each', made by Fenton & Murray [44] (see Hull C C website KINCM:1957.27). She made an excellent reputation on the Hamburg route, sailing in weather that kept other ships in port. However in 1848, for reasons unknown, she was converted to a sailing ship and had her steam apparatus removed [26].

Also in 1835 a Navigation Permit was granted for a steamer called Luxor to sail between Nantes and Orleans. This boat was also from Guibert's stable and had a 12 H.P. apparatus by Fenton Murray and Jackson. The permit describes the engine as having a cylinder diameter of 0.568 m, a stroke of 0.904 m, and working at 30 strokes per minute, operating at 2.5 atmospheres [38].

Drawings survive dated August 1835 of a twin 9 H.P. apparatus supplied to M Guibert at Chalon, along the lines of those supplied for the Soane-Rhone Hirondelles and described as one of two sets [45]. These were probably installed in two steam tugs which were granted their Navigation Permit in 1836 for the Saone, they were named Le Jackson and Le James Fenton [38].

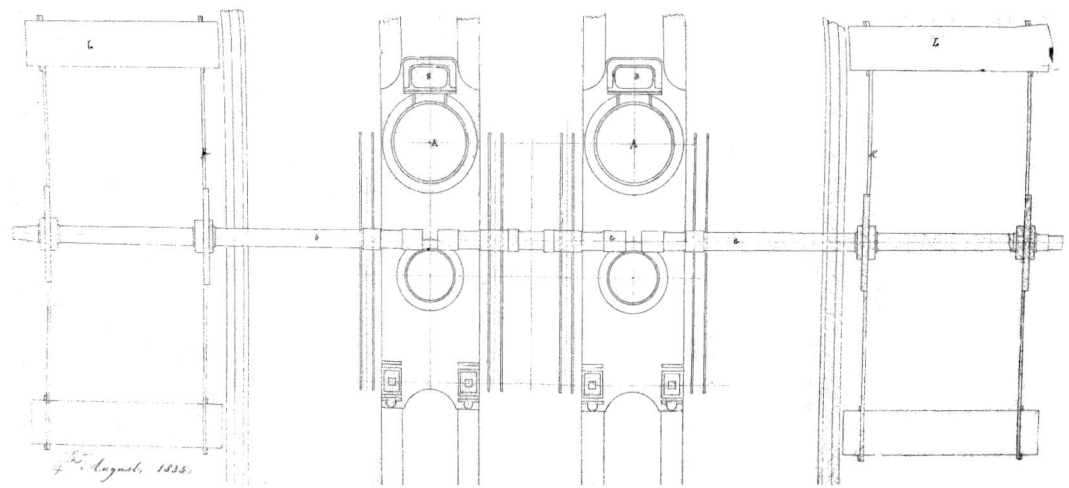

107. Two views of twin 9 H.P. apparatus for Le Jackson and Le James Fenton. Archives du Musee des arts et metiers, CNAM S 156.

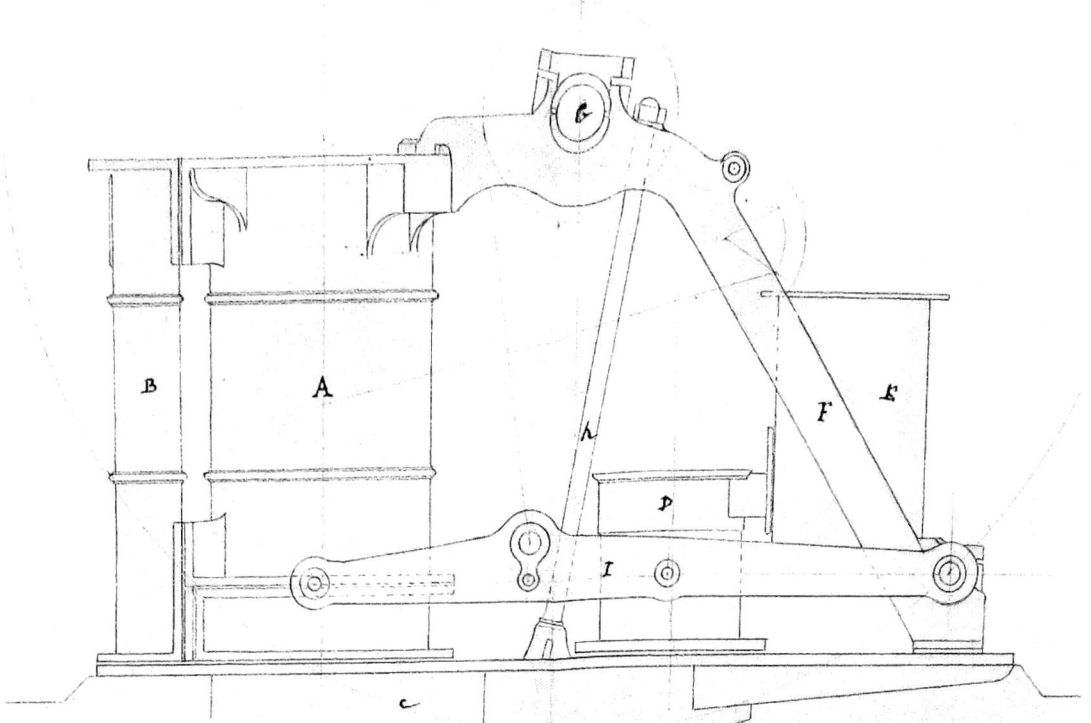

In 1836 the French Government ordered 10 steamboats to operate a mail and passenger service in the Mediterranean [46]. Lycurgue was probably built at Lorient but moved to Indret to have her Fenton Murray & Co engines fitted. The contract between the French Ministry of Finance and Fenton Murray and Co was signed at Leeds on 30 July 1835 with Faveau and Guy Lussac signing for the French Government (subject to ratification). It specified; two double acting low pressure steam engines of 80 H.P. each, paddle wheels of 16 paddles 8 feet long by 2 feet wide with maximum external diameter of 19 feet. The engine cylinders were to be at least 48 inches diameter with a stroke at least 4 foot 6 inches, boilers to be of first class plate iron. It also contained metal specifications for various parts and a comprehensive list of spare parts and tools. Shipment was to be by 1 March and erection on board by 1 July 1836 [47].

In August 1836 a M. Reech published the results of trials conducted using the Papin and Cerbere, both boats to the same plan, but Papin had a Fenton Murray and Co. apparatus and Cerbere an Indret apparatus; based on that in Sphinx but updated. The main findings of the trials were:

1. All other factors being equal Papin was 10 % faster, than Cerbere

2. To equal Papin's speed, the pressure in Cerbere's boilers had to be 12 ½ cms of mercury higher.

3. Cerbere's boilers struggled to turn the paddle wheels at 22 ½ revolutions/minute (the standard). Papin's boilers gave steam equating to a nominal power of 200hp. The major difference was the expansive operation of the Fenton Murray and Co. engines, while the ones manufactured at Indret were non-expansive. The Indret engines were also so badly set up that for 3/100's of the return stroke no pressure acted on the piston and the vacuum was incomplete [48].

Later that year additional trials were made using three of the ships for service in the Mediterranean; Lycurgue, Dante (with engines from Indret modified to work expansively), Minos (with unmodified engines). The results were consistent with the previous trials; the knowledge had been acquired which made it unnecessary to purchase further English engines, and the fears expressed by Faveau in his 1834 report were proved to be ineffective in practice. A large amount of data was collected and calculated for presentation in the report, one of the tables presented is given [48].

Table showing the comparison of steam used and power obtained with the apparatus of several known boats of 160 nominal horse power, from experiment.

Boat	Amount of Stroke before steam cut-off units	Ratio of power obtained units	Effective power HP HP	Comments
Etna	1.000	0.8209	131.34	
Etna	0.988	0.8390	134.24	Control law of Messrs Gengembre and Cave
Minos	0.974	0.8607	137.71	Control law of Messrs Gengembre and Cave
Cerbere and Tartare	0.971	0.8647	138.35	Control law of Messrs Gengembre and Cave
Cocyte	0.933	0.9224	147.58	Mr Gengembre's final control law
Tartare	0.911	0.9543	152.69	Control law as modified at Lorient and Brest
Cerbere	0.905	0.9621	153.94	Control law as modified at Lorient and Brest
Sphinx	0.898	0.9718	155.49	Control law of Sphinx (Mr Fawcett)
Minos	0.892	0.9800	156.80	Control law as modified at Lorient and Indret

Cocyte	0.873	0.9970	159.52	Control law as modified at Lorient and Indret
Papin	0.854	1.0000	160.00	Mr Jackson
	0.850	1.0000	160.00 *	Control law as modified at Lorient and Indret
Lycurgue	0.824	0.9947	159.15	Mr Jackson
Cocyte	0.812	0.9920	158.72	Control law by a Committee at Indret
Castor	0.707	0.9530	152.48 *	Mr Maudslay

* These numbers apply not to the actual engines of the Papin and Castor but to apparatus such as in the Sphinx furnished with control laws the same as in Papin and Castor.

Papin and Lycurgue spent their working lives in the Mediterranean. Papin sank off the coast of Morocco in December 1845 [49]. On 7 November 1854 Lycurgue, who along with some of her sister ships had been taken over by the Messageries Maritime, arrived at Alexandria with Ferdinand De Lesseps on board, at the time of the decision to construct the Suez Canal. She was then stationed at Constantinople and worked the North African coast line. Lycurgue was broken up at Marseille in February 1857 after storm damage [50].

The Messageries Royale received a Navigation Permit for Hirondelle 3 on the Saone in January 1836 with a 12 H.P. apparatus by Fenton Murray & Co. The power plant was described as double working, expansively working at 2 atmospheres pressure and with a cylinder of 0.53 diameter and stroke of 0.79m [38].

Messrs J V G Lauriol were granted a Navigation Permit for the Nantes and Bordeaux in 1837. The ship was built by M. Guibert of Nantes and powered by an 80 H.P. apparatus by Fenton Murray & Co. The engine was described as double cylindered and of low pressure, so no doubt a twin 40 H.P. The pistons were of 53 cm diameter with a stroke of 79 cm, and the engine ran at 32 strokes per minute. The steamer was named after its intended route which was to sail between Nantes and Bordeaux with cargo and passengers [38].

In August 1837 a Lyon newspaper carried a report concerning steam transport on the Upper Rhone. The Upper Rhone Transport Company had announced the delay in starting its new operations from Lyon to Lac du Bourget and Seyssel until April 1838. The reason given, was that while the hull for the first boat was complete, a combination (strike) by workers in England had closed most of the workshops for 4 months and Messrs Jackson of Leeds (ie Fenton Murray and Jackson) had been forced to delay the delivery of the propulsion apparatus until the beginning of 1838. All the machinery for the service had been ordered from the Leeds firm [51]. Three paddle steamers or tugs two of 48 H.P. and one of 20 H.P. are believed to have eventually started a service but were withdrawn after a few years due to the poor demand [52].

In 1840 the Bretmayer Company introduced an iron hulled boat with a 60 H.P. Fenton & Murray engine on the Saone – Lyon route, the Aigle 1. In the same year a new Hirondelle was introduced using earlier engines operating expansively at a pressure of 3 atmospheres, she was shortly followed by yet another Hirondelle in 1841 with new 60 H.P. Fenton & Murray engines and improved accommodation [34]. The latter was probably the same steamer (a Hirondelle) that received a Navigation Permit in 1841 which was described as having boilers and apparatus by Fenton Murray and Co. The paddle wheels were 3.7 metres in diameter. There were two boilers made of 10 mm thick iron plates and each was 3.14 metres by 1.39, having a capacity of 5 cubic metres. The engines were low pressure with expansive working and condensation, the pistons 53 cm with a stroke of 79 cm, 32 strokes per minute and of 50 H.P. (this was probably the assessment of the inspecting engineer) [38]. These engines for a Hirondelle are the last marine engine exports to France by the firm to be uncovered so far.

Notes :

1. Linaker's Patent number 3152 of 1808.
2. Robertson Buchanan, A Practical Treatise on Propelling Vessels by Steam &c (Glasgow 1816) p 40.
3. Science Museum, Goodrich Collection. Letters ref A262.
4. Farey, Rees's Manufacturing Industry Extracts (David & Charles Newton Abbot 1972) Vol. 5 p 143.
5. E A Forward, 'Simon Goodrich and his Work as an Engineer Part 2' Transactions Newcomen Society 1937 p 13.
6. Leeds Intelligencer 21 Jun. 1813 and 26 Jul 1813, Leeds Mercury 26 Jun 1813 and 10 Jul. 1813.
7. 'Early Steam Navigation' Journal of the Society of Arts, 30 Mar. 1877, p 445-447, and 7 Sep. 1877 p 943-944.
8. B.P.P. Reports from Select Committee to enquire into the nuisance caused by the Explosion of Boilers in Steam Vessels 1817
9. Caledonian Mercury 17 Apr 1817.
10. Science Museum, Goodrich Collection. Letters ref A602
11. Francis B Ogden was a nephew of Aaron Ogden an early operator of steam boats around New York (1811). He was aide de camp to General Jackson at the battle of New Orleans in the War of Independence. He took an active interest in steam engines and steam boats. He later became US Consul at Liverpool and subsequently Bristol. He supported John Ericsson, one of the inventors of screw propulsion, and the boat Ericsson used to demonstrate his principles to the Admiralty was the SS Francis B Ogden. Appleton Encyclopaedia of American Biography 1888.
12. Who first introduced the Use of Expansive Steam for Marine Engines? Mechanics Magazine No 944 1842, p 206-208.
13. Mr Ogden's Marine Steam-Engine Mechanics Magazine no 377 1830, p 145-147.
14. University of Virginia Library, Heth Papers, MSS 38-114.
15. M Marestier, Memoire sur Les Bateaux A Vapeur Des Etats-Unis D'Amerique (Paris 1824) p 159
16. Mechanics' Magazine No 994 1842, p 208.
17. The Yorkshire Gazette 22 January 1820.
18. Leeds Intelligencer 22 October 1821.
19. Purdon's directory of 1839.
20. Stephenson's directory of 1842.

21. Humber Packet Boat Website and the Report of the 23rd Meeting of the British Association for the Advancement of Science held at Hull in September 1853, pages 45 to 52.

22. Encyclopaedia Metropolitana, Editors E Smedley, Hugh Rose & Henry Rose, (London 1845) Volume 8 part 1 p 236.

23. Science Museum Library. Goodrich Collection, Journals & Memoranda Volume B47.

24. B.P.P. Reports from Committees, Six Reports on Holyhead Roads; Mr Telford's Report, Holyhead Harbour; Shrewsbury and Bangor Ferry Road &c. Session 5 February to 6 August 1822 Volume VI, p 151.

25. BPP, Holyhead Roads 1822, Appendix No 8, p 214.

26. The Early History of Hull Steam Shipping, F.H.Pearson (1896, Reprint Goole 1984), p 16.

27. Yorkshire Gazette 24 August 1822.

28. Newcastle Courant 28 February and 7 March 1829.

29. National Archives, Admiralty Papers ADM 106 2154 Navy Board In-Letters Digest, Aug. to Dec. 1822.

30. Leeds Mercury 24 Sept. 1825.

31. Trent and Humber Steam Boat Companion by J Greenwood, 1833 lists Calder as working to Goole, as does Stephenson's 1842 directory.

32. The Fouling & Corrosion of Iron Ships; their causes and means of prevention. Charles F T Young. (London 1867) Chapter 3.

33. Science Museum Library. Goodrich Collection, Journals & Memoranda Volume B66.

34. M W Manes, 'Sur la navigation a la vapeur de la Saone et du Rhone' Annales des Pont et Chaussees 2nd serie 1843 1st semestre (Paris 1843), p 16-17. and M Valentin-Smith, Monographie de Saone (Lyon 1852) p 111-112.

35. Morgan and Another v Seaward and Others. 1Web.Pat.cas 167. Reported in The Repertory of Patent Inventions (New Series Vol iv 1835) p 349.

36. National Archives Records of Aire & Calder Rail 800/10, 97

37. Annales de la Societe Academique de Nantes, et du Departement de la Loire-Inferieure. Volume 9. (Nantes 1838), 92-93.

38. Permis de navigation F/14/4229, Archives nationales – Paris.

39. Archives du Musee des arts et metiers - CNAM, Paris, item S87.

40. Paul Schule 'dynastie de « mecaniciens » et d'inventeurs : « les Paul »'.Revue scientifique des Musees d'art et d'histoire (Geneva) XXIX (2009)148.

41. The records of Fawcett and Preston in the Liverpool Maritime Museum.

42. 7DD1-37 Papin, Service historique de la Marine de Vincennes, Paris.

43. C.V.D Brisou, Accueil, Introduction et Developpment de l'energie vapeur dans La Marine Militaire Francaise au XIX Siecle (Paris 2001), and for more information on the background.

44. Hull Packet 3 Jul. 1835.

45. Archives du Musee des arts et metiers - CNAM, Paris, item S156.

46. At Toulon: Mentor, engines by Edward Bury of Liverpool. At Cherbourg: Sesostris, engines by Hallette of Arras; Rhamses, engines by Cave of Paris. At Brest: Tancrede & Leonidas,engines by Miller & Ravenhill of London. At Rochefort: Scamandre & Eurotas, engines by Maudslay,sons and Field of London. At Lorient: Dante & Minos, engines from Indret; Lycurgue, engines Fenton Murray and Co. 1845 return of French Government Finances via Gallica (website of the Bibliotheque nationale de France).

47. Archives du Musee des arts et metiers - CNAM, Paris, item S145.

48. M Reech, Memoire sur Les Machines A Vapeur et Leur Application a Navigation (Paris 1844).

49. The fate of the Papin was the subject of a painting by the well known French maritime artist Auguste Mayer, Naufrage du bateau a Vapeur 'Le Papin'. She is depicted breaking in two on a sandbank.

50. P.Bois, La Grand Siecle Des Messageries in Maritime Histoire de Commerce et L'Industrie de Marseille XIX-XX siecles, (Marseilles 3rd Edition in print), p 162. And so the brief tantalising pointer included in Mr Tyas's Centenary Appreciation of Matthew Murray read at the Science Museum on 24 February 1926 acquires a little more substance. He stated that the firm built some engines for the post packet service between France and Constantinople.

51. Le Censeur, Journal de Lyon. 5 August 1837.

52. Aubert Jean. Historique de la navigation sur le Haut-Rhone francais. In : Les Etudes rhodaniennes. Vol 15 no 1-3, 1939. Pp 181-190. Also Schiff Jean-C. la fin de la navigation sur le Haut-Rhone. In : Les Etudes rhodaniennes. Vol 12 no 2, 1936 pp 259-272.

Chapter 19.

Conclusions and Endings.

Are there any conclusions that may be drawn about Matthew Murray, having looked at his life and work? It remains difficult to be totally positive as the passing of time leaves only glimpses of the man.

The son of a man, who in all probability underwent clearance from an agricultural village in remote Northumberland village, achieved a great deal in becoming the most celebrated engineer in his chosen hometown of Leeds during his life and leaving behind the title of 'The Father of Leeds Engineering', especially when it is considered how much of an engineering centre Leeds became in its manufacturing heyday.

Murray was probably given a positive attitude by his parents. His father went from being a mariner to a whitesmith, a move that he probably could not have managed without marrying into one of the major families of blacksmiths in Newcastle upon Tyne, the Prices. But he showed a determination to improve his situation and rise above the social strictures of the age. And in this he was joined by others, as 'mechanicals' were often men of varied and poor backgrounds determined to pursue their interests. His mother had received an education of sorts in that when she married Reynard Murray she was able to sign the marriage certificate, 'margrat price', whereas Reynard could but place his mark.

Murray learnt a trade initially as a whitesmith as attested by his self-description on his first patent in 1790 and whitesmith and mechanic on his second patent in 1793. By 1799 he styled himself as engineer. While the master under whom he served articles is unknown, and his move to Stockton is obscure, it is probable that wherever he worked in Stockton, one of the products was textile machines.

His move to work for John Marshall, a self-made and determined man, with whom Murray developed a close relationship can only have encouraged him in self-belief in his ability. However the evidence is that he probably did not feel comfortable in dealing with the upper society of manufacturers in other than his technical capacity, or that he found the realms of socio-political activity a severe distraction from his main work. No doubt at the behest of John Marshall, at least twice Murray appears on committees, once for Marshall's first school and then for the initial setting up of the Leeds Mechanics Institute, but Murray seems to have removed himself from these activities. His place in the Mechanics Institute seems to have been taken very quickly by his son-in-law Jo March, who along with Charles Maclea invested considerable effort and time in becoming a part of the socio-political scene in Leeds with both men going on to serve their turn as Mayor of Leeds.

Matthew Murray was however a sociable man, who was obviously in his early to middle life quite partial to drinking in ale houses. This seems to have been undertaken as a means of gathering technical data, sales pitching, meeting suppliers, employees and celebrating. Boulton and Watt made much of Murray's fondness for ale, but as observed earlier, he probably had a very strong head. They also wished to see the man as a dissolute and bad character to justify their pursuit of him. Murray must have had a strong head which would have been necessary to complete business and also possibly to outdrink the likes of Murdoch, which has to remain a possibility in regard to the famous visit Murdoch made to the Round Foundry along with Storey and as to who plied whom with a lot of ale and obtained the others best practice.

As a pure technical innovator in grand ideas Murray was not in the forefront with men such as James Watt senior or Richard Trevithick. Where Murray's genius lay was in recognising the possibilities presented by an invention or an idea and in either improving the concept to make it really practical or to make it viable as a manufacture.

He understood technology and worked at its application. Frank Whittle, talking about the jet engine, once said:

"The invention was nothing. The achievement was making the thing work."
And if this is coupled with another well-known comment – that innovation is 5 per cent inspiration and 95 per cent perspiration perhaps the latter part of both statements summarise Murray's genius.

Arkwright had the inspiration and spun cotton, Kendrew and Porthouse managed to apply the principles to rough spinning of flax. Murray and Marshall shed the perspiration to make a viable system for mass production for linen which with continual process improvement made the early Marshall fortune.

The most well-known example of Murray operating in this fashion is the rack-rail locomotive. Trevithick was the father of locomotion, Murray was capable of making the locomotive a practical paying proposition, adding in his own idea of using two pistons to obviate the need for a fly wheel to navigate the dead spots on the cylinder. The evolution probably being driven by his need to make the locomotive lighter so that it could operate on the rail technology of 1812. It then took Stephenson's commercial vision, along with others, to apply the early work and develop the concept of railways as a passenger transport towards its modern idea.

On the technical side possibly one of the best summaries may be the words written by Bryan Donkin to his wife in December 1816 to the effect that he was going to order his engine from Fenton Murray and Wood as he expected that he would get one as good as one from Boulton and Watt and at a much lower price. On the technical side the engines were good, but as has hopefully been shown Murray was not only good in the product design sphere but was also a forerunner in production engineering; producing very well finished machines at a very reasonable price. It is possible to read into this side that Murray sought the ability to standardise production with repeatable quality. Murray made important improvements and his D valve lasted as a norm for decades. He applied other principles such as those incorporated in the Hypercycloidal movement, but seeing that it had an inherent weakness he did not use the system for long.

Having become a leading manufacturer of textile machinery and steam engines, Murray and the firm entered the realm of mill building and as far as can be determined were the first to offer an investor a package of a near fully equipped mill. They appear at the least to have offered mill plans (based on some form of standardised Bage mill), assistance in the building (Bell's Mill in Dundee has recorded that the foreman was a man from Yorkshire), and provision of the cast iron work, steam engine and some if not the majority of the flax machinery.

The practicality of the hydraulic press, which was certainly not invented by Murray, was vastly improved by his introduction of a method of making 'the top and bottom parts of the press to move or approach one another', whereas previous presses had one part move only. This enabled the pressed cloth to be removed at a more convenient height for packing etc.

Murray was a thinker as well as a doer. An interesting snippet survives in a self-help book published in Boston in 1877:

"My good father, who never rested until he got to heaven, used to tell me how old Mr Murray of Leeds, who was one of the pioneers of the new industries of England, used to push aside every drawing and tool, and to put his work away, when he was beaten by some problem, and go two or three days into a stillness, where no man or work could come near him, and then the problem would, as it were, solve itself" [1].

Matthew Murray was probably fairly convinced of his course of action once his mind was made up, and this may have made him difficult to deal with. There is a feeling of mutual respect for the technical skills but underlying frustration in the dealings between Murray and Marc Brunel. Murray would not jump or put Brunel's orders first and Brunel did not like it, but used Fenton Murray & Co while Maudslay was unavailable or otherwise engaged (Chatham Sawmill) because he could trust their work.

476

Murray's attitude to patents and data rights remains an enigma. There must be little doubt that the legal cases involving his spinning patent (Kendrew v Marshall) and his steam valve and other matters patent (Boulton and Watt) must have made him wary. Although the first was initiated as a case for patent fees against Marshall, when he made the defence that he no longer used Kendrew's machines, the machines were examined and found to be no different in principle. However it was clearly apparent that small changes in distances, speeds and tensions and indeed the addition of leather pieces changed an unpromising machine into one Marshall could use. In his case with Boulton and Watt, he knew he did not have the resources either financially or technically to combat them and so withdrew at the last minute. He also knew by then that his air pump was not worth a fight. In later times he probably sent his heckling machine to the Royal Society to establish his design prior to Hives taking out a patent, and he subsequently patented his improved hydraulic press. He could see the power of patents and possibly their publicity for him and his firm. But if as seems highly likely he had a planing machine of some kind operating very early on, as J.O. March told us, that was so important it was kept in a locked room with only a chosen few allowed access!

Murray did not achieve success by himself and had able partners to manage aspects of the business. David Wood probably shared Murray's keenness to improve manufacturing engineering as he appears to have spent more time in the factory. After Murray stops appearing in Marshall's experiment books on a regular basis, there are references to Wood and it is likely that he undertook the job of keeping their major customer, Marshalls, happy. He was obviously a capable and well liked man. Boulton and Watt considered offering him a job to attract him away from the partnership. He had his method of cleaning linen written up by no less a man than Robert Owen.

James Fenton was the man of business. He had previously been in partnership with John Marshall, which was followed by a leather business in London. However he obviously found his niche at Fenton Murray & Co, where he ended up as a sole trader after the death of David Wood and Matthew Murray. In preparation for his own demise he did introduce managers and then other partners. He it was who re-organised the business to separate the running of the machine business, and introduced Samuel Exley. In his obituary he was referred to as an iron-founder.

William Lister is another shadowy figure in the early days of the partnership having joined at the end of 1803 or early 1804. He came from a fairly local family of iron-founders, and while no doubt making valuable inputs in that field seems to have primarily been a source of finance and has been termed a sleeping partner. His early death in 1811 seems to support his role as a financial partner as some of the partnerships correspondence from that period refers to their shortage of funds due to the necessity of repaying Lister's capital.

Samuel Exley was introduced as a partner by James Fenton in 1832. Reading between the lines, and particularly the correspondence among the Marshall's around 1826 onwards, it would appear that Murray's protégé and son in law Richard Jackson was not enthusiastic about the machine business (textile). So Fenton introduced Exley, who seems to have run the business at a profit until he petitioned to have the partnership concerning that business wound up. Exley does not seem to have made any radical inventions, but the firm remained one of the leading suppliers in Leeds until its demise. This can be gathered by Exley's inclusion in the deputation of Leeds flax and tow machine makers that petitioned the Board of Trade in February 1843.The other notable known members included Peter Fairbairn, J Wordsworth and Charles Maclea. Exley appears to have recovered enough of his capital to set himself up in business as a consultant and agent for selling mills and machinery. He remained in Holbeck for the rest of his life where he died on the 15th December 1879 with an estate valued at under £4000.

Just after he introduced Exley into the business, James Fenton also elevated Richard Jackson to be a partner in both the machine business and the foundry business. Jackson's interest seems to have been the steam engine, particularly the locomotive and the marine engine. With James Fenton's death Jackson appears to have become the business manager for the partnership, although during William Fenton's (a lawyer) term he was probably not in charge of the financing or contracts. Jackson had the misfortune to be in control of the company through the difficult times that were the lot of manufacturing towards the end of the 1830s. Mary Murray, Matthew's widow, died on the 18th December 1836. On December 31st her matrimonial home contents were advertised for auction, and by May 1838 the very materials of the house were also up for auction, so the residence had been dismantled. Richard Jackson and William Fenton had decided to construct a new shop which was to be dedicated to locomotive engines. As has been seen 1837 had been a bad year for strikes and the firm had been forced to delay the shipment of marine engines to France due to inability to produce. Money was tight and the new investment was probably the item that tipped the company over the edge. The later letters to the G.W. Railway are often asking for payment and refer to large sums being due to suppliers, something that Exley also refers to in his submissions for winding up. Fenton Murray & Co invested in premises, unbeknown to them at the time, when many major railway companies were seeking to open up their own locomotive manufacturing facilities and the French Industry was on the cusp of becoming self-sufficient in marine engines. By May 1840 the premises were mortgaged to Beckett's Bank for £4000. The circumstances could not survive Exley's submissions to wind-up the machine side, or the expiration of patience in Jackson's personal or the firm's creditors and in the middle of 1844 all was finished. Jackson seems to have been a good engineer in that under his management improvements and changes were introduced into the products, but reading the submissions made by Exley to the Courts he does not appear to have been able to manage the firm's finances, and indeed can be said to have been less than scrupulous in his use of Exley's capital to bail out the foundry business. After the bankruptcy Richard Jackson moved to London, where by 1846 he was in residence at 35 Tredegar Square off the Mile End Road. In March 1846 he re-married to one Helen or Ellen Cox then resident at 1 Tredegar Square. Jackson's witness was his son in law Andrew Faulds. The married couple moved afterwards to Kings Norton in Birmingham where they were at the time of the 1861 census. Jackson was describing himself as a Civil Engineer at the time, this census for the first time gives us a clue to his origins giving his place of birth as Eccles in Lancashire. A possible birth is one on the 4 July 1790 to William Jackson of Worsley, Eccles a cabinet maker and his wife Ellen [2].

While both David Wood and Matthew Murray each had a son, neither of the sons seems to have wished to take up a permanent position in the firm. Matthew Murray junior, having worked for the firm for a while opted to make a career in Russia. He set up a workshop but no foundry using other facilities. In his last surviving letter home (1827) he is fairly optimistic about winning a contract to put in a water supply to Moscow. He died at Moscow on the 22 July 1835. David Wood junior also had a penchant for working abroad, no doubt in mechanical matters. His death in May 1830 at Dusseldorf, on his journey back to England, was reported in the Leeds newspapers in early June that year. David Wood's other known son died at the age of 16 in 1815. James Fenton never married, and his heir was his brother who was a lawyer. Matthew Murray had three daughters who lived to maturity and all married. Margaret the eldest married Richard Jackson in July 1815.

Ann, the second daughter, married Charles Gascoigne Maclea in January 1816. Mary, the youngest, married Joseph March in July 1825. Jackson became a manager under Murray and is found selling steam engines and meeting customers fairly early on. Maclea and March, deciding not to vie for position at the Round Foundry left the firm and set up on their own in 1825. The Maclea's did not have children, and it would appear that Ann had some issues that classed her as a lunatic under the then laws. In the 1841, 1851, 1861 and 1871 censuses, Ann is found residing in various private asylums in York, and it was in one of these that she died aged 83 in 1874. She was then added in commemoration to Charles Maclea's stone in St Marks Woodhouse cemetery, where he had been buried in 1864. Charles was Mayor of Leeds in 1846/7 but did not serve his full term due to a severe illness. After leaving Maclea and March he held various directorships with major insurance companies and railways. Charles was also on the panel of judges for Mechanical Exhibits for the Great Exhibition in 1851.

Joseph and Mary March had at least eight children, three of whom died young. After Charles Maclea left the business, his place was eventually taken by George March the eldest son. Mary lived until 1864 and is commemorated by a stained glass window in their parish church, the former St Marks Woodhouse (now the Gateway Church). The inscription reads "To the Glory of God, in Memory of Mary the wife of J.O. March of Beech Grove House, who died January 18th 1864, aged 66 years". J. O March survived until 1888, having himself served as Mayor in 1862/3. The premises of Maclea and March were up for rent shortly afterwards. The surviving part of J. O March's beloved Beech Grove House is now part of the offices of Leeds University.

108. Commemorative window of Mary March, Murray's youngest daughter at Gateway Church Leeds, formerly St Marks Woodhouse [3].

Jackson's two sons both served as managers in the firm. William remained in Leeds after the bankruptcy and certainly still had descendants living in the area in 1900, so there are probably still some there. Matthew Murray Jackson went to work for Escher Wyss in Switzerland. The firm developed large marine engines to power the steamboats it built to work the Swiss lakes. The official history of the firm credits the development of the engines to Albert Escher and the two English managers Lloyd and Jackson [4]. M M Jackson's Yorkshire wife Mary died in Zurich in September 1849. He subsequently remarried to an Anna Catharina Tobler. M. M. Jackson left Escher Wyss in 1868 and went to work for the Austrian Danube Navigation Company at Budapest as Engineer in Chief. He retired to Hove where he died on 4th November 1892, his wife having possibly retired separately to Trier in Germany. In his will he passed Matthew Murray's Gold Medal from the Emperor of Russia to his son William, who along with the other main legatee Charles seem to be sons of his second marriage. He had 5 children by his first marriage one of whom Richard, born in 1842, appears to be present in the 1871 census in Hunslet along with his sister Margaret T. He is a candidate for the Richard Jackson who was a senior draughtsman at Kitson & Co.

Richard Jackson senior's daughter Margaret married an Edward King in Zurich, who worked with her brother M. M. Jackson. She returned to England, having presumably outlived her spouse, with her brother and they were living in the same house in Hove in the 1871 census. She returned to the Leeds area and descendants still survive there.

Richard Jackson senior's youngest daughter Susanna Matilda, never married and died at her brother Williams's home in Water Hall Holbeck in June 1864.

It has already been noted that Richard Yale the partner holding the Fenton interests at the time of the firm's demise met a tragic end at the time. His mother Mary and grandmother Isabella Fenton, who had presumably lost almost everything in Richard Yale's bankruptcy, ended up living with cousins in Ripon. Isabella died in the last quarter of 1854 and Mary her daughter in 1886.

The players are all long gone, but hopefully it has been shown that the firm made an impact. The main areas are not those for which the reputation had survived, as the impact of mill building, which appears to have been overlooked by both Trevor Turner and Kilburn Scott has possibly left more of a lasting, if unattributed, impression than any other activity of the firm. The importance of the rack locomotive in the development of rail traction remains impressive. Murray's reputation probably lived as a result of the wide breadth of subjects he tackled and the notable people he worked for and with, Brunel - father and son, Goodrich and Bentham, John Buddle and John Watson the coal viewers and not least John Marshall.

Notes:

1. 'The Simple Truth A Home Book', by Robert Collyer.

2. Thanks to Sheila Bye for this information and to her great help in tracking Richard Jackson.

3. Stained Glass window in the Gateway Church Leeds (formerly St Marks Woodhouse), dedicated to Mary March, nee Murray by her husband Joseph March.

4. Escher Wyss 1805-1955, 150 years of Development. Published by Escher Wyss 1955.

Appendix 1

List of Matthew Murray's patents.

1. Patent number 1752 of A D 1790 Machinery for spinning fibrous materials. (Paid for by Marshalls).

2. Patent number 1971 of A D 1793 Machinery for preparing and spinning flax, hemp, tow, wool and silk. (Paid for by Marshalls). The machinery from this patent appears to be that involved in case between Kendrew and Marshall.

3. Patent number 2327 of A D 1799 Steam Engines. This patent included a mechanism for controlling the fire/steam production, and proposed the use of cylinders lying in a horizontal plane, as became the norm for mill engines etc. some time later.

4. Patent number 2531 of A D 1801 Steam Engines. This is the patent given up due to the action of Boulton and Watt. It included a new type of air pump (which did not work well), changes to ways of packing cylinder tops, changes to valves (B&W's main bone of contention).

5. Patent number 2632 of A D 1802 Combining steam engines to obtain Circular Power &c. This included the famous and long lasting D valve with its use of flat surfaces (and begs the question of when Murray began to use the planing machine), and also the hypercycloidal principles applied to a steam engine.

6. Patent number 3792 of A D 1814 Hydraulic presses for pressing cloth and paper &c. Which introduced a press where both the top and bottom parts moved.

Part 1 Stationary Engines

Category a - sourced from Newspapers

Items advertised may well have been secondhand and only the first advert is
mentioned as some mills were repeatedly up for sale.

Date of advert	Eng. Size	Given Owner	Location	Use	Paper		
Nov 1806	20 H.P	Mersey Wire Mill Co.	Warrington	Wire Mill	various	sale	
June 1807	6 H.P		Hunslett	Flax Mill	Leeds	sale	
Oct. 1809	4 H.P	Bradley & Coxen	Southwark	Foundry	London	sale	
Feb 1810	36 H.P	C S Millward	Bromley Ldn	?	London	sale	
Mar 1810	8-10 H.P	Mr Drew	St George's Fields	Bakery	London	want	
Mar 1811	20 H.P	Scott, Brown & Co	Broadford Mill Aberdeen	Flax Mill	Aberdeen	sale	
Mar 1811	25 H.P	Brown & Co	West Ward Mill Dundee	Flax Mill	Aberdeen	sale	
Mar 1811	6-8 H.P	Mr Lloyd	Uley, Glocs	Fulling ?	London	want	F & M, B& W
Mar 1812	12 H.P	Chrichton	Tay Street Dundee	Flax Mill	Dundee	sale	
May 1812	8 H.P	Donaldson	Lochty Mill nr Dundee/ Arbroath	Flax Mill	Dundee	sale	
July 1812	16 H.P	Robertson & Souter	Townhead Arbroath	Flax Mill	Dundee	sale	
Mar 1813	10 H.P	John Rand	Bradford	textile mill	Leeds	sale	
April 1814	10 H.P	George Cobb	Steander Leeds	Flax Mill	Leeds	sale	
Feb 1815	8 H.P	Robinson, Dearlove	Knaresborough	Flax Mill	Leeds	sale	
April 1816	10 H.P	Stables bros	Horsforth, Leeds	Scribbling Mill	Leeds	sale	
April 1817	12 H.P	Moore, Tennant	Bishop Monkton nr Ripon	Flax Mill	Leeds	sale	
May 1817	12 H.P	Thomas George	St Peters Leeds	Gig Mill	Leeds	sale	
June 1817	25 & 12 H.P	James Ford	Montrose	Flax Mill	Edinburgh	sale	
Sept 1817	30 H.P	James Tennant	Hill-House Bank, Leeds	Flax Mill	various	sale	
Dec. 1817	16 H.P	T Tennant	Holbeck Moor Mill, Leeds	Cloth Mill wool	Leeds	sale	
June 1818	4 H.P	John Knowles	Bent Mills nr Bingley	textiles	Leeds	sale	
Jan 1819	8 H.P		52 Cork St Dublin	manufacture	Dublin	sale	
April 1819	20 H.P		Trottick Mills nr Dundee	Flax Mill	Edinburgh	sale	
Nov 1822	20 H.P	Turney and Bates	Salterhebble nr Halifax	Worsted Mill	Leeds	sale	
May 1824	14 H.P	Mr Atkinson	Bradford	Worsted Mill	Leeds	sale	
Sept 1824	20 H.P	Forster & Co?	Union Mills Holbeck Leeds	Scribbling Mill	Leeds	sale	
March 1826	14 H.P	Richard Jackson	Murray's Mill Marshall St	Textiles?	Leeds	sale	
Oct 1826	24 H.P		Bradford		Leeds	sale	
Jan 1827	26 H.P		Fawdon Colliery, N'berland	railway	Newcastle	sale	
Sept 1827	20 H.P		Rawcliffe, Yorks	Flax Mill	various	sale	
Dec.1827	8 H.P	Naylor & Brearley ?	Milner Royd Mill, Sowerby Br	Textiles ?	Leeds	sale	
Aug 1829	14 H.P	apply at the Leeds Mercury offices			Leeds	sale	
Oct 1829	16 H.P		Redcross St Cripplegate	manufacture	London	sale	
Oct 1829	8 H.P	Linsley ?	Leeds	mustard mill	Leeds	sale	
Jan 1830	8 H.P	Marshall & Co	Holbeck Leeds	Flax Mill	Leeds	sale	
April 1830	10 H.P		Beaver Hole nr Barnsley		various	sale	
May 1830	13 H.P		Shropshire or Manchester		Manchester	sale	
Sept 1830	12 H.P	E & J Taylor	Huddersfield		Leeds	sale	
Jan 1831	10 H.P		Birmingham?		Birmingham	sale	
Nov 1832	30 H.P	Halliley, Stanley	Low Close Mill, Leeds	textile mill	Leeds	sale	
July 1833	24 H.P	London Dock Co	London Docks	construction?	London	sale	
Aug 1833	2 x 6 H.P	Wm Cubitt	Lowestoft navigation	construction	Norfolk	sale	
Feb 1834	30 H.P	Samuel & Darwin	Sheffield	steel works	Sheffield	sale	
June 1834	2 H.P	apply to Leeds Mercury			Leeds	sale	
July 1834	8 H.P		nr Gottenburg Sweden	brewery	London	sale	
Jan 1835	5 H.P	Faviel & Dyson	Fleet Mills	construction?	Leeds	sale	
Mar 1835	10 H.P	North Level Comm's	Tid Gote nr Wisbech	construction	Stamford	sale	
Aug 1835	4 H.P	Thos Cadman & Co	Lady Lane Leeds	textiles	Leeds	sale	
Dec 1835	75 H.P	Messrs Benyons	Castle Foregate, Shrewsbury	Flax Mill	Shrewsbury	sale	
April 1836	20 H.P	Buck and Kershaw	Southowram Bank, Halifax	Worsted Mill	Leeds	sale	
May 1836	c. 20 H.P		Sheffield	manufacture	Sheffield	sale	
June 1836	12 H.P	Forbes Low & Co	Poynernook, Aberdeen	Cotton Mill	Aberdeen	sale	
April 1837	30 H.P	C & R Parker	Darlington	Flax Mill	various	sale	
Sept 1837	12 H.P	South Holland IDB	Sutton St Edmunds, Wisbech	pumping eng.	Leeds	sale	
Sept 1837	30 H.P	Robinson & Watson	Campion Rd Leeds	Scribbling etc	Leeds	sale	
Nov 1838	8 H.P	Josh Batty	Leeds	building	Leeds	sale	
April 1839	20 H.P	Lundy Foot & Co	Minories, London	Tobacco & Snuff	Liverpool	sale	
April 1839	10 H.P	Wm Morris & Co	Wheatley, Halifax	Wire Mill	Bradford	sale	
July 1839	24 H.P	John Howard	Low Fold Mill, Bank, Leeds	Carpet Mill	Leeds	sale	
Oct 1840	20 H.P	Pigou & Co	Dartford Kent	Gunpowder Mill	Newcastle	sale	
May 1841	12 H.P	Wm & John Bairstow	Keighley		Leeds	sale	
Oct 1841	16 H.P	George Popple	Hull	Oil pressing	Leeds	sale	
Nov 1841	6 H.P		Bristol		Bristol	sale	
Mar 1842	16 H.P	Robert Campion	Campionville nr Whitby	Flax Mill	Leeds	sale	
April 1842	10 H.P		Kidacre Mill Leeds		Leeds	sale	
April 1842	30 H.P	Marshall & Co	Water lane Leeds	Flax Mill	Leeds	sale	
Jan 1843	40 H.P	Wm Bruce	Wellington Mills Leeds	Woolen Mill	Leeds	sale	
Apr 1843	20 H.P		Steam Mills Spalding	Corn, Cake & Bone	Stamford	sale	
May 1843	20 H.P	Mr & Mrs Land	Water Lane Holbeck, Leeds		Leeds	sale	
July 1843	6 H.P	W & J Slingsby	Bell Busk Mill nr Skipton		Leeds	sale	
Oct 1843	40 H.P parts	Clarke Plummer Co	N'umberland Mill Newcastle	Flax Mill	Newcastle	sale	
Jan 1844	12 or 17 H.P	Barnsley Calender	Barnsley	Cloth finishing	Leeds	sale	
April 1846	10 H.P		Glasgow	working hammers	Glasgow	sale	

Feb 1850	2.5 H.P	The Baths	Wellington Street Leeds	pumping eng.	Leeds	sale
Mar 1850	2.5 & 12 H.P		Th Parker Hunslett Leeds		Leeds	sale
Aug 1850	40 H.P	?ex Sth London Water	London	pumping eng.	London	sale
Mar 1852	12 H.P	apply Sam Exley	Jack Lane Mill, Hunslett, Leeds	Flax Mill	Leeds	sale
Sept 1852	40 H.P		Wortley Leeds	Flax Mill	Leeds	sale
June 1853	30 H.P	Joseph Bell & Co	Whitehaven	Flax Mill	Preston	sale
June 1854	24 H.P	Clarke Plummer Co	N'umberland Mill Newcastle	Flax Mill	Newcastle	sale
Aug 1856	6 H.P	George Bucknell	nr Taunton	Tanning & bark mill	Exeter	sale
May 1857	60 H.P	Leys Masson & Co	Grandholm nr Aberdeen	Flax Mill	Glasgow	sale
Nov 1859	16 H.P	apply E Shaw	Leeds		Leeds	sale
Mar 1861	12 H.P	apply W Short	Bristol		Bristol	sale
Dec 1862	25 H.P		Isle Mill Holbeck Leeds		Leeds	sale
May 1875	20 inch cyl	Wm D Adlington	Skegby Mill Mansfield		Sheffield	sale

Category b - archive and book sources.

Date	H.P	Name	Location	Use	Source	Note
1795		Murray & Wood ?	reported by Hawkes to B&W	foundry work	B & W	
1797	40 H.P	Fisher & Nixon	Holbeck	textiles	B & W	
	10 H.P	Charles Shipman	Hull	sail cloth	B & W	
	?		in B&W letters.		B & W	
1799	40 H.P	Benyon Marshall Bage	Ditherington, Shrewsbury	Flax Mill	B & W	
	30 H.P	Benyon Marshall Bage	Holbeck Leeds	Flax Mill	B & W	on sale Apr 1842.
	?	G & J Wright	Manchester		B & W	
	8 H.P	Mr Close	Leeds	Dyers	B & W	
	?	Brookes family	Honley Mill nr Huddersfield		Crump	
1801	20 H.P		by the bridge Newcastle	Corn Mill	B & W	
	63" double	Wm King	Jarrow Colliery	pumping eng.	B & W	
	2 nd hand	Messrs Fishwick	Newcastle	forge work	B & W	
	?	Messrs Hawkes & Co	Gateshead	forge work	B & W	
	30 H.P	Ard Walker	Hunslett Leeds	Cotton Mill	WYAS	
1802	15 H.P		Rochdale		B & W	
	4 H.P	Mr Brewin	Bermonsey	bark mill	Farey	
	40 H.P	Mr Swann	London		B & W	
	40 H.P	Coupland Wilkinson	Leeds	Cotton Mill	Warner	
1803	?		Bolton		B & W	
	5 H.P	Milton Mill	nr Huddersfield		B & W	
	?	Heptonstall	Doncaster	Ropery or Flax Mill	B & W	
	10 H.P	Carrs	Swinegate Leeds	Dyers	B & W	
	10 H.P	Wilsons	Water Lane Holbeck Leeds	Buckram Manuf	B & W	
1804	50 H.P	Benyon & Bage	Meadow Lane Leeds	Flax Mill	Goodrich	
	6 H.P	Meux Brewery	London	Brewers	Farey	
	10 H.P	Edelcrantz	sold to Dannemora mine Sweden	pumping eng.	Gustavson	
	10 H.P ?	via Edelcrantz	Stockholm Sweden	distillery	and	
	10 H.P ?	via Edelcrantz	Stockholm Sweden	distillery	Turner	
	?	Mr Cliff	W. Bromwich	Iron and Steel Wks	Goodrich	
c. 1805	100 H.P	Matthew Bell	Willington Colliery, Tyneside	Pumping eng.	NEIMME	
	20 H.P	John Buddle	Elswick Colliery	drawing eng.	NEIMME	
1806	25 H.P	James Brown	Bell or West Ward Mill Dundee	Flax Mill	various	on sale Mar 1811
	15 H.P	Czar of Russia	Russia	Forge Works	Goodrich	
	6 H.P	Czar of Russia	Russia	Forge Works	Goodrich	
	4 H.P	Henry Maudslay	75 Margaret St London	workshops	Sc Mus	
	30 H.P	Admiralty	Block Mills Portsmouth	w'shop & Pumping	Goodrich	
	20 H.P	Samuel Bentham	St Petersburg	? & pumping	Goodrich	
	6 H P	Marc Brunel	Battersea	saw mill	NMM	
	20 H.P.	Edelcrantz	Kungsholmsbron , Stockholm	Corn Mill	{ Daedalus	
	?	Hoganas Stenkolsgruvor	Skane Sweden	pumping eng.	{ 1961	
1807	2 x 20 H.P	West Middx Water Works	Hammersmith	pumping eng.	Lon.Met. A	
1808	?	Hoganas Stenkolsgruvor	Byd Pit Skane Sweden	colliery eng.	Turner +	
	20 H.P	Scott & Brown	Broadford Mill Aberdeen	Flax Mill		on sale Mar 1811
1809	6 H.P	Admiralty	Sheerness Dockyard	pumping eng.	Goodrich	
1810	10 H.P	Admiralty	Deptford Victualling Yard	brewey	Goodrich	
	16 H.P	Marc Brunel	Battersea	saw mill	NMM	
1810 to 1812	160 H.P	John Curwen	Isabella Pit Workington	colliery pumping	Cumbria A	
1811	?	Wills tobacconists	Bristol	work mach'y	Wills A	
	60 H.P	Marshalls	Ditherington, Shrewsbury	Flax Mill	Rimmer	
	20 H.P					
1813	? 60 H.P	Russia	? Alexandrovsky Works	Flax Mill +	NEIMME	
	?	John Buddle	Percy Main Colliery	drawing eng.	? Loco	NEIMME
	50 H.P	Titley Tatham & Walker	Water Hall Mill, Water lane	Flax Mill	Tyas.	
	30 H.P	Admiralty / M Brunel	Chatham Saw Mill	Saw Mill	Nat Arch	
1814	56 H.P	Marshalls	Mill B, Water Lane Holbeck	Flax Mill	Lindley	
1815	?		Killingworth Colliery Tyneside	drawing eng.	Losh's patent	
1816	70 H.P	Marshalls	Mill C Marshall Street	Flax Mill	Lindley	
1818	15 H.P	Cord Factory	Ardwick, manchester	Cord manu.	Griscom	
1819	2 H.P.	Pauper Lunatic Asylum	Wakefield Yorks	Pumping etc	Sylvester	
	10 H.P	Sugar plantation	Demerera Guyana	Via Fawcett & Liddesdale	F & Preston A	
	24 H.P	Grout Bayliss & Co	Yarmouth	Silk Mill	H of Norfolk	
1820	?8 H.P	Newton Chambers	Sheffield	Iron Works	Sheff Arch	
	30 H.P	North Level Drainage	Borough Fen nr Crowlnd Lincs	Pumping eng.	Hinde	
	40 H.P	Sth London Water Wks	Vauxhall, London	Pumping eng.	Lon.Met A	
1821/2	40 H.P	Marshalls	Mill A, Water Lane	Flax Mill	Rimmer	
	56 H.P	Marshalls	Ditherington Shrewsbury	Flax Mill	Rimmer	

1822	10 H.P		Lisbon	Via Fawcett & Liddesdale	F & Preston A
1823	c 4 H.P	Westminster Gas Works	London	plant operation	Mech Mag
1824	60 H.P	Deeping Fen Drainage Bd	Pode Hole nr Spalding	Fen drainage	various
1827	70 H P	Marshalls	Mill D Marshall Str.	Flax Mill	Marshall
1831	26 H.P	Mr Gascoigne	Garforth Colliery, Yorks	Drawing eng.	NEIMME
1834	40 H.P	Sth London Water Wks	Vauxhall, London	Pumping eng.	Lon.Met A sale Aug.1850
1835	8 H.P	Kirkstall Forge	Kirkstall Leeds	Tilt hammer	Kirkstall Fge
1838	40 H.P nom.	Thos Lenty & Co	Castleton Mill Leeds	Flax Mill	The Eng'r
1839	?	Low Moor	Oakenshaw Lift nr Bradford	Pumping	Ind.Arch
1839-1840	20 H.P	Dublin & Kingston Rlwy	Dublin	workshop	Whishaw

Category c. Lindley List made in 1824 of Leeds Engines (duplicates some of the engines in a & b, where known duplicates are marked*)

Hereunder is the Fenton Murray & Co content of the list (70% by number of engines)

Woollen Trade	40 H.P	Jas Hargreave & Son
	32 H.P	R Glover
	30 H.P}	
	15 H.P}	Wm Hirst
	30 H .P	Hirst Bramley & Co
	40 H.P	D Willans & Sons
	30 H.P	T & B Hogg
	20 H.P	L Forster & Co
	30 H.P	P Ingham & Co
	20 H.P	Chorleys & Appleby
	24 H.P	J T Coward
	18 H.P	York & Sheepshankes
	16 H.P	B Blackburn & Sons
	18 H.P	Shann Driver & Co
	18 H.P	I Young & Co
	10 H.P	Outerwood & Co
	12 H.P	J Pearson
	12 H.P	Lord & Robinson
	7 H P	T & J Bischoff & Co
	6 H.P	Thos Lee & Brother
	6 H.P	John Wilson
	8 H.P	Wm Lupton & Co
Stuff Manufacture	36 H.P	Stansfield & Co
Cotton Spinning	35 H.P	R & F Coupland*
Flax Spinning	70 H.P}	
	56 H.P}	
	40 H.P}	
	31 H.P}	
	8 H.P}	Marshall & Co*
	60 H.P}*	
	27 H.P}	Benyon & Co
	60 H.P	Hives & Atkinson
	40 H.P	Moses Atkinson
	20 H.P	Geo Hammond
	24 H.P	I Hunton
	50 H.P	Titley & Co*
	14 H.P	C Charnock & Co
	16 H.P	Thos Land*
	8 H.P	Proctor & Son
	4 H.P	S Grimshaw & Co
	16 H.P}	
	2 H.P}	R Busk
Dying Trade	20 H.P	P & G Sayner
	10 H.P}	
	3 H.P}	W G Scarth
	12 H.P	Josh Wood & Son
	10 H.P	Wm Carr*
	9 H.P	W T Watson
	9 H.P	Banks & Goodman
	8 H.P	Aldam & Co
	8 H.P	Thomas Prince
	7 H.P	Eastburn & Co
	6 H.P	B Musgrave
	6 H.P	W Johnson
Seed Crushing	45 H.P	Hudson & Co
	64 H.P	Josh Medley
	20 H.P	J & J Armistead
	20 H.P	Wm Hainsworth
	12 H.P}	
	10 H.P}	Jonathan Lupton
	15 H.P	J & G Smith
Machine Makers	10 H.P}*	
	10 H.P}	
	6 H.P}	
	6 H.P}	Fenton Murray & Co
	8 H.P	Taylor Wordsworth & Co
	7 H.P	M Cawood & Co
	16 H.P	Thos Brown & Co
	4 H.P	David Blakey

	6 H.P	T & H Briggs
Buckram	40 H.P	Hartley Greens & Co
Cudbear	10 H.P	Jno Wilson & Sons

Part 2 Dredging Equipment

1808	16 or 20 H.P	Hughes Bough & Mills	re-build of a Trevithick dredger
1812	8 H.P ?	Port of Dublin	
1817	10 H.P	Port of Stettin	supplied by Bryan Donkin
1818	4 H.P		enquiry by Donkin, outcome unknown
1820	6 or 10 H.P	Aire & Calder Navigat'n.	
1833	5 H.P	Commiss's Southwold Harbour. Having cleared the harbour it was put up for sale in 1836.
1834	10.H.P	Ouse Navigation (Yorkshire) to a design by Thomas Rhodes.
1837/8	10 H.P	City of Newcastle - Tyne Dredger	
1839	?	Berwick on Tweed Dredger	
1840	15 H.P	Victoria - Shannon Navigation	
	15 H.P	Albert - Shannon Navigation	
1842	?	Tyne Dredger 2 possibly had Fenton, Murray & Co equipment.

Part 3 Marine Engines & Boilers

1808	2 H.P	Linaker - experimental boat at Portsmouth.
1813	8 H.P	Mr Wright's 'L'Actif' / 'Experiment' Scheduled passenger service Norwich / Yarmouth engine was later transferred to the' Courier'.
1817/8	?	Mr Ogden for 'Roanoke' in the US State of Virginia
1818/9	?	Second engine to Ogden's design for New Orleans. Purchased by Ogden in 1824 and used in a tug on the Mississippi.
1820/1	30 H.P	'Leeds' Steam Packet to work Hull / Selby.
1821	2 x 50 H.P	'Hero' London Margate Steam Packet.
1822	2 x 40H.P	Yorkshireman' Weddle & Brownlow, Hull/London re-engined in 1829 with Butterley engines.
1826	34 H.P	'Calder' Hull/Selby packet, later Hull/Goole.
1829	?	Iron paddle steamer for Wm Gravatt.
1830	2 x 8 H.P	'Hirondelle' Saone/ Rhone Messageries Co.
1831	2 x 70 H.P.	S.P.' Transit' Hull/Hamburg and London , St Petersburg
	10 H.P	Iron Steam Boat by Jas Livingstone for Castleford/Goole.
	?	'Liberal' owned by Bronley & Co of Goole.
	?	2 Aire & Calder Navig'n tugs Castleford/Goole (1 By Livingstone?)
1832	8 H.P	'Hirondelle 2' on the Loire, M Guibert.
	20 H.P	for M.Paul probably for the Saone/Rhone.
1835	2 X 80 H.P	'Papin' for the French Navy.
	2 x 120 H.P	'William Darley' for E Gibson of Hull, Hull/Hamburg.
	12 H.P	'Luxor' (M Guibert) on the Loire, Nantes/Orleans.
	2 x 9 H.P.	'Le Jackson' Saone tug.
	2 x 9 H.P.	'Le James Fenton' Saone Tug.
1836	2 X 80 H.P	'Lycurgue' French Gov't Postal Service.
	(2 x?)12 H.P.	' Hirondelle 3' Saone/Rhone Messageries Co.
1837	80 H.P	'Nantes & Bordeau' M Guibert. Nantes/Bordeau.
1838	48 or 20 H.P	Upper Rhone.
1840	60 H.P	'Aigle' M Bretmayer, Saone/Rhone. (? 2x 30)
	60 H.P	'Hirondelle 4 or 5' Saone/Rhone, Messageries Co. (? 2x30).

Boilers only

1817		'Morning Star' of Alloa, nr Stirling.
1834	for sale	'Salamander' in Norfolk.

Part 4 Locomotive engines.

1812-1816	0-4-0	7 off Rack Locomotives.	
1831	2-2-0	Vulcan	Liverpool & Manchester Rlwy
	2-2-0	Fury,	Liverpool & Manchester Rlwy
1832	0-4-0	Constance	Andrezieux-Roanne Rlwy
	0-4-0	Jackson	Andrezieux-Roanne Rlwy
1833	2-2-0	Leeds	Liverpool & Manchester Rlwy
1834	2-2-0	Columbia	Charleston & Hamburg Rlwy
	2-2-0 ?	Nelson	Leeds & Selby Rlwy
1835	0-4-0	Jackson	St Etienne-Lyon Rlwy
1836	2-2-0	Jackson	Paris-St Germain Rlwy
	2-2-0	Denis Papin	Paris-St Germain Rlwy
?	2-4-0	Anson	Leeds & Selby Rlwy
1837	0-4-0	Leeds	London & Birmingham Rlwy
	2-4-0	Hawke	Leeds & Selby Rlwy
1838	2-2-0	Notre-Dame-des-Tables	Montpellier & Cette (Sete) Rlwy
	2-2-0	L'Herault	Montpellier & Cette (Sete) Rlwy
	2-2-0	La Montpellieraine	Montpellier & Cette (Sete) Rlwy
1838 ?	2-4-0	?	Leeds & Selby Rlwy
	2-4-0	?	Leeds & Selby Rlwy
1839	2-2-0	La Cettoise	Montpellier & Cette (Sete) Rlwy
	2-2-0	La Rosine	Montpellier & Cette (Sete) Rlwy

	2-2-2	?	Leeds & Selby Rlwy	
	2-2-2	?	Leeds & Selby Rlwy	
	2-2-2	Agilis	Sheffield & Rotheram Rlwy	
	2-2-2	Leeds	London & South Western Rlwy	
	2-2-2	Eclipse	London & South Western Rlwy	
	2-2-2	Midland Rlwy no. 40	North Midland Rlwy	
	2-2-2	Midland Rlwy no. 41	North Midland Rlwy	
	2-2-2	Midland Rlwy no. 42	North Midland Rlwy	
1839/40	2-2-2	Firefly	Belgian State Rlwy.	
	2-2-0	Mathieu Murray	Paris-Versailles Left Bank	
	2-2-0	? Denis Papin	Paris-Versailles Left Bank	
	2-2-2	Vulkan	Munich & Augsburg Rlwy	
	2-2-2	Mars	Munich & Augsburg Rlwy	
	2-2-2	Jackson	Paris-Orleans Rlwy	
	2-2-2	Newton	Paris-Orleans Rlwy	
1840	2-2-2	Versailles	Paris-Versailles Right bank	
	2-2-2	Kingston	Hull & Selby Rlwy	
	2-2-2	Exley	Hull & Selby Rlwy	
	2-2-2	Selby	Hull & Selby Rlwy	
	2-2-2	Andrew Marvel	Hull & Selby Rlwy	
	2-2-2	Collingwood	Hull & Selby Rlwy	
	2-2-2	Wellington	Hull & Selby Rlwy	
	2-2-2	Phoenix	London & South Western Rlwy	
	2-2-2	Crescent	London & South Western Rlwy	
	2-2-2	Midland Rlwy no.13	North Midland Rlwy	
	2-2-2	Midland Rlwy no.14	North Midland Rlwy	
	2-2-2	Midland Rlwy no.15	North Midland Rlwy	
	2-2-2	Midland Rlwy no.16	North Midland Rlwy	
5/1840	2-2-2	Charon, Firefly	Great Western Rlwy	
9/1840	2-2-2	Cyclops, Firefly	Great Western Rlwy	
3/1841 ?	2-2-2	Stromboli, Leo	Great Western Rlwy	
4/1841	2-2-2	Hecla, Leo	Great Western Rlwy	
5/1841	2-2-2	Cerberus, Firefly	Great Western Rlwy	
6/1841	2-2-2	Etna, Leo	Great Western Rlwy	
7/1841	2-2-2	Harpy, Firefly	Great Western Rlwy	
	2-2-2	Pluto, Firefly	Great Western Rlwy	
9/1841	2-2-2	Minos, Firefly	Great Western Rlwy	
	2-2-2	Ixion, Firefly	Great Western Rlwy	
10/1841	2-2-2	Hecate, Firefly	Great Western Rlwy	
11/1841	2-2-2	Gorgon, Firefly	Great Western Rlwy	
12/1841	2-2-2	Vesta, Firefly	Great Western Rlwy	
	2-2-2	Prosperine, Firefly	Great Western Rlwy	
1/1842	2-2-2	Acheron, Firefly	Great Western Rlwy	
2/1842	2-2-2	Erebus, Firefly	Great Western Rlwy	
	2-2-2	Styx, Firefly	Great Western Rlwy	renamed Phlegethon
3/1842	2-2-2	Hydra, Firefly	Great Western Rlwy	
4/1842	2-2-2	Lethe, Firefly	Great Western Rlwy	
5/1842	2-2-2	Medea, Firefly	Great Western Rlwy	
	2-2-2	Medusa, Firefly	Great Western Rlwy	
7/1842	2-2-2	Ganymede, Firefly	Great Western Rlwy	
8/1842	2-2-2	Argus/Firefly	Great Western Rlwy	

Sale.
Great North of England Rlwy purchased a F M & J '2-2-2 from the Leeds & Selby Rlwy in 1840
and the surviving evidence indicates that it was converted to a ballast engine either '0-4-0 or '0-6-0 (by FM&J).

Appendix 3.

Draft Agreement Fenton Murray & Company and West Middlesex Water
Works Co

Agreement for Two Steam Engines

James Fenton, Matthew Murray, David Wood, & William Lister of
Leeds in the County of York Steam Engine Manufacturers, and Partners
under the Firm or Name of Fenton Murray & Company on the One Part & the
Company of Proprietors of the West Middlesex Water Works on the other part
mutually agree as follows: The above named James Fenton, Matthew Murray,
David Wood, & William Lister do undertake to make, complete, deliver & fix
and set up ready to work at Hammersmith in the County of Middlesex 2
Steam Engines of 20 Horses Power each consisting of the Materials specified
in the annexed List, & agreeable to the Plans delivered to the said Company
of Proprietors and in consideration of the sum hereinafter agreed to be paid &
that they the said James Fenton, Matthew Murray, David Wood, & William
Lister will warrant the said Engines for the term of Four Years & repair any
Accidents that may happen to be caused by any defect of the Materials,
Proportion, Workmanship fixing or fitting up of the said Engines during the
above term, & will appoint and provide an experienced Engineer to examine,
& give Directions for the Management of the said Engines, at least once a
Fortnight during the aforesaid Term of Four Years, the Engines are so
warranted, & the said James Fenton, Matthew Murray, David Wood, & William
Lister, do further agree to construct the moveable Parts of the said Engines to
fit each other so far as to answer every useful purpose by putting or
substituting like parts from one Engine in the place of any Decayed part in the
other in case of Accident. And that they will ship the above two Engines on
board vessels at Leeds the one on or before the first day in June 1807 & the
other on or before the first day in July following. And will fix, & erect the two
Engines in a state fit for working with all convenient speed after the arrival of
the respective materials in London. ~ And the said Company of Proprietors
promises, & agrees on Condition of the said work being done & performed as
aforesaid to pay or cause to be paid to the said James Fenton, Matthew
Murray, David Wood, & William Lister the sum of two thousand pounds in
manner following that is to say, One Thousand Pounds on the Arrival in
London of the whole of the Materials specified in the annexed list, &
composing the aforesaid 2 Steam Engines, & the remaining One Thousand
pounds when the said Engines are set to work – and will further pay unto the
said James Fenton, Matthew Murray, David Wood, & William Lister the sum of
Twenty Five Pounds per Annum for & during the Term of 4 Years in
consideration of their funding an Engineer to examine, & direct the working of
these Engines as aforesaid for and during the Term. And the said Company
of Proprietors do further agree to find & put up with all convenient speed, the
Stone, Brick &Timber Work necessary for the Reception of the said Engines,
& will cause the said Engines, to be properly found with the necessary
Materials Attendants & Repairs for the aforesaid Term of Four Years
excepting such Accidents as may arise from Imperfections in the Engines or

in the fitting or fixing thereof, as before mentioned which are to be repaired at the Expense of the said James Fenton, Matthew Murray, David Wood, & William Lister & stipulations on the part of ye aforesaid James Fenton, Matthew Murray, David Wood, & William Lister They bind themselves to the said and for the true & strict performance of the above conditions in the Penal Sum of One thousand Pounds.

In Witness whereof the said James Fenton, Matthew Murray, David Wood, & William Lister have respectively set their Hands & Seals & the said Company of Proprietors have affixed their Common Seal this   Day of in the year of Our Lord 1807.

And it is further agreed & understood by and between the Parties hereunto that the said James Fenton, Matthew Murray, David Wood, & William Lister shall with all convenient speed upon <u>notice</u> being given unto them or to their agent or Superintendent of any accident or Damage as aforesaid renew repair and restore the said Engine or Engines to a complete & good State of Working at their own <u>Costs</u> & <u>Charges</u> but that they shall not be liable to make good to the Company any loss arising from the said Engine or Engines being out of work for any time in respect of such Accident or Damage unless they shall neglect or delay such renewal repairs & restoration as aforesaid.